Emilio H. Satorre
Gustavo A. Slafer
Editors

Wheat
Ecology and Physiology of Yield Determination

Pre-publication REVIEWS, COMMENTARIES, EVALUATIONS . . .

"This book brings together scientists from many countries to write individual chapters on ecological and physiological factors that determine wheat yields. The authors have done an excellent job of taking broad subject matter and summarizing what is known in understandable terminology. The authors also have presented their views on the potential for future yield gains through technology advances in their subject area. As a soil fertility research-extension professional, I found the genetic discussion very informative and could visualize the need for multidisciplinary teams to move to the next yield level. The editors should be congratulated for assembling a diverse group of scientists to write an excellent book on the challenges facing scientists for future yield improvement in wheat."

David A. Whitney, PhD
Professor,
Department of Agronomy,
Kansas State University,
Manhattan

More pre-publication
REVIEWS, COMMENTARIES, EVALUATIONS . . .

"The authors from South America, the United States, Europe, Oceania, and Africa have dedicated this book to reviewing the major issues involved in the determination of wheat yield. The book certainly fulfills its promise by giving comprehensive information on yield formation as well as future perspectives for increasing wheat yields. By dividing the book into four major sections—physiology, ecology, production systems, and breeding—the editors have made it very reader friendly.

This book skillfully reviews different aspects of yield formation in wheat. It gathers together information on physiology, ecology, production systems, and breeding in a way that helps readers understand the complexity of yield formation. Thus, it is a valuable reference for graduate students as well as cereal researchers and plant breeders. This book is an updated continuation of the world-famous books written on yield formation of wheat in earlier decades."

Pirjo Peltonen-Sainio, PhD
Professor,
Department of Plant Production,
University of Helsinki,
Finland

CRC Press is an imprint of the
Taylor & Francis Group, an **informa** business

Wheat
Ecology and Physiology of Yield Determination

Wheat
Ecology and Physiology of Yield Determination

Emilio H. Satorre
Gustavo A. Slafer
Editors

CRC Press is an imprint of the
Taylor & Francis Group, an **informa** business

CRC Press
Taylor & Francis Group
6000 Broken Sound Parkway NW, Suite 300
Boca Raton, FL 33487-2742

Visit the Taylor & Francis Web site at
http://www.taylorandfrancis.com

and the CRC Press Web site at
http://www.crcpress.com

CONTENTS

ABOUT THE EDITORS

Emilio H. Satorre, PhD, is Professor of Crop Production in the Faculty of Agronomy at the University of Buenos Aires, Argentina. He is the author or co-author of nearly 100 professional papers, abstracts, and proceedings on various aspects of crop and weed physiology and ecology. Dr. Satorre is a member of the National Council for Scientific and Technical Research (CONICET) in Argentina, the Institute of Physiology and Ecology related to agriculture, the Crop Society of America, the Weed Science Society of America, and the Argentine Society of Ecology, among others. In addition, he is Agriculture Coordinator of the Farmer's Agricultural and Extension System in Argentina (AACREA).

Gustavo A. Slafer, PhD, is Adjunct Professor of Crop Production in the Faculty of Agronomy at the University of Buenos Aires, Argentina. He is the author or co-author of over 100 professional papers, abstracts, and proceedings on various aspects of crop ecophysiology and is editor of the book *Genetic Improvement of Field Crops* (Marcel Dekkar, Inc., 1994). Dr. Slafer is an Editorial Board member of the *Journal of Crop Production* (The Haworth Press, Inc.) and speaks internationally on wheat physiology, particularly in relation to yield generation and breeding. In addition, he is a member of the National Council for Scientific and Technical Research (CONICET) in Argentina, the Institute of Physiology and Ecology related to agriculture, the Argentinian Society of Plant Physiology, and the Argentinian Society of Genetics, among others.

Contributors

Edmundo H. Acevedo, PhD, is Professor, Laboratorio de Relación Suelo-Agua, Departamento de Producción Agrícola, Facultad de Ciencias Agrarias y Forestales, Universidad de Chile, Santiago, Chile.

Jose L. Araus, PhD, is Professor, Department de Biología Vegetal, Facultat de Biología, Universitat de Barcelona, Barcelona, Spain.

Philip J. Bauer, PhD, is Research Agronomist, USDA-ARS, Coastal Plain Soil, Water, and Plant Research Center, Florence, South Carolina.

Marta M. Casa Blum, MSc, is Professor, Universidade de Passo Fundo, Faculdade de Agronomia e Medicina Veterinária, Passo Fundo, RS, Brazil.

Basilio Borghi, PhD, is Researcher, Istituto Sperimentale per la Cerealicoltura, S. Angelo Lodigiano, Italy.

Daniel F. Calderini is Research and Teaching Agronomist, Departamento de Producción Vegetal, Facultad de Agronomía, Universidad de Buenos Aires, Buenos Aires, Argentina.

Mike D. Dennett, PhD, is Researcher and Lecturer, Department of Agricultural Botany, School of Plant Sciences, The University of Reading, Whiteknights, Reading, United Kingdom.

Alberto Fereres, PhD, is Researcher, Departamento de Producción Vegetal, Instituto Nacional de Investigaciones Agrarias, Madrid, Spain.

Ralph A. Fischer, PhD, is Research Program Coordinator, Australian Centre for International Agricultural Research, Canberra, Australia.

James R. Frederick, PhD, is Associate Professor, Department of Agronomy, Clemson University, Pee Dee Research and Education Center, Florence, South Carolina.

Robert J. Froud-Williams, PhD, is Researcher and Lecturer, Department of Agricultural Botany, School of Plant Sciences, The University of Reading, Whiteknights, Reading, United Kingdom.

Jeffory A. Hattey, PhD, is Assistant Professor, Department of Plant and Soil Sciences, Oklahoma State University, Stillwater, Oklahoma.

Gordon V. Johnson, PhD, is Professor, Department of Plant and Soil Sciences, Oklahoma State University, Stillwater, Oklahoma.

Jori P. Jordaan, PhD, is Assistant General Manager and Researcher, SENSAKO, Bethlehem, South Africa.

Robert M. D. Koebner, PhD, is Researcher, John Innes Centre, Norwich Research Park, Colney, United Kingdom.

Carlos A. Medeiros, MSc, is Professor, Universidade de Passo Fundo, Faculdade de Agronomia e Medicina Veterinária, Passo Fundo, RS, Brazil.

Daniel J. Miralles, DrSci, is Researcher and Teaching Agronomist, Departmento de Producción Vegetal, Facultad de Agronomía, Universidad de Buenos Aires, Buenos Aires, Argentina.

Roger H. Ratcliffe, PhD, is Researcher, USDA-ARS, Crop Production and Pest Control Research Unit, Department of Entomology, Purdue University, West Lafayette, Indiana.

William R. Raun, PhD, is Associate Professor, Department of Plant and Soil Sciences, Oklahoma State University, Stillwater, Oklahoma.

Erlei Melo Reis, PhD, is Professor, Universidade de Passo Fundo; Faculdade de Agronomia e Medicina Veterinária; Passo Fundo, RS, Brazil.

Matthew P. Reynolds, PhD, is Head of Physiology, Wheat Program, CIMMYT, Mexico DF, Mexico.

Richard A. Richards, PhD, is Researcher, CSIRO, Division of Plant Industry and CRC for Plant Science, Canberra, ACT, Australia.

Victor O. Sadras, PhD, is Researcher, CSIRO, Division of Plant Industry, Locked Bag 59, Narrabri, NSW, Australia.

Santiago J. Sarandon is Adjunct Professor, Departamento de Producción Vegetal, Facultad de Ciencias Agrarias y Forestales, Universidad Nacional de La Plata, La Plata, Argentina.

Roxana Savin, PhD, is Researcher and Teaching Agronomist, Departamento de Producción Vegetal, Facultad de Agronomía, Universidad de Buenos Aires, Buenos Aires, Argentina.

Hernán R. Silva is Researcher, Laboratorio de Relación Suelo-Agua, Departamento de Producción Agrícola, Facultad de Ciencias Agrarias y Forestales, Universidad de Chile, Santiago, Chile.

Paola C. Silva is Researcher, Laboratorio de Relación Suelo-Agua, Departamento de Producción Agrícola, Facultad de Ciencias Agrarias y Forestales, Universidad de Chile, Santiago, Chile.

John W. Snape, PhD, is Head of the Cereals Research Department, John Innes Centre, Norwich Research Park, Colney, United Kingdom.

Boris R. Solar is Researcher, Laboratorio de Relación Suelo-Agua, Departamento de Producción Agrícola, Facultad de Ciencias Agrarias y Forestales, Universidad de Chile, Santiago, Chile.

Peter J. Stone, PhD, is Researcher, New Zealand Institute for Crop and Food Research, Hawke's Bay Research Centre, Hastings, New Zealand.

Robert L. Westerman, PhD, is Professor and Head, Department of Plant and Soil Sciences, Oklahoma State University, Stillwater, Oklahoma.

Preface

New discoveries and novel interpretations of previous findings have produced changes in plant sciences, as well as in most other scientific areas, at a continuously increasing pace. Wheat is (and has likely been for approximately the past 15,000 years) the main crop in the world (regarding both cultivated area and production), a relative importance that is magnified if protein rather than total grain production is considered. That explains why it is among the most studied crop plants worldwide.

Despite isolated examples in several books of the physiological-ecological bases determining changes in yield there does not appear to be a single volume condensing the latest information on these particular fields of wheat science. This book attempts to provide a physiological-ecological approach to understanding wheat yield and its determining processes at the crop level of organization, through bringing together much of the relevant information from many different journals, emphasizing the latest and widely accepted developments conforming the state of the art in each particular field. Its nineteen chapters have been organized in four main parts of *Wheat Physiology*, *Wheat Ecology*, *Wheat Production Systems*, and *Breeding to Further Raise Wheat Yields*, which follow an introductory chapter.

The book is mainly addressed to agronomists, breeders, crop ecologists, and physiologists working in different fields related to wheat yield and its determining processes. Although it has not been written as a textbook on a particular subject, we expect it to become a useful reference for advanced undergraduate and postgraduate students in courses that center their objectives in some chapters of this book. Each author has intended his or her chapter to be a summary description and critical update of the knowledge in each field, with particular reference to developments in the past decade.

The views of scientists from eleven different countries ranging in interests and expertise from geneticists, physiologists, ecologists, and breeders to agronomists have, we believe, enriched the whole work. However, as in most multi-authored books, the trade-off of this approach is that some repetition among chapters as well as occasional (mostly subtle) disagreement among authors emerge in reading the whole book. Since the book is mainly addressed to postgraduate students and researchers, we have left the chapters this way to highlight similarities and differences in the perspectives of the authors on particular effects or processes.

All authors have to put part of their duties or their free time aside to accomplish this task, and we would like to express our sincere gratitude to each of them. We also thank our colleagues in the Cereal Production Unit, within the Department of Plant Production, and particularly scientists in our wheat physiology (R. Savin, D. J. Miralles, D. F. Calderini, M. F. Dreccer, L. G. Abeledo, and E. Whitechurch) and crop ecology labs (A. Guglielmini, F. Rizzo, S. Poggio, G. Gonzalo, and R. Ruiz), who understood the extra time demanded by the editing of this book.

We gratefully acknowledge the permission to reprint figures and tables under copyright: Entomological Society of America, *Environmental Entomology*; Association of Applied Biologists, *Annals of Applied Biology*; American Society of Agronomy, *Agronomy Journal*; and Crop Science Society of America, *Crop Science*.

Finally, we are most grateful to our families for their tolerance and support during the preparation of this book.

Despite the extra time it required, we have really enjoyed editing this work. We do hope the authors have enjoyed writing their chapters too, and that readers will share this feeling with us.

Emilio H. Satorre
Gustavo A. Slafer

PART I: WHEAT PHYSIOLOGY

Chapter 1

An Introduction to the Physiological-Ecological Analysis of Wheat Yield

Gustavo A. Slafer
Emilio H. Satorre

WORLDWIDE IMPORTANCE OF WHEAT

There are approximately 350,000 botanically acknowledged plant species, but only 24 (i.e., 0.007 percent of them) are used as crops to satisfy most human requirements for food and fiber (Wittwer, 1980). Due to our strong dependence upon a limited number of plant species, the future welfare of humankind is strongly linked to the degree of understanding we achieve about their potential productivity and adaptability to environmental constraints (Evans, 1975). Not only are these species remarkably few, but their contributions to the total production are not evenly distributed among crops. Bread wheat (*Triticum aestivum*, L.) undoubtedly plays a major role among the few species widely grown as food sources, and likely was central to the beginning of agriculture (Harlan, 1981).

Wheat-Growing Areas and Production

Bread wheat is the most widely grown crop in the world (Briggle and Curtis, 1987; Kent and Evers, 1994; Slafer, Sattore, and Andrade, 1994). Approximately one sixth of the total arable land in the world is cultivated with wheat. For the 1986-1995 decade this area was ca. 223 million hectares (ha) and the production was ca. 545 million tons (t) (data from FAO, 1988-1995). Not only do these figures represent the largest area and largest production, compared with any other crop, but also the relative difference between wheat and rice and maize (the other cereals widely grown worldwide) becomes even greater if judged by the total amount of protein rather than just by the dry matter they produce (Fischer, 1984). To reach these end-of-century figures, both wheat area and production have increased substantially from their values at the beginning of this century (Slafer, Sattore, and Andrade, 1994).

Despite the fact that both the growing area and the production of wheat worldwide have increased since 1900, the relationship between them during the century has not been linear (see Figure 1.1a). The lack of linearity is strongly based on the facts that (1) wheat production has increased continuously while the growing area has clearly increased only during the initial 50 to 60 years of this century, and (2) wheat yields have been slowly increased during the first half of the century but substantially (ca. 10 times) faster afterward (Slafer, Sattore, and Andrade, 1994). That is why a strong linear relationship can be found for the initial period (1903-1955), with a slope slightly higher than 1 indicating that wheat yield increases contributed steadily but slightly (ca. 21 percent) to increases in wheat production during this period, most of which were, thus, attributable to increases in the growing area (Slafer, Sattore, and Andrade, 1994).

From 1955 onward wheat production increased at a much higher rate than in the previous period and the increase was strongly related to the substantial improvements in yield (see Figure 1.1a inset). Although the relationship has a strong linear component, there is a trend to a curvilinear relationship reflecting the reduction in growing area experienced in most regions of the world (with exceptions, particularly in parts of Asia) during the past decades (Slafer, Sattore, and Andrade, 1994).

Despite that, in general, yields have increased strongly during the period from 1955 to 1995, it has also been noticeable that they have been leveling off to an apparent ceiling during the past decade (Slafer, Sattore, and Andrade, 1994). Although the number of years considered is too small to draw general conclusions, it is remarkable that worldwide average yields have not shown an increase from 1990 to 1995 (see Figure 1.1b). In fact, production during these years declined from ca. 600 to 550 million tons, while the world population has kept its rate of increase of ca. 85 to 90 million people per year (see Figure 1.1b).

If it is confirmed during the next few years that wheat yields are actually reaching a ceiling, it is urgent that we gain a much more comprehensive understanding of its generation, in order to devise new strategies for further increasing yields either through management, breeding, or both. This is increasingly urgent if we are to meet the requirements of a burgeoning

FIGURE 1.1a. Relationship Between Worldwide Wheat Production and Its Growing Area for the Period From 1903 to 1995

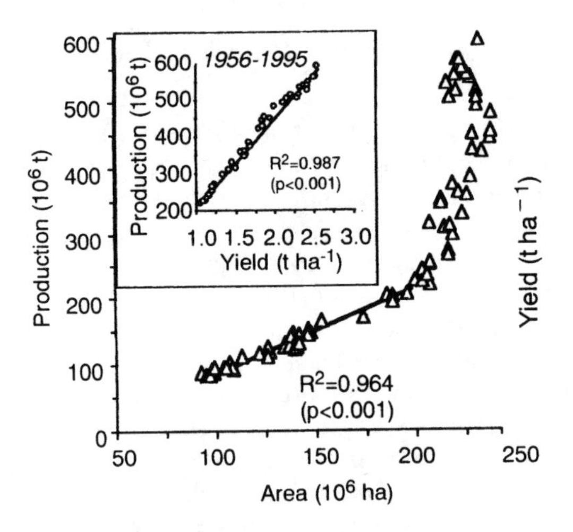

The line, fitted by linear regression for the period 1903-1955, has a slope of 1.21 t ha^{-1}. Inset is the relationship between production and yield for the period 1956-1995, and the line was also fitted by linear regression. Raw data from Annuaires del Institut International D'Agriculture (1914-1947) and FAO Yearbooks (FAO, 1988-1995).

FIGURE 1.1b. Average Yield (□) and Population of the World (●) from 1990 to 1995

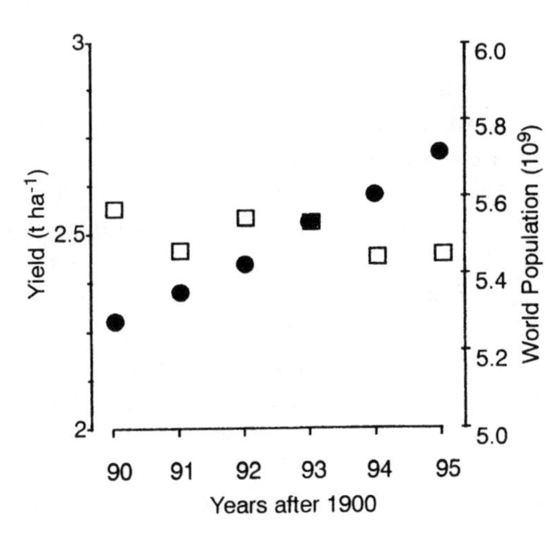

population estimated to reach 8 to 10 billion people during the early decades of the twenty-first century.

Wheat Yield in Different Areas and Historical Trends

Wheat is grown in most regions of the globe. Due to its importance as a food (and feed) source, and to its enormous genetic variability in phenological response to photoperiod and temperature, including vernalization (Slafer and Rawson, 1994), it is scarcely surprising that wheat is grown from almost 60°N in Northern Europe right to 40°S in South America, passing through the equator, in locations ranging in altitude from a few meters to more than 3,000 m above sea level. Some of the countries with the largest wheat-growing areas each year are shown in Figure 1.2. This list includes only countries with more than a million hectares grown per year (averaged for 1990 to 1995), which are spread over all continents.

Average yields for these countries are quite different, ranging from less than 1 to more than 7 t ha^{-1}. Differences in yield reflect differences in the level of inputs and agricultural sophistication as well as in the quality of the edaphic-climatic environmental conditions.

Disregarding the reasons for the differences in average yield, it is noteworthy that almost all countries exhibited a similar trend in yield for the whole century (Slafer, Satorre, and Andrade, 1994; Calderini and Slafer, 1998). Figure 1.3a summarizes six examples of countries differing in their levels of environmental appropriateness for growing wheat, their agricultural sophistication, and their usage of inputs in wheat crops. To highlight the similarities in the trends, avoiding the differences in actual yields, all values were calculated as a proportion of the earliest yield record available for this century.

In all cases yield remained practically unchanged by either genetic or management improvements during the first 50 years of this century or so (Figure 1.3a). For many countries (not only those exemplified in Figure 1.3), the slope for the initial half of the century revealed no yield gains whatsoever (they were not significantly different from zero; Calderini and Slafer, 1998). Therefore, in countries profoundly different in environmental and economic conditions and with quite different agricultural experience, there was a rather poor ability to improve yields through either genetic or management improvement during the first half of this century.

Thereafter, yield increased substantially, though not equally, in all countries (Figure 1.3a). For the examples in Figure 1.3a, the slopes for the United Kingdom, France, Egypt, the former USSR, the United States and Argentina were respectively for the initial 50 years 0.37, 0.29, 0.34, 0.52,

FIGURE 1.2. Harvested Area in Several Countries with More than a Million Hectares Cultivated Yearly

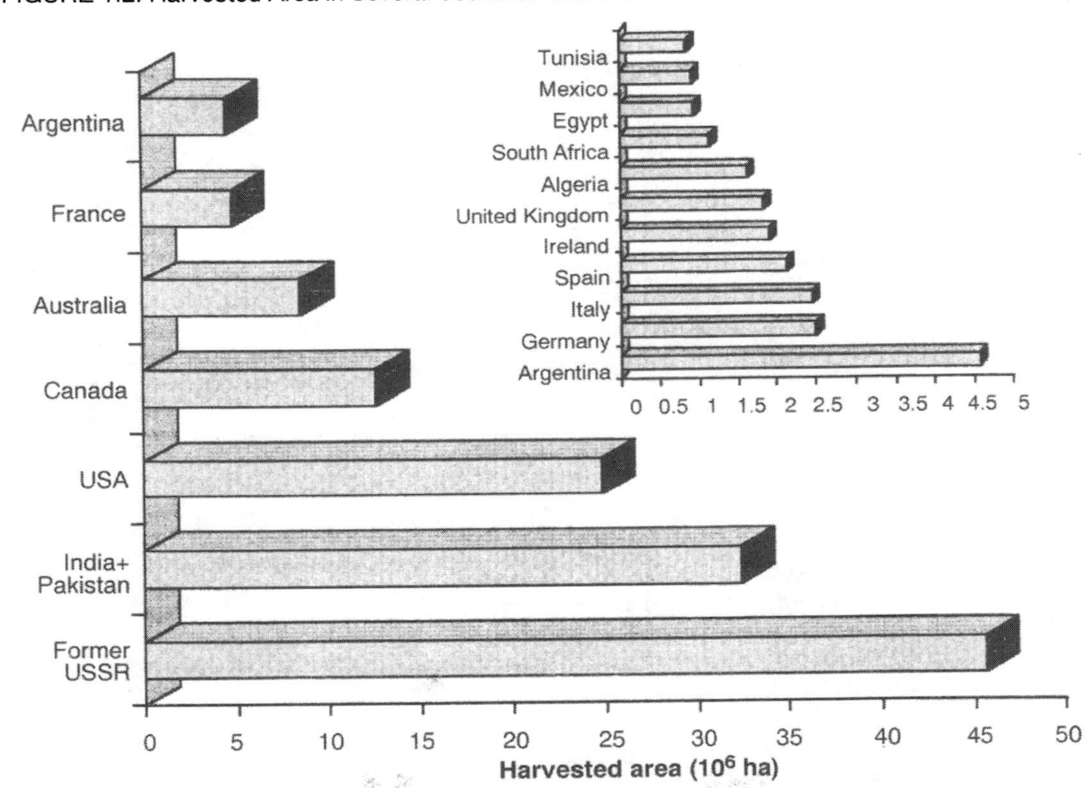

Data are averages for the 1991-1995 period taken from FAO Yearbooks (FAO, 1988-1995).

FIGURE 1.3. Average National Yields for Contrasting Wheat Regions During the Century (a) and During the 1990-1995 Period (b).

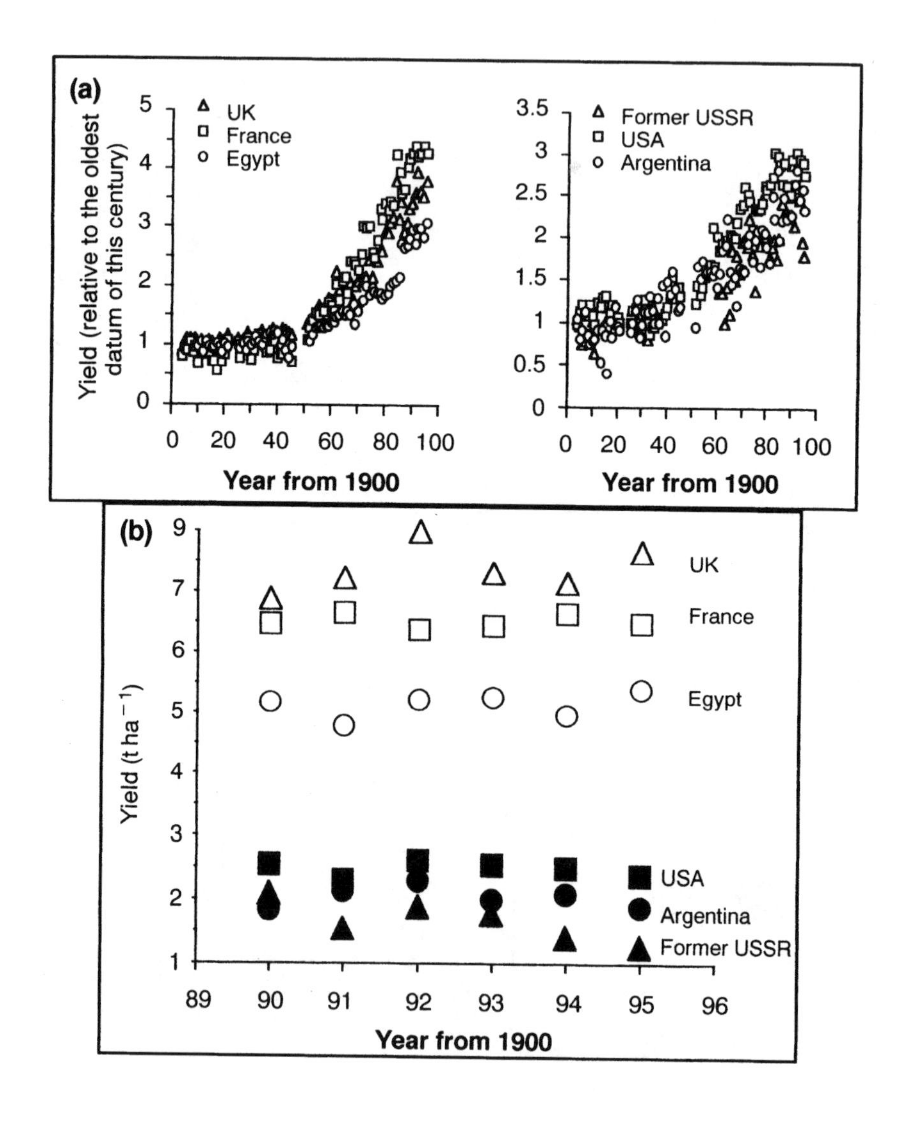

Data in (a) are yields relative to the oldest yield figure available (1903 for all countries except Egypt, 1909). Raw data taken from Annuaires del Institut International D'Agriculture (1914-1947) and FAO Yearbooks (FAO, 1988-1995).

0.25, and 1.30 percent y^{-1} and from then onward 5.56, 7.80, 4.20, 3.01, 3.61, and 2.62 percent y^{-1}.

Despite these large yield gains for the period 1950 to 1995, a detailed analysis of the 1990s highlighted the lack of a trend to increase yields since 1990 (Figure 1.1b), suggesting that they might be asymptotically approaching a ceiling. This was not an artifact and proved certain in many specific countries, beyond the profound differences evidenced by their level of average yields (Figure 1.3b, and see Calderini and Slafer, 1998 for a more comprehensive analysis).

The apparent remarkable lack of success of breeding management to keep increasing yields during the 1990s at a pace similar to that of the 1950s to the 1980s (or even increase them at a slower but significant rate) shown by the latest trends (Figures 1.1b and 1.3b) must be taken cautiously due to the reduced information on which it is based. Unfortunately there is a conflict between waiting to obtain more data to have a proper statistical certainty and taking actions to avoid the consequences of a shortage of wheat availability, which would push the prices up and make it even less available for those who need it most. If this incipient leveling off is reflecting what we can expect rather than "a normal series of years with no increases in yield," new strategies for breaking the apparent barriers of yield should be envisaged. The likelihood of succeeding in this complex task will, we believe, depend strongly upon our improved understanding of the physiological and ecological bases of yield generation in wheat.

Yield Stability

Although most breeders, physiologists, ecologists, and agronomists would agree on the relevance of yield stability, there has been notoriously less effort devoted to understanding it than yield itself (beyond that, understanding yield responses to environmental factors together with the genetic variation in these responses may help substantially to understand yield stability, or lack of it).

Observing yield trends during the century, it becomes obvious that countries differ strongly in their yield stability (see for example Figure 1.3a). Most differences in year-to-year yield among countries likely depend upon the regularity of the climate (mainly rainfall) in each country's wheat region.

Despite its importance, our knowledge of how yield stability has been modified as yield has increased during this century is rather poor and speculative. It has been concluded, mostly from contrasts made between two particular periods (Hazell, 1984; Anderson et al., 1988) that yield stability has been decreasing as yield has increased. The method has the inconveniences, however, that (1) it does not illustrate trends in yield stability, (2) it does not separate variability due to different rates of yield improvements from that due to differences in actual yield instability, and (3) any "abnormal" year occurring in one of these periods may lead to wrong conclusions.

Studies analyzing trends for the whole century (rather than the comparison of the coefficients of variation of average yield for a relatively early and a relatively recent period) for yield stability are infrequent. Slafer and Kernich (1996) have reported trends in yield stability for Australian wheat (and other cereals) since 1900 showing that yield stability did not decrease during the century. More recently, Calderini and Slafer (1998) expanded the analysis of wheat yield stability during the present century to many other countries. These countries were selected to represent a wide range of edaphic-climatic situations in which wheat is produced and to include a wide range of farming systems, yield levels, and geographical distribution.

Trends in absolute terms (i.e., assessing yield stability in t ha^{-1}, disregarding the yield) during the present century revealed a decrease in yield stability in 14 of the 21 countries, but the whole change was relatively small compared with increases in yield. Therefore, yield stability assessed as a percentage of yield indicated increases or at least no changes for most countries during the century (for details see Calderini and Slafer, 1998).

A NICHE FOR THIS BOOK

That wheat is the most important crop and that its yield needs to be continuously increased to match world population growth are good reasons for dedicating books to reviewing major issues involved in yield determination. The most conspicuous early example is the book by Percival, *The Wheat Plant,* written during the first decades of this century (Percival, 1921, reprinted in 1975). In 1967 the ASA-CSSA-SSSA released, as part of their Agronomy Monographs, a multiauthored book that became a recognized source of information for many scientists, *Wheat and Wheat Improvement* (Quisenberry and Reitz, 1967). The well-earned reputation of this book justified not only a reprint but also a second edition, 20 years later (Heyne, 1987). In the meantime, a special book was edited in 1981 to update the information available in some important fields of wheat science (Evans and Peacock, 1981).

To the best of our knowledge, no books dedicated exclusively to wheat and reaching a wide readership have been published since the second edition of *Wheat and Wheat Improvement* (Heyne, 1987). There are, we believe, two major reasons to edit a book on the physiological-ecological approach to wheat yield that makes it worthwhile: updated information and the concentration on the crop level of organization.

OBJECTIVES

Since the early 1950s worldwide wheat production has increased almost exclusively due to increases in yield. If wheat-growing areas are not expected to expand significantly, future production increases will depend more and more on our ability to keep increasing yields. As the world population keeps multiplying we should be able, at least, to maintain the rate of yield increase experienced in most of the second half of this century to satisfy the expected requirements.

As superficially discussed above, there are some indications that wheat yields not only are not increasing lately as they did during the 1950s to the 1980s, but also that they may be leveling off, at least in many countries (see also Calderini and Slafer, 1998). In this context, it may be more necessary than ever to analyze physiological-ecological processes at the crop level to help identify realistic opportunities for future breeding and management.

In this volume, we aim to bring together the views and ideas of recognized scientists from many different countries on specific issues of wheat science, concentrating on the physiological and ecological determination of wheat yield.

REFERENCES

Anderson, J.R., Dillon, J.L., Hazell, P.B.R., Cowie, A.J., and Wang, G.H. (1988). Changing variability in cereal production in Australia. *Rev. Market. Agricultural Economics,* 56, 270-286.

Annuaires del Institut International D'Agriculture (1914-1947). *Annuaires Internationales de Statistiques Agricoles.* Institut International D'Agriculture, Rome.

Briggle, L.W. and Curtis, B.C. (1987). Wheat worldwide. In *Wheat and Wheat Improvement* (Ed. E.G. Heyne). ASA-CSSA-SSSA Publishers, Madison, WI, pp. 1-32.

Calderini, D.F. and Slafer, G.A. (1998). Changes in yield and yield stability in wheat during the 20th century. *Field Crops Research,* 57, 335-347.

Evans, L.T. (1975). Crops and world food supply, crop evolution and the origins of crop physiology. In *Crop Physiology* (Ed. L.T. Evans). Cambridge University Press, Cambridge, UK, pp. 1-22.

Evans, L.T. and Peacock, W.J. (1981). *Wheat Science Today and Tomorrow.* Cambridge University Press, Cambridge, UK.

FAO (1988-1995). *FAO Yearbooks* and *FAO Processed Statistics Series 1*. Food and Agriculture Organisation of the United Nations, Rome.

Fischer, R.A. (1984). Wheat. In *Symposium on Potential Productivity of Field Crops Under Different Environments* (Eds. W.H. Smith and S.J. Banta). IRRI, Los Baños, Philippines, pp. 129-153.

Harlan, J.R. (1981). The early history of wheat. In: *Wheat Science Today and Tomorrow* (Eds. L.T. Evans and W.J. Peacock). Cambridge University Press, Cambridge, UK, pp. 1-19.

Hazell, P.B.R. (1984). Sources of increased instability in Indian and U.S. cereal production. *American Journal of Agricultural Economics*, 66, 302-311.

Heyne, E.G. (1987). *Wheat and Wheat Improvement,* Second Edition. ASA-CSSA-SSSA, Madison, WI.

Kent, N.L. and Evers, A.D. (1994). *Technology of Cereals,* Fourth Edition. Elsevier Science Ltd., Oxford, UK.

Percival, J. (1921). *The Wheat Plant.* E.P. Dutton, New York. (Reprint 1975.)

Quisenberry, K.S. and Reitz, L.P. (1967). *Wheat and Wheat Improvement.* ASA-CSSA-SSSA, Madison, WI.

Slafer, G.A. and Kernich, G.C. (1996). Have changes in yield (1900-1992) been accompanied by a decreased yield stability in Australia cereal production? *Australian Journal of Agricultural Research*, 47, 323-334.

Slafer, G.A. and Rawson, H.M. (1994). Sensitivity of wheat phasic development to major environmental factors: A re-examination of some assumptions made by physiologists and modellers. *Australian Journal of Plant Physiology*, 21, 393-426.

Slafer, G.A., Satorre, E.H., and Andrade, F.H. (1994). Increases in grain yield in bread wheat from breeding and associated physiological changes. In *Genetic Improvement of Field Crops* (Ed. G.A. Slafer). Marcel Dekker, Inc., New York, pp. 1-68.

Wittwer S.H. (1980). The shape of the things to come. In *The Biology of Crop Productivity* (Ed. P.S. Carlson). Academic Press, Inc., New York, pp. 413-459.

Chapter 2

Wheat Development

Daniel J. Miralles
Gustavo A. Slafer

INTRODUCTION

To understand the physiological and ecological aspects of wheat yield, it is necessary to study the physiology of crop development, as the effects of certain environmental factors on crop growth and yield differ depending upon the developmental stages when these factors act (see for example Fischer and Aguilar, 1975; Fischer, 1985; Thorne and Wood, 1987; Savin and Slafer, 1991; Slafer et al., 1994). In other words, grain yield is more sensitive to environmental factors during certain developmental phases than others (Landes and Porter, 1989; Slafer, Calderini, and Miralles, 1996).

Main Developmental Stages and Phases

Crop development is defined as a sequence of phenological events controlled by external factors, determining changes in the morphology and/or function of some organs (Landsberg, 1977). The development of a wheat plant can be thus described in different stages or phases which, in turn, can be defined in terms of internal or external morphological changes. However, crop development is a continuity of vegetative, reproductive, and grain-filling phases through which the crop initiates and grows its organs and completes its life cycle. The duration of each phase, and the number of

We thank Dr. J. F. Angus (CSIRO, Canberra) for critical reading of the manuscript. D. J. Miralles was on leave from the Department of Plant Production, Faculty of Agronomy, University of Buenos Aires and acknowledges the support given by CONICET (Consejo Nacional de Investigaciones Cientificas y Tecnicas, Argentina).

primordia initiated, is determined by interactions between genetic and environmental factors.

Several scales have been made to describe major developmental events. Some of them deal with development growth stages visible without dissection of the shoot apex (e.g., Large, 1954; Haun, 1973; Zadoks, Chang, and Konzak, 1974), while others describe the morphological changes in the apical meristem (e.g., Moncur, 1981; Waddington, Cartwright, and Wall, 1983; Gardner, Hess, and Trione, 1985). External morphological stages allow a nondestructive identification of developmental progress but provide no information about the sequence and timing of events in the shoot apex, where development actually occurs. Many important developmental events are not observable without apex dissection and any reported relationship between external and internal developmental events is hard to extrapolate due to the different responses of leaf appearance (and tillering) and apex development to major environmental factors, as is discussed later.

Figure 2.1 shows a simplified schematic diagram of developmental progress on an arbitrary time scale using some easily recognizable developmental features as delimiters of phases. These features include crop emergence, double ridge, which occurs some time after floral initiation (Kirby, 1990; Delécolle et al., 1989), terminal spikelet initiation, when all spikelets have been initiated, heading, anthesis, maturity, and crop harvest (see Figure 2.1).

These stages mark changes in phasic development. There are three major phases: the vegetative phase, when the leaves are initiated; the reproductive phase, when floret development occurs until the number of fertile florets (virtually the number of grains) is determined; and the grain-filling phase, when the grain first develops the endosperm cells and then grows to determine the final grain weight. These phases are delimited by sowing-floral initiation (which is only recognizable a posteriori as discussed below), floral initiation-anthesis, and anthesis-maturity (see Figure 2.1).

Before anthesis the number of grains is determined and after anthesis the grains are actually filled and the individual grain weight is established. The number of grains per unit land area and the averaged individual grain weight are the two major yield components (see Figure 2.1).

Although they do not delimit major developmental phases, the initiation of both the first double ridge and the terminal spikelet are important early reproductive markers. The former is the first visible (through a microscope) sign that the plant is undoubtedly reproductive, while the latter marks the end of the spikelet initiation phase when the final number of spikelets per spike is determined, but also under commercial field conditions it almost invariably coincides with the beginning of stem elongation, which raises the shoot apex above the soil surface.

FIGURE 2.1. Schematic Diagram of Wheat Growth and Development

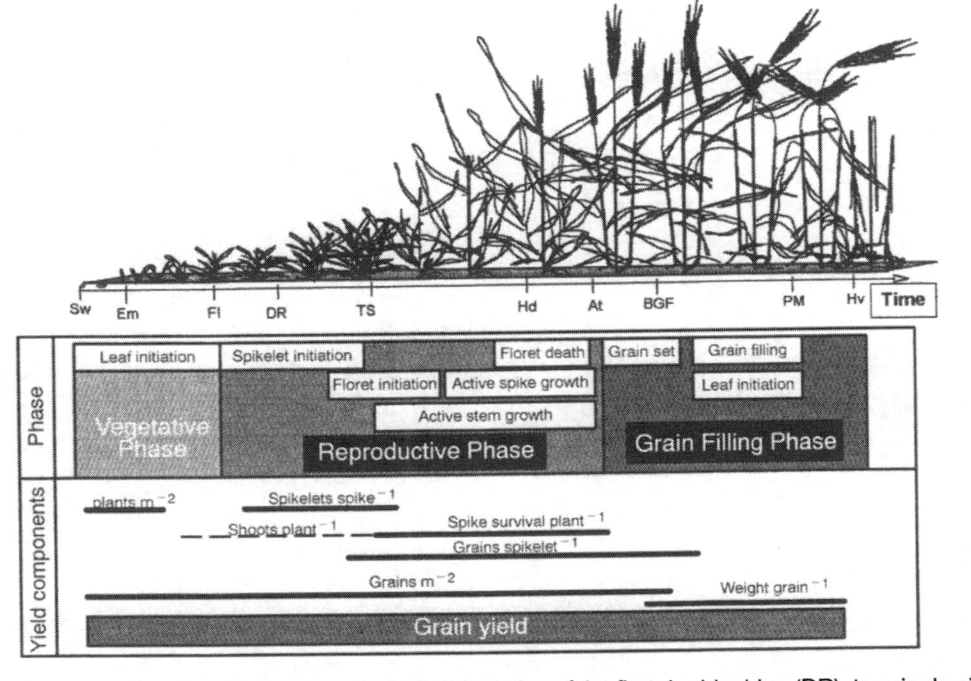

Diagram shows the stage of sowing (Sw), emergence (Em), initiation of the first double ridge (DR), terminal spikelet initiation (TS), heading (Hd), anthesis (At), beginning of the grain-filling period (BGF), physiological maturity (PM), and harvest (Hv). Boxes indicate the periods of differentiation or growth of some organs within the vegetative, reproductive, and grain-filling phases and timing when different components of grain yield are produced. Adapted from Slafer and Rawson (1994b).

Relationship with Yield Components

Figure 2.1 also shows associations between the development of the apex and the formation of the components of yield. Formally, yield components are being formed at any time during the life cycle of the crop, but undoubtedly some phases are more important in determining yield potential than others. It has been generally concluded that the period between terminal spikelet initiation and anthesis is of paramount importance (Fischer, 1984, 1985; Kirby, 1988; Siddique, Kirby, and Perry, 1989; Slafer, Andrade, and Satorre, 1990; Savin and Slafer, 1991). This is because grain yield appears to be better related to the number of grains per unit land area than to individual grain weight (Fischer, 1985; Thorne and Wood, 1987; Slafer and Andrade, 1989; Slafer and Savin, 1994).

Inspecting with more detail the preanthesis period, some authors have found associations between developmental stages and yield components (e.g., Rawson, 1970, 1971; Rawson and Bagga, 1979). If there are associations between the duration of the phases and absolute yield, then it becomes important to determine which factors do modify the duration of these phases. This is further discussed later when effects of major factors on developmental phases are analyzed.

Aim

The main objective of this chapter is to provide an updated, general revision of the dynamics of initiation and appearance of main organs during crop development, and of how genotype and environment affect the timing of developmental events. For this purpose, the effects of the major factors controlling the duration of different developmental phases are analyzed. This chapter is based in a review by Slafer and Rawson (1994b), which we refer to frequently throughout. We attempt not only to update the information from that review, but also discuss other aspects such as leaf initiation, leaf appearance, and the interrelationship between leaf appearance and primordia initiation in the apex, as well as the phasic development discussed by Slafer and Rawson (1994b).

DYNAMICS OF INITIATION AND APPEARANCE OF VEGETATIVE AND REPRODUCTIVE ORGANS

Leaf and Spikelet Initiation

Seed imbibition reinitiates the intense metabolic activity in the nondormant seed and, following the imbibition of the seed, leaf initiation is resumed in the shoot apex. The apex appears as a dome with leaf primordia being formed in its base. Depending on the depth of sowing and on the rate of leaf initiation corresponding to the cultivar, ca. one to three more leaf primordia are initiated before seedling emergence. These primordia add to those already initiated in the mother plant (the embryo in the mature grain has already initiated ca. three to four leaf primordia; Kirby and Appleyard, 1987; Hay and Kirby, 1991; see Figure 2.1). Then by seedling emergence, the shoot apex has five to seven leaf primordia, and this is considered to be the range in minimum final leaf number observable when plants are grown under the most inductive environmental conditions and/or in the least sensitive cultivars.

After seedling emergence, the shoot apex retains the shape of a dome during a period whose length depends strongly on the genotype and the environmental conditions. Then it elongates and the initiation of leaf primordia may continue as single ridges around the elongated apex (Fisher, 1973), until the onset of floral initiation.

Despite the morphological changes in the apical meristem during the vegetative phase, it appears that leaf primordia are initiated at a single rate on a thermal time basis (which implies it is highly sensitive to temperature, see Figure 2.2). The reciprocal of this rate, the thermal time between the initiation of two consecutive leaf primordia, or simply the plastochron (Wilhelm and McMaster, 1995), is thus a single value for any environmental condition (i.e., it is mostly insensitive to photoperiod or vernalization). Although there is genetic variation for plastochron (Evans and Blundell, 1994), a common value in the literature might be ca. 50°C d, above a base temperature of 0°C (Kirby et al., 1987; Delécolle et al., 1989).

Leaf initiation ceases, and the maximum number of leaves in the main shoot is determined, when the apex changes from vegetative to reproductive development at floral initiation. As mentioned above, there is no single morphological change that allows an unequivocal determination of timing of floral initiation. It is only possible to determine it a posteriori, simply by subtracting final leaf number from the accumulated number of leaf and spikelet primordia. When this is done, mostly (but not always; see Delécolle et al., 1989) the change in slope of total number of primordia vs. thermal time coincides with floral initiation (see Figure 2.2).

Some time later than floral initiation the first double ridge is initiated, when leaf and spikelet primordia appear as double ridges around the shoot apex. The lower ridge is the leaf primordium, which does not develop further, and the upper ridge is the spikelet primordium (Bonnett, 1966; Gardner, Hess, and Trione, 1985). The double-ridge stage has been frequently used as an indicator of the end of the vegetative phase, but it has

FIGURE 2.2. Leaf and Spikelet Primordia versus Thermal Time After Sowing in a Hypothetical Wheat Crop

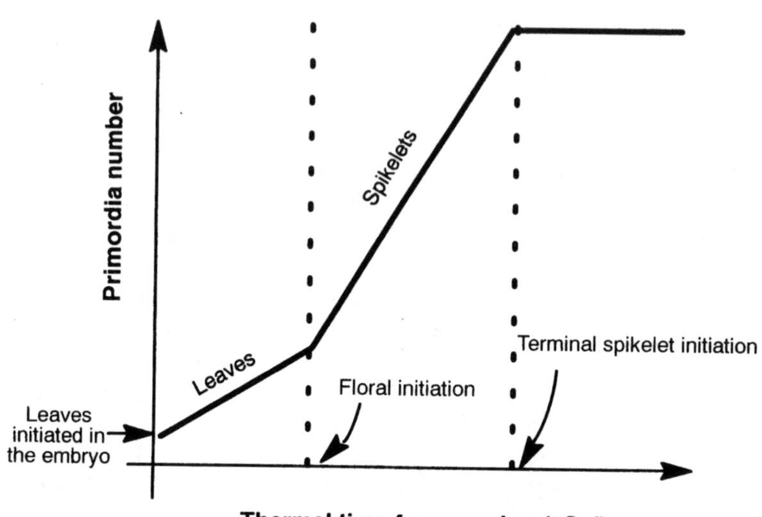

long been recognized that the first spikelet primordium may be initiated before the first double ridge appears (e.g., Delécolle et al., 1989; Kirby, 1990).

The phase of spikelet initiation that commenced at floral initiation continues until the initiation of the terminal spikelet in the apical meristem, when the maximum number of spikelets is fixed.

Leaf Appearance

Seedling emergence is frequently taken as the appearance of the tip of the first leaf through the coleoptile. This occurs very shortly after the coleoptile has emerged through the soil surface. As mentioned above, several leaves have been already initiated in the apex by the time the tip of the first-initiated leaf is appearing. This ensures a minimum duration for reproductive development given by the number of primordia that have to appear until flag leaf appearance and the length of the phyllochron (period between the appearance of two successive leaves), plus the period between flag leaf appearance and anthesis. Moreover, as the rate of leaf initiation is normally much faster than that of leaf appearance, the phyllochron has been frequently reported to be approximately twice the plastochron.

Therefore, the longer the period from seedling emergence to floral initiation, the higher the number of leaf primordia that have to appear after floral initiation (Hay and Kirby, 1991), and consequently the longer the reproductive phase from floral initiation to anthesis.

Both genetic and environmental factors have been shown to affect the rate of leaf appearance. Syme (1974), for example, concluded that for the set of cultivars included in his study semidwarf wheats produce leaves more quickly throughout growth than standard-height cultivars. During the last two decades many examples have been published exhibiting a large genetic variation in phyllochron (e.g., Halloran, 1975; Allison and Daynard, 1976; Halloran, 1977; Rahman and Wilson, 1977; Rawson et al., 1983; Stapper, 1984; Rawson, 1986; Stapper and Fischer, 1990; Kirby, 1992; Rawson, 1993; Slafer, Halloran, and Connor, 1994; Slafer and Rawson, 1997).

Effects of environmental conditions on rate of leaf appearance are evident when crops are sown at different dates or locations (Baker, Gallagher, and Monteith, 1980; Kirby, Appleyard, and Fellowes, 1982a; Delécolle et al., 1985; Kirby and Perry, 1987; Hay and Delécolle, 1989; Masle, Doussinault, and Sun, 1989; Stapper and Fischer, 1990; Cao and Moss, 1991a). Temperature has the most significant influence on rate of leaf appearance, and thus the concept of thermal time is frequently used to analyze the dynamics of leaf appearance (Gallagher, 1979; Baker, Gallagher, and Monteith, 1980; Kirby, Appleyard, and Fellowes, 1982a; Baker et al., 1986; Bauer et al., 1988; Hay and Delécolle, 1989; Kirby et al., 1989; Cao and Moss, 1991a, 1991b; Hay and Kirby, 1991; Kirby, 1992; Slafer et al., 1994; Calderini, Miralles, and Sadras, 1996; Slafer and Rawson, 1997).

Once the effect of temperature has been taken into account by using thermal time, two general behaviors have been described in the literature. While some authors have found that the rate of leaf appearance changes with plant ontogeny (Wiegand, Gerbermann, and Cuellar, 1981; Baker et al., 1986; Hay and Delécolle, 1989; Calderini, Miralles, and Sadras, 1996), many others found this rate to be constant throughout development (Gallagher, 1979; Baker, Gallagher, and Monteith, 1980; Kirby, Appleyard, and Fellowes, 1982a, 1985; Rawson et al., 1983; Bauer, Frank, and Black, 1984; Delécolle et al., 1985; Baker et al., 1986; Hunt and Chapleau, 1986; Bauer et al., 1988; Hay and Delécolle, 1989; Kirby et al., 1989; Masle, Doussinault, and Sun, 1989; Cao and Moss, 1991a, 1991b; Krenzer, Nipp, and McNew, 1991; Longnecker, Kirby, and Robson, 1993; Slafer, Halloran, and Connor, 1994; Cone, Slafer, and Halloran, 1995; Slafer and Rawson, 1995c, 1995d). Slafer and Rawson (1997) found that the apparent conflict may be reflecting a response to photoperiod, resulting in a nonlinear relationship between leaf number and thermal time. Thus, a single cultivar would present a single phyllochron or not depending on its sensitivity to photoperiod and on the daylength under which it is grown.

Disregarding the actual shape of the relationship between the rate of leaf appearance and thermal time, both the absolute photoperiod (Cao and Moss, 1989; Masle, Doussinault, and Sun, 1989; Slafer and Rawson, 1995d, 1995e) and its rate of change at seedling emergence (Baker, Gallagher, and Monteith, 1980; Kirby, Appleyard, and Fellowes, 1985; Kirby and Perry, 1987) have been suggested as factors affecting phyllochron in wheat. Photoperiod, when manipulated artificially, appears to have a general though small impact on phyllochron (Slafer and Rawson, 1995d, 1997), and a less frequent, but more marked response, when very sensitive cultivars are grown under photoperiods sufficiently short (see below). In studies where the rate of change of photoperiod was artificially manipulated and imposed as a treatment under field conditions, phyllochron became insensitive to this factor (Kirby, Appleyard, and Fellowes, 1982b; Ritchie, 1991; Slafer, Halloran, and Connor, 1994; Kernich, Slafer, and Halloran, 1995). It is likely that what was believed to be an effect of the rate of change of photoperiod from time-of-sowing experiments was simply an effect of differences in air vs. meristem temperature among sowing dates while always using air temperature for calculating thermal time (Jamieson et al., 1995). Ritchie and NeSmith (1991) discussed other inconsistencies in the use of rate of change of photoperiod at seedling emergence to predict rate of leaf appearance in wheat.

Interrelationship Between Primordia Initiation and Leaf Appearance

As both the initiation of leaf and spikelet primordia and the appearance of leaves are mostly sensitive to temperature, these processes may be analyzed as interrelated, correlating the dynamics of internal and external developmental processes. Kirby (1990) observed that, regardless of environmental and genetic differences, the relationship between leaf initiation and leaf appearance could be well described by a single, linear regression. However, when spikelet primordia are related to emerged leaves, the rate of spikelet initiation appears to be a decreasing function of total leaf number (Kirby, 1990). This evidence demonstrates a clear relationship between spikelet initiation, leaf emergence, and total number of leaves on a shoot. Therefore, this relationship allows describing the leaf and spike initiation phases in terms of the number of emerged leaves and final number of leaves initiated in the main shoot. For a detailed description of this model see Kirby (1995).

Tillering

Besides the bud corresponding to the main shoot apex, axillary tiller buds are developed in each phytomer (the basic morphological unit of the shoot consisting of leaf, internode, node, and axillary bud below the point of sheath attachment). Each of these buds has the potential to further develop into leafy tillers.

In the absence of any restriction in availability of assimilates (due to either inter- or intraplant competition), the emergence of tillers is closely related to leaf emergence: the appearance of the first primary tiller coincides with that of the fourth leaf, i.e., approximately three phyllochrons after seedling emergence. The subsequent primary tillers appear at regular intervals of one phyllochron (Masle, 1985; Porter, 1985), and thus the relationship between the number of primary tillers and the number of visible leaves on the main shoot is linear with a slope close to one and an abscissa intercept corresponding to the number of leaves appeared before the onset of tillering (normally three to four). However, tillers of different higher order (secondary, tertiary, etc.) may eventually appear from the axillary buds developed in each tiller phytomer, with a similar relationship to tiller leaf number than that described for the main shoot. Thus the overall pattern of potential tiller emergence follows a Fibonacci series (Masle, 1985). The coleoptile tiller only appears under favorable growing conditions, when it is the first tiller in the plant (Baker and Gallagher, 1983).

However, most wheat crops grow with virtually unlimited resources (in relation to their demands) only for a short initial period of their growing season, and therefore the relationship described above only holds for that short period (whose actual length is quite variable in response to the availability of resources per individual plant). Immediately after resources become limiting not all the tillers that were potentially expected to appear do so, and the rate of tiller appearance, though still positive, becomes increasingly slower than that predicted from the potential explained by the Fibonacci series. Later, not only are there not enough resources to allow new tillers to appear, but their availability is insufficient to maintain growth of all tillers and some die, in reverse of the order they appeared.

Although there is not a mechanistic relationship between the onset of tiller mortality and developmental progress, it generally coincides with the beginning of stem elongation. This likely reflects the sharp increase in the demand for assimilates and nutrients by the vegetative internodes that elongate from then on. As the beginning of stem elongation is frequently related to the initiation of the terminal spikelet on the main-shoot apex, the length of the tillering period is therefore indirectly related to plant development (see Hay and Kirby, 1991) for a particular plant density.

Floret Development and Survival

Shortly after the initiation of spikelets, floret initiation begins in the spikelets first initiated, those in the lower region of the middle third of the spike, continuing toward both ends of the spike. The development of different floret pieces begins in the basal positions of each spikelet and progresses from there toward the distal positions (e.g., Sibony and Pinthus, 1988). By the time of terminal spikelet initiation, three to five florets are commonly initiated in the central spikelets. Floret initiation within each spikelet continues approximately until ca. 200 to 300°C d before the appearance of the flag leaf ligule, when apparently no further florets are initiated (Kirby, 1988), but those initiated keep progressing in their development. In the short period from booting to heading/anthesis, many of the initiated florets abort, and only a few of the many floret primordia set fertile florets at anthesis. The maximum number of floret primordia per spikelet normally ranges between 6 and 12 (Sibony and Pinthus, 1988; Youssefian, Kirby, and Gale, 1992), mostly depending on the spikelet position. From those, only one to four (rarely five) florets complete their development to produce a fertile floret.

Kirby (1988) observed that "floret death" was coincident with the onset of rapid growth of stems and spikes, suggesting that competition for assimilates would determine the rate of floret mortality, and then the final number of fertile florets. Although floret "death" or "mortality" is the most common term in the literature, the actual process is that after the onset of rapid growth of stems and spikes relatively few primordia keep developing normally as the rate of subsequent development decreases progressively from those florets placed in positions proximal to the rachis relative to those in distal positions (Langer and Hanif, 1973; Miralles, 1997). The proportion of florets that maintain a normal rate of development after the onset of rapid spike and stem growth is related to the availability of assimilates to the growing spikes, due to an improved crop growth during this period (Fischer, 1985; Thorne and Wood, 1987; Savin and Slafer, 1991). Studies that used genotypes with different partitioning between spikes and stems also demonstrated that the higher the spike/stem ratio at anthesis, the larger the number of fertile florets and grains (Brooking and Kirby, 1981; Siddique, Kirby, and Perry, 1989; Slafer and Andrade, 1991; Slafer and Miralles, 1993; Calderini, Dreccer, and Slafer, 1995; Miralles, 1997).

Grain Filling

Wheat is a kleistogamus plant, i.e., pollination and fertilization occur prior to the extrusion of the anthers. The dynamics of fertilization follow the same pattern as spikelet differentiation and floret development across the spike (Rawson and Evans, 1970). Once the fertile florets are fertilized they become potential grains.

Grain set, the proportion of fertile florets actually producing a "normal" grain, is normally less than 100 percent, likely because of competition for assimilates (Savin and Slafer, 1991). The period of grain set is characterized by substantial grain development with virtually no grain growth, and is therefore described as the "lag phase" when the grains slowly accumulate dry matter up to ca. 5 to 10 percent of their final weight (e.g., Loss et al., 1989). During this phase, which lasts for ca. 20 to 30 percent of the total postanthesis period (Gebeyehou, Knott, and Baker, 1982) and most of the endosperm cells are developed, the potential size of the grains is defined. Endospermatic cell formation can continue during part of the subsequent grain growth phase (Nicolas, 1985).

After the lag phase, grain growth accelerates until reaching a maximum rate. For most of this period, grain weight is frequently assumed to have a lineal relationship with thermal time, and grain growth would proceed at a single, maximum rate. Toward the end of this phase, grain growth declines and grains reach their maximum dry weight, at physiological maturity. Bilineal models, assuming a single rate of grain growth and a sharp end at physiological maturity, fit the dynamics of grain growth as well as a logistic model does (e.g., Miralles, Dominguez, and Slafer, 1996).

Unless grain filling is terminated by an external environmental factor, such as severe water stress (Nicolas, Gleadow, and Dalling, 1984) or sudden heat shock (Savin, Stone, and Nicholas, 1996), the period from anthesis to physiological maturity is quite conservative in terms of degree days (and thus almost exclusively sensitive to temperature; see below).

EFFECTS OF MAJOR FACTORS ON THE DURATION OF DEVELOPMENTAL PHASES

Changes in sowing dates (Angus et al., 1981; Kirby, Appleyard, and Fellowes, 1985; Hay, 1986) or location (Bauer et al., 1988) can strongly modify the duration of different developmental phases. The major components of the environment that affect development are temperature (both low temperature associated with the vernalization requirements and temperature per se) and photoperiod (Pirasteh and Welsh, 1980; Fischer, 1984; Davidson et al., 1985; Hay and Kirby, 1991; Slafer and Rawson, 1994b). Other factors, such as level of nutrition (Rodriguez, Santa Maria, and Pomar, 1994) and water availability, plant density, and radiation (Evans

1987; Rawson, 1993) and CO_2 concentration (Rawson, 1992) may have a small effect on development when treatments are extreme (Slafer, 1995).

Of the three major factors, temperature alone has a universal impact on the rates of wheat development (Aitken, 1974). Contrasting with responses to photoperiod and vernalization, there are no developmental phases nor cultivars insensitive to temperature (Slafer and Rawson, 1995a, 1995b). Thus, in general terms, the higher the temperature the faster the rate of development and consequently the shorter time to complete a particular developmental phase (Slafer and Rawson, 1994b). Phases such as sowing-seedling emergence or anthesis-maturity in wheat are undoubtedly insensitive to both photoperiod and vernalization, and many cultivars have low sensitivity to these two factors at any phenophase.

Photoperiod and vernalization are generally considered to account for most of the differences in development rate among cultivars, and any "residual" difference after the vernalization and photoperiod requirements were satisfied would be the consequence of differences in "basic development rate" or "intrinsic earliness" (Major, 1980; Flood and Halloran, 1984; Masle, Doussinault, and Sun, 1989; but see Slafer, 1996).

Developmental Responses to Temperature

Developmental responses to temperature start as soon as the seed is imbibed (Roberts, 1988) and continues until maturity (Angus et al., 1981; Del Pozzo et al., 1987; Porter et al., 1987; Slafer and Savin, 1991). From the various models that have been proposed to predict the timing of development as affected by temperature, the most widely accepted is the thermal time (with units of degree days, °C d; Monteith, 1984), which has been widely used in simulation models (Stapper, 1984; Weir et al., 1984; Ritchie, 1991; Ritchie and NeSmith, 1991; Porter, Jamieson, and Wilson, 1993). In practice, it is taken as the summation of differences between daily mean temperature and a base temperature, which in fact is the calendar time weighted by the thermal conditions of each day.

The concept of thermal time is mechanistically based on a linear relation between rate of development and temperature, from base to optimum temperature (see Figure 2.3). Then the calculated intercept on the abscissa and the thermal conditions when the rate of development becomes maximum are the base and optimum temperatures, respectively, while the reciprocal of the slope is the thermal time. Therefore, for the thermal time to be a single value at any thermal environment the relationship described in Figure 2.3 must be linear, otherwise thermal time would have different values for different temperatures.

FIGURE 2.3. General Model for the Relationship Between Rate of Development and Temperature with a Strong Linearity Between Base (Tb) and Optimum (To) Temperatures, As Implicit in the Concept of Thermal Time

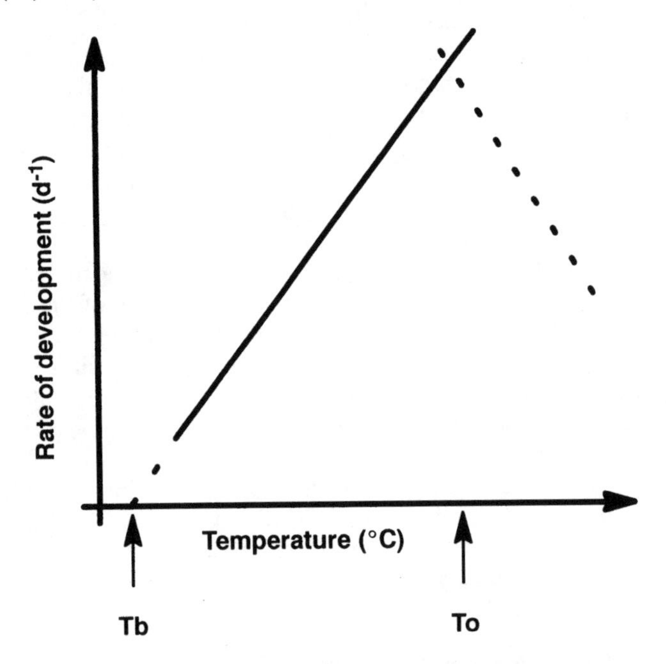

The slope ([1/d] $°C^{-1}$ = $[°C\ d]^{-1}$) represents the reciprocal of the thermal time (°C d) required for that phase using the estimated Tb.

These parameters are, however, specific for particular combinations of genotype and phenophase. Although all wheat genotypes respond to temperature by increasing the rate of development in any particular phenophase, it is possible to find differences among genotypes and phenophases in the sensitivity to temperature (Rawson and Bagga, 1979; Angus et al., 1981; Del Pozzo et al., 1987; Porter et al., 1987; Hunt, van der Poorten, and Pararajasingham, 1991; Slafer and Savin, 1991; Richards, 1992; Rawson and Richards, 1993; Slafer and Rawson, 1994b, 1995a). In fact, when curvilinear rather than linear associations are reported for the relationship between rate of development and temperature (see an example in Angus et al., 1981) it may be that the curvilinearity stems from considering a long developmental phase (e.g., from seedling emergence to anthesis) during which there were several shorter phenophases, that may have linear responses but with different sensitivities to temperature (for details see Slafer and Rawson, 1995a).

Furthermore, what has been frequently defined as "intrinsic earliness" to explain cultivar differences in rate of development in absence of photoperiod and vernalization effects could well be the result of the interaction between the genotype and temperature (for details see Slafer, 1996). Furthermore, the accuracy of the application of the thermal time concept to predict phenological development would dependend on the accuracy of the parameters chosen for its calculation (Morrison, McVetty, and Shaykewich, 1989).

The rates of leaf initiation and appearance are also linearly related to temperature (e.g., Slafer and Rawson, 1995c). The slope of the relationship is the rate of these processes per degree day, and the reciprocal of those slopes are the plastochron and phyllochron. Once again for accepting single thermal times for plastochron and phyllochron, the relationships between the rates of leaf initiation and appearance versus temperatures must be linear.

As temperature affects the rates of development during leaf initiation and the actual rate of leaf initiation in a similar way, there is virtually no effect of temperature on the number of primordia initiated and final leaf number is hardly affected by mean temperatures (Rawson and Zajac, 1993; Slafer and Rawson, 1994a).

General Models for Responses to Photoperiod and Vernalization

Vernalization and photoperiod responses provide the bases for the wide adaptability of wheat. Vernalization response is recognized as an acceleration of development following the exposure to low temperature, which is perceived directly by the apex once the seed starts the imbibition process or even during seed development in the mother plant (Purvis, 1961). Vernalization temperatures range from 0 to 15°C, being generally most effective in the range from 3 to 10°C (Weir et al., 1984). It is reasonably well established that wheat genotypes vary widely in their sensitivity to vernalization, ranging from insensitive to an obligate requirement (Slafer and Rawson, 1994b). Genes responsible for this response (Vrn/vrn) have been identified (Hoogendoorn, 1985; Flood and Halloran, 1986).

Regarding the photoperiod response, wheat is a long-day plant (the longer the daylength, the faster the rate of development). The photoperiod stimulus is perceived by the leaves and the signal is transmitted to the apex (Evans, 1987). Consequently, the plants cannot respond to photoperiod until the first leaf emerges. There is also a wide range of genetic variation in response to photoperiod and the major genes controlling the sensitivity to this factor (Ppd/ppd) have also been identified and mapped (Börner et al., 1993). Both in photoperiod and vernalization responses the dominant alleles confer insensitivity. Thus the presence of recessive (vrn and/or ppd) alleles in a noninductive environment (warm temperatures or long photoperiod) determines a longer growing cycle.

A simple model for describing the general effects of vernalization and photoperiod on the duration of a sensitive phenophase or its reciprocal, the rate of development for that phase, include two ranges for each factor, dividing the response into two types of reaction (see Figure 2.4).

One of the phases is responsive (for the below-optimum values of photoperiod and vernalization) while the other is unresponsive (for the above-optimum values of photoperiod and vernalization; Slafer, 1996). For plants developing under the first range of values, the longer the daylength or vernalization pretreatment, the shorter the length of the phase (Figure 2.4a), due to an acceleration of the rate of development (Figure 2.4b). When the optimum for photoperiod and vernalization is reached the plant shows the minimum time or the fastest rate of development for a particular developmental phase (Slafer, 1996), and no further response to these factors is found (Figure 2.4).

It must be noted that the term "optimum" is exclusively used here regarding the responses of the rate of development to either photoperiod or vernalization, i.e., values of these factors that saturate the responses reaching the maximum possible rate, which does not mean that these conditions optimize yield, time to anthesis, or any other agronomic response (Slafer, 1996). These optimum values as well as the slopes for the responses to below-optimum levels are specific for each cultivar and phenophase.

Developmental Responses to Vernalization

Although vernalization was recognized as early as 1837, and demonstrated experimentally in 1928 by Lysenko (see Flood and Halloran, 1986), little is yet known about the dynamics of its response (Hay and Kirby, 1991; Slafer and Rawson, 1994b). Vernalization mostly affects the length of the sensitive phases but does not appear to affect the rate of primordium initiation (e.g., the plastochron). Therefore, responses in length of the vegetative phase are mostly paralleled by those in final leaf number, and the latter is commonly used as an indicator of vernalization sensitivity (e.g., Levy and Peterson, 1972; Halloran, 1975, 1977; Kirby, 1992).

FIGURE 2.4. Schematic Diagram of the Changes in Duration (a) and in Rate of Development (b) for a Sensitive Phenophase Under Different Photoperiod and Vernalization Treatments

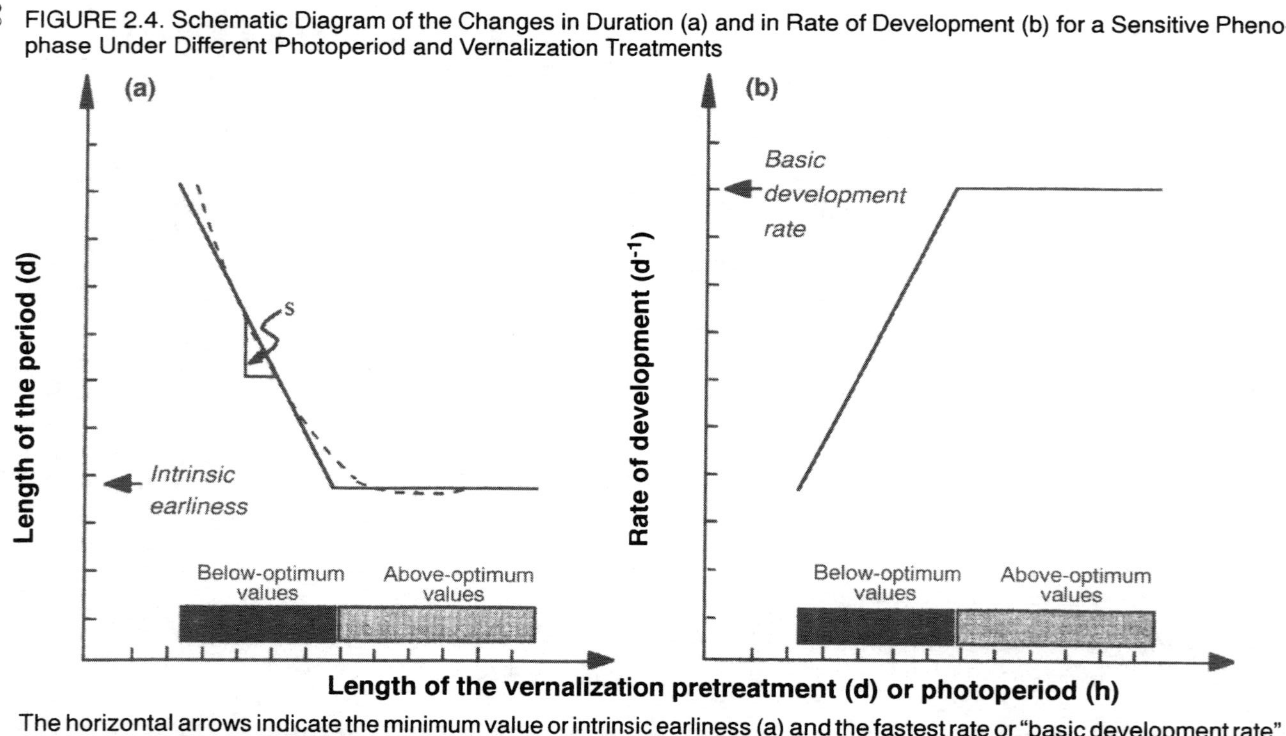

The horizontal arrows indicate the minimum value or intrinsic earliness (a) and the fastest rate or "basic development rate" (b) under above-optimum photoperiod and vernalization treatments. The dark and shaded bars in both diagrams indicate the values of below- and above-optimum photoperiods and vernalizations, while the slope of the relationship under below-optimum values represents the sensitivity (S) to these factors. Solid line in (a) represents the linear fit—by hand—of the curve (tagged line) frequently reported (adapted from Slafer, 1996).

Although it has been stated that vernalization response is characteristic of winter wheats, the fact that spring cultivars may also respond has been widely recognized (e.g., Levy and Peterson, 1972; Halloran, 1977; Jedel, Evans, and Scarth, 1986; Kirby, 1992). The difference between the spring and winter wheats may be restricted to the magnitude of vernalization response, being generally small in most spring compared with most winter types (e.g., Levy and Peterson, 1972), which is in agreement with the definition of spring and winter wheats adopted by Stapper (1984).

Most studies designed to describe general patterns of vernalization response have disregarded different parameters such as the optimum vernalization length for reaching the minimum duration of the phase, and the sensitivity to vernalization for any length of the pretreatment below the optimum (Figure 2.4). Similarly, the effects of vernalization on different phases of development have rarely been considered, mostly because it is widely accepted that vernalization affects only the length of the vegetative period to floral initiation (Halse and Weir, 1970; Flood and Halloran, 1986; Roberts et al., 1988; Ritchie, 1991). However, several authors (e.g., Halloran and Pennell, 1982; Fischer, 1984; Stapper, 1984) have recognized that vernalization also affects the duration of the following period to terminal spikelet initiation, though the impact is less than on the vegetative phase. Furthermore, Slafer and Rawson (1994b) showed (using data reported by Rahman, 1980) that vernalization affected the length of the period from double ridge to terminal spikelet initiation more than the length of the vegetative phase, at least in some cultivars. Manupeerapan and colleagues (1992) also found strong responses to vernalization during the spikelet initiation phase. It is possible that influences of vernalization extend beyond terminal spikelet initiation. For example, Masle, Doussinault, and Sun (1989) found that vernalization was associated with earlier anthesis, for a similar date of lemma appearance, which might indicate effects of this factor on the rate of reproductive development even beyond the possible indirect effect due to the increase in the number of leaves (Slafer, Connor, and Halloran, University of Melbourne, 1995, unpublished), which deserves further study.

Only very limited information is available in the literature on possible effects of vernalization on rate of leaf appearance. Cao and Moss (1991b), working with only one cultivar and a single vernalization treatment, have found no differences in phyllochron between vernalized and unvernalized plants, at least up to the time of anthesis in the vernalized plants. However, they reported a reduction in rate of leaf appearance in the unvernalized plants, supporting the idea of a direct effect of vernalization on the rate of development of reproductive phases. Clearly much more information is required before any generalization may be attempted.

A peculiarity of the response to vernalization is that the acceleration produced by exposing sensitive cultivars to cool temperatures may be reversed if the period of low temperature is interrupted, an effect known as "devernalization." This was experimentally proven by Gregory and Purvis (1948) and Purvis and Gregory (1952) using temperatures >30°C, a temperature that has since become adopted as a threshold for devernalization in wheat (Loomis and Connor, 1992). More recently, however, Dubert et al. (1992) have shown devernalization at ca. 20°C, implying that the process could be common in the field.

Developmental Responses to Photoperiod

Immediately after seedling emergence many crops exhibit a juvenile phase of insensitivity to photoperiod (also termed the "photoperiod-insensitive preinductive phase"; Ellis et al., 1992) that imposes a lower limit for the length of the vegetative phase and, thus, for the final number of leaves. The existence of a juvenile phase has been demonstrated for at least some cultivars of soybean (Collison et al., 1993), maize (Kiniry et al., 1983), barley (Roberts et al., 1988) and sunflower (Villalobos, Hall, and Ritchie, 1990). As wheat does not appear to possess a juvenile phase before it becomes sensitive to photoperiod (Hay and Kirby, 1991; Slafer and Rawson, 1995d), plants would perceive the photoperiod stimulus immediately after seedling emergence and the minimum number of leaves may coincide with the number of leaf primordia initiated by seedling emergence.

General trends in the literature suggest that a long photoperiod reduces the time to anthesis due to an acceleration of development primarily from seedling emergence to floral initiation (indirectly assessed as double ridges), or sometimes to terminal spikelet initiation (e.g., Rawson, 1971; Rahman and Wilson, 1977; Major, 1980; Rahman, 1980; Davidson et al., 1985; Rawson, 1993; Rawson and Richards, 1993). Most results in the literature also agree that the acceleration of development is associated with a reduced final number of leaf and spikelet primordia (Rawson, 1971; Wall and Cartwright, 1974; Halloran, 1977; Rahman and Wilson, 1977; Rahman, 1980; Pinthus and Nerson, 1984; Miglietta, 1991; Kirby, 1992; Rawson, 1993; Rawson and Richards, 1993; Evans and Blundell, 1994; Slafer et al., 1994; Slafer and Rawson, 1995d, 1995e). This association implies that the rate of primordium initiation on the apex is not, or is just slightly, affected by photoperiod (Miglietta, 1989; Rawson, 1993; Rawson and Richards, 1993; Evans and Blundell, 1994; Slafer and Rawson, 1995d).

Although it is usually considered that photoperiod primarily affects the rate of vegetative development (Porter and Delécolle, 1988), data from Allison and Daynard (1976), Rahman and Wilson (1977), Angus et al. (1981), Masle, Doussinault, and Sun, (1989), Connor, Theiveyanathan, and Rimmington (1992), and Manupeerapan et al. (1992) revealed that the length of the phase from terminal spikelet initiation to anthesis may be in some cultivars, even more responsive to photoperiod than vegetative or spikelet initiation phases.

It is clear from the reappraisal done by Slafer and Rawson (1994b) that all phases (vegetative, spikelet initiation, and stem growth) considered in the studies conducted by Allison and Daynard (1976) and Rahman and Wilson (1977) were sensitive to photoperiod, the general model described in Figure 2.4 being applicable to all of them. However, the phenophases considered as well as the cultivars analyzed strongly differed in the parameters of that model. For example, in some genotypes the value of the optimum photoperiod was similar for the vegetative and spikelet initiation phases, while in others this value differed slightly or strongly. The responses to photoperiod in the phase from terminal spikelet initiation to anthesis ranged from the lowest sensitivity to a strong response that may even be qualitative for very short photoperiods in some cultivars.

In a more recent work of Slafer and Rawson (1996), using wheat cultivars with different photoperiod sensitivity, it was shown that the sensitivity to photoperiod was stronger in the winter than in the spring wheats used (and intermediate in a semiwinter Australian wheat) and that the difference was even more noticeable for the later than the earlier phases of development. Furthermore, cultivars differed for most parameters of the responsiveness to photoperiod. Optimum photoperiod differed significantly among cultivars (from ca. 15 to 21 h). Calculations of optimum photoperiod for wheat derived from data by Rawson (1971) showed values from 12 to longer than 16 h for the both the vegetative and spikelet initiation phases. Of remarkable interest is that the parameters of the response to photoperiod appear to be independent of each other, and might be therefore open to independent genetic manipulation (Major, 1980; Slafer and Rawson, 1996).

Exposure to long photoperiods almost invariably reduces the time to heading in a positive relationship with reductions in the final number of leaves, but also with a frequently slight effect on the rate of leaf appearance (Stapper, 1984; Cao and Moss, 1989; Rawson, 1993; Cao and Moss, 1994; Slafer and Rawson, 1995d).

Furthermore, Slafer and Rawson (1997) have found that a change in phyllochron during plant ontogeny may be part of a response to photoperiod. In their work, it appeared that there was a unique rate of leaf appearance throughout plant development when plants were grown under relatively long photoperiods (the actual value depending on the sensitivity of the cultivar). However, when sensitive cultivars were exposed to sufficiently short photoperiods a single value of phyllochron was not appropriate for leaves at all insertions, when the relationship between number of leaves and time was better fitted by a bilinear than a linear model, implying that at some stage there is a reduction in phyllochron in response to photoperiod (Slafer and Rawson, 1997). The evidence of Slafer and Rawson (1997) in accordance with the observations of Stapper and Fischer (1990) showed that the change in phyllochron occurred at approximately the time when the sixth leaf had appeared, not associating the change in slope with any particular stage of development of the apex.

This kind of response to short photoperiods would operate mainly on the length of the reproductive phases (beyond the indirect effect that results from the increased number of leaves to appear after floral initiation). Plants with this sensitivity would extend their duration to heading well beyond what would be expected from changes in the final number of leaves. There may be room to speculate that, if genetic differences in sensitivity of phyllochron to photoperiod can be characterized, this trait could be used in breeding for higher yield potential (see Slafer, Calderini, and Miralles, 1996).

Interactions Between Factors

In general, it is accepted that if the duration of the phases of development are expressed in degree days, rather than calendar time, the response to photoperiod and vernalization would be the same for any temperature regime (Major, 1980; Major and Whelan, 1985; Evans and Blundell, 1994). However, some evidence in the literature suggests that interactions between these factors might be important (see Slafer and Rawson, 1994b). For example, all genotypes in a study carried out by Wall and Cartwright (1974) were sensitive to photoperiod and temperature, but just one (Sonora 64) showed similar responses to temperature at two levels of photoperiod. In other genotypes under long photoperiod the effect of temperature on the duration from sowing to heading was less (Tokwe, Yecora 70, and Siete Cerros) or more (Mexico 120) than under short photoperiod, clearly revealing a strong temperature/photoperiod interaction depending on the cultivar. In a more detailed study, Slafer and Rawson (1996) found that photoperiod response, like photoperiod sensitivity and optimum photoperiod, were affected by temperature and this effect varied with cultivar and developmental phase. These interactions could reflect some effects of photoperiod on the slope of the relationship between rate of development and temperature, but also there is the possibility of an effect of photoperiod on the intercept of this relationship and thus on base temperature (for details see Slafer and Rawson, 1995e).

Interactions between photoperiod and vernalization are also quite common. Davidson and colleagues (1985) examined the photoperiod and vernalization effects in a data set of 117 genotypes of wheat. A reanalysis of Davidson's data (Slafer and Rawson, 1994b) demonstrated that several cultivars did not respond to vernalization when they were grown under short photoperiods and did not respond to photoperiod unless their vernalization requirements (or part of them) were satisfied. Meanwhile most genotypes that were unresponsive to only one of these factors were highly responsive to a period of vernalization and long photoperiod in combination. Not only do photoperiod and vernalization interact strongly to determine the time to heading in wheat, but also the magnitude and direction of that interaction is cultivar dependent. As shown by Slafer and Rawson (1994b), the response to these factors can be very simple as in the cultivar Sunset, which was insensitive to photoperiod and vernalization, as well as in Steinwedel or Bezostaya, which showed an important response to one of these factors but were practically insensitive to the other. Other genotypes exhibited strong interactions between photoperiod and vernalization, including cultivars that behave as short-long day plants (Evans, 1987; Manupeerapan et al., 1992), which reach heading earlier under short than under long photoperiod if unvernalized (for details see Slafer and Rawson, 1997).

CONCLUSIONS

In this chapter developmental process and interactions with environmental variables have been described. The diversity of responses, as a consequence of the interaction between genotype and environment, makes it difficult sometimes to summarize and integrate many of these developmental processes. The information in this review could be particularly useful to incorporate into the development routine in some simulation models to more accurately predict development events. In addition, we believe that some critical crop developmental phases such as spike-growth period, which determines the sink size, could be exploited in the future to increase the number of grains per square meter and thus yield potential. Some of the developmental crop phases could be genetically manipulated by breeders as an alternative way to increase yield (Slafer, Calderini, and Miralles, 1996). However, much work must be done on the possibility of development as a way to increase yield potential, especially understanding which genes could be used in the future to modulate growth to achieve this goal.

REFERENCES

Aitken Y. (1974). *Flowering Time, Climate and Genotype.* Melbourne University Press, Melbourne.

Allison C.J. and Daynard T.B. (1976). Effect of photoperiod on development and number of spikelets of a temperate and some low-latitude wheats. *Annals of Applied Biology* 83, 93-102.

Angus J.F., Mackenzie D.H., Morton R., and Schafer C.A. (1981). Phasic development in field crops. II. Thermal and photoperiodic responses of spring wheat. *Field Crops Research* 4, 269-283.

Baker C.K. and Gallagher J.N. (1983). The development of winter wheat in the field. 1. Relation between apical development and plant morphology within and between season. *Journal of Agricultural Science* 101, 327-335.

Baker C.K., Gallagher J.N., and Monteith J.L. (1980). Daylength change and leaf appearance in winter wheat. *Plant Cell and Environment* 3, 285-287.

Baker J.T., Pinter P.J., Reginato R.J., and Kanemasu E.T. (1986). Effects of temperature on leaf appearance in spring and winter wheat cultivars. *Agronomy Journal* 78, 605-613.

Bauer A., Frank A.B., and Black A.L. (1984). Estimation of spring wheat leaf growth rates and anthesis from air temperature. *Agronomy Journal* 76, 829-835.

Bauer A., Garcia R., Kanemasu E.T., Blad B.L., Hatfield J.L., Major D.J., Regginato R.J., and Hubbard K.G. (1988). Effect of latitude on phenology of "Colt" winter wheat. *Agricultural and Forest Meteorology* 44, 131-140.

Bonnett O.I. (1966). Inflorescences of maize, wheat, rye, barley and oats: Their initiation and development. Illinois College of Agriculture, *Agricultural Experiment Station Bulletin* 721, 105 pp.

Börner A., Worland A.J., Plaschke J., Schumann E.Y., and Law C.N. (1993). Pleiotropic effects of genes for reduced height (Rth) and daylength insensitivity (Ppd) on yield and its components for wheat grown in Middle Europe. *Plant Breeding* 111, 204-216.

Brooking I.R. and Kirby E.J.M. (1981). Interrelationships between stem and ear development in winter wheat: The effects of a Norin 10 dwarfing gene Gai/Rht2. *Journal of Agricultural Science* 97, 373-381.

Calderini D.F., Dreccer M.F., and Slafer G.A. (1995). Genetic improvement in wheat yield and associated traits: A re-examination of previous results and latest trends. *Plant Breeding* 11, 108-112.

Calderini D.F., Miralles D.J., and Sadras V.O. (1996). Appearance and growth of individual leaves as affected by semidwarfism in isogenic lines of wheat. *Annals of Botany* 77, 583-589.

Cao W. and Moss D.N. (1989). Daylength effect on leaf emergence and phyllochron in wheat and barley. *Crop Science* 29, 1021-1025.

Cao W. and Moss D.N. (1991a). Phyllochron change in winter wheat with planting date and environmental changes. *Agronomy Journal* 83, 396-401.

Cao W. and Moss D.N. (1991b). Vernalization and phyllochron in winter wheat. *Agronomy Journal* 83, 178-179.

Cao W. and Moss D.N. (1994). Sensitivity of winter wheat phyllochron to environmental changes. *Agronomy Journal* 86, 63-66.

Collison S.T., Summerfield R.J., Ellis R.H., and Roberts E.H. (1993). Durations of the photoperiod-sensitive and photoperiod-insensitive phases of development to flowering in four cultivars of soybean [*Glycine max* (L.) Merrill]. *Annals of Botany* 71, 389-94.

Cone A.E., Slafer G.A., and Halloran G.M. (1995). Effects of moisture stress on leaf appearance, tillering and other aspects of development in *Triticum tauschii. Euphytica* 86, 55-64.

Connor D.J., Theiveyanathan S., and Rimmington G.M. (1992). Development, growth, water-use and yield of a spring and a winter wheat in response to time of sowing. *Australian Journal of Agricultural Research* 43, 493-516.

Davidson J.L., Christian K.R., Jones D.B., and Bremner P.M. (1985). Responses of wheat to vernalization and photoperiod. *Australian Journal of Agricultural Research* 36, 347-359.

Del Pozzo A.H., Garcia-Huidobro J., Novoa R., and Villaseca S. (1987). Relationship of base temperature to development of spring wheat. *Experimental Agriculture* 23, 21-30.

Delécolle R., Couvreur F., Pluchard P., and Varlet-Grancher C. (1985). About the leaf-daylength model under French conditions. In *Wheat Growth and Modelling* (Eds. W. Day and R.K. Atkin.). Plenum Press, New York, pp. 25-31.

Delécolle R., Hay R.K.M., Guerif M., Pluchard P., and Varlet-Grancher C. (1989). A method of describing the progress of apical development in wheat based on the time course of organogenesis. *Field Crops Research* 21, 147-160.

Dubert F., Filek M., Marcinska I., and Skoczowski A. (1992). Influence of warm intervals on the effect of vernalization and the composition of phospholipid fatty acids in seedlings of winter wheat. *Journal of Agronomy and Crop Science* 168, 133-141.

Ellis R.H., Collison S.T., Hudson D., and Patefield W.M. (1992). The analysis of reciprocal transfer experiments to estimate the durations of the photoperiod-sensitive and photoperiod-insensitive phases of plant development: An example in soya bean. *Annals of Botany* 70, 87-92.

Evans L.T. (1987). Short day induction of inflorescence initiation in some winter wheat varieties. *Australian Journal of Plant Physiology* 14, 277-286.

Evans L.T. and Blundell C. (1994). Some aspects of photoperiodism in wheat and its wild relatives. *Australian Journal of Plant Physiology* 21, 551-562.

Fischer R.A. (1984). Wheat. In *Symposium on Potential Productivity of Field Crops Under Different Environments* (Eds. W.H. Smith and S.J. Banta). IRRI, Los Baños, pp. 129-153.

Fischer R.A. (1985). Number of kernels in wheat crops and the influence of solar radiation and temperature. *Journal of Agricultural Science* 100, 447-461.

Fischer R.A. and Aguilar I.M. (1975). Yield potential in a dwarf spring wheat and the effect of carbon dioxide fertilization. *Agronomy Journal* 68, 749-755.

Fisher J.E. (1973). Developmental morphology of the inflorescence in hexaploid wheat cultivars with and without the cultivar Norin 10 in their ancestry. *Canadian Journal of Plant Science* 53, 7-15.

Flood R.G. and Halloran G.M. (1984). Basic development rate in spring wheat. *Agronomy Journal* 76, 260-264.

Flood R.G. and Halloran G.M. (1986). Genetics and physiology of vernalization response in wheat. *Advances in Agronomy* 39, 87-125.

Gallagher J.N. (1979). Field studies of cereal leaf growth. I. Initiation and expansion in relation to temperature and ontogeny. *Journal of Experimental Botany* 30, 625-636.

Gardner J.S., Hess W.M., and Trione E.J. (1985). Development of a young wheat spike: A SEM study of Chinese Spring wheat. *American Journal of Botany* 72, 548-559.

Gebeyehou G., Knott D.R., and Baker R.J. (1982). Rate and duration of grain filling in durum wheat cultivars. *Crop Science* 22, 337-340.

Gregory F.G. and Purvis O.N. (1948). Reversal of vernalization by high temperature. *Nature* 161, 859-860.

Halloran G.M. (1975). Genotype differences in photoperiodic sensitivity and vernalization response in wheat. *Annals of Botany* 39, 845-851.

Halloran G.M. (1977). Developmental basis of maturity differences in spring wheat. *Agronomy Journal* 69, 899-902.

Halloran G.M. and Pennell A.L. (1982). Duration and rate of development phases in wheat in two environments. *Annals of Botany* 49, 115-121.

Halse N.J. and Weir R.N. (1970). Effects of vernalization, photoperiod and temperature on phenological development and spikelet number of Australian wheat. *Australian Journal of Agricultural Research* 21, 383-393.

Haun J.R. (1973). Visual qualification of wheat development. *Agronomy Journal* 65, 116-119.

Hay R.K.M. (1986). Sowing date and the relationships between plant and apex development in winter cereals. *Field Crops Research* 14, 321-337.

Hay R.K.M. and Delécolle R. (1989). The setting of rates of development of wheat plants at crop emergence: Influence of the environment on rates of leaf appearance. *Annals of Applied Biology* 115, 333-341.

Hay R.K.M. and Kirby E.J.M. (1991). Convergence and synchrony—A review of the coordination of development in wheat. *Australian Journal of Agricultural Research* 42, 661-700.

Hoogendoorn J. (1985). The basis of variation in date of ear emergence under field conditions among the progeny of a cross between two winter wheat varieties. *Journal of Agricultural Science* 104, 493-500.

Hunt L.A. and Chapleau A.M. (1986). Primordia and leaf production in winter wheat, triticale, and rye under field conditions. *Canadian Journal of Botany* 64, 1972-1976.

Hunt L.A., van der Poorten G., and Pararajasingham S. (1991). Postanthesis temperature effects on duration and rate of grain filling in some winter and spring wheats. *Canadian Journal of Plant Science* 71, 609-617.

Jamieson P.D., Brooking I.R., Porter J.R., and Wilson D.R. (1995). Prediction of leaf appearance in wheat: A question of temperature. *Field Crops Research* 41, 35-44.

Jedel P.E., Evans L.E., and Scarth R. (1986). Vernalization responses of a selected group of spring wheat (*Triticum aestivum* L.) cultivars. *Canadian Journal of Plant Science* 66, 1-9.

Kernich G.C., Slafer G.A., and Halloran G.M. (1995). Barley development is unaffected by rate of change of photoperiod. *Journal of Agricultural Science* 124, 379-388.

Kiniry J.R., Ritchie J.T., Musser R.L., Flint E.P., and Iwig W.C. (1983). The photoperiod sensitive interval in maize. *Agronomy Journal* 75, 687-690.

Kirby E.J.M. (1988). Analysis of leaf, stem and ear growth in wheat from terminal spikelet stage to anthesis. *Field Crops Research* 18, 127-140.

Kirby E.J.M. (1990). Co-ordination of leaf emergence and leaf and spikelet primordium initiation in wheat. *Field Crops Research* 25, 253-264.

Kirby E.J.M. (1992). A field study of the number of main shoot leaves in wheat in relation to vernalization and photoperiod. *Journal of Agricultural Science* 118, 271-278.

Kirby E.J.M. (1995). Factors affecting rate of leaf emergence in barley and wheat. *Crop Science* 35, 11-19.

Kirby E.J.M and Appleyard M. (1987). *Cereal Development Guide.* NAC Cereal Unit, Stoneleigh, UK, 95 pp.

Kirby E.J.M. and Perry M.W. (1987). Leaf emergence rates of wheat in a Mediterranean environment. *Australian Journal of Agricultural Research* 38, 455-464.

Kirby E.J.M., Appleyard M., and Fellowes G. (1982a). Effect of sowing date on the temperature response of leaf emergence and leaf size in barley. *Plant, Cell and Environment* 5, 477-484.

Kirby E.J.M., Appleyard M., and Fellowes G. (1982b). Rate of change of day-length and leaf emergence. *Annual Report of the Plant Breeding Institute for 1982,* Plant Breeding Institute, Cambridge, UK, 115.

Kirby E.J.M., Appleyard M., and Fellowes G. (1985). Variation in development of wheat and barley in response to sowing date and variety. *Journal of Agricultural Science* 104, 383-396.

Kirby E.J.M., Porter J.R., Day W., Adam J.S., Appleyard M., Ayling S., Baker C.K., Belford R.K., Biscoe P.V., Chapman A., Fuller M.P., Hampson J., Hay R.K.M., Matthews S., Thompson W.J., Weir A.H., Willington V.B.A., and Wood D.W. (1987). An analysis of primordium initiation in Avalon winter wheat crops with different sowing dates and at nine sites in England and Scotland. *Journal of Agricultural Science* 109, 107-121.

Kirby E.J.M., Siddique K.H.M., Perry M.W., Kaesehagen D., and Stern W.R. (1989). Variation in spikelet initiation and ear development of old and modern Australian wheat varieties. *Field Crops Research* 20, 113-128.

Krenzer E.G. and Nipp T.L. (1991). Mainstem leaf development and tiller formation in wheat cultivars. *Agronomy Journal* 83, 667-670.

Krenzer E.G., Nipp T.L., and McNew R.W. (1991). Winter wheat mainstem leaf appearance and tiller formation vs. moisture treatment. *Agronomy Journal* 83, 663-667.

Landes A. and Porter J.R. (1989). Comparison of scales used for categorising the development of wheat, barley, rye and oats. *Annals of Applied Biology* 115, 343-360.

Landsberg J.J. (1977). Effects of weather on plant development. In *Environmental Effects on Crop Physiology* (Eds. J.J. Landsberg and C.V. Cutting). Academic Press, London, pp. 289-307.

Langer R.H.M. and Hanif M. (1973). A study of floret development in wheat (*Triticum aestivum* L.). *Annals of Botany* 37, 743-751.

Large E.C. (1954). Growth stages of cereals. Illustration of Feecks scale. *Plant Pathology* 3, 128-129.

Levy J. and Peterson M.L. (1972). Response of spring wheats to vernalisation and photoperiod. *Crop Science* 12, 487-490.

Longnecker N.E., Kirby E.J.M., and Robson A.D. (1993). Leaf emergence, tiller growth and apical development of nitrogen-deficient spring wheat. *Crop Science* 33, 154-160.

Loomis R.S. and Connor D.J. (1992). *Crop Ecology: Productivity and Management in Agricultural Systems.* Cambridge University Press, Cambridge.

Loss S.P., Kirby E.J.M., Siddique K.H.M., and Perry M.W. (1989). Grain growth and development of old and modern Australian wheats. *Field Crops Research* 21, 131-146.

Major D.J. (1980). Photoperiod response characteristics controlling flowering of nine crop species. *Canadian Journal of Plant Science* 60, 777-784.

Major D.J. and Whelan E.D.P. (1985). Vernalization and photoperiod response characteristics of a reciprocal substitution series of Rescue and Cadet hard red spring wheat. *Canadian Journal of Plant Science* 65, 33-39.

Manupeerapan T., Davidson J.L., Pearson C.J., and Christian K.R. (1992). Differences in flowering responses of wheat to temperature and photoperiod. *Australian Journal of Agricultural Research* 43, 575-584.

Masle J. (1985). Competition among tillers in winter wheat: Consequences for growth and development of the crop. In *Wheat Growth and Modelling* (Eds. W. Day and R.K. Atkin). Plenum Press, New York, pp. 33-54.

Masle J., Doussinault G., and Sun B. (1989). Response of wheat genotypes to temperature and photoperiod in natural conditions. *Crop Science* 29, 712-721.

Miglietta F. (1989). Effect of photoperiod and temperature on leaf initiation rates in wheat (*Triticum* spp.). *Field Crops Research* 21, 121-130.

Miglietta F. (1991). Simulation of wheat ontogenesis. II. Predicting dates of ear emergence and main stem final leaf number. *Climate Change* 1, 151-160.

Miralles D.J. (1997). Determinantes fisiologicos del crecimiento, el rendimiento y la generacion del numero y el peso de los granos en lineas isogenicas de trigo. PhD thesis, University of Buenos Aires, Buenos Aires.

Miralles D.J., Dominguez C.F., and Slafer G.A. (1996). Relationship between grain growth and postanthesis leaf area duration in dwarf, semidwarf and tall isogenic lines of wheat. *Journal of Agronomy and Crop Science* 177, 115-122.

Moncur M.W. (1981). Floral initiation in field crops. An atlas of scanning electron micrographs. Division of Land Use Research, CSIRO Canberra, Australia.

Monteith J.L. (1984). Consistency and convenience in the choice of units for agricultural science. *Experimental Agriculture* 20, 105-117.

Morrison M.J., McVetty P.B.E., and Shaykewich C.F. (1989). The determination and verification of a baseline temperature for the growth of Westar summer rape. *Canadian Journal of Plant Science* 69, 455-464.

Nicolas M.E. (1985). Effects of post-anthesis drought on wheat. PhD thesis, School of Agriculture and Forestry, University of Melbourne, Melbourne, Australia.

Nicolas M.E., Gleadow R.M., and Dalling M.J. (1984). Effects of drought and high temperature on grain growth in wheat. *Australian Journal of Plant Physiology* 11, 553-556.

Pinthus M.J. and Nerson H. (1984). Effects of photoperiod at different growth stages on the initiation of spikelet primordia in wheat. *Australian Journal of Plant Physiology* 11, 17-22.

Pirasteh B. and Welsh J.R. (1980). Effect of temperature on the heading date of wheat cultivars under a lengthening photoperiod. *Crop Science* 20, 453-456.

Porter J.R. (1985). Approaches to modelling canopy development in wheat. In *Wheat growth and modelling* (Eds. W. Day and R.K. Atkin). Plenum Press, New York, 69-81.

Porter J.R. and Delécolle R. (1988). Interaction of temperature with other environmental factors in controlling the development of plants. In *Plants and Temperature* (Eds. S.P. Long and F.I. Woodward). The Company of Biologists Limited, Cambridge, UK, pp. 133-156.

Porter J.R., Jamieson P.D., and Wilson D.R. (1993). Comparison of the wheat simulation models AFRCWHEAT2, CERES-Wheat and SWHEAT for non-limiting conditions of crop growth. *Field Crops Research* 33, 131-157.

Porter J.R., Kirby E.J.M., Day W., Adam J.S., Appleyard M., Ayling S., Baker C.K., Beale P., Belford R.K., Biscoe P.V., Chapman A., Fuller M.P., Hampson J., Hay R.K.M., Hough M.N., Matthews S., Thompson W.J., Weir A.H., Willington V.B.A., and Wood D.W. (1987). An analysis of morphological development stages in Avalon winter wheat crops with different sowing dates and at ten sites in England and Scotland. *Journal of Agricultural Science* 109, 107-121.

Purvis O.N. (1961). The physiological analysis of vernalisation. *Encyclopedia of Plant Physiology* 16, 76-122.

Purvis O.N. and Gregory F.G. (1952). Studies in vernalization in cereals XII. The reversibility by high temperature of the vernalized condition in Petkus winter rye. *Annals of Botany* 1, 1-21.

Rahman M.S. (1980). Effect of photoperiod and vernalization on the rate of development and spikelet number per ear in 30 varieties of wheat. *Journal of the Australian Institute of Agricultural Science* 46, 68-70.

Rahman M.S. and Wilson J.H. (1977). Determination of spikelet number in wheat. I. Effects of varying photoperiod on ear development. *Australian Journal of Agricultural Research* 28, 565-574.

Rawson H.M. (1970). Spikelet number, its control and relation to yield per ear. *Australian Journal of Biological Sciences* 23, 1-5.

Rawson H.M. (1971). An upper limit for spikelet number per ear in wheat as controlled by photoperiod. *Australian Journal of Agricultural Research* 22, 537-546.

Rawson H.M. (1986). High temperature tolerant wheat: A description of variation and a search for some limitations to productivity. *Field Crops Research* 14, 197-212.

Rawson H.M. (1992). Plant responses to temperature under conditions of elevated CO_2. *Australian Journal of Botany* 40, 473-490.

Rawson H.M. (1993). Radiation effects on development rate in a spring wheat grown under different photoperiods and high and low temperatures. *Australian Journal of Plant Physiology* 20, 719-727.

Rawson H.M. and Bagga A.K. (1979). Influence of temperature between floral initiation and flag leaf emergence on grain number in wheat. *Australian Journal of Plant Physiology* 6, 391-400.

Rawson H.M. and Evans L.T. (1970). The pattern of grain growth within the ear of wheat. *Australian Journal of Biological Science* 23, 753-764.

Rawson H.M., Hindmarsh J.H., Fischer R.A., and Stockman Y.M. (1983). Changes in leaf photosynthesis with plant ontogeny and relationships with yield per ear in wheat cultivars and 120 progeny. *Australian Journal of Plant Physiology* 10, 503-14.

Rawson H.M. and Richards R.A. (1993). Effects of high temperature and photoperiod on floral development in wheat isolines differing in vernalisation and photoperiod genes. *Field Crops Research* 32, 181-192.

Rawson H.M. and Zajac M. (1993). Effects of higher temperatures, photoperiod and seed vernalisation on development in two spring wheats. *Australian Journal of Plant Physiology* 20, 211-222.

Richards R.A. (1992). The effect of dwarfing genes in spring wheat in dry environments. II. Growth, water use and water use efficiency. *Australian Journal of Agricultural Research* 43, 529-539.

Ritchie J.T. (1991). Wheat phasic development. In *Modelling Plant and Soil Systems* (Eds. J. Hanks and J.T. Ritchie). American Society of Agronomy, Madison, WI, pp. 31-54.

Ritchie J.T. and NeSmith D.S. (1991). Temperature and crop development. In *Modelling Plant and Soil Systems* (Eds. J. Hanks and J.T. Ritchie). American Society of Agronomy, Madison, WI, pp. 5-29.

Roberts E.H. (1988). Temperature and seed germination. In *Plants and Temperature* (Eds. S.P. Long and F.I. Woodward). Symposia of the Society for Experimental Biology Number 42, The Company of Biologists Limited, Cambridge, UK), pp. 109-132.

Roberts E.H., Summerfield R.J., Cooper J.P., and Ellis R.H. (1988). Environmental control of flowering in barley (*Hordeum vulgare* L.). I. Photoperiod limits to long-day responses, photoperiod-insensitive phases and effects of low-temperature and short-day vernalization. *Annals of Botany* 62, 127-144.

Rodriguez D., Santa Maria G.E., and Pomar M.C. (1994). Phosphorus deficiency affects the early development of wheat plants. *Journal of Agronomy and Crop Science* 173, 69-72.

Savin R. and Slafer G.A. (1991). Shading effects on the yield of an Argentinian wheat cultivar. *Journal of Agricultural Science* 116, 1-7.

Savin R., Stone P.J., and Nicolas M.E. (1996). Responses of grain growth and malting quality of barley to short periods of high temperature in field studies using portable chambers. *Australian Journal of Agricultural Research* 47, 465-477.

Sibony M. and Pinthus M.J. (1988). Floret initiation and development in spring wheat (*Triticum aestivum* L.). *Annals of Botany* 62, 473-479.

Siddique K.H.M., Kirby E.J.M., and Perry M.W. (1989). Ear-to-stem ratio in old and modern wheats; relationship with improvement in number of grains per ear and yield. *Field Crops Research* 21, 59-78.

Slafer G.A. (1995). Wheat development as affected by radiation at two temperatures. *Journal of Agronomy and Crop Science* 17, 249-263.

Slafer G.A. (1996). Differences in phasic development rate amongst wheat cultivars independent of responses to photoperiod and vernalization. A viewpoint of the intrinsic earliness hypothesis. *Journal of Agricultural Science* 126, 403-419.

Slafer G.A. and Andrade F.H. (1989). Genetic improvement in bread wheat (*Triticum aestivum*) yield in Argentina. *Field Crops Research* 21, 289-296.

Slafer G.A. and Andrade F.H. (1991). Changes in physiological attributes of the dry matter economy of bread wheat (*Triticum aestivum*) through genetic improvement of grain yield potential at different regions of the world. A review. *Euphytica* 58, 37-49.

Slafer G.A. and Miralles D.J. (1993). Fruiting efficiency in three bread wheat (*Triticum aestivum*) cultivars released at different eras. Number of grains per spike and grain weight. *Journal of Agronomy and Crop Science*, 170, 251-260.

Slafer G.A. and Rawson H.M. (1994a). Does temperature affect final numbers of primordia in wheat? *Field Crops Research* 39, 111-117.

Slafer G.A. and Rawson H.M. (1994b). Sensitivity of wheat phasic development to major environmental factors: A re-examination of some assumptions made by physiologists and modellers. *Australian Journal of Plant Physiology* 21, 393-426.

Slafer G.A. and Rawson H.M. (1995a). Base and optimum temperatures vary with genotype and stage of development in wheat. *Plant Cell & Environment* 18, 671-679.

Slafer G.A. and Rawson H.M. (1995b). Intrinsic earliness and basic development rate assessed for their response to temperature in wheat. *Euphytica* 83, 175-183.

Slafer G.A. and Rawson H.M. (1995c). Rates and cardinal temperatures for processes of development in wheat: effects of temperature and thermal amplitude. *Australian Journal of Plant Physiology* 22, 913-926.

Slafer G.A. and Rawson H.M. (1995d). Development in wheat as affected by timing and length of exposure to long photoperiod. *Journal of Experimental Botany* 46, 1877-1886.

Slafer G.A. and Rawson H.M. (1995e). Photoperiod × temperature interactions in contrasting wheat genotypes: Time to heading and final leaf number. *Field Crops Research* 44, 73-83.

Slafer G.A. and Rawson H.M. (1996). Responses to photoperiod change with phenophase and temperature during wheat development. *Field Crops Research* 46, 1-13.

Slafer G.A. and Rawson H.M. (1997). Phyllochron in wheat as affected by photoperiod under two temperature regimes. *Australian Journal of Plant Physiology* 24, 151-158.

Slafer G.A. and Savin R. (1991). Developmental base temperature in different phenological phases of wheat (*Triticum aestivum*). *Journal of Experimental Botany* 42, 1077-1082.

Slafer G.A. and Savin R. (1994). Source-sink relationships and grain mass at different positions within the spike in wheat. *Field Crops Research* 37, 39-49.

Slafer G.A., Andrade F.H., and Satorre E.H. (1990). Genetic-improvement effects on pre-anthesis physiological attributes related to wheat grain yield. *Field Crops Research* 23, 255-263.

Slafer G.A., Calderini D.F., and Miralles D.J. (1996). Yield components and compensation in wheat: Opportunities for further increasing yield potential. In *Increasing Yield Potential in Wheat: Breaking the Barriers* (Eds. M.P. Reynolds, S. Rajaram, and A. McNab), CIMMYT, Mexico City, pp. 101-133.

Slafer G.A., Calderini D.F., Miralles D.J., and Dreccer M.F. (1994). Preanthesis shading effects on the number of grains of three bread wheat cultivars of different potential number of grains. *Field Crops Research* 36, 31-39.

Slafer G.A., Halloran G.M., and Connor D.J. (1994). Development rate in wheat as affected by duration and rate of change of photoperiod. *Annals of Botany* 73, 671-677.

Stapper M. (1984). SIMTAG: A simulation model of wheat genotypes. In *Model Documentation.* University of New England, Armidale, Australia, and ICARDA, Aleppo, Syria.

Stapper M. and Fischer R.A. (1990). Genotype, sowing date and planting spacing influence on high-yielding irrigated wheat in southern New South Wales. I. Phasic development, canopy growth and spike production. *Australian Journal of Agricultural Research* 41, 997-1019.

Syme J.R. (1974). Leaf appearance rate and associated characters in some Mexican and Australian wheats. *Australian Journal of Agricultural Research* 25, 1-7.

Thorne G.N. and Wood D.W. (1987). Effects of radiation and temperature on tiller survival, grain number and grain yield in winter wheat. *Annals of Botany* 59, 413-426.

Villalobos F.J., Hall A.J., and Ritchie J.T. (1990). Oilcrop-sun: a crop growth and development simulation model of the sunflower. Phenology model calibration and validation. In *Proceedings of the Inaugural Congress* (Ed. A. Scaife), European Society of Agronomy, Paris, pp. 2-35.

Waddington S.R., Cartwright P.M., and Wall P.C. (1983). A quantitative scale of spike initial and pistil development in barley and wheat. *Annals of Botany* 51, 119-130.

Wall P.C. and Cartwright P.M. (1974). Effects of photoperiod, temperature and vernalization on the phenology and spikelet number of spring wheats. *Annals of Applied Biology* 72, 299-309.

Weir A.H., Bragg P.L., Porter J.R., and Rayner J.H. (1984). A winter wheat crop simulation model without water or nutrient limitations. *Journal of Agricultural Science* 102, 371-382.

Wiegand C.L., Gerbermann A.H., and Cuellar J.A. (1981). Development and yield of hard red winter wheats under semitropical conditions. *Agronomy Journal* 73, 29-37.

Wilhelm W.W. and McMaster G.S. (1995). Importance of the phyllochron in studying development and growth in grasses. *Crop Science* 35, 1-3.

Youssefian S., Kirby E.J.M., and Gale M.D. (1992). Pleiotropic effects of the G.A. insensitive Rht dwarfing gene in wheat. 2. Effects on leaf, stem and ear growth. *Field Crops Research* 28, 191-210.

Zadoks J.C., Chang T.T., and Konzak C.F. (1974). A decimal code for the growth stage of cereals. *Weed Research* 14, 415-421.

Chapter 3

Physiological and Numerical Components of Wheat Yield

James R. Frederick
Philip J. Bauer

INTRODUCTION

Wheat (*Triticum aestivum* L.) grain yields have increased substantially since the 1950s (Slafer, Satorre, and Andrade, 1993; Simmons, 1987; Schmidt, 1984). About half of this yield improvement has resulted from genetic increases in grain yield and the other half is due to improved production technologies and practices. Genetic improvement in wheat yield can be attributed to selection for improved agronomic characteristics conferring either higher yield potential or greater stress tolerance (Slafer, Satorre, and Andrade, 1993; Simmons, 1987; Schmidt, 1984). The exact physiological basis for the genetic gain in grain yield potential is unknown. However, it stands to reason that the physiological processes controlling wheat yield have been altered during the course of yield improvement, and that new cultural practices have changed how these processes affect wheat growth and development.

Crop physiologists have historically taken a reductionist approach when trying to identify the factors that control yield, usually assuming that improvements in only one or a few processes will result in substantial yield increases. Processes identified as having a significant effect on wheat growth and development include nutrient uptake and metabolism, photosynthesis and respiration, carbon partitioning, leaf senescence, and plant water relations. It is now generally believed that wheat grain yield is a function and integration of all these processes, each of which can be altered by the climatic conditions during the growing season and the cultural practices used to produce the crop. Identifying and understanding how these processes interact to regulate yield will aid in the development and selection of higher-yielding cultivars for specific production systems. Knowledge of how the different physiological processes interact to control yield will also allow scientists to more accurately predict wheat yield responses to management practices.

In this chapter, we will discuss the physiological traits associated with wheat yield and how these traits affect one another. For wheat, these physiological traits affect the number of kernels per m^2 and/or individual kernel weight. A more thorough discussion of how biotic and abiotic stresses affect wheat growth and development is given in Chapters 2, 4, 5, 8, 9, and 10. We recognize that the various classes of wheat grown throughout the world differ in many important physiological and agronomic characteristics, such as vernalization requirement (see Chapter 2) and grain quality (see Chapter 5). However, there are many physiological traits are common to all of these classes of wheat, which will be the focus of our discussion. As will be pointed out, many of these physiological processes occur not only in wheat, but also in a similar manner in other grain crops, such as soybean *(Glycine max* (L.) Merr.) and corn (*Zea mays* L.).

FACTORS AFFECTING KERNEL NUMBER

The number of kernels per m^2 is determined by the number of kernel-bearing tillers per m^2 and the number of kernels per spike. Many factors affect tiller initiation and survival, such as genotype, class of wheat (winter versus spring), cultural practices used (planting date, seeding rate, and soil fertility), and growing conditions (air and soil temperatures, soil water conditions). The effects of these factors are more thoroughly reviewed in other chapters. One aspect of tiller development we would like to discuss briefly is the relationship between timing of tiller appearance and tiller fertility. In general, only those early-formed tillers that are initiated when fewer than four to six leaves are on the main stem will survive to produce a fertile spike (Kirby, 1983). The exact physiological control of this tiller regulation is not known. However, this synchrony in tiller age contributes to a similar anthesis date among tillers on the same plant. Spikelet initiation has been found to occur earlier with respect to leaf appearance and at a faster rate on tillers than on the main stem (Stern and Kirby, 1979), which should also contribute to a synchronized anthesis date among tillers. A similar anthesis date among tillers results in the various stages of grain fill and kernel dry down occurring at about the same time for all spikes on the plant.

The other component of kernel number per m^2 is the number of kernels per spike. For bread wheat, results from the comparison of old and modern cultivars suggest that yield improvement has primarily been the result of an increase in the number of kernels per spike and, since spike number has changed little, kernel number per m^2 (see review by Slafer, Satorre, and Andrade, 1993). A very close relationship between grain yield and kernel number per m^2 also occurs in soft red winter wheat (see Figure 3.1). Compared to older cultivars, the higher kernel number per spike for newer cultivars is associated with a greater ratio of spike to total aboveground dry weight at anthesis and a greater partitioning of assimilate to spike formation (and less to vegetative structures) during the three-week period prior to anthesis (Slafer, Andrade, and Satorre, 1990; Siddique, Kirby, and Perry, 1989). This period is also when the flag and penultimate (leaf below

FIGURE 3.1. Soft Red Winter Wheat Grain Yield As a Function of Kernel Number per Square Meter for Four Wheat Cultivars

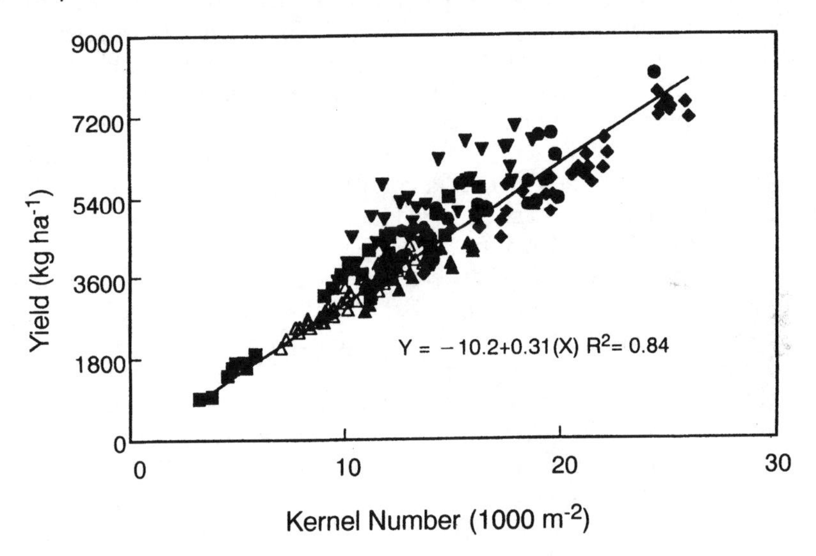

Note: Grown with two levels of spring-applied N in 1994 (●) and 1995 (▼) as reported by Frederick (1997); wheat grown with different rates of spring-applied N under irrigated and nonirrigated conditions in 1992 (◆) and 1993 (■) as reported by Frederick and Camberato (1995a); and wheat produced with different combinations of surface tillage and deep tillage in 1994 (▲) and 1995 (Δ) as reported by Frederick and Bauer (1997) and in 1996 (○) (unpublished data). Each data point is a plot value for each treatment. Regression equation was developed using individual plot values.

flag) leaves are formed, the two most important leaves with regard to photosynthate production during grain fill. Therefore, the occurrence of photosynthesis-reducing stress during this time should decrease both sink demand (number of potential kernels per head) and source supply (leaf area index) for photosynthate.

If genetic yield improvement is the result of a greater partitioning of assimilate to spike formation prior to anthesis, then factors which increase photosynthate supply during that time, such as a high leaf area index and a high leaf photosynthetic rate, should be important in establishing a high kernel number per m^2 in newer cultivars. Research results indicate these associations occur in wheat. Reductions in leaf photosynthesis and leaf area index due to drought occurring prior to anthesis are correlated with reductions in the number of kernels per spike (Frederick and Camberato, 1995a, 1995b). The uptake rate of most essential wheat nutrients is maximum during stem elongation (Karlen and Sadler, 1990). For example, between 70 and 90 percent of the total aboveground N (nitrogen) is accumulated prior to anthesis (Oscarson et al., 1995; Dalling, 1985; Cregan and van Berkum, 1984). Therefore, nutrient deficiencies during this period should reduce both vegetative development and spike formation. In support of this, we found the application of below recommended rates of spring-applied N to result in lower leaf N concentrations and photosynthetic rates prior to anthesis, lower leaf area indices near anthesis, and fewer kernels per m^2 at maturity, compared to instances when higher N rates are applied (Frederick and Camberato, 1995a, 1995b).

The observed relationship between leaf photosynthesis and kernel number per spike has led to the speculation that spike development is directly regulated by photosynthate availability. Mohaptra, Aspinall, and Jenner (1982) and Scott, Dougherty, and Langer (1975) reported that sucrose supply appears to be important for spikelet development. Shading during floret development has been found to result in fewer florets and fewer kernels per spikelet (Stockman, Fischer, and Brittain, 1983). Loss of leaf area prior to anthesis due to insect feeding and/or foliar diseases reduces leaf photosynthate production and the number of kernels per m^2. As stated above, drought stress reduces both leaf photosynthesis (Frederick and Camberato, 1995b) and kernel number per spike (Frederick and Camberato, 1995a). However, the loss of a water potential gradient in growing spike tissue, rather than reduced carbohydrate supply, may be the primary cause for reduced spike growth under drought stress, as has been reported for vegetative growth (Nonami and Boyer, 1989).

For soft red winter wheat, leaf area index at the boot growth stage (Feekes growth stage 10.0, as described by Large, 1954), total aboveground dry weight near anthesis, and the number of kernels per m^2 at maturity are highly correlated with one another (see Figure 3.2). The good correlation between leaf area and kernel number per m^2 suggests that the number of potential kernels formed and the plant's potential to produce photosynthate to fill these kernels are either directly or indirectly coupled in wheat. The relationship between leaf area index and kernels per m^2 was linear over the entire range of leaf area indices measured. The lack of a plateau in this relationship suggests that additional yield improvement could be made for high-yielding environments by selecting for higher leaf area indices, at least for soft red winter wheat grown in the southeastern United States. However, a point of diminishing return would eventually be obtained with this approach as the wheat leaves begin to shade one another. When this point of increased shading is obtained, Austin et al. (1976) proposed that a more vertical leaf orientation prior to anthesis should be selected for. The relationship between leaf area index and kernel number per m^2 shown in Figure 3.2 also suggests that yield improvement for cultivars to be grown in lower-yielding environments should involve selection for increased stress tolerance, so that leaf area production and spike formation will continue during periods of adverse growing conditions.

FACTORS AFFECTING INDIVIDUAL KERNEL WEIGHT

In contrast to kernel number per m^2, there appears to be little relationship between individual kernel weight and grain yield in soft red winter wheat (see Figure 3.3). Reductions in kernel weight generally have less effect on wheat yield than kernel number. For example, shading wheat during the grain-filling period reduces individual kernel weight, but has much less effect on grain yield (see discussion by Slafer, Satorre, and Andrade, 1993). Slafer, Satorre, and Andrade (1993) reported that several studies have shown that kernel weight has been reduced during the course of yield improvement in bread wheat. The lack of association between kernel weight and yield indicates that there has been little success at simultaneously increasing both kernel number per m^2 and kernel weight in wheat. The major limitations to increasing kernel weight under conditions of high kernel numbers are not known, although these limitations must affect the rate and/or duration of grain fill. It seems likely that photosynthate production during grain fill somehow limits kernel weight, as 70 to 90 percent of the grain dry weight comes from photosynthate made during the grain-filling period (Austin et al., 1977; Bidinger, Musgrave, and Fischer, 1977). Only when prolonged periods of photosynthesis-reducing stress occur during grain fill does photosynthate made prior to anthesis

FIGURE 3.2. Vegetative Dry Weight As a Function of Leaf Area Index, Kernel Number per m^2 As a Function of Vegetative Dry Weight, and Kernel Number per m^2 As a Function of Leaf Area Index for Soft Red Winter Wheat

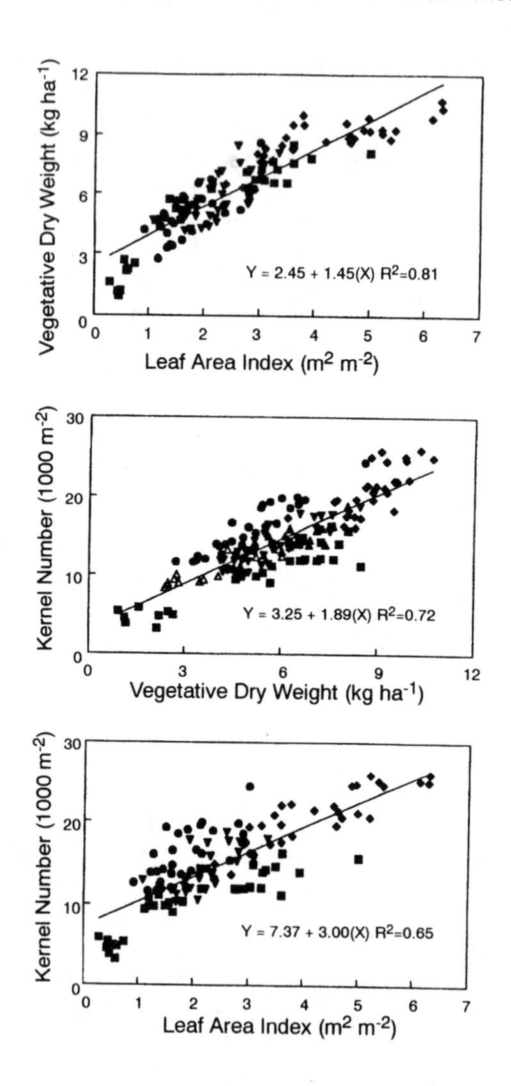

Note: Leaf area index was measured near the boot growth stage and vegetative dry weight was measured near anthesis. Data points are from studies described and referenced in Figure 3.1. Each data point is a plot value for each treatment. Regression equations were developed using individual plot values.

FIGURE 3.3. Soft Red Winter Wheat Grain Yield As a Function of Individual Kernel Weight

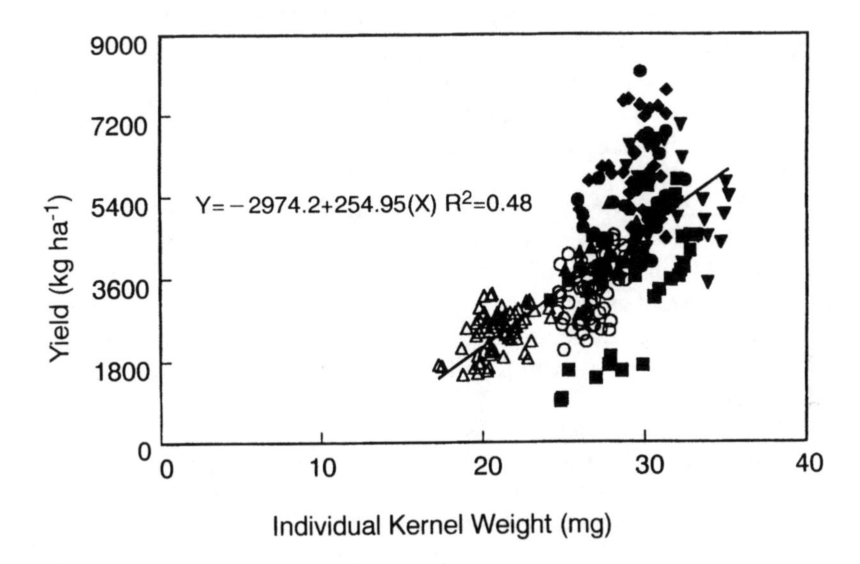

Note: Data points are from studies described and referenced in Figure 3.1. Each data point is a plot value for each treatment. Regression equation was developed using individual plot values.

become a significant part of grain yield (Bidinger, Musgrave, and Fischer, 1977). Under normal growing conditions, about half of the photosynthate moved into the grain originates from the flag leaf, with the remainder coming from the spike, leaf sheaths, and penultimate leaf (Rawson et al., 1983). Therefore, the rate and/or duration of kernel growth probably is at least partially limited by the rate and/or duration of photosynthate production by the flag leaf during grain fill.

Factors Affecting Kernel Growth Rate

Grain dry weight accumulation is linear during most of the grain-filling period (see reviews by Simmons, 1987 and Egli, 1994). This linear increase is not specific to wheat, but also occurs in other grain crops, such as soybean and corn (Egli, 1994). The rate of increase in wheat kernel weight appears to be under genetic control (Hunt, van der Poorten, and Pararajasingham, 1991; Darroch and Baker, 1990; Bruckner and Frohberg,

1987), with endosperm cell number having a positive effect on kernel growth rate (Brocklehurst, 1977). Individual kernels on the spike differ in rate of growth (Simmons, Crookston, and Kurle, 1982; Simmons and Crookston, 1979; Simmons and Moss, 1978). However, there is little evidence to suggest the growth rate of a kernel at a specific location on the spike and the average kernel growth rate of all kernels on the spike differ in response to a given treatment. In soybean, these two methods for estimating seed growth characteristics are frequently correlated (Egli, 1994).

When comparing genotypes, only a weak association has been found between kernel growth rate and wheat yield (Frederick, 1997; Van Sanford, 1985). The lack of consistent associations between kernel weight and yield and between the rate of kernel growth and kernel weight suggest that little genetic yield improvement can be made by selecting for higher kernel growth rates. Little association between seed growth rate and seed weight or yield has also been found in soybean (see review by Frederick and Hesketh, 1993).

Several researchers have proposed that increases in leaf photosynthetic rate during grain fill should result in higher wheat kernel growth rates (Frederick and Camberato, 1994; Simmons, Crookston, and Kurle, 1982). However, evidence indicates that this may not be true. Frederick and Camberato (1995b) found that increasing the rate of spring-applied N increased flag-leaf photosynthesis of irrigated wheat during the early stages of grain fill, but had little effect on kernel growth rate (Frederick and Camberato, 1995a). Drought stress during grain fill reduces leaf photosynthetic rate, but has little effect on kernel growth rate (Frederick and Camberato, 1995a, 1995b; Egli, 1994). Selection for higher rates of leaf photosynthesis during grain fill has been found to result in little increase in grain yield (Austin, 1989). The lack of correlation between single leaf photosynthesis and yield across genotypes may be caused by other differences between genotypes that affect leaf photosynthetic rate, such as leaf age, leaf area and thickness, and assimilate demand by reproductive structures (see discussion by Frederick and Hesketh, 1993). Within a genotype, the lack of increase in kernel growth rate when leaf photosynthesis is increased indicates photosynthate supply may not be limiting kernel growth rate, at least during the early stages of grain fill before rapid leaf senescence and loss of photosynthetic activity occur. With respect to obtaining high kernel weights, the duration of leaf photosynthesis during grain fill may be just as important as the rate of leaf photosynthesis.

Increases in leaf photosynthesis during the early portion of grain fill are associated with increases in vegetative dry weight (Frederick and Camberato, 1995b; Kiniry, 1993). Nonstructural carbohydrate concentrations also increase during this time (Kiniry, 1993; Davidson and Chevalier, 1992; Ford et al., 1979) and probably account for most of the vegetative dry-weight increases. These carbohydrate reserves are found primarily in the stem and are utilized for maintenance respiration and grain fill (Austin et al., 1977). The occurrence of drought during grain fill results in smaller increases in vegetative dry weight (Frederick and Camberato, 1995b) and less stored carbohydrate accumulation (Davidson and Chevalier, 1992) during the early stages of grain fill. However, drought increases the proportion of the grain weight originating from stem reserves, with values ranging from near 10 percent under normal conditions to greater than 40 percent when drought or heat stress occur (Davidson and Chevalier, 1992; Austin et al., 1977; Bidinger, Musgrave, and Fischer, 1977; Rawson and Evans, 1971). Stem reserves may serve to maintain a linear rate of kernel growth when photosynthate production declines during the latter portion of grain fill (Simmons, 1987). Under normal growing conditions, the lag period at the end of grain fill may be caused by the depletion of carbohydrate reserves. Little research has been conducted to determine whether the depletion of stored carbohydrates during the latter portion of grain fill directly affects the termination of grain fill.

There have been several reports of an inverse relationship between kernel number per m^2 and kernel weight (Frederick and Camberato, 1995a; Frederick and Camberato, 1994; Slafer and Andrade, 1993). Two theories involving kernel growth rate have been put forth to explain this phenomenon (Slafer, Sattore, and Andrade, 1993). First, kernels within the spike differ in rate of dry weight accumulation, with kernels in the proximal location of the spikelet and the central section of the spike tending to have higher kernel growth rates than those further out (Simmons, Crookston, and Kurle, 1982; Simmons and Crookston, 1979; Simmons and Moss, 1978). Therefore, increases in kernel number per spike would result in more kernels produced at locations more distal from the center of the spikelet and spike, locations having slower rates of kernel growth. Second, it has been proposed that, if photosynthate production is not increased proportionally to the increase in kernel number, increases in kernel number per spike should result in more competition between kernels for assimilate, resulting in a lower average rate of kernel growth. The latter theory is not consistent with the fact that photosynthate supply during the early stages of grain fill appears to be in excess of grain demand. Therefore, there should be sufficient assimilate to support the growth of more kernels per spike, at least during the first portion of grain fill when canopy photosynthetic rates and vegetative carbohydrate reserves are high. On the other hand, an increase in kernel number without an increase in photosynthate production would probably result in a faster depletion of vegetative carbohydrate reserves during the latter portion of grain fill when leaf photosynthetic rates are low. This situation would result in a shorter duration of grain fill and lower kernel weights.

One other explanation may exists for the inverse relationship found between kernel number and kernel weight. The high leaf area index and vegetative biomass associated with a high kernel number per m^2 (Figure 3.2) can result in greater soil water depletion during the growing season and more severe plant water deficits during the grain-filling period. The greater severity of plant water deficits under these conditions results in lower leaf photosynthetic rates and smaller kernels, compared to times when less vegetative growth is produced (Frederick and Camberato, 1995a, 1995b, 1994).

When comparing cultivars, we found an inverse relationship between kernel growth rate and the duration of grain fill (see Figure 3.4). Others have found a similar negative association between these two variables (May and Van Sanford, 1992; Van Sanford, 1985). Frederick and Hesketh (1993) proposed that the high demand for assimilate with high seed growth rates increases leaf senescence, shortens the duration of seed fill, and results in smaller seed in soybean. Similar relationships may also occur in wheat, thus explaining the inverse relationship between the rate and duration of kernel growth shown in Figure 3.4. On the other hand, wheat genotypes with high rates of kernel growth may just proceed through the various stages of kernel development at a faster rate, rather than having a higher kernel growth rate at each stage of kernel development. Breeders must also be aware that selecting for genotypes with high average kernel growth rates can concomitantly cause a selection for fewer kernels per spike (see previous discussion).

Factors Affecting the Duration of Grain Fill

Physiological maturity date is controlled by anthesis date, the rate of kernel growth, and the duration of grain fill. The duration of grain fill, in return, is determined by such factors as plant health and nutrient status, reproductive sink demand for assimilate, and air temperature. The duration of grain fill may also be determined by the capacity of the grain to utilize available assimilates, as determined shortly after anthesis by the number of endosperm cells and starch granules formed (Egli, 1994; Brocklehurst, 1977). However, kernel removal (Simmons, Crookston, and Kurle, 1982) and CO_2 enrichment during grain fill (Fischer and Aguilar, 1976; Krenzer and Moss, 1975) have been found to increase mean kernel weight, indicating that maximum kernel weight is usually not obtained

FIGURE 3.4. Relationship Between Average Kernel Growth Rate and the Effective Filling Period for the Soft Red Winter Wheat Cultivars

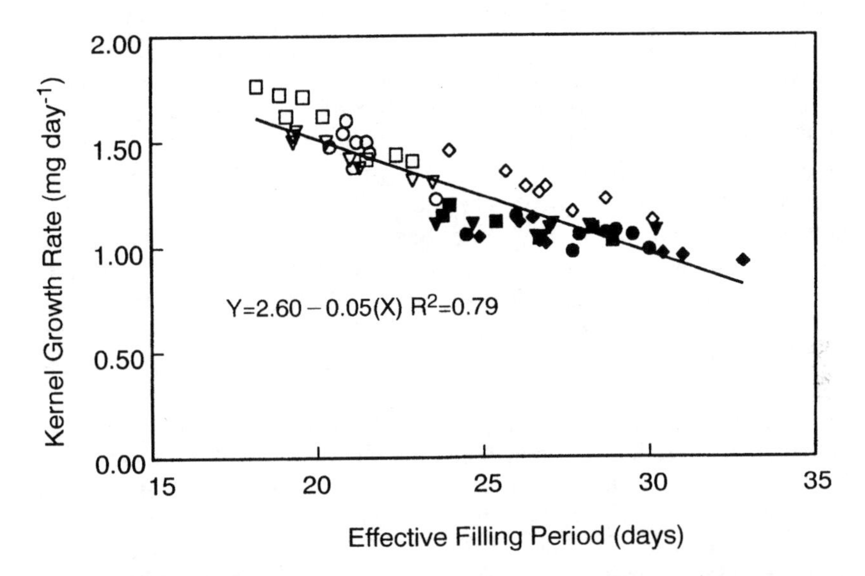

Note: Cultivars 'Northrup King Coker 9835' (●), 'Northrup King Coker 9803' (▼), 'Andy' (◆), and 'Gore' (■) in 1994 and 'Northrup King Coker' 9835 (○), 'Northrup King Coker' 9803 (▽), 'Andy' (◇), and 'Gore' (□) in 1995. Data taken from Frederick (1997). Each data point is a plot value for each treatment. Regression equation was developed using individual plot values.

under field conditions. Since carbohydrate supply appears to have little effect on kernel growth rate, kernel removal and CO_2-enrichment treatments must increase kernel weight by increasing the amount of stored carbohydrates available for kernel growth during the latter part of grain fill, thus extending the duration of grain fill. Large-seeded genotypes appear to come closer to obtaining their maximum kernel size than smaller-seed genotypes (Blade and Baker, 1991; Simmons, Crookston, and Kurle, 1982), suggesting that extending the duration of grain fill should result in greater kernel weight increases in small-seeded genotypes.

Since kernel growth rate is linear during most of the grain-filling period, the duration of grain fill can be estimated by dividing kernel dry weight at maturity by the rate of kernel growth. This estimate of grain fill has been termed the effective filling period. The shorter the lag period before and after the period of linear growth, the more accurate the effective filling period is for estimating the actual grain-filling period.

The duration of grain fill may be controlled by the rate of leaf senescence which, in turn, may be regulated by the N status of the plant and the rate of N demand by the developing grain (see review by Frederick and Hesketh, 1993). As for grain dry weight accumulation, the rate of grain N accumulation is linear for most of the grain-filling period (Sofield et al., 1977). Most of the N in the grain originates from N taken up prior to anthesis (Oscarson et al., 1995; Van Sanford and MacKown, 1987; Waldren and Flowerday, 1979). Wheat retains the capacity to take up N after anthesis (Oscarson et al., 1995; Van Sanford and MacKown, 1987), and late applications of N (at boot stage or later) usually increase leaf N concentration and may delay leaf senescence (Tindall, Stark, and Brooks, 1995; Banziger, Feil, and Stamp, 1994). These observations have led researchers to believe that rapid leaf senescence during grain fill is the result of reduced N uptake caused by the depletion of available soil N, in the absence of late N fertilizer applications (Dalling, 1985). Although little is known about root development and activity under field conditions, it is likely that root growth diminishes and root density decreases during grain fill, which may also contribute to the decrease in N uptake during this time. In any case, these observations suggest that the linear rate of grain N accumulation can be met by either the active uptake of N during grain fill and/or from the remobilization of N from vegetative tissue. Nitrogen remobilization probably is the most important source of grain N under most production systems (Simmons, 1987).

Since N is an important component of the chlorophyll molecules and enzymes associated with photosynthesis, the remobilization of vegetative N during grain fill results in the loss of leaf area and photosynthetic activity (Frederick, 1997; Frederick and Camberato, 1995b; Hunt and van der Poorten, 1985). Frederick (1997) found that the photosynthetic activity of wheat flag leaves decreases rapidly when the developing grain reaches about half of its final size. Therefore, once the flag leaf has senesced, remobilization and spike photosynthesis would be the main (and probably only) sources of carbohydrate for grain fill.

Drought stress reduces wheat N accumulation. Therefore, if the rate of N deposition in the grain is not reduced by drought occurring during grain fill, an even greater proportion of the grain N would originate from remobilized N. This increase in N remobilization and loss of photosynthetic activity would at least partially explain why drought stress during grain fill accelerates leaf senescence and shortens the duration of grain fill (Frederick and Camberato, 1995a, 1995b). Drought during grain fill results in higher grain protein concentrations, and grain N is usually inversely related to grain yield (Slafer, Satorre, and Andrade, 1993; Terman, 1979). The occurrence of drought during grain fill should have little effect on final grain N content if most of the grain N originates from the remobilization of N accumulated prior to anthesis, and drought has little effect on the rate of N remobilization. On the other hand, if most of the grain carbohydrate originates from photosynthate produced during grain fill, the occurrence of photosynthesis-reducing drought during that time would reduce overall carbohydrate supply (current photosynthate and stem reserves). Therefore, the duration of grain fill would be shortened if the rate of grain dry-weight accumulation remained linear. The maintenance of grain N accumulation, but a reduction in total grain carbohydrate accumulation, would explain why grain protein concentration is higher with drought.

Responses of wheat to management practices designed to extend the duration of grain fill and increase kernel weight may partially depend on the climatic conditions during the grain-filling period. There have been many reports of late N fertilizer applications increasing leaf N concentration and grain protein concentration, but having little effect on kernel weight or grain yield (see discussions by Tindall, Stark, and Brooks, 1995 and Slafer, Satorre, and Andrade, 1993). These responses may partially be due to adverse growing conditions during grain fill. For example, if drought occurs during that time, photosynthesis would be reduced even if leaf N concentrations are high. Consequently, there would be more vegetative N for remobilization to the grain when N is applied near anthesis, but little additional photosynthate produced. This situation would result in a higher grain protein concentration with late N applications, but little increase in kernel weight or grain yield. On the other hand, cool air temperatures and good soil moisture conditions during the latter portion of grain fill would favor delayed leaf senescence, an extended period of photosynthate production, and a longer grain-filling period (Wiegand and Cuellar, 1982), especially with late N applications (Banziger, Feil, and Stamp, 1994). Therefore, under these conditions, the uptake of N during grain fill may result in higher kernel weight and grain yield (Evans, Wardlaw, and Fischer, 1975).

SUMMARY AND STRATEGIES FOR FUTURE YIELD IMPROVEMENT

Research results indicate that genetic yield gains in wheat have been due to a greater partitioning of assimilate to reproductive development, and less to vegetative dry matter production, during the period of spike formation prior to anthesis (see review by Slafer, Satorre, and Andrade,

1993). These changes have resulted in modern wheat cultivars having a high kernel number per spike and kernel number per m^2 (Figure 3.1; Slafer, Satorre, and Andrade, 1993). Less partitioning of assimilate to vegetative tissue may have resulted in modern cultivars having less leaf area per plant and/or less vegetative carbohydrate reserves prior to anthesis than older cultivars. Therefore, newer cultivars are probably more yield sensitive than older cultivars to stresses that reduce photosynthetically active leaf area and leaf photosynthetic rates during the period of spike formation and grain set. This greater reliance of newer cultivars on photosynthate production during the period of spike formation prior to anthesis would explain why yield differences between old and new cultivars are greater in high-yielding (high-photosynthate producing) environments than in low-yielding environments (Slafer and Andrade, 1993; Austin, Ford, and Morgan, 1989). Obviously, there is a limit to the magnitude of yield improvement that can occur by increasing kernel number per m^2 without increasing leaf area, as sink demand would eventually exceed photosynthate production. The linear relationship between leaf area index and kernel number shown in Figure 3.2 suggests that further increases in kernel number of wheat cultivars to be produced in high-yielding environments could be accomplished by selecting for a higher leaf area index. The amount of yield improvement that could be achieved with this approach would depend on the increase in leaf shading at higher leaf area indices.

Selecting for higher kernel numbers in wheat would also take advantage of the better growing conditions that usually occur prior to anthesis, compared to during the grain-filling period. In most wheat-producing areas, air temperatures generally increase and soil water levels decrease during the spring growing season. Good growing conditions prior to and at anthesis would allow the genetic potential for high kernel numbers to be expressed. There appears to be less potential for increasing grain yield by selecting for heavier kernels (Figure 3.3), especially under conditions of high kernel number per m^2. Neither a high kernel growth rate nor a long duration of grain fill have been consistently found to be closely associated with a high kernel weight. The kernel growth rate of wheat cultivars appears to be controlled by the physiological and/or anatomical characteristics of the kernel and/or spike, rather than carbohydrate supply (Thornley, Gifford, and Bremner, 1981; Jenner and Rathjen, 1978; Simmons and Moss, 1978). With respect to kernel demand, it appears that wheat will usually produce photosynthate in excess of grain demand or respirational needs during the first part of the grain-filling period and possibly not enough photosynthate during the latter part of grain fill. Carbohydrate produced in excess of grain needs is stored primarily in wheat stems. The accumulation of carbohydrate reserves appears to have at least an evolutionary advantage, serving as a source of carbohydrate to maintain grain fill during periods of drought stress (Kiniry, 1993; Gallagher, Biscoe, and Hunter, 1976).

Leaf N concentrations, leaf photosynthesis, and stem carbohydrate reserves all decrease to low levels during the latter portion of grain fill (Frederick and Camberato, 1995b; Kiniry 1993; Ford et al., 1979). Little research has been conducted to determine whether physiological maturity and the termination of grain growth under normal growing conditions in the field are caused by the lack of available photosynthate near the end of grain fill or whether the depletion of carbohydrate reserves is timed to coincide with physiological maturity. Plant water stress probably occurs to some degree during grain fill in nonirrigated wheat. Water stress accelerates leaf senescence, reduces leaf photosynthesis, and results in lower levels of vegetative carbohydrate reserves. These reductions in photosynthate supply are associated with a shorter grain-filling period, indicating that the duration of kernel growth may depend at least partly on carbohydrate availability near the end of grain fill under field conditions.

Further selection for higher kernel numbers per m^2 in wheat, without increasing leaf area index, may result in an earlier remobilization of vegetative C and N during grain fill due to the higher sink-to-source ratio. If the rate of kernel growth is not altered, then a faster rate of leaf senescence and shorter duration of grain fill would result due to the increased demand for C and N. These relationships between kernel number, leaf area, and kernel weight would at least partially explain why kernel weight has decreased as breeders have selected for higher kernel numbers during the course of yield improvement (see review by Slafer, Satorre, and Andrade, 1993).

One possible solution to maintaining the duration of grain fill and kernel weight under conditions of high assimilate demand (high kernel number per m^2) may be to select for a stay-green (delayed senescence) characteristic in wheat, as is found in modern corn hybrids. Most leaves of higher-yielding, modern U.S. corn hybrids remain green and retain chlorophyll until physiological maturity, whereas the leaves of lower-yielding, older U.S. corn hybrids senesce prior to physiological maturity, as in wheat. This stay-green characteristic is different from the delayed leaf senescence associated with a low assimilate demand by developing grain (see Frederick and Hesketh, 1993). In corn, the stay-green characteristic has been reported to be associated with a greater lodging and biotic stress resistance (Duvick, 1992), a greater drought tolerance (Nissanka, Dixon, and Tollenaar, 1997; Frederick et al., 1989), a higher kernel number per ear (Frederick et al., 1989), a greater total (Tollenaar, McCullough, and Dwyer, 1993) and vegetative (Frederick, Below, and Hesketh, 1990) biomass during the latter portion of grain fill, a greater N uptake during grain fill under both irrigated and nonirrigated conditions (Frederick, Below, and Hesketh, 1990), and a longer duration of kernel fill (Cavalieri and Smith, 1985). Genetic yield improvement in corn appears to have occurred without a large change in leaf area index at anthesis (Tollenaar, McCullough, and Dwyer, 1993). Therefore, modern corn hybrids probably have a higher sink-to-source ratio than older hybrids, as has been proposed for wheat (Slafer, Satorre, and Andrews, 1993). Sustained photosynthetic activity and the high levels of vegetative carbohydrate reserves late into the grain-filling period may allow the duration of grain fill to be extended (or at least maintained) in modern corn hybrids under conditions of high assimilate demand (high kernel number per ear). If this is true, then selecting for a similar stay-green characteristic in wheat should increase the duration of kernel fill and kernel weight of genotypes having a high kernel number per m^2 to leaf area index ratio (high sink-to-source ratio for photosynthate), assuming genes for this trait can be found in wheat.

Carbohydrate reserves have been shown to be of value in maintaining kernel growth during periods of photosynthesis-reducing drought stress in wheat (Kiniry, 1993). On the other hand, vegetative carbohydrate reserves are low during the latter portion of grain fill in wheat (Kiniry, 1993), which should reduce the plant's ability to buffer itself during periods of drought occurring during that time. In addition, carbohydrate reserves used to support grain growth during drought occurring later in the grain-filling period probably cannot be replenished in wheat once the stress is relieved because of the decline in leaf area and photosynthate capacity during that time. The earlier utilization of stem reserves and the loss of photosynthetic capacity when drought occurs late in the grain-filling period would limit carbohydrate supplies, which, in addition to accelerated leaf senescence, may explain why the duration of wheat grain fill is shortened by drought. For corn, the duration of grain fill may not be shortened by drought because of the high levels of stored carbohydrates and the capacity to continue photosynthesis late into the grain-filling period once the stress is relieved. These traits would account for the greater drought tolerance of modern corn hybrids. It seems very likely that a similar trait would benefit wheat produced in regions of the world where drought stress is common during grain fill.

The stay-green characteristic may be of little value for improving wheat yield in most wheat-producing areas if reproductive demand for assimilate is not also increased. Without a concurrent increase in kernel number per

m^2 (increase in assimilate demand), incorporating a stay-green characteristic in wheat would theoretically result in a lengthening of the grain-filling period, as has been proposed for the late application of N fertilizer near anthesis. If growing conditions are warm and dry throughout grain fill (conditions that favor accelerated leaf senescence and a shortened grain-filling period), the stay-green characteristic would have little benefit. However, in wheat-producing areas where climatic conditions are conducive to a long grain-filling period, a stay-green characteristic may enhance photosynthate production during the latter portion of grain fill, resulting in heavier kernels.

In conclusion, there appears to be significant opportunity to further increase the genetic yield potential of wheat. The most promising approach appears to be for breeders to continue selecting for a higher kernel number per m^2, especially for areas where growing conditions become warm and dry during the grain-filling period. There appears to be less opportunity for genetic yield improvement by selecting for heavier kernels either by way of higher kernel growth rates and/or a longer duration of grain fill. Data indicate there has been little success at increasing kernel weight, especially when kernel number is high. However, we have identified several possible limitations to kernel-weight increases under these conditions. Solutions to these limitations may partially depend on the climatic conditions normally encountered during the growing season.

REFERENCES

Austin, R.B., 1989. Genetic variation in photosynthesis. *J. Agric. Sci.,* 112: 287-294.

Austin, R.B., Edrich, J.A., Ford, M.A., and Blackwell, R.D., 1977. The fate of the dry matter, carbohydrates and ^{14}C lost from leaves and stems of wheat during grain fill. *Ann. Bot.* (London), 41: 1309-1321.

Austin, R.B., Ford, M.A., Edrich, J.A., and Hooper, B.E., 1976. Some effects of leaf posture on photosynthesis and yield in wheat. *Ann. Appl. Biol.,* 83: 425-446.

Austin, R.B., Ford, M.A., and Morgan, C.L., 1989. Genetic improvement in the yield of winter wheat: A further evaluation. *J. Agric. Sci.,* 112: 295-301.

Banziger, M., Feil, B., and Stamp, P., 1994. Competition between nitrogen accumulation and grain growth for carbohydrates during grain filling of wheat. *Crop Sci.,* 34: 440-446.

Bidinger, F., Musgrave, R.B., and Fischer, R.A., 1977. Contribution of stored pre-anthesis assimilate to grain yield in wheat and barley. *Nature* (London), 270: 431-433.

Blade, S.F., and Baker, R.J., 1991. Kernel weight response to source-sink changes in spring wheat. *Crop Sci.,* 31: 1117-1120.

Brocklehurst, P.A., 1977. Factors controlling grain weight in wheat. *Nature* (London), 266: 348-349.

Bruckner, P.L., and Frohberg, R.C., 1987. Rate and duration of grain fill in spring wheat. *Crop Sci.,* 27: 451-455.

Cavalieri, A.J., and Smith, O.S., 1985. Grain filling and field drying of a set of maize hybrids released from 1930 to 1982. *Crop Sci.*, 25: 856-860.

Cregan, P.B., and van Berkum, P., 1984. Genetics of nitrogen metabolism and physiological/biochemical selection for increased grain crop productivity. *Theor. Appl. Genet.*, 67: 97-111.

Dalling, M.J., 1985. The physiological basis of nitrogen redistribution during grain filling in cereals. In J. Hasper, L. Scrader, and R. Howel (Editors), *Exploitation of Physiological and Genetic Variability to Enhanced Crop Productivity.* Weverly Press, Madison, WI, p. 55-71.

Darroch, B.A., and Baker, R.J., 1990. Grain filling in three spring wheat genotypes: Statistical analysis. *Crop Sci.,* 30: 525-529.

Davidson, D.J., and Chevalier, P.M., 1992. Storage and remobilization of water-soluble carbohydrates in stems of spring wheat. *Crop Sci.*, 32: 186-190.

Duvick, D.N., 1992. Genetic contributions to advances in yield of U.S. maize. *Maydica,* 37: 69-79.

Egli, D.B., 1994. Seed growth and development. In K.L. Boote, J.M. Bennett, T.R. Sinclair, and G.M. Paulsen (Editors), *Physiology and Determination of Crop Yield.* American Society of Agronomy, Madison, WI, pp. 127-148.

Evans, L.T., Wardlaw, I.F., and Fischer, R.A., 1975. Wheat. In L.T. Evans (Editor), *Crop Physiology.* Cambridge University Press, Cambridge, UK, pp. 101-149.

Fischer, R.A., and Aguilar, I., 1976. Yield potential in a dwarf spring wheat and the effect of carbon dioxide fertilization. *Agron. J.*, 68: 749-752.

Ford, M.A., Blackwell, R.D., Parker, M.L., and Austin, R.B., 1979. Associations between stem solidity, soluble carbohydrate accumulation and other characters in wheat. *Ann. Bot.*, 44: 731-738.

Frederick, J.R., 1997. Winter wheat leaf photosynthesis, stomatal conductance, and leaf N concentration during reproductive development. *Crop Sci.*, 37: 1819-1826.

Frederick, J.R., and Bauer, P.J., 1997. Winter wheat responses to surface and deep tillage on the southeastern Coastal Plain. *Agron. J.*, 88: 829-833.

Frederick, J.R., Below, F.E., and Hesketh, J.D., 1990. Carbohydrate, nitrogen and dry matter accumulation and partitioning of maize hybrids under drought stress. *Ann. Bot.,* 66: 407-415.

Frederick, J.R., and Camberato, J.J., 1994. Leaf net CO_2-exchange rate and associated leaf traits of winter wheat grown with various spring nitrogen fertilization rates. *Crop. Sci.,* 34: 432-439.

Frederick, J.R., and Camberato, J.J., 1995a. Water and nitrogen effects on winter wheat in the southeastern Coastal Plain: I. Grain yield and kernel traits. *Agron. J.*, 87: 521-526.

Frederick, J.R., and Camberato, J.J., 1995b. Water and nitrogen effects on winter wheat in the southeastern Coastal Plain: II. Physiological responses. *Agron. J.*, 87: 527-533.

Frederick, J.R., and Hesketh, J.D., 1993. Genetic improvement in soybean: Physiological attributes. In G.A. Slafer (Editor), *Genetic Improvement of Field Crops*. Marcel Dekker, Inc., New York, pp. 237-286.

Frederick, J.R., Hesketh, J.D., Peters, D.B., and Below, F.E., 1989. Yield and reproductive trait responses of maize hybrids to drought stress. *Maydica*, 34: 319-328.

Gallagher, J.N., Biscoe, P.V., and Hunter, B., 1976. Effects of drought on grain growth. *Nature* (London), 264: 541-542.

Hunt, L.A., and van der Poorten, G., 1985. Carbon dioxide exchange rates and leaf nitrogen contents during ageing of the flag and penultimate leaves of five spring-wheat cultivars. *Can. J. Bot.*, 63: 1605-1609.

Hunt, L.A., van der Poorten, G., and Pararajasingham, S., 1991. Postanthesis temperature effects on duration and rate of grain filling in some winter and spring wheats. *Can. J. Plant Sci.*, 71: 609-617.

Jenner, C.F., and Rathjen, A.J., 1978. Physiological basis of genetic differences in the growth of grains of six varieties of wheat. *Aust. J. Plant Physiol.*, 5: 249-262.

Karlen, D.L., and Sadler, E.J., 1990. Nutrient accumulation rates for wheat in the southeastern Coastal Plain. *Comm. Soil Sci. Plant Anal.*, 21: 1329-1352.

Kiniry, J.R., 1993. Nonstructural carbohydrate utilization by wheat shaded during grain growth. *Agron. J.*, 85: 844-849.

Kirby, E.J.M., 1983. Development of the cereal plant. In D.W. Wright (Editor), *The Yield of Cereals*. Royal Agricultural Society of England, London, pp. 1-3.

Krenzer, E.G., and Moss, D.N., 1975. Carbon dioxide enrichment effects upon yield and yield components in wheat. *Crop Sci.*, 15: 71-74.

Large, E.C., 1954. Growth stages in cereals: Illustrations of the Feekes scale. *Plant Pathol.*, 3: 128-129.

May, L., and Van Sanford, D.A., 1992. Selection of early heading and correlated response in maturity of soft red winter wheat. *Crop Sci.*, 32: 47-51.

Mohapatra, P.K., Aspinall, D., and Jenner, C.F. 1982. The growth and development of the wheat apex: The effects of photoperiod on spikelet production and sucrose concentration in the apex. *Ann. Bot.*, 49: 619-626.

Nissanka, S.P., Dixon, M.A., and Tollenaar, M., 1997. Canopy gas exchange response to moisture stress in old and new maize hybrid. *Crop Sci.*, 37: 172-181.

Nonami, H., and Boyer, J.S., 1989. Turgor and growth at low water potentials. *Plant Physiol.*, 89: 798-804.

Oscarson, P., Lundborg, T., Larsson, M., and Larsson, C.M., 1995. Genotypic differences in nitrate uptake and nitrogen utilization for spring wheat grown hydroponically. *Crop Sci.*, 35: 1056-1062.

Rawson, H.M., and Evans, L.T., 1971. The contribution of stem reserves to grain development in a range of wheat cultivars of different heights. *Aust. J. Agric. Res.,* 22: 851-863.

Rawson, H.M., Hindmarsh, J.H., Fischer, R.A., and Stockman, Y.M., 1983. Changes in leaf photosynthesis with plant ontogeny and relationships with yield per ear in wheat cultivars and 120 progeny. *Aust. J. Plant Physiol.,* 10: 503-514.

Schmidt, J.W., 1984. Genetic contributions to yield gains in wheat. In W.H. Fehr (Editor), *Genetic Contributions to Yield Gains of Five Major Crop Plants.* Crop Science Society of America Special Publication No. 7, Madison, WI, pp. 89-101.

Scott, W.R., Dougherty, C.T., and Langer, R.H.M., 1975. An analysis of a wheat yield depression caused by high sowing rate with reference to the pattern of grain set within the ear. *New Zealand J. Agric. Res.,* 8: 209-214.

Siddique, K.H.M., Kirby, E.J.M., and Perry, M.W., 1989. Ear-to-stem ratio in old and modern wheats; relationship with improvement in number of grains per ear and yield. *Field Crops Res.,* 21: 59-64.

Simmons, S.R., 1987. Growth, development, and physiology. In E.G. Heyne (Editor), *Wheat and Wheat Improvement.* Agron. Series No. 13, Second Edition, American Society of Agronomy, Madison, WI, pp. 77-113.

Simmons, S.R., and Crookston, R.K., 1979. Rate and duration of growth of kernels formed at specific florets in spikelets of spring wheat. *Crop Sci.,* 19: 690-693.

Simmons, S.R., Crookston, R.K., and Kurle, J.E., 1982. Growth of spring wheat kernels as influenced by reduced kernel number per spike and defoliation. *Crop Sci.,* 22: 983-988.

Simmons, S.R., and Moss, D.N. 1978. Nitrogen and dry matter accumulation by kernels formed at specific florets in spikelets of spring wheat. *Crop Sci.,* 18: 139-143.

Slafer, G.A., and Andrade, F.H., 1993. Physiological attributes related to the generation of grain yield in bread wheat cultivars released at different eras. *Field Crops Res.,* 31: 351-356.

Slafer, G.A., Andrade, F.H., and Satorre, E.H., 1990. Genetic-improvement effects on pre-anthesis physiological attributes related to wheat grain yield. *Field Crops Res.,* 23: 255-263.

Slafer, G.A., Satorre, E.H., and Andrade, F.H., 1993. Increases in grain yield in bread wheat from breeding and associated physiological changes. In G.A. Slafer (Editor), *Genetic Improvement of Field Crops.* Marcel Dekker, Inc., New York, pp. 1-68.

Sofield, I., Wardlaw, I.F., Evans, L.T., and Zee, S.Y., 1977. Nitrogen, phosphorus and water contents during grain development and maturation in wheat. *Aust. J. Plant Physiol.,* 4: 799-810.

Stern, W.R., and Kirby, E.J.M., 1979. Primordium initiation at the boot apex in four contrasting varieties of spring wheat in response to sowing date. *J. Agric. Sci.,* 93: 203-215.

Stockman, Y.M., Fischer, R.A., and Brittain, E.G., 1983. Assimilate supply and floret development within the spike of wheat (*Triticum aestivum* L.). *Aust. J. Plant Physiol.*, 10: 585-594.

Terman, G.L., 1979. Yields and protein content of wheat grain as affected by cultivar, N, and environmental growth factors. *Agron. J.*, 71: 437-440.

Thornley, J.H.M., Gifford, R.M., and Bremner, P.M., 1981. The wheat spikelet-growth response to light and temperature-experiment and hypothesis. *Ann. Bot.*, 47: 713-725.

Tindall, T.A., Stark, J.C., and Brooks, R.H., 1995. Irrigated spring wheat response to topdress nitrogen as predicted by flag leaf nitrogen concentration. *J. Prod. Agric.*, 8: 46-52.

Tollenaar, M., McCullough, D.E., and Dwyer, L.M., 1993. Physiological basis of the genetic improvement of corn. In G.A. Slafer (Editor), *Genetic Improvement of Field Crops.* Marcel Dekker, Inc., New York, pp. 183-236.

Van Sanford, D.A., 1985. Variation in kernel growth characters among soft red winter wheats. *Crop Sci.*, 25: 626-630.

Van Sanford, D.A., and MacKown, C.T., 1987. Cultivar differences in nitrogen remobilization during grain fill in soft red winter wheat. *Crop Sci.*, 27: 295-300.

Waldren, R.P., and Flowerday, A.D., 1979. Growth stages and distribution of dry matter, N, P, and K in winter wheat. *Agron. J.*, 71: 391-397.

Wiegand, C.L., and Cuellar, J.A., 1982. Duration of grain filling and kernel weight of wheat as affected by temperature. *Crop Sci.*, 21: 95-101.

Chapter 4

Nitrogen As Determinant of Wheat Growth and Yield

Basilio Borghi

INTRODUCTION

Nitrogen accounts for only a small portion of total plant weight; however, it plays a crucial role in plant metabolism because more than 90 percent of the plant N is in protein. There is plenty of nitrogen in the earth but plants can use only specific forms of it, which are generally available in the soil only in limited amounts during the growing season. As a consequence, N could be the main factor limiting yield potential in numerous wheat-growing areas of the world.

The low N supply coupled with its fundamental role in plant metabolism make it necessary to optimize the management of N resources to increase the efficiency of N use in crop systems (Novoa and Loomis, 1981). Basically, this can be achieved in two ways: first, by increasing the proportion of soil N absorbed by the crop and, second, by increasing the accumulation of N compounds in the edible part of the crop.

A better understanding of N evolution in the soil and of the physiological, biochemical, and genetic aspects of N metabolism are necessary for the optimization of N nutrition. In this chapter only the general principles of N economy in wheat, i.e., nitrogen evolution in the soil, uptake, assimilation, translocation, remobilization and storage, will be treated. The recommended fertilizer modalities for the different wheat-growing areas of the world are described in other sections of this book.

NITROGEN IN THE SOIL

In general N is present in nature in forms not accessible to plants: about 98 percent is found in primary rocks and 2 percent in the earth's atmosphere, mostly as N_2 gas. Only a very small fraction is present as organic matter in living or dead organisms. The process of organic matter mineralization, which takes place in the soil, gradually releases the nitrate (NO_3^-) or ammonium (NH_4^+) ions, the only N compounds the roots of wheat plants are able to uptake.

Soil organic matter has a relatively constant 10:1 C/N ratio and is present in different amounts, ranging from 1 to 5 percent of dry weight, in the different wheat-growing areas of the world. The process of N mineralization annually involves about 2 to 3 percent of the total N present in organic matter. Assuming that in the soil layer explored by the wheat roots, 2000 to 4000 kg of N ha^{-1} in organic form is present, only 40 to 80 kg of N ha^{-1} will be available to the crop during the life cycle. Since the N needed to produce 100 kg of wheat grain can roughly be estimated at 3 kg, it is clear that, in most growing conditions, a wheat crop needs more N than is made available by the mineralization process. In some of the low-input wheat-growing areas of the world, such as those in Australia, legume-based pastures in a crop rotation provide the main if not the only source of N for wheat, but in general the gap between the N provided by the soil and the amount needed for wheat is bridged by distributing N fertilizers. On a worldwide scale the use of N fertilizers increased rapidly in the 1960s and in the 1970s, reaching a peak at the end of the 1980s. It then diminished but is now expected to increase slowly in the second part of the 1990s to a level, in the year 2000, 4 to 7 percent higher than in 1989 (Bumb, 1995). In the long term, N prices tend to decrease. In the past this trend was favored by the development of new energy-saving processes. In several countries fertilizer prices, but also wheat prices, are subsidized and therefore a farmer's decision about N fertilizer may depend more on the ratio between fertilizer and wheat prices rather than on the absolute price of the fertilizer. The ratio between the two prices is generally low, ranging from a maximum of 7 in Australia to values below 1.5 in many developing countries. Assuming that only 3 kg of N is needed to produce 100 kg of grain, it is clear that farmers tend to apply all the N needed to maximize the yield, which means that the economic yield, in most cases, coincides with the maximum yield.

Nitrate fertilizer is the form of N immediately available to plants, provided it has been dispersed in a wet soil or distributed to the soil surface prior to rain or artificial irrigation. *Ammonium fertilizer* has a reduced mobility as the ammonium cation (NH_4^+) is absorbed by negatively charged colloids. In well-aerated soils with adequate moisture (40 to 60 percent of field capacity) and a pH close to 7, it can be rapidly transformed by *nitrosomonas* and *nitrobacter* microorganisms to nitrate form. The

N conversion is strongly affected by temperature, the minimum for the process being around 6°C. At 20°C, 50 percent of the ammonium is converted to nitrate in two weeks; at 30°C the same process takes only one week. *Urea* is generally considered a slow-acting N fertilizer when applied to the soil because the conversion to ammonium is carried out by urease, the temperature-dependent microbial enzyme. Urea can be distributed as a foliar spray and, in this case, is immediately absorbed by the plant.

Apart from N released during organic matter mineralization and N from mineral fertilizers and organic manure, other minor sources of soil N are represented by wet and dry deposition from the atmosphere. In the arable soils of Western Europe net mineralization in the growing season amounts to 0.5 to 1.5 kg ha^{-1} day^{-1}, while N deposition from the atmosphere amounts to 10 to 20 kg ha^{-1} $year^{-1}$. The trade-off of the N in the soil is synthetically reported on the left side of Figure 4.1.

NITROGEN METABOLISM IN THE PLANT

The wheat plant takes up N preferentially as NO_3^- ions, NH_4^+ uptake being confined to abnormal soil conditions (acid soils, waterlogged field) or to fields heavily fertilized with ammonium or organic manure (Austin and Jones, 1975). The uptake of soil N depends on the size and activity of the roots. Root-to-shoot ratio is greatly affected by the N concentration in the soil solution and roots have the greatest efficiency and also the first priority on N use.

The absorbed NO_3^- can be reduced to ammonium directly in the roots, or preferentially in the shoots. It is a two-step process catalyzed by the nitrate and nitrate reductase enzymes and the overall reaction is called nitrate reductase (NR) (see Figure 4.1). Nitrate reduction, and therefore protein synthesis, may be limited by the amount of NR enzyme present in the plant or by the level of the enzyme's activity. NR activity has therefore been proposed as a potential grain yield and protein content indicator (see Cregan and van Berkum, 1984).

When NO_3^- is abundant in the soil and/or the need of N for plant growth is reduced (i.e., at the beginning of the growing season), a limited amount of NO_3^- can be stored in the vacuoles, particularly at the base of the tillers. The concentration of NO_3^- in the sap extracted from the plant can be regarded as an indicator of the N nutritional status of the plant (Gate, 1995).

After reduction, N is combined with C to produce an ample spectrum of organic compounds, mainly proteins, playing important structural and functional roles. Synthesis of these compounds proceeds via the formation of

FIGURE 4.1. Nitrogen in the Soil and in the Plant

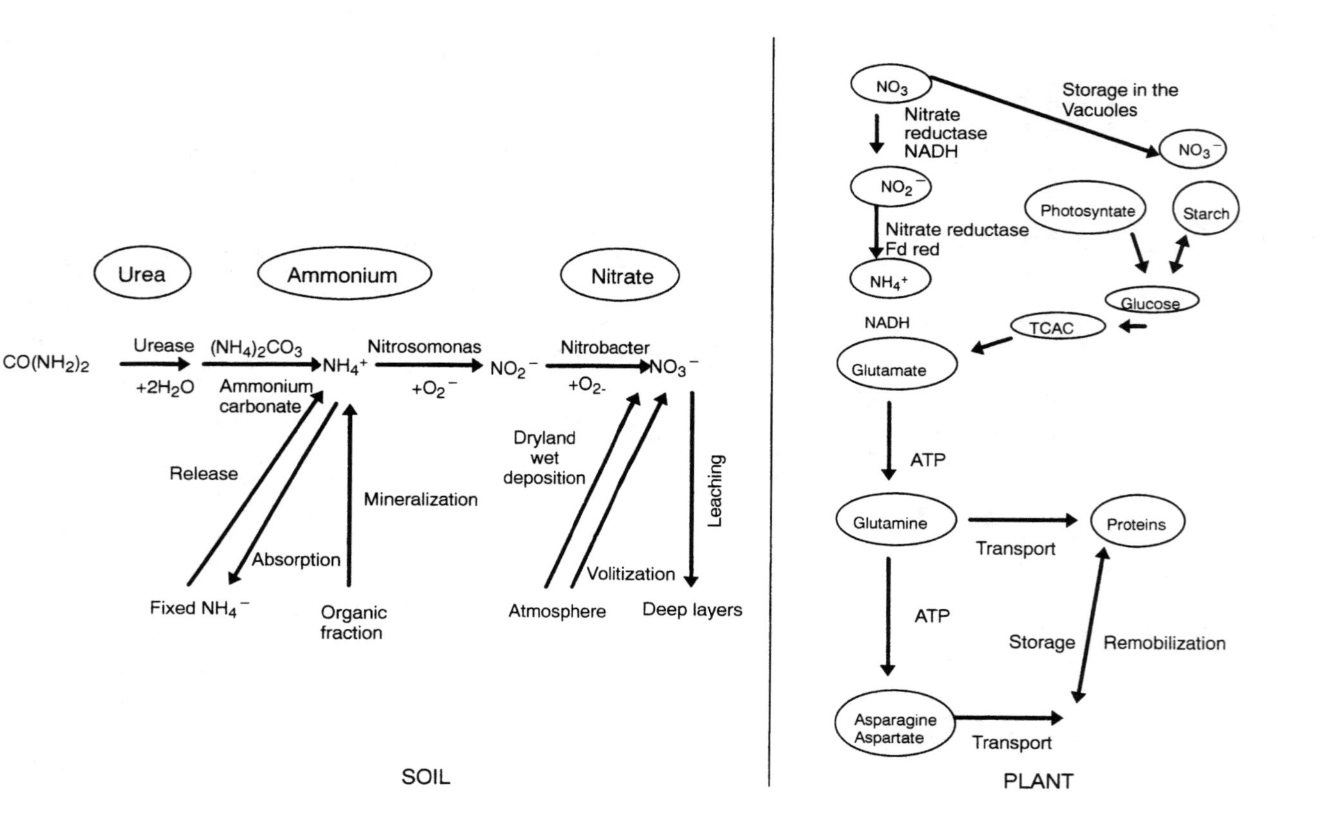

glutamate and glutamine which, together with asparagine and aspartate, are the most common organic N compounds transported in the phloem. The direct transfer of NO_3^- and NH_4^+ accounts for an insignificant amount of the N transported in the phloem (Simpson, Lambers, and Dalling, 1983); glutamine is the prevailing form of xylematic transport and its relative importance increases during senescence of the phloematic system (see Figure 4.1).

During the vegetative phase, N undergoes a continuous movement within the plant from roots to shoots and vice versa, but also between different tillers of the same plant. After flowering, the N flux is mainly oriented toward accumulation in the developing kernels. This process is largely sustained by N remobilization from vegetative organs because, at that stage, N uptake, in most cases, is very limited (Blum, 1988).

Photosynthesis and N metabolism are closely connected. On one hand, the functional activity of the photosynthetic apparatus is largely dependent on N availability in the plant, because N contributes to the formation of the structural and functional proteins of the chloroplast. On the other hand, N organication is an energy-demanding process requiring an efficient photosynthetic apparatus. N deficiency negatively affects the development of an efficient photosynthetic apparatus (number of units) and reduces total biomass. In practice, the former effect can be monitored in terms of variations in chlorophyll content or, more easily, of variations in the intensity of the green color of the leaves and, as such, serves as an indicator of the N nutritional status of the crop (Peltonen, Virtanen, and Haggrén, 1995).

NITROGEN EFFECTS AT THE PLANT AND CROP LEVEL

Response to N will initially be illustrated with reference to an ideal situation where wheat plants can grow without serious environmental constraints, apart from N availability. In this condition, the N fertilizer distributed on the surface of the soil will rapidly become available to the roots thanks to frequent light rainfalls. Something close to this ideal situation is found in the wheat-growing areas of Central-Western Europe where grain yield potential approaches 10 t ha^{-1}.

The amount of dry matter produced per unit of land increases linearly with the increase of N fertilizer applied in an ample range of N rates up to a level where a plateau is reached (see Figure 4.2). At plant level, the biomass increase is associated with larger leaves that stay green longer, taller stem, and larger number of tillers surviving to maturity and bearing fertile spikes. The rate of photosynthesis is normally not increased by N fertilizer; excess fertilizer may decrease the net assimilation rate per unit of green area

FIGURE 4.2. Effects of Nitrogen Fertilizer Rates on Biological Yield, Harvest Index, Grain Yield, Protein Content, and Breadmaking Quality

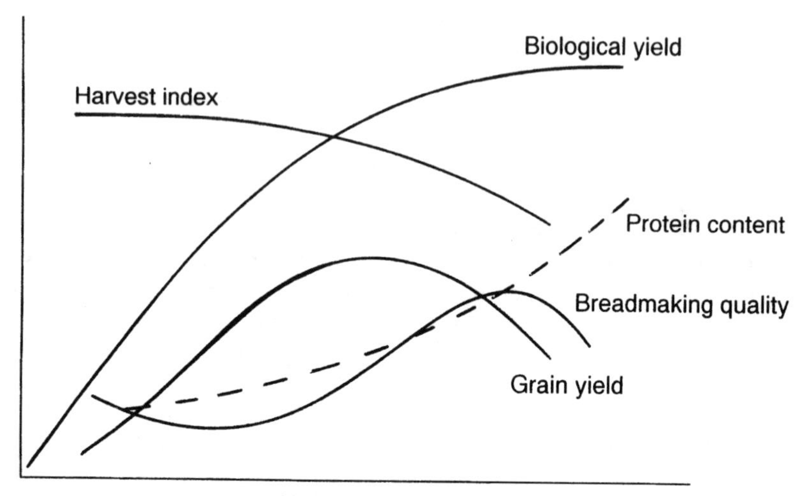

because of the reciprocal shading caused by luxurious crop growth (Pearman, Thomas, and Thorne, 1979). The grain yield response to N parallels that of biomass with three significant differences: the slope of the linear growth phase is lower; it reaches a plateau at lower N rates; and, soon after the ceiling, grain yield decreases. The different reaction to N of biomass and grain yield is well described by the harvest index, which decreases with the increase of N rates. The negative effects of high N rates on grain yield are due to weakening of the vegetative organs, which make the crop more prone to lodging and the tender tissues more vulnerable to parasites. Among the yield components, the number of kernels m^{-2} is the best indicator of wheat response to N (Meynard, 1987). Yield components are affected not only by the rate of N fertilization but also by the timing of N application or, conversely, by N scarcity at critical stages. For instance, a deficiency at shooting time decreases the number of bearing-ear tillers; spikelets per spike and kernels per spike are negatively affected by N deficiency during the period of spike initiation, which corresponds to the rapid increase of N demand. At flowering time, N deficiency may reduce seed setting. During grain filling, only a limited amount of N is taken up from the soil (about 20 percent of the total N) and, in most situations, the N provided by the mineralization of organic matter appears sufficient for plant needs.

Nitrogen remobilization from senescent vegetative tissues to reproductive organs is the dominant process during grain filling. Senescence is a highly organized process and follows a common pattern, the speed of which can be modified by environmental conditions but not altered. Senescence starts from the lower leaves, the flag leaf being the last. Yellowing begins at the top of the leaf and gradually reaches the leaf sheath. Culms and spikes (glumes and awns) remain green longer and, besides producing the energy for nitrogen remobilization throughout photosynthesis, are the last source of protein accumulation in the grain (Simpson, Lambers, and Dalling, 1982, 1983).

Wheat has a great capacity to accumulate in the grain most of the N absorbed by the plant or present in its vegetative organs, as is shown by the linear increase of protein content in response to N fertilizer (Figure 4.2). The wheat plant possesses the ability to store in the developing grain all the nitrogen potentially translocable: in an experiment in which the sink was halved by eliminating one spikelet out of two at heading time, grain yield per spike dropped to 65 percent of the control while protein content rose from 11.8 percent to 16.9 percent and the total amount of N per spike did not significantly differ from that present in the intact spikes (Borghi et al., 1986).

Breadmaking quality increases with the N supply and reaches a peak at an N supply level above that needed to achieve maximum yield (Figure 4.2). Thereafter protein quality decreases with increase of protein content because the extra N accumulated in the grain is represented by gliadins or nonprotein nitrogen, which depress breadmaking quality (Borghi et al., 1986).

OPTIMIZATION OF NITROGEN FERTILIZATION

It is recognized that the optimal combination of new genotypes and higher N rates was the driving force of the Green Revolution and that the future increases in total grain production necessary to satisfy the demands of an increasing population must be achieved essentially by crop intensification because of the lack of new land. On the other hand, there is, particularly in the developed countries, increasing alarm concerning the potential negative effects on the environment of excessive use of N fertilizer. Thus the supply of sufficient N fertilizer to achieve optimum grain yield and quality while reducing the risk of pollution caused by inappropriate N applications is a crucial task in wheat cultivation.

Three basic factors should be considered with regard to the identification of the optimum N fertilization rate: (1) the total amount of N needed by the crop for optimum grain yield in a given situation, (2) the amount of

N available from sources other than N fertilizer, and (3) the efficiency of the N applied in satisfying N demand during the cropping seasons (Viets, 1965). The difficulties lie in the unpredictability of, or in the difficulty of measuring, each of these factors. For instance, the total N absorbed by the plant includes not only the N stored in the grain but also the N left in the straw or in the vegetative parts (leaves, glumes, etc.), which is lost during threshing; the N accumulated in the roots; and the N volatilized from the plant tissues or extruded from the roots. Moreover, "optimal grain yield" is a vague concept and its determination with sufficient accuracy requires, for each homogeneous growing area, the organization of complex experiments repeated over several years. Furthermore, the N available from sources other than applied fertilizer is difficult to ascertain because of the complexity of the factors involved, such as soil type, previous crop, quantity and nature of soil organic matter, climatic trends, etc. The efficiency of the N applied in satisfying the N demand of the crop depends on the type of fertilizer, timing of application, seasonal trends, etc.

These difficulties have not prevented agronomists from developing N rate recommendations that are widely and successfully applied by farmers. In the past recommendations for a given wheat-growing area were based on fixed N rates inferred from N rate studies and were considered valid for all situations. A further refinement included the classification of available soils based on the expected N mineral content inferred from the previous crops and management practices. Another step forward was represented by the determination of the amount of mineral N present in the soil at the moment of N fertilization, usually at the end of the winter. Finally a more complete method, which took into account also the N made available by the mineralization process during the whole plant life cycle and the expected yield, was established. The approach is based on the predicted "N balance," which represents a simplification of the complex balance of the numerous N inputs and outputs. For instance, N losses caused by volatilization are assumed to be equivalent to the N input from rains and nitrogen symbiotic fixation (Hebert, 1969). The "N balance" is opened at the beginning of the vegetation period. At that stage the initial amount of N available from sources other than fertilizer is represented by the N already accumulated in the plant plus the mineral N present in the soil profile explored by the roots. This initial N amount can be analytically determined; however, about ten soil samplings per field in the layers from 30 to 120 cm (according to soil depth) are recommended to achieve a sufficiently precise N estimate ($\pm$ 10 percent) (Gate, 1995); in the Mediterranean climate, to achieve the same level of precision, 20 samplings are needed (Grignac and Perret, 1987).

The amount of N made available by the mineralization process during the vegetation period is difficult to determine because of the complexity of the factors involved and calls for fine "tuning" at the local level or refined modeling (Allison, 1966). The different elements to be considered for soil N estimation have been analytically described for the French conditions and summarized in synoptic tables that include the estimate of N released from an ample range of past management practices involving different previous crops, modalities of organic manure distribution, and so on (Gate, 1995).

To complete the N balance three further elements are needed: (1) the yield potential of the site, (2) the amount of N needed to sustain the potential yield, and (3) the actual yield expected in a given field and year. Yield potential can be estimated with adequate accuracy by means of N rate studies carried out over several years. To determine the optimal N rate for achieving maximum yield, the trials should include at least five N levels (Needham, 1984). The amount of N absorbed by the crop and the corresponding maximum yield achieved in different N dose experiments are linked by a linear regression with a *b* coefficient close to 3. This means that a wheat crop requires 3 kg of N per 100 kg of grain produced (Gate, 1995). These two elements allow prediction of the expected N demand of a crop if it is to achieve the maximum yield, which is usually the farmer's target. This yield target can be adjusted before the final N distribution to take into account the specific events of the year, which are likely to result in higher or lower grain yields.

Once the optimum N rate has been determined, there is the problem of distribution timing to be considered. The N demand of the wheat crop is minimal during the winter, rises rapidly after shooting, and gradually decreases after flowering; on the other hand, N fertilizer is very mobile in the soil, with consequent leaching hazards. Dose splitting and correct timing in relation to the variable demand during the life cycle is therefore an important aspect of N fertilization. N optimization can be achieved by monitoring the crop nutritional status directly on the plant during the life cycle. Several factors made this assay problematic: optimal values of plant N concentration differ at different growth stages or in different plant organs, within the same tissue the values fluctuate during the day, and it is difficult to relate the observed nitrate content values to specific N fertilizer rates. From the practical point of view, the nitrate content of the sap extracted from the basal tillers seems to be the best indicator of the crop nutritional status. This test is recommended in France in the case of split applications to optimize the second or third dressing (Gate, 1995).

WATER SUPPLY AND NITROGEN ASSIMILATION

In most of the rainfall regions of the world wheat is subject to water stress that may occur at different stages during the life cycle. For instance, the Mediterranean climate regions are characterized by a relatively abundant water availability in the early stages, followed by a progressive water deficit after flowering caused by a decrease of rainfall associated with an increase in air temperature, which increases evapotranspiration. In such circumstances the plants undergo an acceleration of senescence with important effects on dry matter and N accumulation in the grain. Normally the total amount of N fertilizer is given to the crop at sowing or it is split into two doses, the second being distributed before shooting. When grain filling occurs in conditions of water deficit, early N application increases the preanthesis contribution of dry matter and N to grain basically in two ways: by increasing early vigor and by improving the remobilization process. Rapid growth in the early stages of development, favored by N application, results in a larger and deeper root system able to explore a deeper soil layer, thus increasing the amount of water available to the plant. The parallel increase of aboveground biomass, and particularly an early cover of the soil, reduces water evaporation from the soil and increases the waterflow throughout the plant, i.e., the proportion of precipitation that passes through the crop, with a consequent improvement of water use efficiency. A negative aspect is represented by the increase of transpiration caused by the larger biomass, which may cause an early depletion of the soil moisture and severe water stress after anthesis, leading, in the worst situations, to lack of grain formation, the so-called "haying off" of the crop (Fischer and Kohn, 1966). Larger biomass before anthesis also means a larger amount of carbon and nitrogen compounds available in the plant to sustain grain filling in a period when climatic conditions (high temperatures and water deficit) severely hamper N uptake and carbohydrate synthesis. The important role of relocation from vegetative to reproductive organs in dry environments has long been recognized (Blum, 1988) but only recently has the beneficial role of N application in dry environments been described in physiological terms. Operating in a Mediterranean climate region of Western Australia characterized by postanthesis water deficit, Palta and Fillery (1995a, 1995b) found that an N application of 60 kg ha^{-1}, compared with a smaller application of 15 kg ha^{-1}, increased preanthesis contribution of dry matter to grain yield more than 10-fold, while the N remobilized into grain increased 2.3-fold, leading to a 54 percent increase in N despite a 5-fold reduction in postanthesis uptake. This result can be explained assuming that the demand for early stored N was stimulated by the increase of the size of the sink induced by the larger availability of N at the highest rate. Another effect of the high rate of N was the reduction of N losses from the plant during grain filling. In fact the anticipated closure of the life cycle induced by the more rapid development of postanthesis water deficit decreased the N losses peculiar to the late phases of grain filling.

The traditional strategy to optimize grain yield in dry environments is based on the economization of the soil water resources during the vegetative growth in order to save enough water for the critical stage of grain filling. One of the consequences is the limited use of N fertilizer because N fertilizer must be applied early in the season when the amount of rainfall is difficult to predict. In these circumstances farmers prefer to avoid the risk of luxurious vegetative growth in order to reduce the initial biomass and the consequent transpiratory water losses.

The optimal N rate for rainfall areas can be experimentally determined by a two-step approach. In the first step the achievable yields with limited water supply are deduced from experimental trials including different levels of water supply under nonlimiting fertility conditions. The second step includes N rate experiments aimed at optimizing N rate for a given water supply. Available water for a specific crop and the relative optimum N supply can finally be calculated on the basis of the water stored in the soil at the time of N application, the water that can be provided by irrigation, and the expected rainfall of the area (Prihar, Gajri, and Arora, 1985). A supplementary late application can be recommended in a wet year (Turner and Begg, 1981).

In conclusion, the results available in the literature indicate that in drought-prone areas the wheat plant is able to translocate efficiently to the grain most of the dry matter and N accumulated before anthesis in the vegetative organs, including the infertile tillers, and point to the opportunity of maximizing the biomass at flowering through the use of N fertilizer. A problem may arise from high temperatures seldom associated with post-anthesis water stress. In fact the combination of thermal and water stress can accelerate senescence to the point where the remobilization process is severely hampered (Fischer and Kohn, 1966).

GENETIC ASPECTS OF NITROGEN NUTRITION

At the beginning of this century, when the important contribution of N nutrition to grain yield became evident and the synthesis of ammonium from atmospheric N made fertilizers available on a large scale, it was found that the available wheat varieties were not able to exploit the extra fertility induced by N fertilizer because of their intrinsic susceptibility to lodging. Breeders were therefore stimulated to develop a new plant idiotype more "tolerant" to N fertilization. The attempt to improve lodging resistance by improving stem stiffness while keeping plant height constant did not yield satisfactory results, above all because this approach was essentially based on selection within the old local populations.

A real breakthrough was achieved in Italy by Nazareno Strampelli (responsible for the world's first green revolution), who introduced into the Italian germplasm the short-straw trait from the Japanese cultivar Akagomughi (Vallega, 1974). According to Law (1983), Akagomughi was a good source of earliness and dwarfism because of the strong genetic linkage between the Ppd1 gene for earliness and the dwarfing gene Rht 8. The best progenies turned out to be both early and with short straw. The plant model of these progenies was completely different from that prevailing at the time in the cultivated varieties. Apparently the breeders were rather skeptical about the potential values of this idiotype because the plants were thin and had small heads (Michahelles, personal communication). More enthusiasm was shown by agronomists who, by using heavier N fertilizer rates, succeeded in increasing the yield potential and thereby ensured the success of the semidwarf high-input high-yielding varieties.

Oil crises and a growing concern with environmental aspects of wheat cultivation in the 1970s led plant breeders to breed wheat cultivars more efficient in N utilization. The problem was initially tackled by searching for genetic variability for grain protein content. Johnson, Schmitt, and Mattern (1968) identified the variety Atlas 66 as a valuable source of high protein because the trait appeared to be controlled by a relatively small number of genes located in the chromosomes of the homeologous group 5. Johnson, Mattern, and Schmitt (1967) and Johnson, Dreier, and Grabouski (1973) provided evidence that the high protein content of Atlas 66-derived lines was the consequence of a more efficient and complete N translocation from foliage to grain and succeeded in incorporating Atlas 66 "protein genes" in hard winter wheat varieties.

Starting from the evidence that the grain protein content depends on the amount of vegetative biomass per unit of grain biomass, Kramer (1979) estimated, on the basis of the data available in the literature, a 0.13 percent decrease of grain protein content for each 1 percent increase in harvest index. He used these figures to compare varieties with different biomass and harvest index and observed that, after leveling to the common harvest index of 43 percent, the grain protein content differences between varieties disappeared. This was true also of the variety Atlas 66, characterized by a very low harvest index, and Kramer therefore questioned the use of this variety as a source of protein genes.

Several studies have been reported in the literature concerning attempts to modify genetically the N metabolism of wheat plants. Most of the papers published in the 1970s report genetic variability for some metabolic steps in nitrogen organication and, in particular, nitrate reductase (NR) activity. The reason for investigating NR activity was that, in conditions of maximal photosynthesis, actual NR activity may be a limiting factor for grain yield and that breeding for increased NR activity, as measured in vitro, should serve to enhance grain yield or N content (Hewitt, 1979). Under certain conditions the measurement of NR may be useful to predict yield capacity, but selection for superior genotypes on the basis of NR activity has produced variable results (Cregan and van Berkum, 1984). In fact it has been shown that NR activity significantly differs among wheat genotypes, and is a heritable trait responding to selection (reviewed by Clark, 1990). However, attempts to demonstrate a crucial role of NR activity in the N nutrition of wheat have been unsuccessful. The complexity of N metabolism, the interactions with carbon metabolism, and the complex sink-source relationships tend to suggest that direct genetic control of N efficiency is unlikely (Blum, 1988).

In recent years, increasing attention has been paid to the possibility of enhancing the efficiency of N use, i.e., increasing the yield per unit of N supplied to the crop. Efficiency is the combination of uptake efficiency (the portion of N taken up per unit of N supplied) and translocation efficiency (the portion of absorbed N recovered in the grain). Genotypic differences for both traits have been reported by several authors (reviewed by Beninati and Busch, 1992). Rapid early root growth and deeper rooting can reduce N leakage caused by winter rains (Fischer, 1981). Cox, Qualset, and Rains (1985) observed that genetic variation in N assimilation plays a role in determining grain yield and protein concentration in wheat and found genetic variability in N assimilation after anthesis. The lines that assimilated more N than required for their yield level and revealed high N translocation turned out to have a higher protein concentration (Cox, Qualset, and Rains, 1986). Clarke et al. (1990), on the contrary, did not find sufficient genetic variability for N utilization parameters among the cultivars they tested and concluded by questioning the convenience of selection for N efficiency. A similar opinion was expressed by Bertholdsson and Stoy (1995), who showed that the high protein concentration of some American hard wheat cultivars was a direct consequence of their low dry-matter production in the grain-filling period.

In conclusion, the possibility of selecting more N-efficient cultivars has difficulties already emphasized by Fischer in 1981 and confirmed in the most recent literature. In particular it has been seen that the efficiency mechanisms become important at very low levels of N availability when yield level is below the economic threshold and that further improvement of the nitrogen harvest index is problematic because of the high values already achieved in modern varieties. Furthermore, higher values could probably be achieved through a faster removal of N from the green organs, with expected negative effects on yield. It is a fact that modern short-straw, high-yielding varieties are more efficient in N use in the sense that they are able to produce more grain per unit of N absorbed. The negative side of this peculiarity is represented by the lower grain protein concentration. The negative effects of the lower protein percentage on end-use properties (i.e., breadmaking or pasta-making characteristics) have been largely counterbalanced by intensive selection for a more favorable composition of the protein subunits (high and low molecular weight glutenins) accumulated in the grain, as shown in Table 4.1.

Recently the countries belonging to the European Union, faced by problems of wheat surpluses and of risks of pollution of phreatic water as a result of heavy N fertilization, have started to impose restrictions on the use of N fertilizer. Investigations have been undertaken to evaluate the possibility of breeding wheat cultivars not only able to use N fertilizer more efficiently, but able also to tolerate N deficiency. Experiments in which sets of cultivars are grown at optimum N nutrition and under N stress have indicated the existence of genetic variability for tolerance of N deficiency

TABLE 4.1. Agronomic and Qualitative Characteristics of Six Groups of Varieties Bred in Italy During This Century

Varietal group	Year of release	Plant height (cm)	H.I. (%)	Grain yield (t/ha)	Protein (%)	Alveograph W ($J \times 10^{-4}$)	Quality* score
1	1900	111	34	3.64	12.0	131	5.5
2	1920	117	36	3.56	11.6	65	5.6
3	1930	108	38	4.11	11.5	102	7.2
4	1950	94	40	4.20	10.9	110	7.3
5	1970	89	43	4.63	10.7	126	8.4
6	1980	80	44	4.92	11.2	160	10.9
l.s.d. (p=0.05)		2	2	0.27	0.4	35	—

Source: Canevara et al., 1994.
*This score reflects the effects of each high molecular weight subunit on gluten quality.

(Peltonen, Virtanen, and Haggrén, 1995). For instance, in Germany the Federal Office of Plant Cultivars recently certified that the cultivar Pegassos possesses an "improved N efficiency" and a "pronounced low-input suitability" (El Bassam, 1998).

It is likely that modern high-yielding cultivars might differ in tolerance of N deficiency and consequently that specific breeding programs can produce new cultivars more adapted to a low-input agriculture. It is a fact that the yield potential accumulated in modern cultivars through breeding confers an advantage over the old ones which is expressed in a wide spectrum of N availability as exemplified in Figure 4.3).

FIGURE 4.3. Grain Yield of Varieties of Different Eras Cultivated at Three Levels of Nitrogen Supply

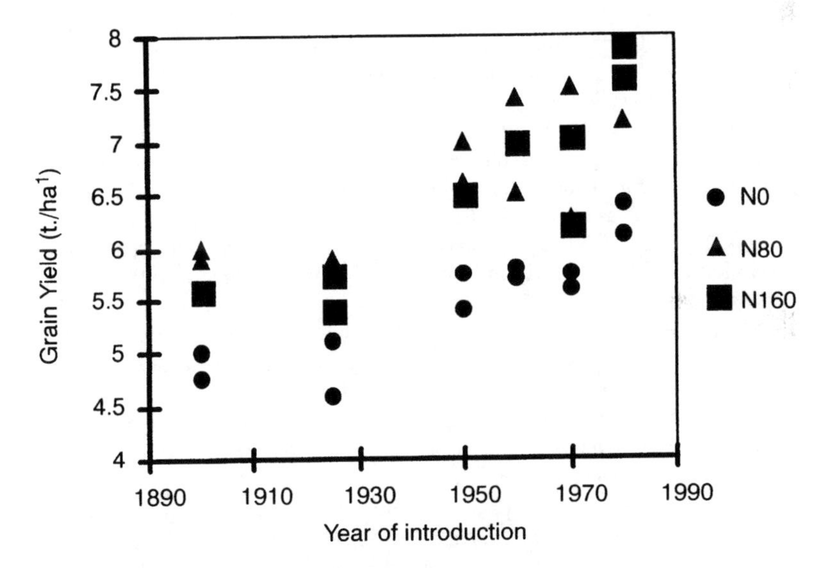

Source: Data kindly provided by Guarda and colleagues.

REFERENCES

Allison, F.E., 1966. The fate of nitrogen applied to soils. *Adv. Agron.*, 18: 219-258.

Austin, R.B. and Jones, H.G., 1975. The physiology of wheat. pp. 20-73. In *Plant Breeding Institute, Annual Report, 1974.* Plant Breeding Institute, Cambridge, UK.

Beninati, N.F. and Busch, R.H., 1992. Grain protein inheritance and nitrogen uptake and redistribution in a spring wheat cross. *Crop Sci.,* 32: 1471-1475.

Bertholdsson, N.-O. and Stoy, V., 1995. Yields of dry matter and nitrogen in highly diverging genotypes of winter wheat in relation to N-uptake and N-utilization. *J. Agron. & Crop Sci.,* 175: 285-295.

Blum, A., 1988. *Plant Breeding for Stress Environment.* CRC Press, Boca Raton, FL, p. 223.

Borghi, B., Corbellini, M., Cattaneo, M., Fornasari, M.E., and Zucchelli, L., 1986. Modification of the sink/source relationships in bread wheat and its influence on grain yield and grain protein content. *J. Agron. & Crop Sci.,* 157: 245-254.

Bumb, B.L., 1995. World nitrogen supply and demand. An overview. In P.E. Bacon (Editor), *Nitrogen Fertilization in the Environment.* Marcel Dekker, New York, pp. 1-40.

Canevara, M.G., Romani, M., Corbellini, M., Perenzin, M., and Borghi., B., 1994. Evolutionary trend of morphological, physiological, agronomical, and qualitative traits in *Triticum aestivum* L. cultivars bred in Italy since 1900. *Eur. J. Agron.,* 3(3):175-185.

Clark, R.B., 1990. Physiology of cereals for mineral nutrient uptake, use, and efficiency. In V.C. Baligar and R.R. Duncan (Editors), *Crop As Enhancers of Nutrient Use.* Academic Press, New York, pp. 131-209.

Clarke, J.M., Campbell, C.A., Cutforth, H.W., De Pauw, R.M., and Winkelman, G.E., 1990. Nitrogen and phosphorus uptake, translocation, and utilization efficiency of wheat in relation to environment and cultivar yield and protein levels. *Can. J. Plant Sci.,* 70: 965-977.

Cox, M.C., Qualset, C.O., and Rains, D.W., 1985. Genetic variation for nitrogen assimilation and translocation in wheat. I. Dry matter and nitrogen accumulation. *Crop Sci.,* 25: 430-435.

Cox, M.C., Qualset, C.O., and Rains, D.W., 1986. Genetic variation for nitrogen assimilation and translocation in wheat. III. Nitrogen translocation in relation to grain yield and protein. *Crop Sci.,* 26: 737-740.

Cregan, P.B. and van Berkum, P., 1984. Genetics of nitrogen metabolism and physiological/biochemical selection for increased grain crop productivity. *Theoretical and Applied Genetics,* 67: 97-111.

El Bassam, N., 1998. A concept of selection for "low input" wheat varieties. In H.J. Braun, F. Altay, W.E. Kronstad, S.P.S. Beniwal, and A. McNab (Editors), *Proc. 5th Int. Wheat Conference,* Ankara, Turkey, 9-14 June 1996. Kluwer Academic, Dordrecht, Germany.

Fischer, R.A., 1981. Optimizating the use of water and nitrogen through breeding of crops. In J. Monteith and C. Webb (Editors), *Soil Water and Nitrogen in Mediterranean-Type Environments.* Martinus Nijhoff & W. Junk. The Hague, pp. 249-278.

Fischer, R.A. and Kohn, G.D., 1966. The relationship of grain yield to vegetative growth and post-flowering leaf area in the wheat crop under conditions of limited soil moisture. *Aust. J. Agric. Res.,* 17: 281-295.

Gate, P., 1995. *Ecophysiologie du Blé: De la Plante à la Culture.* Tec & Doc. Paris, p. 429.

Grignac, B. and Perret, V., 1987. On the French hard wheat production and breeding. In B. Borghi (Editor), *Hard Wheat: Agronomic, Technological, Biochemical and Genetic Aspects.* Commission of the European Communities, Report EUR 11172 EN, Brussels, Luxembourg, pp. 151-158.

Hebert, J., 1969. La fumure azotée du blé tendre d'hiver. *Bull. Tech. Inf.,* 244: 755-766.

Hewitt, E.J., 1979. Primary nitrogen assimilation from nitrate with special reference to cereals. In J.H.H. Spiertz and Th. Kramer (Editors), *Crop Physiology and Cereal Breeding.* Proceedings of a Eucarpia Workshop, 14-16 November 1978. Pudoc, Wageningen, The Netherlands, pp. 139-155.

Johnson, V.A., Dreier, A.F., and Grabouski, P.H., 1973. Yield and protein responses to nitrogen fertilizer of two winter wheat varieties differing in inherent protein content of their grain. *Agron. J.,* 65: 259-263.

Johnson, V.A., Mattern, P.J., and Schmidt, J.W., 1967. Nitrogen relations during spring growth in varieties of *Triticum aestivum* L. differing in grain protein content. *Crop Sci.,* 7: 664-667.

Johnson, V.A., Schmidt, J.W., and Mattern, P.J., 1968. Cereal breeding for better protein impact. *Econ. Bot.,* 22: 16-25.

Kramer, Th., 1979. Environmental and genetic variation for protein content in winter wheat (*Triticum aestivum* L.). *Euphytica.,* 28: 209-218.

Law, C.N., 1983. Prospects for directed genetic manipulation in wheat. *Genet. Agr.,* 37: 115-132.

Meynard, J.M., 1987. L'analyse de l'élaboration du rendement sur les essais de fertilisation azotée. *Perspectives Agric.,* 115: 76-83.

Needham, P. 1984. *The Basis of Current Nitrogen Recommendations for Cereals.* Reference Book, Ministry of Agriculture, Fisheries and Food. UK No. 385, ADAS, Great Westminster House, London.

Novoa, R., and Loomis, R.S., 1981. Nitrogen and plant production. *Plant Soil.,* 58: 177-204.

Palta, J.A. and Fillery, I.R.P., 1995a. N application increase pre-anthesis contribution of dry matter to grain yield in wheat grown on a duplex soil. *Aust. J. Agric. Res.,* 46: 507-518.

Palta, J.A. and Fillery, I.R.P., 1995b. N application enhances remobilization and reduces losses of pre-anthesis N in wheat grown on a duplex soil. *Aust. J. Agric. Res.,* 46: 519-531.

Pearman, I., Thomas, S.M., and Thorne, G.N., 1979. Effect of nitrogen fertilizer on photosynthesis of several varieties of winter wheat. *Ann. of Bot.,* 43(5): 613-621.

Peltonen, J., Virtanen, A., and Haggrén, E., 1995. Using a chlorophyll meter to optimize nitrogen fertilizer application for intensively-managed small-grain cereal. *J. Agron. Crop Sci.,* 174: 309-318.

Prihar, S.S., Gajri, P.R., and Arora, V.K., 1985. Nitrogen fertilization of wheat under limited water supplies. *Fertilizer Res.,* 8(1): 1-8.

Simpson, R.J., Lambers, H., and Dalling, M.J., 1982. Translocation of nitrogen in a vegetative wheat plant (*Triticum aestivum* L.). *Physiol. Plant.*, 56: 11-17.
Simpson, R.J., Lambers, H., and Dalling, M.J., 1983. Nitrogen redistribution during grain growth in wheat (*Triticum aestivum* L.). *Physiol. Plant.*, 71: 7-14.
Turner, N.C. and Begg, J.E., 1981. Plant-water relations and adaptation to stress. In J. Monteith and C. Webb (Editors), *Soil Water and Nitrogen in Mediterranean-Type Environments.* Martinus Nijhoff & W. Junk. The Hague, pp. 97-131.
Vallega, J., 1974. Historical perspective of wheat breeding in Italy. *Proceedings of Fourth Wheat Seminar,* 21 May-2 June 1973. FAO/Rockfeller Foundation. Tehran, Iran, pp. 115-126.
Viets, F.G. Jr., 1965. The plant's need for and use of nitrogen. In W.V. Bartholomew and F.E. Clark (Editors), *Soil Nitrogen.* Agronomy 10, American Society of Agronomy, Madison, WI, pp. 503-549.

Chapter 5

Grain Quality and Its Physiological Determinants

Peter J. Stone
Roxana Savin

INTRODUCTION

Wheat is the world's most important agricultural product. It is the most-produced crop in the world, with an annual harvest of ca. 540 million tons. Approximately 90 percent of this production is consumed directly by humans, and consequently wheat supplies ca. 20 percent of the world's total plant-derived edible dry matter (Evans, 1993).

The primacy of wheat among the cereals can be further demonstrated by examining its place in history. The Bible states of bread that "of wheaten flour thou shalt make it" (Exodus 29:2), and for a considerable time since, wheat has been the preferred grain for bread making. Indeed, wheat was valued so highly that ca. 970 B.C. King Solomon (Song of Solomon 7:2) serenaded one of his beaux with the words, "Thy navel is like a round goblet, which wanteth not liquor: thy belly is like a heap of wheat set about with lilies. . . ."

Wheat is the primary crop because it is grown across an exceptionally diverse range of environments, from the arid plains of Africa to the humid valleys of Vietnam, and from the cold of Nepal (Rao and Whitcombe, 1977) to the heat of India (Bagga and Rawson, 1977). However, it would be a mistake to think that the extreme popularity of wheat is merely the result of its wide range of adaptation. Indeed, it is probably more logical to argue that wheat has become adapted to such a wide range of environments because humans have found it so useful that they attempt to cultivate it under even the most unlikely conditions.

Why Is Wheat Such a Popular Crop?

One of the most significant properties of wheat is the extremely wide variety of uses it serves. Quite apart from the more "predictable" wheat products, such as:

- bread
- pasta
- crackers
- cookies
- noodles
- pretzels
- cakes
- breakfast cereals
- puddings

Wheat (or part thereof) is also widely used for the following:

- beer
- MSG
- vitamins (thiamine, riboflavin, niacin)
- smallgoods (manufactured meat products)
- binder and filler (paper and textiles)
- sugars (glucose, maltose, fructose, dextrose)
- food thickeners (soups, stews, sauces, gravies)
- extruded snack foods
- starch
- cardboard
- couscous
- bran
- semolina
- biodegradable films
- gluten
- confectionery
- burghul
- wheat germ

While the other major grains (rice and maize) also have important secondary and industrial functions, neither matches wheat for sheer versatility.

How Does Wheat Differ from Other Cereals?

Wheat is supreme among cereals largely because its grain contains proteins with unique chemical and physical properties. When ground wheat grain (flour) is mixed with water the storage proteins bind in such a way that a coherent mass called gluten is formed. Gluten exhibits plasticity, strength, and elasticity, which means that it is capable of deformation when placed under pressure, that it tends to resist that deformation, and that it tends to reassume its original form when pressure is released. These properties enable wheat flour to form a cohesive dough that can expand to accommodate gas, and yet resist stretching to the bursting point. Without these characteristics, leavened breads would not be possible. Furthermore, the cohesive properties of gluten make it relatively simple to separate wheat protein from other constituents of the grain, which helps to explain the widespread use of the secondary products of wheat (e.g., starch) as industrial and food additives. It is largely because of the unique properties of gluten that wheat is the world's most widely consumed source of food.

Objective

We shall examine the science behind the unique properties of wheat as a food item, and will endeavor to demystify the physiological processes by which these properties are conferred upon the humble wheat grain. It is only through understanding the processes that contribute to wheat quality that we can hope to manipulate them to our full advantage, and obtain wheat with the quality attributes that we desire.

WHAT MAKES WHEAT SO SPECIAL?— THE WHEAT PROTEINS

What Is Gluten, and How Does It Work?

Before the unique physical properties of gluten can be fully appreciated, an understanding of the basics of wheat protein chemistry is necessary. A comprehensive description of the wheat proteins in a single chapter would not be possible—there are simply too many of them. The gluten protein alone consists of more than 100 different polypeptide components, which can be classified and subclassified on the basis of their amino acid sequence, amino acid composition, or by their size (Müller and Wieser, 1995). We shall not concern ourselves with excessive intricacy here: a number of functional classifications can be conveniently discerned, and their interrelationships are most easily explained diagrammatically (Figure 5.1).

Since the pioneering days of cereal chemistry, the different fractions of wheat protein have been classified on the basis of their solubility. Sequential extraction of protein in solutions of increasing strength enabled four basic classes of protein to be discerned: (1) albumins, which are soluble in water; (2) globulins, which are insoluble in water but dissolve in dilute salt solution; (3) gliadins, which do not dissolve in either water or saline water, but are soluble in 70 percent aqueous ethanol; and (4) glutenins, which are not soluble in any of the solutions above, but which dissolve in dilute acids or alkalis. Fractionation on this basis, followed by Kjeldahl nitrogen analysis of each fraction, enabled early workers to accurately determine grain protein composition and relate it to flour and dough properties (Osborne, 1907; Woodman, 1922; Woodman and Engledow, 1924).

This simple method of separation, classification, and measurement forms the basis of many of today's more complex methods of protein chemistry, which have enabled further subdivision of these basic protein classes and consequently a greater understanding of their role in product quality.

The metabolic proteins (albumin and globulin) typically make up 20 to 30 percent of the total grain protein (Figure 5.1) (Woodman and Engledow, 1924; Jennings and Morton, 1963; Martín del Molino et al., 1988),

FIGURE 5.1. Protein Composition of a Typical Wheat Grain

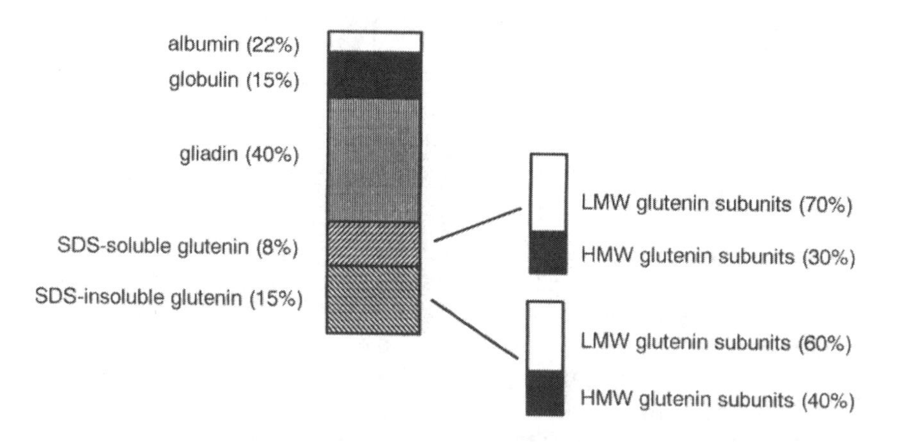

Note: Bars show the percentage of each protein fraction. SDS is sodium dodecyl sulfate and LMW and HMW are low and high molecular weight, respectively.

although this varies with genotype (Singh, Donovan, and MacRitchie, 1990; Stone and Nicolas, 1994), nitrogen nutrition (Abrol et al., 1971; Doekes and Wennekes, 1982; Stenram, Heneen, and Olered, 1990), and perhaps most importantly, method of measurement (Stone and Nicolas, 1996a). Despite their abundance in the grain, albumin and globulin have a minimal impact on dough strength and breadmaking quality, as they play only a minor role in the protein interactions that are required for the formation of cohesive gluten. This is partly because they are not chemically disposed toward protein-protein interactions, and also because they reside primarily in the embryo region of the grain, a fraction that is deliberately excluded from the white flours commonly used in bread and pastry making. While this means that we shall pay relatively little attention to them here, the reader should be reminded that the albumins and globulins are essential for the growth and development of the wheat seedling (Dell'Aquila et al., 1983), which relies on the energy and nutrients made available by their hydrolytic and proteolytic capacity.

The storage proteins compose the remaining 70 to 80 percent of wheat grain protein, and are primarily responsible for determining the physical properties of dough and consequently many aspects of grain quality. We should keep in mind that their primary role is as a store of energy and nutrients for the germination and growth of wheat seedlings—it is mere fortuity that they have secondary properties useful for humans.

The storage proteins can be subdivided into classes on a number of bases, but size and solubility are the most usual distinctions: each is related to a considerable extent, and both mechanistically influence dough properties. In general, the proteins of higher molecular mass are less soluble and contribute more to dough strength than the smaller and more soluble proteins, which largely determine the extensibility of a dough (Gupta et al., 1995; Schropp and Wieser, 1996). It is on this basis that we shall discuss the role of the storage proteins (gliadin and glutenin) in determining dough properties. (The authors beg the forgiveness of cereal chemists for endeavoring to simplify what has become something of a field for specialists. A detailed analysis of the "true" composition of wheat storage proteins is beyond the scope of this discussion, but interested readers are encouraged to refer to a number of excellent reviews (Shewry, Halford, and Tatham, 1989; MacRitchie, du Cros, and Wrigley, 1990; Weegels, Hamer, and Schofield, 1996) which discuss the genetics and physicochemistry of gluten proteins).

The storage or gluten proteins fall into two major classes: the gliadins and the glutenins, each of which makes up approximately 30 to 40 percent of the total protein in the wheat grain. Gliadins are the smaller of the two, and range in size from about 30,000 to 80,000 kDa (which means that they are still "large" molecules, by normal biological standards). Gliadins are generally nonaggregating, which means that they do not join together to form large polymer chains. For this reason, the gliadins are unlikely to contribute strength to gluten. Rather, gliadin is largely responsible for extensibility in doughs, or the ability to stretch without breaking (Branlard and Dardevet, 1985; Gupta, Khan, and MacRitchie, 1993). The extensible properties of gliadin can be graphically illustrated by purifying it and mixing it with water, under which conditions it develops a honey-like consistency (Simmonds, 1989).

This is in marked contrast with the behavior and properties of glutenin, which when treated in a like manner exhibits the consistency of hard rubber. The resistance of glutenin to stretching and compression can be explained largely by its molecular properties, not the least important of which is its size. Glutenin is a family of individual proteins that vary in size from 12,000 to over 130,000 kDa (MacRitchie, du Cros, and Wrigley, 1990), but it is not usually as individuals that glutenin molecules exert their particular influence on dough strength. In contrast with the gliadins, glutenins are aggregating proteins, which means that they can form strong bonds with each other. By these interglutenin associations, glutenin polypeptides have the ability to form macromolecules with molecular weights of up to ten million (Grimwade et al., 1996), which makes them among the largest proteins in the natural world (Wrigley, 1996). It is the macromolecule of glutenin that is central to the formation of the cohesive gluten network required for a dough strong enough to withstand the stresses of breadmaking.

The ability of glutenin to form a macromolecule is largely determined by the proportion of the different classes of glutenin present in the wheat grain. Just as wheat proteins can be broadly classified by their solubility in water, salts, ethanol, and acids, the glutenins can be further divided into subclasses on the basis of their solubility in dilute detergent, particularly sodium dodecyl sulfate (SDS) (Figure 5.1) (Gupta, Khan, and MacRitchie, 1993). The glutenin proteins that are soluble in SDS tend to be smaller and contribute less to dough strength than those that are not SDS-soluble, and this appears to be related to the size of the protein subunits that are the building blocks of each glutenin molecule (Gupta, Khan, and MacRitchie, 1993; Gupta et al., 1995). Generally, the greater the proportion of high molecular weight glutenin subunits in a given glutenin molecule, the greater its ability to form a macromolecule, and the greater its contribution to dough strength (Gao and Bushuk, 1993; Gupta, Khan, and MacRitchie, 1993). SDS-soluble glutenin contains a smaller proportion of HMW glutenin subunits than SDS-insoluble glutenin, which consequently helps to explain the latter's greater contribution to dough strength (Gupta, Khan, and MacRitchie, 1993).

As with most things, however, size is not everything: a series of classical experiments showed that a combination of LMW and HMW glutenin subunits was required for maximum dough strength, and that removal of certain subunits from glutenin (by breeding subunit-deletion biotypes) had a proportionally greater impact on dough strength than on the amount of polymer in the protein (Gupta and MacRitchie, 1994; Gupta et al., 1994; Gupta et al., 1995). Evidently, while the size of the building blocks (subunits) is important in determining the size of the building (glutenin molecule), their size alone is not sufficient to determine the size of the city (macromolecule).

Chemical and structural differences in glutenin subunit composition are also likely to influence the ability of the resulting glutenin molecule to form macromolecules. The most important chemical and structural property affecting macromolecule formation appears to be the number and presentation of cysteine residues on the subunit, although direct evidence is difficult to obtain (Shewry, Halford, and Tatham, 1992). Nevertheless, cysteine residues are essential for macropolymer formation, acting as the chemical "glue" between individual glutenin molecules.

This point marks our departure from wheat protein chemistry, and we are now free to discuss some fundamentals of practical wheat quality.

How Do Glutenin Macropolymers Form, and How Does Their Formation Relate to Wheat Quality?

As stated in the introduction, wheat is valuable very largely because gluten enables its dough to develop the plasticity, strength, and elasticity needed to make bread rise. How do these properties emanate from the molecules that have been described above?

As for protein composition, the story of dough formation is both complex and incomplete, and much of the evidence is inconclusive or contradictory (see review by Weegels, Hamer, and Schofield, 1996). As such, we will here provide a simplified model, which nevertheless explains the essential links between the chemistry and the physics of dough, and consequently bread and wheat quality.

The conversion of flour to dough requires two essential inputs: water and energy (in the form of mixing). When flour is first mixed with water, the first major event is the unfolding of the gluten proteins to give a randomly arranged suspension of molecules. As we have seen, each glutenin molecule (ca. 30 percent of the total protein) has a number of cysteine residues attached along its length. When two cysteine residues come into close contact, a strong covalent (disulfide) bond may form, effectively gluing the glutenin molecules together. Mixing facilitates this disulfide bonding by moving the glutenin molecules past one another, thereby increasing the chance that two cysteine residues will meet. Every time that two glutenin molecules bind together in this way, their molecular mass increases, and so too does the average molecular mass of the dough. Of course, more than two glutenin molecules can bind together, and so it does not require a lot of mixing before the average molecular mass of the dough becomes very high (over 1 million kDa). Once the molecules begin to join into these macromolecules, the effects of disulfide bonding are supplemented by "entanglement." Just as the probability of string becoming tangled and knotted increases with its length, so too does the probability of protein entanglement: long molecules are more likely to make longer molecules when mixed.

As with any polymer system (from oil to polythene to dough), as the average molecular mass of the mixture increases, so too does its viscosity. As the dough is mixed, it becomes "thicker" or more viscous in feel, and consequently more energy is required to mix it at a given rate. In other words, the dough becomes stronger as it is mixed because its average molecular mass increases.

Further mixing increases the degree of interglutenin disulfide bonding to the point where the maximum potential number of disulfide bonds is reached. Cereal chemists and bakers call this point "peak dough development," because further mixing will not impart any more strength to the dough.

In fact, further mixing tends to reduce dough strength, because rather than assisting in the formation of new disulfide bonds, additional mixing energy can only break those that already exist. The result is a reduction in the average molecular mass of the dough, with a concomitant reduction in viscosity, or strength, and an increase in extensibility and stickiness. Cereal chemists and bakers simply call this phenomenon overmixing, and the latter definitely attempt to avoid it, as it reduces dough and bread quality, and the increased dough stickiness makes cleaning mixing equipment tedious.

It should be clear, then, that grain protein composition is fundamental in determining the physical properties of a dough. In general, the greater the molecular mass of the proteins in the grain, the greater the strength of the resulting dough. Later, we will discuss how different varieties of wheat and environmental variation (climate and cultural practice) affect protein composition, and consequently grain quality.

Glutenin Is Important, But Is It Enough?

Thus far, our discussion has implicitly assumed that dough strength tends to be inadequate, and that the potential to develop strength limits wheat quality. This is true for most of the world wheat crop, and explains why a premium is usually paid for breadmaking wheats: the expense of adulterating baker's flours by adding gluten (as a means of increasing strength to an acceptable level) is one that most millers and bakers would rather avoid.

This is not to say that doughs cannot be too strong for breadmaking. The best breads are produced from doughs that have a mix of strength, elasticity, and plasticity, properties largely determined by the balance between the two main classes of gluten protein: gliadin and glutenin. As mentioned earlier, gliadin tends to impart plasticity and elasticity to a dough, and glutenin, strength. What happens when the balance between these two protein fractions is altered?

When an optimum balance between strength and elasticity is achieved, the resulting dough does not require excessive mixing energy to reach peak development. The dough is strong enough to trap the thousands of tiny gas pockets that will expand during proving and baking, and keep them separate from each other, yet sufficiently elastic to enable the cells to expand, giving rise to a light-textured loaf with evenly distributed cells, and a moderate loaf volume (Figure 5.2). Such a loaf requires flour with the correct balance between gliadin and glutenin, as well as skilled baking.

If the ratio of gliadin:glutenin is too low, the dough is generally excessively strong and relatively inextensible. The result is a dough that requires a high (and costly) energy input before peak dough development is reached. Furthermore, the high strength and low elasticity of the dough prevents expansion of the gas bubbles during proving and baking, resulting in a loaf of heavy texture and low volume (high density) (Figure 5.2).

Doughs made from flours with a high gliadin:glutenin ratio tend to require low mixing energies, and because most bakers use a fixed energy input, weak doughs such as this are frequently overmixed, thereby further weakening an initially weak dough. Apart from being sticky and difficult to handle, a dough made from flour with an excessive gliadin:glutenin ratio will tend to be too weak to hold gas in discrete cells during proving and baking. As a result, the gas cells tend to merge, giving bread that has large holes in it, and an excessive or "blown out" loaf volume (Figure 5.2).

Clearly, dough strength is necessary, but not in itself sufficient for making acceptable loaves of bread; the balance between strength and elasticity plays a vital role in determining the quality of the end product. This "secret" is not new—as if to mark the hundredth anniversary of the

FIGURE 5.2. A Diagrammatic Representation of the Effect of Protein Composition on Grain Quality, As Determined by Baking

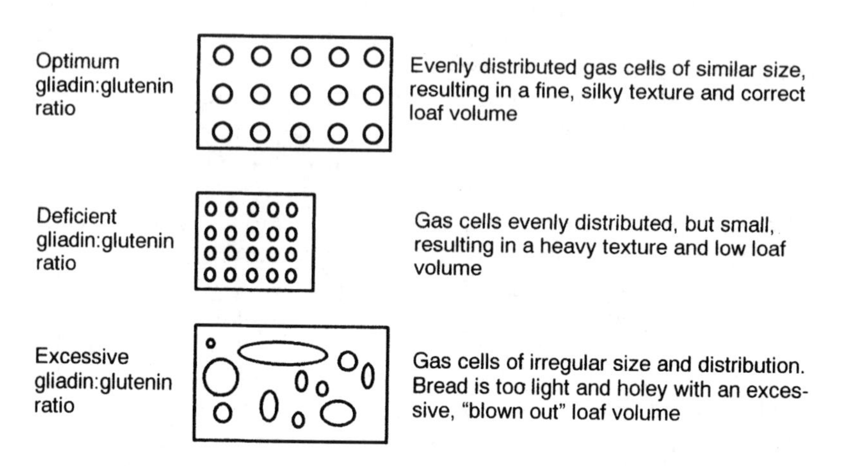

Source: Adapted from Simmonds, 1989.

event, Weegels, Hamer, and Schofield (1996) communicated that the Académie des Sciences, Paris, heard in 1896 that flours with a high gliadin:glutenin ratio held gas well during proving but not baking, and that flours with a low gliadin:glutenin ratio did not hold gas well during either proving or baking. The correct balance between the two gluten fractions was required for an acceptable loaf (Fleurent, 1896). The gluten proteins of wheat, although long invisible to the prying human eye, have for some time been evident enough in our loaves of bread.

Proteins and the Measurement of Quality

Thus far, our discussion has largely focused on the physicochemistry of gluten, and its influence on the quality of wheat for breadmaking. We have used bread as a "model" product, a tool for demonstrating the importance of dough strength and some of the chemistry of wheat proteins that contribute to it. In doing so, we have largely ignored the myriad end uses of wheat, a number of which were outlined in the introduction. Furthermore, we have not yet considered how to arrive at an estimation of wheat quality, even for a single product, such as bread. This is a deficiency we should endeavor to correct, if only because the value of a grain of wheat is directly proportional to its quality.

Because wheat has such a wide variety of potential end uses, it would be difficult to describe a given batch of wheat as simply "good" or "bad" in quality; it is really just best suited to one or more of these end uses. So grain quality is perhaps less an index of the inherent properties of a grain than the specific requirements of those who wish to mill, process, or eat them.

It would appear then that definition of grain quality is something of a moving target. The difficulties of the multiple uses of wheat are compounded by the fact that even for a single use, such as breadmaking, a wide range of products are produced (e.g., pan, flat, and steamed breads). Furthermore, even for the manufacture of a single product, grain is exposed to a wide variety of processes, each with its own peculiar requirements. For example, wheat for pan breadmaking is sequentially stored, milled, mixed, proved, baked, stored, and consumed. Characteristics of wheat that are highly valued for one of these processes may be considered undesirable for another.

The most practicable means of overcoming the need for these specific quality requirements is to use an index of quality that in some way integrates the most important quality parameters, to give an "on average" guide to quality. That which is most commonly used is a combination of cultivar identification and protein percent, which together give a good indication of protein composition and concentration. These two parameters give an indication of likely best end use, such that for a given (hypothetical) cultivar, the optimum end use might change with protein content, as shown in Figure 5.3.

While relationships such as this are generally assumed, there is currently no means of predicting accurately how the properties of a given cultivar will change with protein content, other than the use of empirical relationships or commercial experience, which are used to estimate grain "quality" for a given cultivar/protein content combination. To understand why it is that measures of protein content and genotype can be used as indicators of grain quality, we need to return briefly to our knowledge of dough physicochemistry.

In essence, protein percentage and cultivar serve merely as a surrogate measure for dough strength and associated characteristics, such as loaf volume. In Figure 5.3 we could substitute grain protein percentage for some measure of dough strength, and there would be little change in the order or distribution of wheat products on the graph. It follows then that grain protein percentage and dough strength are usually strongly and positively correlated (MacRitchie, 1978; Branlard and Dardevet, 1985; Camp-

FIGURE 5.3. Indicative End Use for Wheat of a Given Protein Content

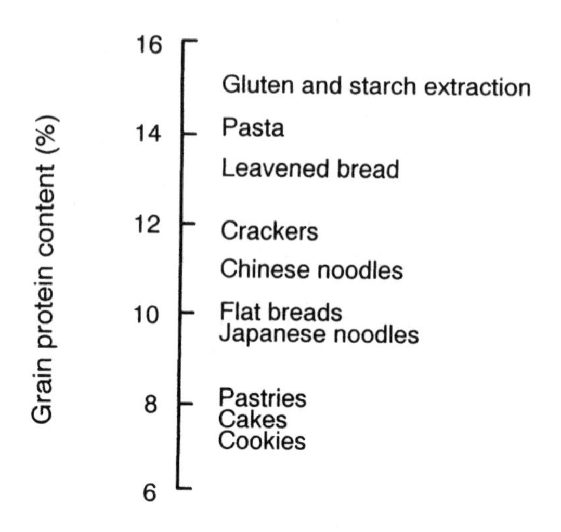

bell et al., 1987; Hay 1993; Khatkar, Bell, and Schofield, 1996), especially within a cultivar.

From the discussion in the "What Is Gluten, and How Does It Work?" section, it should be clear that the composition of wheat proteins is fundamental in determining dough strength. However, it was not explained at that point that the balance of grain proteins is largely genetically determined. In fact, the classes of gluten present in the grain are so genotype-dependent that they can be used to identify individual cultivars (Wrigley, 1992), in much the same way that fingerprints can be used to uniquely identify humans. From what we have learned thus far, it should be obvious that because wheat cultivars vary in grain protein composition, the inherent strength characteristics of their dough will also vary.

Protein percentage is also important because the amount of protein in a dough affects the extent to which protein macromolecules can hold the dough together. The higher the concentration of protein, the greater its ability to impart strength to a dough. Protein percentage is determined by a combination of agronomics and genetics, which is to say that there is a general tendency for cultivars of wheat to produce grains of high, medium, or low protein percentage, which can be altered by factors such as fertilizer application and sowing time.

Together then, protein percentage and cultivar give a good general indication of the type of dough strength that might be expected from a given batch of wheat, and from this experienced grain handlers and processors can frequently determine the best end use for that wheat. As we know, wheats for leavened breadmaking need to make strong doughs, but wheat destined for cookies should have weak dough properties (if they had gluten strong and extensive enough to hold gas, they might rise when cooked, and would not make the hard, dense, and flat products we expect cookies to be). Intermediate strength characteristics are required for noodles and crackers, as illustrated in Figure 5.3. Large-scale handlers of wheat and flour usually find that with experience, they can match specific mixes of cultivar and protein percentage to their best end use, thereby adding to the value of both the raw and end products.

THE NONPROTEIN COMPONENTS OF THE WHEAT GRAIN

Thus far, we have only looked at the influence of protein content and composition on wheat quality, and have conveniently ignored the fact that we have concerned ourselves with only about 12 percent of the grain. Although this seems remiss, an apology is not quite in order—even though it is in a minority, the influence of protein on wheat quality, and on bread in particular, cannot be underestimated. Having said that, we shall now consider the influence of the remainder of the wheat grain on bread quality.

Just as the balance between the different protein fractions is important in determining the overall properties of wheat protein, the balance between the proteins and the nonprotein components of the grain influences the properties of wheat flour. The composition of the wheat grain is, prima facie, quite simple (Figure 5.4).

Carbohydrate

A typical 50-mg wheat grain is dominated by carbohydrate, which constitutes approximately 70 percent of the total grain mass. Almost all of this carbohydrate is present as starch, with the remainder (less than 3 percent) comprising sucrose, glucose, fructose, and other less abundant sugars (Abou-Guendia and D'Appolonia, 1973; Boyacioglu and D'Appolonia, 1994a). Although these sugars are not present in great quantities, they are important during proving and baking, as they contribute to the sugar supplies required by yeast for the production of leavening gas.

Starch is just thousands of glucose molecules linked together to give a water-insoluble polymer. In wheat, starch occurs in small granules, which are densely packed in the endosperm of wheat and which serve as the

FIGURE 5.4. The Basic Ingredients of a Wheat Grain

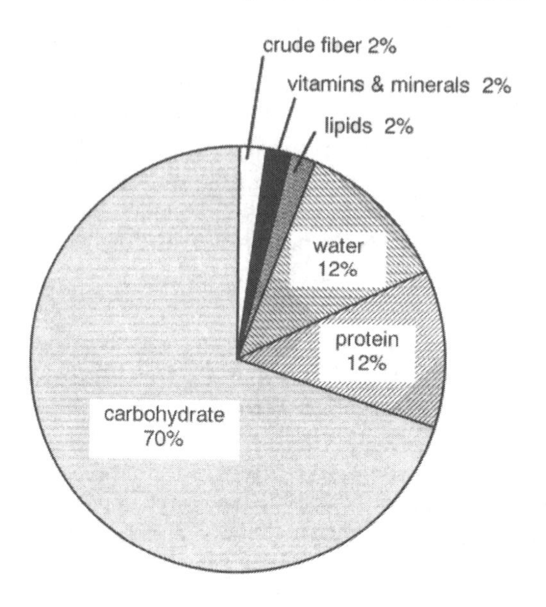

principal supply of energy for the germinating seed and seedling. Starch granules occur in three distinct groups (A, B, and C granules), which are distinguished on the basis of size, although their properties and the physiology of their synthesis are also different. The A-type starch granules are the largest (ca. 10-50 μm diameter) and differ from the B- and C-type granules in that they are lenticular in shape. They typically make up approximately 80 percent of the total starch mass, yet only ca. 10 percent of the total granule number. The B-type granules are spherical, with a diameter of 5 to 10 μm. They constitute about 15 percent of the starch mass and starch granule number. C-type granules are also spherical, and their small size (2 to 5 μm diameter) means that although they comprise up to 75 percent of the total starch granule population, they contribute only about 5 percent of the total starch mass.

In their native state, none of the starch granules are soluble in water, although the breakage that occurs when grain is milled for flour (10 to 35 percent of granules become damaged) makes them prone to swelling and partial solubilization in cold water. The structure of the starch granule is not uniform, but varies from crystalline to amorphous, depending on the chemical form of starch present in the granule.

Wheat starch is made up of two main fractions, amylose and amylopectin, which make up approximately 25 and 75 percent of the total starch mass, respectively (Leloup, Colonna, and Buleon, 1991; Boyacioglu and D'Appolonia, 1994b). The relationship between amylose and amylopectin is somewhat analogous to that between gliadin and glutenin: each of the starch fractions differ in size and structure, and so have different physical and chemical properties, and the balance between these properties affects the functionality of starch as a whole.

Amylose is an essentially linear polymer of 1-4 linked alpha-D-glucose, which in wheat is about five chains long and has slight branching (Noel, Ring, and Whittam, 1992). Amylopectin is much larger and more highly branched, with individual molecules making up tens of thousands of glucose units. The linear structure of amylose means that for a given molecular mass, it is longer than amylopectin, making it more prone to entanglement with starch and other molecules. As a result, amylose is a more effective thickening agent than amylopectin, a property for which it is valued as an additive in the food industry.

One of the most important properties of starch is its propensity to gelatinize. At approximately 60°C, starch granules become soluble in water because the energy in solution becomes sufficient for bonds between starch molecules to be broken and partially replaced by hydrogen bonds between starch and water (McGee, 1991). In this state, starch absorbs water and swells, forming a viscous (gelatinous) fluid. It largely this ability to gelatinize that explains why flours are so useful in sauces and gravies.

In bread, starch serves a number of important roles. In addition to making up the bulk of the loaf, starch reinforces the gluten network by providing a semisolid structure to which gluten can adhere. Second, because starch occurs in discrete granules, it can move around the dough to fill up spaces that are created as the loaf changes shape during baking (McGee, 1991). Starch also plays an important part in regulating the distribution of water in a loaf of bread. As the loaf heats up and rises during baking, water is forced out of the now-coagulating protein structure. Just as this occurs, the starch is more able to absorb water, because it has reached its gelatinization temperature. Protein does not give up as much water as starch can absorb (starch can absorb over ten times its weight in water), and so bread is primarily composed of partially gelatinized starch (McGee, 1991).

In time, the structural changes that occur in the starch granules during baking are partially responsible for the staling of bread. When the loaf is removed from the oven and begins to cool, the partially gelatinized amylose rapidly becomes less viscous, and it is this firming that makes a loaf easier to slice as it cools. Amylopectin is slower to resume its crystalline state, and this slow firming of the loaf is part of staling. Staling is often associated with dryness, and this occurs because as the amylopectin resumes its crystalline structure water is "squeezed" out of the lattice, and it migrates down a concentration gradient and, eventually, out of the loaf.

It should be evident that starch is very important in determining bread quality. Despite this significance, starch properties are not usually seen as a fundamental determinant of wheat quality, especially for breadmaking. This probably has less to do with the inherent importance of starch properties than the fact that starch composition is, relative to protein, very stable between genotypes and environments. Consequently, even though starch properties are important in determining the actual qualities of a grain, the fact that they are not highly variable means that there is not a high priority placed on measuring them to determine the value of wheat—its "quality." This view is changing somewhat with the increased importance of noodle wheats on the world market. The gelatinization properties of starch are of primary importance in determining the "mouth feel" of noodles, and so the value of noodle wheats (Oda et al., 1980; Crosbie 1991; Wang and Seib, 1996).

Lipid

Lipids are a fatty substance which compose only ca. 2 percent of the mass of flour. Despite their relative quantitative obscurity, flour lipids are absolutely essential for breadmaking. Removal of lipid from a dough prevents rising during baking, and there is evidence which suggests that lipids are involved in the binding of gliadin to glutenin within the gluten structure, and of gluten to starch within the dough as a whole. Flour lipids are known to be closely associated with both starch granules (Soulaka and Morrison, 1985; Morrison, 1988; Williams, Shewry, and Harwood, 1994), and gluten proteins (Weegels et al., 1994; Carcea and Schofield, 1996), although it is not clear which play the most important role in influencing dough properties (Kaldy, Kereliuk, and Kozub, 1993). Despite their impact on dough, grain lipid content and composition are not often used as a guide to grain quality, and this is largely because their role in determining quality is poorly understood.

HOW DO WHEAT GRAINS GROW?

Carbohydrate Accumulation

As discussed previously, carbohydrate makes up ca. 70 percent of the mass of a wheat grain, and starch constitutes almost 97 percent of the total mass of carbohydrate. Given the size of its contribution to grain mass, one might think of starch accumulation as the driving force behind grain growth, and it is in this light that we shall examine it.

The growth of a wheat grain is classically sigmoidal in form: a short lag phase is followed by a longer yet more rapid linear period of growth, at the end of which the rate of dry matter accumulation slows, until at the point of physiological maturity, no further additions are made to grain mass (Figure 5.5). This characteristic kernel growth curve is very largely a reflection of the physiology of starch deposition, which is, after all, at least 70 percent responsible for the construction of the curve.

FIGURE 5.5. The Growth of a Wheat Kernel, Showing Key Events in Starch Accumulation

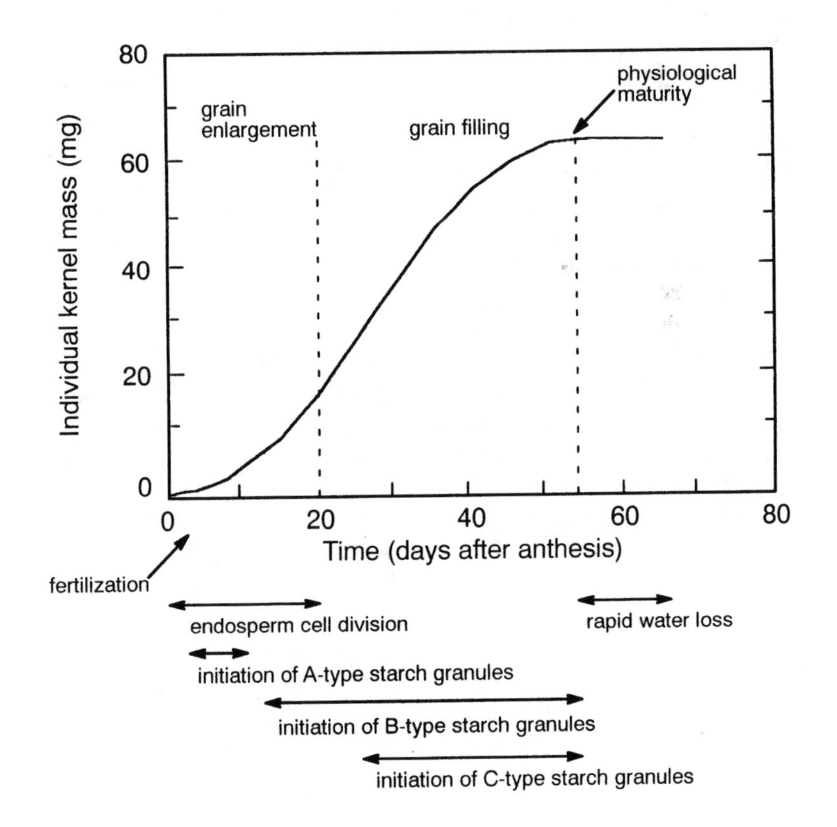

Jenner, Ugalde, and Aspinall (1991) have elegantly described grain growth and development as being the product of two processes: grain enlargement and grain filling. Grain enlargement can be thought of as the process that builds the structures into which starch and other constituents of the endosperm will be added during grain filling. It therefore follows that grain enlargement occurs first.

Grain enlargement commences with the act of fertilization, generally lasts ca. 20 days (Jenner, Ugalde, and Aspinall, 1991), and is the dominant process during the lag phase of grain growth. Within a few hours of fertilization, endosperm cell nuclei begin to multiply via mitotic division, a process that continues synchronously (recurring at the same successive instants of time) until cell walls are formed three to four days later (Wardlaw, 1970). From this point, synchrony ceases, and consequently the rate of cell division slows to ca. 25 percent its previous rate (Jenner, Ugalde, and Aspinall, 1991). For the next 15 days, cell division continues at rate of more than 6000 cells per day, until the final population of ca. 100,000 endosperm cells is reached (Briarty, Hughes, and Evers, 1979).

During this period of intense cell division, cells are simultaneously enlarging, mainly under pressure from water, and will reach an intermediate volume of ca. 2 $\mu m^3 \times 10^{-5}$. At this point, the endosperm may be characterized as a large (0.04 cm^3) bag, filled with 100,000 of these smaller bags, each of which contains a nucleus, a good deal of water, and the first traces of starch and protein. The lag phase of grain growth has finished, and with the appearance of starch and protein, grain filling has begun (Figure 5.5).

The lag phase has been time spent in manufacturing the sites into which starch and other grain constituents will be inserted, but as we have seen, there is some synthetic activity during the lag phase as well: grain enlargement and grain filling overlap to some extent. Sites of starch synthesis, in particular, become active during the lag phase of grain growth. A-type amyloplasts (envelopes in which A-type starch granules are synthesized) are initiated in the period 4 to 12 days after anthesis (Figure 5.5), and begin to fill with starch almost immediately. Each amyloplast is capable of giving rise to a single starch granule (Bechtel et al., 1990), but fewer than half of all A-type amyloplasts initiated will actually give birth to one. At the end of cell division a typical endosperm cell might contain 150 to 200 A-type amyloplasts, yet only 50 to 60 A-type starch granules (Briarty, Hughes, and Evers, 1979). Although these granules are initiated relatively early in the life of the grain, they do not attain their full size until much later; as stated previously, grain filling is largely a process whereby existing structures are enlarged or added to.

B-type amyloplasts are initiated from around 15 days after anthesis, and initiation appears to continue until near physiological maturity, and as a result the initiation and growth of B-type starch granules occurs concomitantly for most of grain filling (Figure 5.5). As many as 900 B-type amyloplasts might be initiated within a single endosperm cell (Briarty, Hughes, and Evers, 1979), but only around 250 produce a granule by maturity (Bechtel et al., 1990). Little is known about the initiation and growth of C-type starch granules, except that they start to appear at ca. 25 days after anthesis (Bechtel et al., 1990). It is not yet clear whether initiation is a discrete or continuous event.

The storage of starch within the wheat grain is therefore a little like a Russian doll (starch granules reside within amyloplasts within endosperm cells within endosperm). Readers interested in how the starch is constructed within the amyloplast are referred to reviews by Hawker, Jenner, and Niemitz (1991) and Preiss (1991).

Protein Accumulation

Despite the long-standing appreciation of the influence of protein composition on wheat quality (Fleurent, 1896; Guess, 1900; Woodman, 1922), there has, until recently, been little quantitative information on the process of accumulation of the different classes of protein (Kaczkowski, Kos, and Moskal, 1986; Kaczkowski, Kos, and Pior, 1988; Huebner, Kaczkowski, and Bietz, 1990; Stenram, Henreen, and Olered, 1990; Gupta et al., 1996; Stone and Nicolas, 1996a).

This paucity of data can largely be ascribed to the technical difficulties involved in obtaining accurate, repeatable, and readily quantifiable measures of the components of wheat protein. The extremely large size range of wheat proteins means that it has been difficult to separate and quantify them using a single method. This difficulty has largely been overcome with the advent of high performance liquid chromatography (HPLC), which has enabled the full complement of proteins to be analyzed simultaneously, repeatably, and with relative ease, making the analysis of fractional protein composition something of a routine task (Huebner, Kaczkowski, and Bietz, 1990; Singh, Donovan, and MacRitchie, 1990; Singh et al., 1990; Gupta, Khan, and MacRitchie, 1993).

In addition to simply satisfying our curiosity, a knowledge of the pattern of fractional protein accumulation in the wheat grain may provide some tangible practical benefits. An improved understanding of the mechanics of protein accumulation may expedite attempts to manipulate the nutritional value and end-use quality of wheat, whether through novel genetic techniques or through classical plant breeding. In addition, knowledge of the stages during grain growth at which the different protein fractions are synthesized is likely to enhance our ability to predict the impact on mature grain quality of stresses such as drought, disease, and high temperature, any of which may occur at specific stages of grain growth. Short-term perturbations during grain filling are likely to affect the balance of proteins, especially if they accumulate at different stages of grain growth.

During grain growth, storage proteins are deposited in discrete membranous bodies within the developing starchy endosperm (Graham and Morton, 1963; Graham, Morton, and Raison, 1963; Graham, Morton, and Simmonds, 1963). Within each protein body, proteins are synthesized in the ribosome and are moved sequentially down a secretory pathway to the endoplasmic reticulum, then to the Golgi apparatus, and are finally deposited in the vacuole (Bednarek and Raikhel, 1992; Noiva and Lennarz, 1992; Robinson et al., 1996). During this process, the protein is converted from a relatively simple, functionless, and linear molecule to a mature protein with a complex three-dimensional structure that fits it for its unique role in cellular and whole plant function (Buchner, 1996; Hendershot et al., 1996). Movement from one site in the secretory pathway to the next depends upon the correct protein conformation being reached, whether by disulfide bonding or some other means of attaining stability and 3-D conformity. At each step, "signal proteins" bind to the nascent protein, to biochemically guide it through the pathway, and to exercise "quality control," thereby ensuring that only correctly manufactured and conformed proteins are present in the vacuole (Bednarek and Raikhel, 1992). But enough of the biochemistry of protein synthesis; needless to say the synthesis and deposition of storage proteins within the protein bodies is a highly integrated and regulated process. It should be noted that the protein bodies do not remain as discrete entities for the duration of grain growth; as the grain matures, the protein bodies lose their integrity so that in the mature grain the storage proteins form a continuous matrix throughout the endosperm (Parker, 1980).

Having examined storage protein synthesis with respect to space, we should now examine the pattern of protein synthesis with respect to time. We shall concentrate on protein accumulation within a single kernel of wheat, which makes it easy to get a feel for the dynamics of protein accumulation, and from which the accumulation at a whole-crop level can be determined simply by multiplication.

Protein accumulation in the wheat grain is generally sigmoidal, and the rate of deposition in the linear phase is about 0.15 to 0.20 mg per day (Figure 5.6), with variation in the rate of deposition resulting from factors such as genotype, fertilization, and temperature (Sofield et al., 1977a; Sten-

FIGURE 5.6. The Accumulation of Protein Fractions in the Developing Grain of Wheat cv. Egret

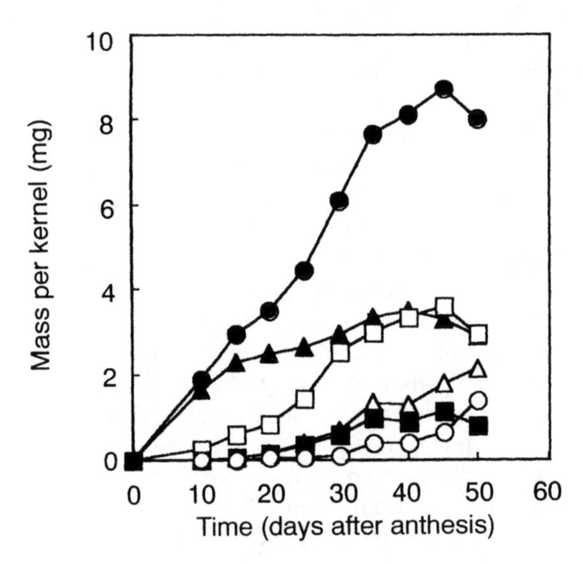

● Total grain protein; ▲ albumin/globulin; □ monomer; Δ total polymer; ■ SDS-soluble polymer; ○ SDS-insoluble polymer. Adapted from Stone and Nicolas (1996a).

ram, Heneen, and Olered, 1990; Johansson et al., 1994; Stone and Nicolas, 1996a). Consequently, it is difficult to say how much protein a typical wheat grain contains, but between 4 and 10 mg would cover most of the range.

The component fractions of the total grain protein must clearly accumulate at a lower rate, and as we shall see, both the rate and duration of synthesis differs for each of the major protein fractions (albumin, globulin, gliadin, and glutenin). For simplicity, the following discussion presents data for an "average" wheat variety grown under an "average" (18°C) temperature.

The metabolic proteins are the first to accumulate in appreciable amounts. Albumin and globulin together make up approximately 90 percent of the total grain protein in the first ten days of grain growth (Figure 5.6). Even though they are deposited in the grain for most of the grain-filling period, the proportion of albumin and globulin in total grain protein declines during grain growth, so that they make up ca. 20 to 30 percent at maturity. This decline occurs because the synthesis of storage protein

(70 to 80 percent of the mature protein mass) occurs somewhat later in grain filling.

Gliadins are the first storage proteins to accumulate in readily measurable amounts, and make their first appearance five to ten days after anthesis (Figure 5.6), at which point they make up approximately 10 percent of the total protein. From then until maturity, monomer typically accumulates at an average rate of about 0.1 mg per day, but its proportion of total grain protein reaches a maximum about halfway through grain growth, and may remain constant at 30 to 40 percent to maturity (Figure 5.6).

Glutenins are frequently the last of the proteins to appear in the grain, and may not be present in significant quantities until as late as 20 days after anthesis. Glutenin tends to accumulate linearly from then until maturity, and makes up a steadily increasing proportion of the total grain protein, reaching a maximum of 30 to 40 percent in the harvestable grain (Figure 5.6). The two solubility classes of glutenin (the SDS-soluble and the SDS-insoluble) do not accumulate synchronously; deposition of SDS-soluble glutenin tends to predominate early (20 to 35 days after anthesis), whereas SDS-insoluble glutenin tends to be synthesized later in grain filling (30 to 50 days after anthesis) (Figure 5.6). Consequently SDS-soluble glutenin may be up to approximately 90 percent of the total glutenin 35 days after anthesis, although it might typically constitute only 40 percent at maturity.

There are three main points to be distilled from this data:

1. The deposition of the different classes of protein is highly asynchronous, and consequently both the balance and amount of the different protein fractions varies throughout grain growth.
2. The average molecular mass of the proteins increases throughout grain growth, as shown by: (a) the declining proportion of smaller proteins (albumins and globulins) during grain filling; (b) the decrease in gliadin:glutenin ratio with time; and (c) the increasing proportion of SDS-insoluble glutenin in the total glutenin protein. We would expect that as the molecular mass of the proteins increases, so would the ability of the flour to form a cohesive dough. Interestingly, these recently obtained chemical data correspond closely with earlier studies that reported an increased ability to form a coherent mass of gluten as grain filling progressed (Woodman and Engledow, 1924).
3. The asynchronicity of fractional protein deposition means that perturbations to grain growth are also likely to affect grain protein composition, whether this is through disruption during a certain phase of grain growth (and so protein deposition) or through a gener-

al truncation of grain filling. A number of studies have shown that short periods of heat stress alter the rate and timing of deposition of particular protein fractions (Stone and Nicolas, 1996b; Stone, Nicolas, and Wardlaw, 1996), and it is probable that drought and disease would also affect protein composition, and consequently quality, although there is presently little supporting evidence. Of course, the different protein fractions may well have differing sensitivities to stress, and so the likely effect of stress on protein composition is far from clear.

This knowledge of fractional protein deposition during grain growth is also helping to link our understanding of the biochemistry, physiology, and genetics of why some varieties of wheat make strong flour, whereas others only make weak flours. Studies to date have been limited to a small number of genotypes, but thus far, it appears that strong varieties produce SDS-insoluble glutenins at a higher rate than weak varieties do (Gupta et al., 1996; Stone and Nicolas, 1996a), and that this can be related to the presence or absence of relatively simple genetic factors (Gupta et al., 1996).

In the near future, we should aim to integrate our knowledge of the physiology, genetics, and biochemistry of protein deposition to enable us to manipulate it through simple agronomic means. It stands to reason that practices such as reducing or even increasing stress at given stages of grain growth, or specifically timed fertilizer application, may be used to intelligently alter grain protein composition and consequently grain quality.

GRAIN QUALITY IS THE RESULT OF INTERACTIONS BETWEEN GRAIN COMPONENTS

Grain quality is more than the sum of the contributions of protein, starch, and lipid to dough strength, loaf volume, or any other measure of quality. As we have shown for the proteins, it is not just the presence of a given constituent that determines quality, but rather the interaction between different constituents. The picture for grain quality becomes even more complex when we consider that changes in protein composition and concentration frequently occur against a background of changing starch composition and concentration. Furthermore, lipids and other factors must also enter the quality equation. It is clear that no single component of the wheat grain can explain more than a fraction of the total variation that occurs between grain samples. We must attempt to understand the nature of these interactions if we are to successfully grow wheat of a given quality.

Physiological Interactions That Determine Grain Quality

Thus far, we have examined factors that determine grain quality in splendid isolation, which is not as they occur in the grain. Using examples, we shall now try to show (1) how grain constituents interact to determine grain quality, and (2) how factors such as agronomic management and climate can affect the physiology of grain quality.

We have shown that protein percentage is frequently used as a primary measure of grain quality ("Proteins and the Measurement of Quality"). Physiologically, however, it should be obvious that protein percentage is itself a secondary product of the interaction of protein with other grain constituents, mainly starch. A change in the quantity of either protein or starch will impact on the protein percentage of the grain, and consequently its apparent quality. It then follows that differences in the physiology of accumulation of starch and protein will be fundamental in explaining the effect of agronomic management and environment on grain protein percentage.

Nitrogen Fertilizer

Nitrogen fertilizer is one of the most-used tools for altering grain yield and quality, based upon the common knowledge that it can increase grain yield, grain protein percentage, or both. When nitrogen availability is lower than that required to maximize grain yield, the well-known hyperbolic response of yield to fertilizer occurs, whereby yield increases asymptotically to a maximum possible for a given environment (Figure 5.7, (2)). Around this point are a number of regions in which starch and protein accumulation respond differentially to nitrogen availability.

Starting from a low level of nitrogen availability, the first increment of N fertilizer increases the amounts of both starch and protein in the grain, but the response of starch is usually the greater. The first increment of N, therefore, tends to increase yield but decrease protein percentage, resulting in the frequently reported negative relationship between grain yield and protein percentage (Malloch and Newton, 1934; Benzian and Lane, 1979; Terman, 1979; Benzian and Lane, 1981; Morris and Paulsen, 1985). Before the critical level of N is attained, the response of starch and protein accumulation enters a second region of response, in which additional N fertilizer will often have a reduced (but still positive) effect on starch

FIGURE 5.7. Diagrammatic Representation of the Response of Yield (▬) and Protein Percentage (—) to Nitrogen Fertilizer

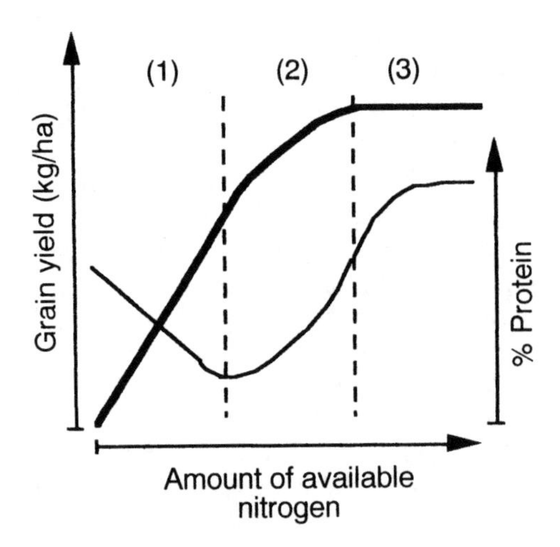

accumulation, and a proportionally greater impact on protein accumulation. The net effect of N in the second region of response is therefore a small increase in yield, and a comparatively large increase in protein percentage. As greater amounts of N are added, the crop may reach the third region of response, where maximum yield has been attained. In this region, additional fertilizer does not affect the amount of starch in the grain, but it does increase the amount of grain protein. As a result, protein percentage is highly responsive to nitrogen in this "luxury consumption" region of N addition.

At the risk of oversimplifying a complex interaction, we may think of the response of yield and protein percentage to N fertilizer in these terms: as the critical level of N for yield (starch) is approached, a proportionally greater amount of N accumulates in the grain as protein. As such, addition of N from a low base level increases yield, whereas that from a higher base level increases protein percentage. It is clear that the effects of N fertilizer on grain protein percentage emanate from the differing responses of starch and protein accumulation to increased levels of N.

We shall build on this simple model by adding to it a response of grain protein composition to N fertilizer. In addition to increasing grain protein percentage, additional N in the second region of response also influences grain protein composition. It is widely reported that as grain protein percentage increases, the proportion of glutenin in the grain decreases (Abrol et al., 1971; Dubetz and Gardiner, 1979; Dubetz et al., 1979; Doekes and Wennekes, 1982; Stenram, Heneen, and Olered, 1990), and that this occurs because the increase in accumulation of gliadin is greater than that of glutenin (Abrol et al., 1971; Dubetz and Gardiner, 1979). This introduces a second level of interactions into our grain quality equation, and opposing effects of N fertilizer on grain quality. If N fertilizer increases both grain protein percentage (which tends to increase dough strength) and the ratio of gliadin:glutenin (which tends to reduce strength), what is the overall effect of N fertilizer on grain quality?

The responses of grain protein percentage and composition to N fertilizer tend to interact such that additional N frequently increases dough strength and consequently grain quality. This occurs because in most instances, the positive influence of protein percentage on strength is sufficiently large to more than compensate for the negative effect of an increased gliadin:glutenin ratio. Clearly, the extent to which fertilizer N increases grain quality is influenced by the differential responses of starch, gliadin, and glutenin to additional N, as represented in the model below.

Figure 5.8a shows the typical response to the first increment of N (that applied in the first region of response). Figure 5.8b shows the typical response to the second increment of N (that applied in the second region of response). The response of dough strength is shown for (1) that resulting from a change in protein percentage alone, (2) that resulting from a change in protein composition alone, and (3) that resulting from the combined effects of protein percentage and composition.

Elevated Temperature

Climate has a profound impact on grain quality, and helps to explain why different parts of the world produce such different types of wheat. Grain yield decreases by approximately 3 to 4 percent for each 1°C rise in temperature above 15°C during grain filling (Wardlaw and Wrigley, 1994), and this occurs primarily because starch accumulation is reduced by temperature. As temperature increases above a mean of 15°C during grain filling, the increase in the rate of starch deposition does not adequately compensate for the reduced duration of starch accumulation (Sofield et al., 1977b; Nicolas, Gleadow, and Dalling, 1984), and consequently grain size and yield are reduced.

Grain protein accumulation also declines as temperature increases, although it is usually less temperature sensitive than starch. As a result, increased temperatures during grain filling tend to increase grain protein percentage (Sosulski, Paul, and Hutcheon, 1963; Kolderup, 1975, 1979;

FIGURE 5.8. A Model Describing the Response of Grain Composition and Dough Strength to Nitrogen Fertilizer Application

A. First increment of N (region 1):

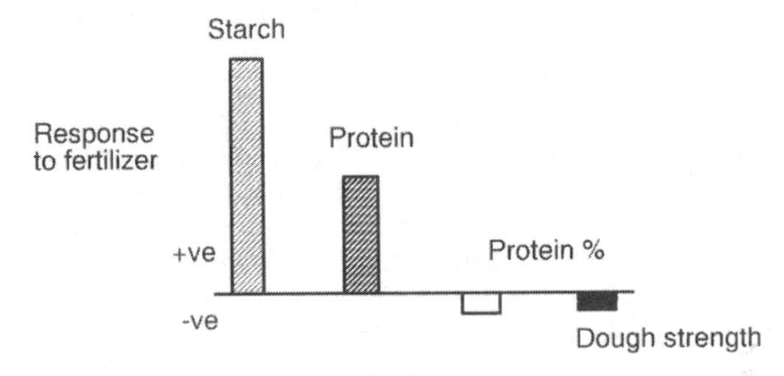

B. Second increment of N (region 2):

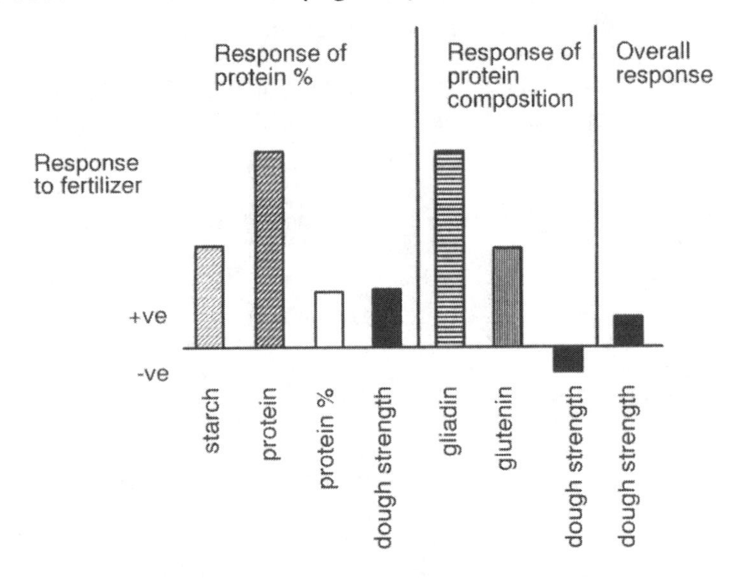

Schipper, Jahn-Deesbach, and Weipert, 1986; Randall and Moss, 1990; Rao et al., 1993). This generally results in an increase in grain quality, as measured by dough strength and related measures (Sosulski, Paul, and Hutcheon, 1963; Schipper, Jahn-Deesbach, and Weipert, 1986; Randall and Moss, 1990), although recent work has shown that as temperature rises above a certain threshold (usually ca. 30°C) the positive relationship between grain protein percentage and dough strength breaks down (Randall and Moss, 1990; Blumenthal et al., 1991; Blumenthal, 1993; Blumenthal, Barlow, and Wrigley, 1994), and may even become negative (Stone, Gras, and Nicolas, 1997). This change in the "normal" relationship between grain protein percentage and dough strength is caused by a temperature-induced change in grain protein composition. In many varieties of wheat, the ratio of gliadin:glutenin increases with temperature above ca. 30°C (Stone and Nicolas, 1994; Stone, Nicolas, and Gras, 1994; Blumenthal et al., 1995a, 1995b; Stone and Nicolas, 1996b; Stone, Nicolas, and Wardlaw, 1996), although this threshold is lower in heat-sensitive cultivars (Stone, Gras, and Nicolas, 1997). It appears that gliadin:glutenin ratio increases because the accumulation of gliadin is reduced less by elevated temperature than that of glutenin (Stone, Nicolas, and Wardlaw, 1996).

As with the effects of fertilizer application, temperature induces a series of changes in the physiology of the components of grain quality, which interact in complex ways to change the quality apparent in the mature grain. The general effects of high temperature on grain quality are summarized in Figure 5.9.

These brief case studies should demonstrate that even on a simple level, the interaction between protein percentage and composition will affect

FIGURE 5.9. A Model Describing the Response of Grain Composition and Dough Strength to Elevated Temperature

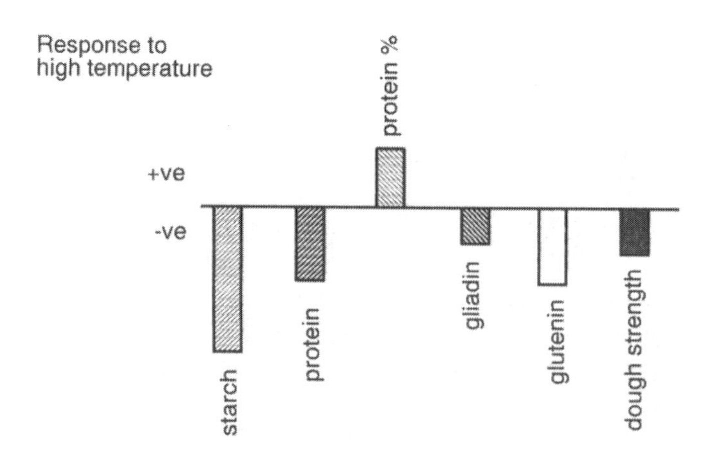

grain quality. Furthermore, the interactive effects of starch (amylose:amylopectin ratio, granule size distribution) and lipid (type and content) add a further layer of complexity to the physiological interactions that affect mature grain quality.

Understanding the way in which the different grain components interact to determine the quality characteristics of wheat has been something of a Holy Grail for cereal chemists. Improved methodology is taking this research into something of a new frontier, by enabling physiologists to probe into the processes that "make" quality. It is likely that by understanding these processes and their role in the determination of mature grain quality we will be able to manipulate quality to our advantage.

Implications for Physiologists and Breeders

The mechanisms by which simple agronomic and genetic factors modify or control grain quality are becoming much more apparent than they were only a few years ago. As a result, seasonal variation in grain quality that was previously considered random is now known to be systematic, and affected by identifiable and quantifiable climatic factors. A more complete mechanistic understanding of this variation should, in time, enable us to predict grain quality, and more important, to manipulate it so that specific quality targets can be attained. Success at either of these aims would significantly increase grain profits, by reducing costs (transport, segregation, and testing) and increasing buyer confidence in product quality.

Part of the challenge for physiologists is to understand the aims and language of grain processors and cereal chemists, so that there can be greater integration of the information concerned with grain quality and the physiology of grain filling. Furthermore, a knowledge of the quality aims of grain processors and end users should help physiologists to tailor their research to meet their needs and those of the cereal-eating public.

The fruits of an improved understanding of the physiology of grain quality are already becoming evident in many wheat breeding programs, where selection for specific physiological traits is being used to select for varieties that produce seasonally stable (predictable) quality characteristics. The greater our knowledge of factors that affect grain quality, the greater the probability of increasing it.

Our understanding of the genetic control of many aspects of wheat quality can be used to resolve many of the physiological riddles that remain unanswered. The use of near-isogenic lines and subunit-deletion biotypes is helping to separate out the effects of genotype and environment on grain quality, so that their interaction might be more clearly understood.

The scope for cooperation between breeders, physiologists, and cereal chemists is clear and wide.

CONCLUSIONS

Grain quality is determined by both the structure and composition of the mature grain, and is therefore a highly complex character resulting from the interaction of many biochemical processes and a large number of genes. At the simplest level, mature grain composition depends strongly on the relative accumulation of carbohydrate and protein (Figures 5.5 and 5.6), which are in turn responsive to environmental factors operating during grain growth, such as nitrogen availability (Figure 5.8) and temperature (Figure 5.9).

Understanding the mechanisms by which genetic and seasonal variation modify grain composition, and hence quality, is an important step in fine-tuning breeding and management strategies aimed at producing products of the quality demanded (or yet to be demanded) by an increasingly discerning market.

REFERENCES

Abou-Guendia, M. and D'Appolonia, B.L. (1973). Changes in carbohydrate components during wheat maturation. II. Changes in sugars, pentosans and starch. *Cereal Chemistry* 50, 723-734.

Abrol, Y.P., Uprety, D.C., Ahuja, V.P., and Naik, M.S. (1971). Soil fertilizer levels and protein quality of wheat grains. *Australian Journal of Agricultural Research* 22, 195-200.

Bagga, A.K. and Rawson, A.M. (1977). Contrasting responses of morphologically similar wheat cultivars to temperatures appropriate to warm climates with hot summers: A study in controlled environment. *Australian Journal of Plant Physiology* 4: 877-887.

Bechtel, D.B., Zayas, I., Kaleikau, L., and Pomeranz, Y. (1990). Size-distribution of wheat starch granules during endosperm development. *Cereal Chemistry* 67, 59-63.

Bednarek, S.Y. and Raikhel, N.V. (1992). Intracellular trafficking of secretory proteins. *Plant Molecular Biology* 20, 133-150.

Benzian, B. and Lane, P. (1979). Some relationships between grain yield and grain protein of wheat experiments in South-east England and comparison with such relationships elsewhere. *Journal of the Science of Food and Agriculture* 30, 59-70.

Benzian, B. and Lane, P. (1981). Interrelationships between nitrogen concentration in grain, grain yield and added fertilizer nitrogen in wheat experiments in South-east England. *Journal of the Science of Food and Agriculture* 32, 35-43.

Blumenthal, C.S., Barlow, E.W.R., and Wrigley, C.W. (1993). Growth environment and wheat quality: The effect of heat stress on dough properties and gluten proteins. *Journal of Cereal Science* 18, 3-21.

Blumenthal, C.S., Batey, I.L., Bekes, F., Wrigley, C.W., and Barlow, E.W.R. (1991). Seasonal changes in wheat-grain quality associated with high temperatures during grain filling. *Australian Journal of Agricultural Research* 42, 21-30.

Blumenthal, C., Bekes, F., Gras, P.W., Barlow, E.W.R., and Wrigley, C.W. (1995a). Identification of wheat genotypes tolerant to the effects of heat stress on grain quality. *Cereal Chemistry* 72, 539-544.

Blumenthal, C., Gras, P.W., Bekes, F., Barlow, E.W.R., and Wrigley, C.W. (1995b). Possible role for the *Glu-D1* locus with respect to tolerance to dough-quality change after heat stress. *Cereal Chemistry* 72, 135-136.

Blumenthal, C.S., Wrigley, C.W., Gras, P.W., Batey, I.L., and Barlow, E.W.R. (1994). Genetic sources of tolerance to quality variation due to growth environment. *Proceedings of the 44th Australian Cereal Chemistry Conference* (J.F. Panozzo and P.G. Downie, Eds.), Royal Australian Chemical Institute: Melbourne, Australia, pp. 62-66.

Boyacioglu, M.H. and D'Appolonia, B.L. (1994a). Characterization and utilization of durum wheat for breadmaking. I. Comparison of chemical, rheological, and baking properties between bread wheat flours and durum wheat flours. *Cereal Chemistry* 71, 21-28.

Boyacioglu, M.H. and D'Appolonia, B.L. (1994b). Characterization and utilization of durum wheat for breadmaking. III. Staling properties of bread made from bread wheat flours and durum wheat flours. *Cereal Chemistry* 71, 34-41.

Branlard, G. and Dardevet, M. (1985). Diversity of grain proteins and bread wheat quality. I. Correlation between gliadin bands and flour quality characteristics. *Journal of Cereal Science* 3, 329-343.

Briarty, L.G., Hughes, C.E., and Evers, A.D. (1979). The developing endosperm of wheat—a stereological analysis. *Annals of Botany* 44, 641-658.

Buchner, J. (1996). Supervising the fold: Functional principles of molecular chaperones. *The FASEB Journal* 10, 10-19.

Campbell, W.P., Wrigley, C.W., Cressey, P.J., and Slack, C.R. (1987). Statistical correlations between quality attributes and grain protein composition for 71 hexaploid wheats used as breeding parents. *Cereal Chemistry* 64, 293-299.

Carcea, M. and Schofield, J.D. (1996). Protein-lipid interactions in wheat gluten—reassessment of the occurrence of lipid-mediated aggregation of protein in the gliadin fraction. *Journal of Cereal Science* 24, 101-113.

Crosbie, G.B. (1991). The relationship between starch swelling properties, paste viscosity and boiled noodle quality in wheat flours. *Journal of Cereal Science* 13, 145-150.

Dell'Aquila, A., Colaprico, G., Taranto, G., and Carella, G. (1983). Endosperm changes in developing and germinating *T. aestivum, T. turgidum* and *T. monococcum* seeds. *Cereal Research Communications* 11, 107-113.

Doekes, G.J. and Wennekes, L.M.J. (1982). Effect of nitrogen fertilization on quantity and composition of wheat flour proteins. *Cereal Chemistry* 59, 276-278.

Dubetz, S. and Gardiner, E.E. (1979). Effect of nitrogen fertilizer treatments on the amino acid composition of Neepawa wheat. *Cereal Chemistry* 56, 166-168.

Dubetz, S., Gardiner, E.E., Flynn, D., and De La Roche, A.I. (1979). Effect of nitrogen fertilizer on nitrogen fractions and amino acid composition of wheat flour protein. *Canadian Journal of Plant Science* 59, 299-305.

Evans, L.T. (1993). *Crop Evolution, Adaptation and Yield.* Cambridge University Press, Cambridge, UK.

Fleurent, E. (1896). Sur une méthode chimique d'appreciation de la valeur boulangére des farines de blé. *Comptes Rendus Hebdomoadaires des Séances de l'Academi des Sciences, Paris* 123, 755-758.

Gao, L. and Bushuk, W. (1993). Polymeric glutenin of wheat lines with varying number of high molecular weight glutenin subunits. *Cereal Chemistry* 70, 475-480.

Graham, J.S.D. and Morton, R.K. (1963). Studies of proteins of developing wheat endosperm: Separation by starch-gel electrophoresis and incorporation of [35S]sulphate. *Australian Journal of Biological Sciences* 16, 357-365.

Graham, J.S.D., Morton, R.K., and Raison, J.K. (1963). Isolation and characterization of protein bodies from developing wheat endosperm. *Australian Journal of Biological Sciences* 16, 375-384.

Graham, J.S.D., Morton, R.K., and Simmonds, D.H. (1963). Studies of proteins of developing wheat endosperm: Fractionation by ion-exchange chromatography. *Australian Journal of Biological Sciences* 16, 350-356.

Grimwade, B., Tatham, A.S., Freedman, R.B., Shewry, P.R., and Napier, J.A. (1996). Comparison of the expression patterns of genes coding for wheat gluten proteins involved in the secretory pathway in developing caryopses of wheat. *Plant Molecular Biology* 30, 1067-1073.

Guess, H.A. (1900). The gluten constituents of wheat and flour and their relation to breadmaking qualities. *Journal of the American Chemistry Society* 22, 262-268.

Gupta, R.B., Khan, K., and MacRitchie, F. (1993). Biochemical basis of flour properties in bread wheats. I. Effects of variation in the quantity and size distribution of polymeric protein. *Journal of Cereal Science* 18, 23-41.

Gupta, R.B. and MacRitchie, F. (1994). Allelic variation at glutenin subunit and gliadin loci, Glu-1, Glu-3 and Gli-1 of common wheats. II. Biochemical basis of the allelic effects on dough properties. *Journal of Cereal Science* 19, 19-29.

Gupta, R.B., Masci, S., Lafiandra, D., Bariana, H.S., and MacRitchie, F. (1996). Accumulation of protein subunits and their polymers in developing grains of hexaploid wheats. *Journal of Experimental Botany* 47, 1377-1385.

Gupta, R.B., Paul, J.G., Cornish, G.B., Palmer, G.A., Bekes, F., and Rathjen, A.J. (1994). Allelic variation at glutenin subunit and gliadin loci, Glu-1, Glu-3 and Gli-1 of common wheats. I. Its additive and interaction effects on dough properties. *Journal of Cereal Science* 19, 9-17.

Gupta, R.B., Popineau, Y., Lefebvre, J., Cornec, M., Lawrence, G.J., and MacRitchie, F. (1995). Biochemical basis of flour properties in bread wheats. II. Changes in polymeric protein formation and dough/gluten properties associated with the loss of low Mr or high Mr glutenin subunits. *Journal of Cereal Science* 21, 103-116.

Hawker, J.S., Jenner, C.F., and Niemietz, C.M. (1991). Sugar metabolism and compartmentation. *Australian Journal of Plant Physiology* 18, 227-237.

Hay, R. L. (1993). Effect of flour quality characteristics on puff pastry baking performance. *Cereal Chemistry* 70, 392-396.

Hendershot, L., Wei, J., Gaut, J., Melnick, J., Aviel, S., and Argon, Y. (1996). Inhibition of immunoglobulin folding and secretion by dominant negative BiP ATPase mutants. *Proceedings of the National Academy of Science of USA* 93, 5269-5274.

Huebner, F.R., Kaczkowski, J., and Bietz, J.A. (1990). Quantitative variation of wheat proteins from grain at different stages of maturity and from different spike locations. *Cereal Chemistry* 67, 464-470.

Jenner, C.F., Ugalde, T.D., and Aspinall, D. (1991). The physiology of starch and protein deposition in the endosperm of wheat. *Australian Journal of Plant Physiology* 18, 211-226.

Jennings, A.C. and Morton, R.K. (1963). Amino acids and protein synthesis in developing wheat endosperm. *Australian Journal of Biological Sciences* 16, 384-394.

Johansson, E., Oscarson, P., Heneen, W.K., and Lundborg, T. (1994). Differences in accumulation of storage proteins between wheat cultivars during development. *Journal of the Science of Food and Agriculture* 64, 305-313.

Kaczkowski, J., Kos, S., and Moskal, M. (1986). Protein fractional composition in developing wheat grains. *Die Nahrung* 30, 437-439.

Kaczkowski, J., Kos, S., and Pior, H. (1988). The level changes of nitrogen containing fractions during the development of wheat *Triticum aestivum* grains. *Acta Physiologia Plantarum* 10, 31-40.

Kaldy, M.S., Kereliuk, G.R., and Kozub, G.C. (1993). Influence of gluten components and flour lipids on soft white wheat quality. *Cereal Chemistry* 70, 77-80.

Khatkar, B.S., Bell, A.E., and Schofield, J.D. (1996). A comparative study of the inter-relationships between mixograph parameters and bread-making qualities of wheat flours and glutens. *Journal of the Science of Food and Agriculture* 72, 71-85.

Kolderup, F. (1975). Effects of soil moisture and temperature on yield and protein production in wheat. *Meldinger Fra Norges Landsbruks Høgskote* 54, 1-18.

Kolderup, F. (1979). Application of different temperatures in three growth phases of wheat. II. Effects on ear size and seed setting. *Acta Agriculturae Scandinavia* 29, 11-16.

Leloup, V.M., Colonna, P., and Buleon, A. (1991). Influence of amylose-amylopectin ratio on gel properties. *Journal of Cereal Science* 13, 1-13.

MacRitchie, F. (1978). Baking quality of wheat flours. *Journal of Food Technology* 13, 187-194.

MacRitchie, F., du Cros, D.L., and Wrigley, C.W. (1990). Flour polypeptides related to wheat quality. *Advances in Cereal Science Technology* 10, 79-145.

Malloch, J.G. and Newton, R. (1934). The relation between yield and protein content of wheat. *Canadian Journal of Research* 10, 774-779.

Martín del Molino, I.M., Rojo, B., Martinez-Carrasco, R., and Pérez, P. (1988). Amino acid composition of wheat grain. I. Changes during development. *Journal of the Science of Food and Agriculture* 42, 29-37.

McGee, H. (1991). *On Food and Cooking: The Science and Lore of the Kitchen.* London, HarperCollins.

Morris, C.F. and Paulsen, G.M. (1985). Development of hard winter wheat after anthesis is affected by nitrogen nutrition. *Crop Science* 25, 1007-1010.

Morrison, W.R. (1988). Lipids in cereal starches: A review. *Journal of Cereal Science* 8, 1-15.

Müller, S. and Wieser, H. (1995). The location of disulphide bonds in a-type gliadins. *Journal of Cereal Science* 22, 21-27.

Nicolas, M., Gleadow, R.M., and Dalling, M.J. (1984). Effects of drought and high temperature on grain growth in wheat. *Australian Journal of Plant Physiology* 11, 553-566.

Noel, T.R., Ring, S.G., and Whittam, M.A. (1992). Physicochemical analysis of wheat starch. In *Seed Analysis.* (H.-F. Linskens and J.F. Jackson, Eds.). Berlin, Springer Verlag, pp. 333-346.

Noiva, R. and Lennarz, W.J. (1992). Protein disulfide isomerase. A multifunctional protein resident in the lumen of the endoplasmic reticulum. *Journal of Biological Chemistry* 267, 3553-3556.

Oda, M., Yasuda, Y., Okazaki, S., Yamauchi, Y., and Yokoyama, Y. (1980). A method of flour quality assessment for Japanese noodles. *Cereal Chemistry* 57, 253-254.

Osborne, T.B. (1907). *The Proteins of the Wheat Kernel.* Publications of the Carnegie Institution Washington, vol. 84. Washington, Judd and Detweiler.

Parker, M.L. (1980). Protein body inclusions in developing wheat endosperm. *Annals of Botany* 46, 29-36.

Preiss, J. (1991). Biology and molecular biology of starch synthesis and its regulation. *Oxford Surveys of Plant Molecular and Cell Biology* 7, 59-114.

Randall, P.J. and Moss, H.J. (1990). Some effects of temperature regime during grain filling on wheat quality. *Australian Journal of Agricultural Research* 41, 603-617.

Rao, A.C.S., Smith, J.L., Jandhyala, V.K., Papendick, R.I., and Parr, J.F. (1993). Cultivar and climatic effects on the protein content of soft white winter wheat. *Agronomy Journal* 85, 1023-1028.

Rao, A.R. and Whitcombe, J.R. (1977). Genetic adaptation for vernalization requirement in Napalese wheat and barley. *Annals of Applied Biology,* 85: 121-130.

Robinson, A.S., Bockhaus, J.A., Voegler, A.C., and Wittrup, K.D. (1996). Reduction of BiP levels decreases heterologous protein secretion in *Saccharomyces cerevisiae. The Journal of Biological Chemistry* 271, 10017-10022.

Schipper, A., Jahn-Deesbach, W., and Weipert, D. (1986). Untersuchungen zum Klimateinfluss auf die Weizenqualitat. *Getreide, Mehl und Brot* 40, 99-103.

Schropp, P. and Wieser, H. (1996). Effects of high molecular weight subunits of glutenin on the rheological properties of wheat gluten. *Cereal Chemistry* 73, 410-413.

Shewry, P.R., Halford, N.G., and Tatham, A.S. (1989). The high-molecular-weight subunits of wheat, barley and rye: Genetics, molecular biology, chemistry and role in wheat gluten structure and functionality. *Oxford Surveys in Plant Molecular and Cell Biology* 6, 163-219.

Shewry, P.R., Halford, N.G., and Tatham, A.S. (1992). High-molecular-weight subunits of wheat glutenin. *Journal of Cereal Science* 15, 105-120.

Simmonds, D.H. (1989). *Wheat and Wheat Quality in Australia.* Melbourne, CSIRO.

Singh, N.K., Donovan, R., Batey, I., and MacRitchie, F. (1990). Use of sonication and size-exclusion high-performance liquid chromatography in the study of wheat flour proteins. I. Dissolution of total proteins in the absence of reducing agents. *Cereal Chemistry* 67, 150-161.

Singh, N.K., Donovan, R., and MacRitchie, F. (1990). Use of sonication and size-exclusion high-performance liquid chromatography in the study of wheat flour proteins. II. Relative quantity of glutenin as a measure of breadmaking quality. *Cereal Chemistry* 67, 161-170.

Sofield, I., Evans, L.T., Cook, M.G., and Wardlaw, I.F. (1977b). Factors influencing the rate and duration of grain filling in wheat. *Australian Journal of Plant Physiology* 4, 785-797.

Sofield, I., Wardlaw, I.F., Evans, L.T., and Zee, S.Y. (1977a). Nitrogen, phosphorus and water contents during grain development and maturation in wheat. *Australian Journal of Plant Physiology* 4, 799-810.

Sosulski, F.W., Paul, E.A., and Hutcheon, W.L. (1963). The influence of soil moisture, nitrogen fertilization, and temperature on quality and amino acid composition of Thatcher wheat. *Canadian Journal of Soil Science* 43, 219-228.

Soulaka, A.B. and Morrison, W.R. (1985). The amylose and lipid contents, dimensions, and gelatinization characteristics of some wheat starches and their A- and B-granule fractions. *Journal of the Science of Food and Agriculture* 36, 709-718.

Stenram, U., Heneen, W.K., and Olered, R. (1990). The effect of nitrogen fertilizers on protein accumulation in wheat (*Triticum aestivum* L.). *Swedish Journal of Agricultural Research* 20, 105-114.

Stone, P.J., Gras, P.W., and Nicolas, M.E. (1997). The influence of recovery temperature on the effects of a brief heat shock on wheat. III. Grain protein composition and dough properties. *Journal of Cereal Science* 25, 129-141.

Stone, P.J. and Nicolas, M.E. (1994). Wheat cultivars vary widely in their responses of grain yield and quality to short periods of post-anthesis heat stress. *Australian Journal of Plant Physiology* 21, 887-900.

Stone, P.J. and Nicolas, M.E. (1996a). Varietal differences in mature protein composition of wheat resulted from different rates of polymer accumulation during grain filling. *Australian Journal of Plant Physiology* 23, 727-737.

Stone, P.J. and Nicolas, M.E. (1996b). Effect of timing of heat stress during grain filling on two wheat varieties differing in heat tolerance. II. Fractional protein accumulation. *Australian Journal of Plant Physiology* 23, 739-749.

Stone, P.J., Nicolas, M.E., and Gras, P.W. (1994). The response of wheat yield and quality to high temperature is affected by cultivar and the timing of heat stress. *Proceedings of the 44th Australian Cereal Chemistry Conference* (J.F. Panozzo and P.G. Downie, Eds.), Royal Australian Chemical Institute: Melbourne, Australia, pp. 52-54.

Stone, P.J., Nicolas, M.E., and Wardlaw, I.F. (1996). The influence of recovery temperature on the effects of a brief heat shock on wheat. II. Fractional protein accumulation during grain growth. *Australian Journal of Plant Physiology* 23, 605-616.

Terman, G.L. (1979). Yields and protein content of wheat grain as affected by cultivar, N, and environmental growth factors. *Agronomy Journal* 71, 437-440.

Wang, L. and Seib, P.A. (1996). Australian salt-noodle flours and their starches compared to U.S. wheat flours and their starches. *Cereal Chemistry* 73, 167-175.

Wardlaw, I.F. (1970). The early stages of grain development in wheat: Response to light and temperature in a single variety. *Australian Journal of Biological Science* 23, 765-774.

Wardlaw, I.F. and Wrigley, C.W. (1994). Heat tolerance in temperate cereals: An overview. *Australian Journal of Plant Physiology* 21, 695-703.

Weegels, P.L., Hamer, R.J., and Schofield, J.D. (1996). Functional properties of wheat glutenin. *Journal of Cereal Science* 23, 1-18.

Weegels, P.L., Marseille, J.P., Bosveld, P., and Hamer, R.J. (1994). Large-scale separation of gliadins and their bread-making quality. *Journal of Cereal Science* 20, 253-264.

Williams, M., Shewry, P.R., and Harwood, J.L. (1994). The influence of the "greenhouse effect" on wheat (*Triticum aestivum* L.) grain lipids. *Journal of Experimental Botany* 45, 1379-1385.

Woodman, H.E. (1922). The chemistry of the strength of wheat flour. *Journal of Agricultural Science* 12, 231-243.

Woodman, H.E. and Engledow, F.L. (1924). A chemical study of the development of the wheat grain. *Journal of Agricultural Science* 14, 563-586.

Wrigley, C.W. (1992). Identification of cereal varieties by gel electrophoresis of the grain proteins. In *Seed Analysis* (H.-F. Linskens and J.F. Jackson, Eds.). Berlin, Springer Verlag, pp. 17-41.

Wrigley, C.W. (1996). Biopolymers—Giant proteins with flour power. *Nature* 381(6585), 539-540.

PART II: WHEAT ECOLOGY

Chapter 6

Effects of Sowing Date and the Determination of Optimum Sowing Date

Mike D. Dennett

INTRODUCTION

The production of wheat is spread over a larger area encompassing a more diverse range of climates and cropping systems than rice is. Variations in national yields reflect the diversity of wheat climates and production systems throughout the world. For example, yields are high in the relatively benign climates of Europe (7.5 t ha^{-1} in the United Kingdom) and much lower in the predominantly water-deficient wheat growing areas of Australia (1.7 t ha^{-1}). In 1997 there were 29 countries, spread through all continents, that sow more than 1 million ha of wheat (FAO, 1998).

Such an extensive spread of production involves a great diversity of climates. However, most wheat production is found in areas with annual rainfall between 200 and 1,000 mm and with mean temperatures over the whole growing period of the crop of less than 18°C (Bunting et al., 1982). This reflects the fact that times of sowing and crop durations are chosen so that the crop can successfully complete its life cycle in the relatively favored climatic conditions of moderate temperatures and sufficient, but not excessive, water supply. In recent years, though, wheat cultivation in warmer areas has been increasing (Fischer and Byerlee, 1991).

The factors dominating the choice of optimum sowing date will differ through this diverse range of climates and cropping systems. In some climates there may be a narrow range of suitable planting dates, while in others the range may be much wider. Nevertheless there are some general considerations about sowing date that can be applied throughout the range of wheat cropping systems. It is convenient to consider these under headings corresponding roughly to three periods in the life of a wheat crop: sowing and establishment, growth and development, and grain maturation and harvest.

SOWING AND ESTABLISHMENT

The main physiological requirement is for adequate soil water to germinate the seed and sustain the developing seedling. A minimum water content of 35 to 40 percent is needed for germination (Evans, Wardlaw, and Fischer, 1975) but the continual availability of soil moisture for the seedling is of prime importance. Thus a decision to sow needs to be based on a combination of knowledge of the amount of water in the soil and the expectation of rainfall being received. Deeper sowing may be appropriate in drier conditions. Waterlogging can substantially reduce plant emergence and this may reduce yield, though to a smaller extent (Hough, 1990).

Temperature affects germination and emergence. Wheat has been shown to germinate over a wide range of temperatures, for example from 4 to 37°C, with an optimum between 20 and 25°C (Evans, Wardlaw, and Fischer, 1975). Genotypic variation in the optimum temperature has been observed (e.g., Ali-Zi et al., 1994). For temperatures up to the optimum, emergence has been related to thermal time, with 150°C d (base temperature 0°C) being quoted for normal seeding depths of winter wheat in England (HGCA, 1997) and for wheat in North America (Cook and Veseth, 1991). Aggarwal and colleagues (1994) suggested 65°C d (base temperature 3.6°C) for wheat in India. Deeper-sown crops obviously take longer to emerge, with delays estimated as 8°C d cm^{-1} (HGCA, 1991), 20°C d cm^{-1} (Cook and Veseth, 1991) and 17.5°C d cm^{-1} (Aggarwal et al., 1994). These estimates are all based on standard meteorological measurements of air temperature, which are convenient to use, though their relationship with the soil temperature experienced by the seed may well differ with location.

Both the growth and development of the young seedling will be influenced by temperature after sowing. For autumn-sown crops it will be necessary to ensure that the plants grow sufficiently, but not excessively, to avoid winter damage, and for spring sown crops it is usually desirable to have a rapid expansion of leaf area.

It is not only physiological factors that are relevant to the choice of sowing date. Depending upon the farming system or crop rotation, cultivation for sowing may be constrained by the harvest of the preceding crop. The ability to cultivate or prepare the land for sowing may well be weather dependent, usually through an excess or a deficiency of soil moisture, as well as possibly being limited by the capacity of machinery and labor. Decreasing the time between the harvest of the previous crop and sowing may make weed control more difficult, increase the risk of disease carry-over, and give an increased workload.

GROWTH AND DEVELOPMENT

The distinction between plant growth (i.e., an increase in size) and plant development (i.e., the initiation of organs and the sequence of changes of form from sowing to maturity) is well established (e.g., Fisher, 1984). It is however worth repeating because satisfactory yields can only be obtained by ensuring that the pattern of development with time is in balance with the opportunities for growth.

The rate of development of a wheat cultivar is primarily a function of temperature and daylength. It seems that the only decisions a wheat grower can make to influence the developmental pattern are the choice of cultivar and the sowing date. Development then depends upon the subsequent patterns of temperature and daylength, and variations in temperature from year to year are the prime cause of differences in the timing of crop development between years. In glasshouse production of horticultural crops, the grower may be able to control developmental patterns by manipulation of temperature and daylength regimes, for example in the production of tomatoes (Bedding, 1981). To a limited extent the use of plastic mulches or floating crop covers can provide some control over the rate of development of field-grown vegetables such as potatoes (Jenkins, 1995). The wheat grower must rely upon his initial choices of sowing date and cultivar. Water shortages at particular stages may decrease tiller number or spikelet number or may hasten the end of grain filling, but these all have detrimental effects on yield and are not used by the farmer to control developmental patterns.

Crop growth involves the capture from the environment of four major resources: photosynthetically active radiation (PAR), water, carbon dioxide, and nutrients (Monteith, 1994). In climates where wheat is grown the availability of PAR and water usually show a strong seasonal variation, but turbulent mixing in the atmosphere ensures relatively constant supplies of carbon dioxide. The presence of nutrients in the soil is partly a consequence of the location and soil type but in practical terms is largely a consequence of agronomic decisions. However, the capture of these resources and their conversion into biomass depends at any time upon the size of the crop canopy and root system as well as temperature and soil water status.

The timing of the initiation of vegetative and reproductive organs and the numbers of such organs therefore depends upon temperature and photoperiod, but the survival and subsequent size of such organs is dependent upon the supply of assimilate. The choice of sowing date is therefore vital to ensure both that sufficient grain sites are initiated and that sufficient assimilate will be available to initially support and subsequently fill these sites. The number of grain sites per unit area of ground depends upon the number of tillers as well as the number of grain sites per ear, though there is often some compensation between these components. The number of tillers will be influenced by the length of period of tiller initiation as well as by sowing density, suggesting that interactions between sowing density and sowing date are likely to influence yield. Tiller survival may well be increased by the availability of nitrogen, so it can be argued that as well as cultivar and sowing date, sowing density and fertilizer application are factors that the grower can use to influence crop development, as expressed by the number of potential grain sites per unit of ground.

Vernalization is one specific aspect of some wheat cultivars, primarily those of midlatitude origin (Evans, 1993). This is the need for a period of low temperatures to stimulate reproductive development. Originally this was an important mechanism for delaying development until the risk of frost damage to the reproductive apex was past but the value of such a mechanism obviously depends upon location and sowing date. Craigon, Atherton, and Sweet (1995) have applied the concept of vernalizing degree days to data on the winter wheat cultivar Norin 27 and suggested the optimum temperature for vernalization was about 5°C. In practical terms in any country cultivars requiring vernalization have latest recommended planting dates to ensure they receive sufficient periods of cold (e.g., NIAB, 1997).

MATURATION AND HARVEST

The choice of sowing date is also influenced by the likely conditions during grain maturation and harvest. A basic principle is that the crop should mature within the climatically determined growing season. Thus the aim is to complete grain filling by the time soil moisture is effectively depleted or low temperatures limit growth. Monteith and Elston (1993) state that the largest mean yield is obtained when the biologically determined growth period of the crop closely matches the length of the physically determined growing season. However, they also point out that this level of yield may not be stable. Stability can be achieved by shortening the growth period, that is, using a quicker-maturing cultivar. The selection of appropriate combinations of cultivar and planting date can be investigated with crop growth models (e.g., Jones and O'Toole, 1987; Aggarwal and Kalra, 1994; Muchow, Hammer, and Vanderlip, 1994).

In a relatively constant environment having a crop mature close to a particular date could be achieved equally by changing sowing dates or choosing cultivars with different maturity requirements. In reality the seasonal variations in temperature and daylength mean that the selection of an appropriate cultivar is often a more effective method than altering sowing date. For example, in England autumn-sown wheat crops differ in maturity by less than two weeks, despite differences in sowing dates of two to three months (HGCA, 1991). A recent study with eight sowing dates of spring wheat in England (Mulholland et al., 1997) showed that cvs. Minaret and Canon required 1392 and 1420° C d (base temperature 1°C) from sowing to maturity. Thus the first four sowing dates (March 17 to April 28) covered a range of 42 days, but all crops matured within a 10-day period, with Canon maturing slightly later for each sowing.

Climatic conditions during grain filling affect grain quality, particularly protein content. Studies have been reviewed by Gooding and Davies (1996) and in general warmer temperatures increase protein content but may also adversely affect protein quality. The effects of water are less clear because of the interactions between water, growth, and nitrogen availability.

Harvesting and storage of grain is easier and cheaper if the grain can ripen before harvest. Typically grain moisture content is reduced from about 45 to 20 percent (HGCA, 1997). Ideally this requires a period of up to two weeks with little rainfall and considerable solar radiation. Large amounts of rain can lead to the sprouting of grain in the ear, especially if the crop has lodged. This gives an increase in alpha amylase activity and a reduction in Hagberg falling number. Cool and prolonged ripening can also decrease Hagberg falling number without visible sprouting (Gooding and Davies, 1996). These aspects of quality emphasize the fact that conditions at the end of the season should be considered when deciding on cultivar and sowing date.

EXPERIMENTAL ESTIMATION OF OPTIMUM SOWING DATES

The classic approach to estimation of optimal sowing dates has been to carry out direct field experiments with a range of sowing dates. Synthesis of results then provides recommendations, usually accompanied by an estimate of the effects of deviations from the optimum on the reduction in yield. Some such experiments for a limited number of environments are discussed later. These experiments are certainly necessary and may well be sufficient to establish practical recommendations. There are two potential difficulties in particular that should be considered in interpreting the results. First, many of these experiments continue for two or three years and, given the multitude of ways in which crops respond to weather during their lives, as outlined above, this short period of experimentation may only represent a small sample of possible weather sequences following sowing. Second, most of the ways in which sowing date can influence yield are likely to be cultivar dependent. Thus the rapid turnover of wheat cultivars in high-intensity cropping systems makes the provision of current information difficult.

WHEAT IN ENGLAND

In England 75 percent of wheat is sown in the autumn (MAFF, 1997). Typically, the wheat-growing soils early in the autumn will be relatively warm and dry. Germination of early sown crops often depends upon rainfall after sowing. Later sowing in the autumn means that plants germinate under conditions of lower soil temperatures and increasing soil moisture content. Temperatures will decrease with time and growth in the winter months is limited by temperature. Soil profiles recharge over winter and are usually relatively full of water when growth accelerates as temperatures increase in the spring. The crop generally grows with an increasing soil moisture deficit as rates of evapotranspiration exceed those of rainfall from April onward. Crops may suffer from shortage of water during grain filling but Weir (1988) concluded that the risk of yield reduction is mostly restricted to shallow-rooting or diseased crops. For healthy, deep-rooting first wheat crops the probability of yield loss due to water stress in any year varied between 0.02 and 0.16 over England.

From a physiological viewpoint, earlier sowing should enable the crop to establish itself more easily and to form a larger leaf canopy and a more extensive root system before growth slows in the winter. The bigger root system should protect against soil heaving caused by frost but a bigger canopy may suffer more frost damage. If development is too advanced frost can damage the reproductive apex, with major effects on yield. As temperatures increase in the spring the larger canopy intercepts more solar radiation and the deeper root system gives access to more stored soil water during the season. Dry matter production of cereals in England is clearly related to the amount of solar radiation intercepted (Gallagher and Biscoe, 1978; Sylvester-Bradley et al., 1990), with the radiation use efficiency

(RUE) being relatively constant. This reflects the findings of Weir (1988) that the risk of yield reduction is relatively small because shortage of water will reduce RUE in wheat (e.g., Giunta, Motzo, and Deidda, 1995). The seasonal increase in temperature lags behind that of solar radiation in the spring and as temperature is a major factor limiting leaf expansion, late-sown crops usually intercept much less radiation than early-sown crops.

These ideas are reflected in current practice, with recommended planting dates being mid-September to mid-October (HGCA, 1991). Indeed, in the autumn of 1997, 75 percent of winter wheat was sown by the end of September (Harris, 1997). Historically, earlier sowing has been considered desirable, being recommended by agricultural pioneers Jethro Tull and William Cobbett in the 1730s and 1820s respectively (Wibberley, 1989). In the 1930s, with long-season cultivars and with wheat often being sown after fallow, optimal sowing dates were in September (Wibberley, 1989). After World War II the availability of manufactured nitrogen fertilizer, the development of new cultivars, and the use of more intensive crop rotations resulted in an optimal sowing date of mid-October (e.g., Eddowes, 1976). This continued until about 1980, when the gradual abandonment of traditional rotations, the substantial increase in the area of oil seed rape (*Brassica napus*), which because of its early harvest provides an excellent entry for wheat, the adaption of reduced or minimal cultivation techniques, and improvements in herbicides, particularly for grass weeds, led to earlier sowings. Green and Irvins (1985) found that in the East Midlands of England the potential yield of wheat declined by 0.35 percent for each day sowing was delayed after September 22. However, studies have shown significant cultivar × sowing date interactions. Early-developing cultivars were unresponsive to early sowing whereas late-developing cultivars with good resistance to lodging do produce higher yields from early sowing (Murphy, Frost, and Evans, 1993).

The place of wheat in cropping patterns and the consequent management decisions may lead to conditions that override the physiological basis of yield benefit from early sowing. A comprehensive set of multifactorial experiments on winter wheat were conducted at Rothamsted Experimental Station in the early 1980s. Early sowing of crops grown after a two-year break from cereals increased yields (Prew et al., 1985), but crops sown after oats showed no benefit and those sown after barley showed a decrease in yield (Prew et al., 1986). It seems that crops sown after potatoes benefited from the greater residual nitrogen in the soil compared to those sown after oats. Crops sown after barley were infected by take-all (*Gaeumannomyces graminis*) and the infection was greater for the early-sown crops, reducing yields below those of late-sown crops (Prew et al., 1986).

In general, drilling cereals under warmer conditions and soon after the harvest of the previous crop has led to additional pest and disease problems. These conditions usually led to an increase in the activity of slugs, which can cause serious damage to young seedlings. Winged aphids may still be active, increasing the chance of transmission of barley yellow dwarf virus from volunteer cereal plants. The risks of infection by mildews and rusts are increased by these conditions and the longer the crop is in the ground the greater the risk of eyespot infection. Late sowing delays the initiation of infections of *Septoria tritici* (Murphy, Frost, and Evans, 1993). Thus earlier sowing requires cultivars with appropriate disease resistances and increased chemical inputs. The optimal sowing date for organic production of wheat is in late autumn to permit time for weed and disease control (Gooding et al., 1993).

One consequence of earlier sowing is that specific developmental stages are likely to occur at earlier dates. Applications of herbicide at inappropriate growth stages can cause crop damage (Chapter 8, this volume). Dennett and Murphy (1983) linked a simple development model to weather-based criteria for the suitability of application of herbicides (spray days) to demonstrate that early sowing considerably reduced the probability of being able to apply herbicides at the correct growth stage. One beneficial environmental consequence of earlier sowing is that the greater autumn and winter growth may lead to a greater amount of nitrogen being recovered from the soil, reducing the amount of nitrogen susceptible to leaching. Cosser et al. (1994) found that sowing in September rather than October increases N uptake by between 15 and 35 kg ha^{-1} depending upon cultivar.

Spring wheats can be sown when soil conditions permit, with yield reductions of about 0.2 t ha^{-1} $week^{-1}$ after mid-March (HGCA, 1991). A recent practice in England has been to sow spring wheats in late autumn to combine high yield with good breadmaking quality. Cultivars that develop rapidly are less suited for this because of the possibility of frost damage.

Provided there is adequate weed and disease control, yields in England are normally greater for early sowing of both winter and spring wheat. Planting recommendations tend to be based on calendar dates rather than on weather events. However, planting early in the autumn may be sometimes be delayed by lack of rain and late planting of winter wheat and early planting of spring wheat may be delayed by the risk of damage to wet soils.

WHEAT IN AUSTRALIA

The agroclimatology of wheat production in Australia was described by Nix (1975). There is a diversity of climates with annual rainfall totals ranging from about 260 to 780 mm. Many climatic constraints are relevant to the choice of sowing date at particular locations, for example frost damage at anthesis, large deficits of rainfall during grain filling, and the uncertainty of receiving adequate planting rains. Thus there is considerable scope for the selection of cultivars and planting dates to optimize production and during recent years the availability of new cultivars, modern herbicides, and minimum tillage equipment has prompted a reassessment of sowing strategies in many areas of Australia. Here we shall explore some of these studies and relate them to the general principles discussed earlier.

The northeastern wheat belt of Western Australia has cool wet winters and hot dry summers and an annual rainfall of 300 to 350 mm. Rainfall in the growing season (May to October) is about 200 to 260 mm and the start varies between late April and mid-June. The end of the season is dry with high rates of potential evapotranspiration (Kerr, Siddique, and Delane, 1992). After an extensive series of experiments over three years Kerr, Siddique, and Delane (1992) concluded that the earliest possible sowing, the timing of which depended upon rainfall, gave the highest yields, but that the type of cultivar which gave the highest yield depended upon the year. Medium- or long-season cultivars performed best with early sowings but gave substantially lower yields with late starts. Short-season cultivars gave the best yields in these seasons. The authors suggest that the best option for farmers is to maintain stocks of several cultivars, and to sow as early as possible in each season using a long-season cultivar if the rains are early and a short-season one if the rains are late. This is similar to the ideas of response farming as applied to millet in West Africa (Stewart, 1991).

The southern wheat belt of Western Australia is wetter (300 to 500 mm annual rainfall) than the northern belt but has a greater risk of frost damage around anthesis. Experiments here involved four sites, 12 to 15 cultivars, and three sowing dates in each of three years (Shackley and Anderson, 1995). The first sowing in each case was carried out as soon as the season started. For long and midseason cultivars highest yields were obtained from the first sowings (usually early May) and declined linearly with delay in sowing, long-season cultivars showing the greatest rate of decline. Short-season cultivars showed no response to early sowing and produced lower yields. There seems to be no advantage in maintaining stocks of several cultivars to suit sowing times; long or midseason cultivars should be suitable in most years. The optimal flowering periods of these cultivars, so that grain filling is not affected by terminal drought, does however mean that some risk of frost damage at anthesis has to be accepted if yields are to be maximized (Shackley and Anderson, 1995). There was also clear evidence in these studies that foliar diseases were important in determining sowing dates for cultivars of different maturities.

In the central wheat belt of Western Australia growers advanced their sowing dates from the traditional mid-June to mid-May with the introduction of high yielding semidwarf wheats and improved weed control and reduced cultivation techniques as demonstrated by Anderson and Smith (1990). Anderson, Heinrich, and Abbots (1996) further concluded that longer-season cultivars could be sown earlier without loss of yield by utilizing the rainfall-based sowing opportunity that occurs before early May in 25 percent of years. However, Gregory and Eastham (1996) found that although sowing at the break of the season produced a larger wheat biomass, compared to sowing later, this was not translated into yield because of disease and drought during grain filling.

In high-yielding conditions in Victoria, late-maturing cultivars gave the highest yields for early (April) sowing. Rainfall records suggest a probability of 0.74 of being able to sow early and use such cultivars. If the season starts late then a short-season cultivar will probably give a slightly higher yield (Coventry et al., 1993). In all these areas of Australia sowing is mostly triggered by rainfall events, rather than being based on calendar dates. In general, highest potential yields are obtained by making full use of the season, the vernalization and photoperiod responses of the late-maturing cultivars making this possible (Coventry et al., 1993).

The importance of matching the developmental pattern of the crop to the thermal environment was demonstrated by Cooper (1992), who grew cultivars with a wide range of maturity under irrigation in New South Wales, with a wide range of sowing dates. Early wheats sown early suffered frost damage and winter cultivars sown late were not fully vernalized. Anthesis in mid-September produced the highest yield for each cultivar. This was the earliest date consistent with avoiding frost damage. Yields decreased by about 1.3 percent per day delay in anthesis, similar to that reported for dryland crops. It seemed that this was due to increased temperatures reducing the length of the grain-filling phase (Cooper, 1992).

OTHER REGIONS

Though the basic ideas outlined above guide sowing dates throughout the world, in many locations specific constraints dominate cropping timetables. In tropical regions avoidance of heat stress, especially at anthesis, is important. In Sudan highest yields were obtained when anthesis occurred at the coolest time of the year (Ishag and Ageeb, 1991). Similar results were obtained in the Indian Punjab where a photothermal quotient, calculated around heading, was closely related to yield (Ortiz-Monasterio, Dhilon, and Fischer, 1994). In some regions avoidance of heat stress at anthesis may mean tolerating heat stress at the seedling stage (Byerlee, 1992). By contrast, sowing date and genotype need to be carefully chosen at high altitude in Nepal to avoid sterility due to cold conditions at anthesis (Subedi, Floyd, and Budhathoki, 1998).

In fallow wheat systems in Nebraska, the aim is to obtain autumn establishment without excessive vegetative growth, which makes inefficient use of stored water and increases the risk of crown and foot rots (Fenster, 1988). In spring wheat areas of Canada, planting dates are related to current soil and weather status (Major, Hill, and Toure, 1996). In dry areas of the Middle East farmers may wait for the rains to be established before deciding whether to plant wheat. Such decisions are amenable to a probabilistic analysis based on weather data (Stewart, 1988).

DETERMINATION OF OPTIMAL SOWING DATES

Optimal sowing dates are usually considered as those that give the greatest yield. The discussion below will continue with this assumption, but these dates may not represent the economic or social optima. The increased costs of fungicides associated with early sowing in England, especially in a wet winter, may make the economic optimum later than the yield optimum. One social reason for early sowing is that the farmer feels more relaxed once the crop is in the ground and is less susceptible to future weather (Wibberley, 1989).

Crop growth models should be able to assist with the choice of planting date, but this has not been common. Muchow, Hammer, and Vanderlip (1994) used crop models to assess the risks associated with different planting dates and cultivars in sorghum. Aggarwal et al. (1994) validated a wheat model (WTGROW) with data from a wide series of experiments in India and then used the model to look at potential yields at 138 sites (Aggarwal and Kalra, 1994). Delay in sowing consistently reduced yields, typically by about 0.75 percent of potential yield per day. By using a long series of climatic data the authors were able to construct cumulative yield distributions for each sowing date, thus showing how yield variability changes with cultivar and sowing date. This seems to be an essential part of any modeling-based assessment of sowing date. There was, however, no explicit validation of the effects of sowing date on yield, though this was not the main aim of the paper. The importance of such validation can be assessed from the work of Porter, Jamieson, and Wilson (1993), who compared three wheat simulation models with field data for crops grown under nonlimiting water and nutrient regimes in New Zealand. The comparison identified weak areas and suggested development of each model. Overall the two best models typically predicted yields to within 12 to 15 percent. However, in this work wheat was planted on three dates in 1986 (May 9, June 13, July 25). All models predicted a decline in yield with sowing date, but in reality yields did not differ between sowing dates. However, a later comparison showed that the models were capable of reasonable predictions of yield for crops with different water regimes (Jamieson et al., 1998b). Hunt, Pararajasingham, and Wiersma (1996) tested whether a wheat model (Cropsim—wheat) could be used to examine the response to sowing date of four recent spring wheat cultivars in Canada. Field data were obtained from 12 sowings at four-day intervals. The simulation model, which calculated grain number as a function of biomass shortly after anthesis, did not give good estimates for the early and late sowings. These predictions were improved by also relating grain number to solar radiation and temperature at anthesis (Hunt, Pararajasingham, and Wiersma, 1996).

These results suggest that the current simulations of development and its links with growth are not sufficient to predict accurately the effects of sowing date. One suggestion is to replace the simulation of development based on the visible state of the stem apex by a scheme based on the production of leaves and leaf primordia (Kirby and Weightman, 1997; Weightman et al., 1997; Jamieson et al., 1998a). Another problem is that the specification of a cultivar in models is often through a small number of genetic coefficients which are characteristic of a group of cultivars and cannot describe small, but important differences between such cultivars (Russell, van Gardingen, and Wilson, 1993). It is necessary to collect information about cultivars in a more detailed way than at present and to make better use of the many cultivar trials currently undertaken (Russell, van Gardingen, and Wilson, 1993). Another limitation is that crop growth models and crop disease models still tend to be separate entities. Rickman and Klepper (1991) suggested ways of coupling such models, but the problems are complex. It seems that at present physiological models can give some guidance as to sowing dates but these need to be interpreted in view of local knowledge of crop rotations, tillage methods, disease risks, and weed control.

The importance of sowing date in determining the environment prior to anthesis, which largely controls the number of potential grain sites, has clearly been demonstrated (e.g., Manupeerapan and Pearson, 1993). It has been shown that in Argentina old cultivars achieved small yields because they were sink limited after anthesis (Calderini, Dreccer, and Slafer, 1997). Similar limitations can occur physiologically; thus, in general sowing dates need to ensure first that sufficient grain sites are provided and second that duration of filling, as determined by temperature and water, is sufficient to utilize as much as possible of this sink capacity.

In conclusion, two points about climate are worth consideration. First, the past few years have seen considerable interest and progress in the study of global teleconnections and seasonal forecasting. The influence of El Niño, or the Southern Oscillation, on patterns of temperature and rainfall in the Southern Hemisphere is a prime example. Stone, Hammer, and Woodruff (1993) considered how climatic prediction could be used for risk assessment of wheat production in northeastern Australia. Meinke, Stone, and Hammer (1996) used a simulation model of peanut crops to show how different phases of the Southern Oscillation would influence not only yields but the timing and number of planting opportunities. If skills in seasonal forecasting continue to develop then optimum planting date and choice of cultivar could become a dynamic concept, changing in advance from year to year. This is much more likely in spring wheat areas with short growing seasons than in winter wheat areas where forecasts would have to cover at least nine months.

Second, it is now reasonably established that humans have affected the global climate through changes in concentrations of greenhouse gases (IPCC, 1996). This is leading to an increase in global atmospheric temperatures, though the effect on weather patterns in any specific locality is much less clear. Sustaining global production of wheat may well mean changes in areas of production or, perhaps more likely, changes in the cultivars grown in any locality. Determination of optimum planting dates will continue to be an ongoing process as climates and cultivars continue to change. The better descriptions we have of both climatic conditions and physiological responses during trials, the easier this task should be. We need to move away from the concept of how a cultivar performs in a location to how it performs in an environment.

REFERENCES

Aggarwal, P.K. and Kalra, N., 1994. Analyzing the limitations set by climatic factors, genotype, water and nitrogen availability on productivity of wheat II. Climatically potential yields and management strategies. *Field Crops Res.*, 38: 93-103.

Aggarwal, P.K., Kalra, N., Singh, A.K., and Singh, S.K., 1994. Analyzing the limitations set by climatic factors, genotype, water and nitrogen availability on productivity of wheat I. The model description, parmetrization and validation. *Field Crops Res.,* 38: 73-91.

Ali-Zi, Mahalasshmi, V., Singh, M., Ortiz-Ferrara, G., and Peacock, J.M., 1994. Variation in cardinal temperatures for germination among wheat (*Triticum aestivum*) genotypes. *Ann. Appl. Biol.,* 125: 367-375.

Anderson, W.K., Heinrich, A., and Abbots, R., 1996. Long season wheats extend sowing opportunities in the central wheat belt of Australia. *Aust. J. Exp. Agric.,* 36: 203-208.

Anderson, W.K. and Smith, W.R., 1990. Yield advantage of two semi-dwarf compared with two tall wheats depends on sowing time. *Aust. J. Agric. Res.,* 8: 811-826.

Bedding, A.J., 1981. *Tomato Production Part 5, Crop Production.* MAFF Book 2248, Ministry of Agriculture, Fisheries and Food, London.

Bunting, A.H., Dennett, M.D., Elston, J., and Speed, C.B., 1982. Climate and crop distribution. In K. Blaxter and L. Fowden (Editors), *Food, Nutrition and Climate.* Applied Science Publishers, London, pp. 43-74.

Byerlee, D., 1992. *Dryland Wheat in India: The Impact of Technical Change and Future Research Challenges.* CIMMYT Economics working paper 92-05, CIMMYT, Mexico City.

Calderini, D.F., Dreccer, M.F., and Slafer, G.A., 1997. Consequences of breeding on biomass, radiation interception and radiation use efficiency in wheat. *Field Crops Res.,* 52: 271-281.

Cook, R.J. and Veseth, R.J., 1991. *Wheat Health Management.* The American Phytopathological Society Press, St. Paul, MN.

Cooper, J.L., 1992. Effect of time of sowing and cultivar on the development and grain yield of irrigated wheat in the Macquarie Valley, New South Wales. *Aust. J. Exp. Agric.,* 32: 345-353.

Cosser, N.D., Thompson, A.J., Gooding, M.J., and Davies, W.P., 1994. Influences of variety and establishment date on nitrogen accumulation during winter in the shoot system of winter wheat. *Aspects Appl. Biol.,* 39, The impact of genetic variation on sustainable agriculture: 195-200. Association of Applied Biologists, Warwick, UK.

Coventry, D.R., Reeves, T.G., Brooke, H.D., and Cann, D.K., 1993. Influence of genotype, sowing date, and seeding rate on wheat development and yield. *Aust. J. Exp. Agric.,* 33: 751-757.

Craigon, J., Atherton, J.G., and Sweet, N., 1995. Modelling the effects of vernalization on progress to final leaf appearance in winter wheat. *J. Agric. Sci., Camb.,* 124: 369-377.

Dennett, M.D. and Murphy, K.J., 1983. Examining the probability of applying herbicides at the correct growth stage from weather records. *Aspects of Applied Biology,* 4, Influence of environmental factors on herbicide performance and crop and weed biology: 521-530. Association of Applied Biologists, Warwick, UK.

Eddowes, M., 1976. *Crop Production in Europe.* Oxford University Press, Oxford, UK.

Evans, L.T., 1993. *Crop Evolution, Adaptation and Yield.* Cambridge University Press, Cambridge, UK.

Evans, L.T., Wardlaw, I.F., and Fischer, R.A., 1975. Wheat. In L.T.Evans (Editor), *Crop Physiology: Some Case Histories.* Cambridge University Press, Cambridge, UK, pp. 101-149.

FAO, 1998. *FAOSTAT Primary Crops Production Database.* http://fao.apps/fao.org

Fenster, C.R., 1988. Fifty years of tillage practices for winter wheat. In *Challenges in Dryland Agriculture: A Global Perspective.* Texas Agricultural Experimental Station, Bushland, TX, pp. 150-151.

Fischer, R.A. and Byerlee, D.R., 1991. Trends of wheat production in the warmer areas: Major issues and economic considerations. In D.A. Sanders (Editor), *Wheat for the Nontraditional Warm Areas.* CIMMYT, Mexico City, pp. 3-27.

Fisher, N.M., 1984. Crop growth and development: The vegetative phase. In P.R. Goldsworthy and N.M. Fisher (Editors), *The Physiology of Tropical Field Crops.* John Wiley, Chichester, pp. 119-161.

Gallagher, J.N. and Biscoe, P.V., 1978. Radiation absorption, growth and yield of cereals. *J. Agric, Sci., Camb.*, 91: 47-60.

Giunta, F., Motzo, R., and Deidda, M., 1995. Effects of drought on leaf area development, biomass production and nitrogen uptake of durum wheat grown in a Mediterranean environment. *Aust. J. Agric. Res.*, 46: 99-111.

Gooding, M.J. and Davies, W.P., 1996. *Wheat Production and Utilization: Systems, Quality, and the Environment.* CAB International, Wallingford, UK.

Gooding, M.J., Davies, W.P., Thompson, A.J., and Smith, S.P., 1993. The challenge of achieving breadmaking quality in organic and low input wheat in the UK—a review. *Aspects of Applied Biology,* 36, Cereal quality III: 189-198. Association of Applied Biologists, Warwick, UK.

Green, C.F. and Irvins, J.D., 1985. Time of sowing and the yield of winter wheat. *J. Agric. Sci., Camb.*, 104: 235-238.

Gregory, P.J. and Eastham, J., 1996. Growth of shoots and roots, and the interception of radiation by wheat and lupin crops on a shallow duplex soil in response to time of sowing. *Aust. J. Agric. Res.,* 47: 427-447.

Harris, R., 1997. Cereal drilling three quarters done already. *Farmers Weekly,* 127(15): 57.

HGCA, 1991. *Physiology in the Production and Improvement of Vereals.* HGCA Research Review 18, Home-Grown Cereals Authority, London.

HGCA, 1997. *The Winter Wheat Growth Guide.* Home-Grown Cereals Authority, London.

Hough, M.N., 1990. EUR 13039—*Agrometeorological Aspects of Crops in the United Kingdom and Ireland.* A review for sugar beet, oilseed rape, peas, wheat, barley, oats, potatoes, apples and pears. European Communities Commission, Luxembourg.

Hunt, L.A., Pararajasingham, S., and Wiersma, J.V., 1996. Effects of planting date on the development and yield of spring wheat: simulation of field data. *Can. J. Plant Sci.*, 76: 51-58.

IPCC, 1996. *Climate Change 1995. The Science of Climatic Change.* Cambridge University Press, Cambridge, UK.

Ishag, H.M. and Ageeb, O.A.A., 1991. The physiology of grain yield in wheat in an irrigated tropical environment. *Expl. Agric.*, 27: 71-77.

Jamieson, P.D., Brooking, I.R., Semenov, M.A., and Porter, J.R., 1998a. Making sense of wheat development: a critique of methodology. *Field Crops Res.*, 55: 117-127.

Jamieson, P.D., Porter, J.R., Goudriaan, J., Ritchie, J.T., van Keulen, H., and Stol, W., 1998b. A comparison of the models AFRCWHEAT2, CERES-Wheat, Sirius, SUCROS2 and SWHEAT with measurements of wheat grown under drought. *Field Crops Res.*, 55: 23-44.

Jenkins, P.D., 1995. Effects of plastic film covers on dry-matter production and early tuber yield in potato crops. *Ann. Appl. Biol.*, 127: 201-213.

Jones, C.A. and O'Toole, J.C., 1987. Application of crop models in agro-ecological characterization: Simulation models for specific crops. In A.H. Bunting (Editor), *Agricultural Environments: Characterization, Classification and Mapping.* CAB International, Wallingford, UK, pp. 199-209.

Kerr, N.J., Siddique, K.H.M., and Delane, R.J., 1992. Early sowing with wheat cultivars of suitable maturity increases grain yield of spring wheat in a short season environment. *Aust. J. Exp. Agric.*, 32: 717-733.

Kirby, E.J.M. and Weightman, R.M., 1997. Discrepancies between observed and predicted growth stages in wheat. *J. Agric. Sci., Camb.*, 129: 379-384.

MAFF, 1997. *Agricultural Census of England: Results for 1997.* Ministry of Agriculture, Fisheries and Food, London.

Major, D.J., Hill, B.D., and Toure, A., 1996. Prediction of seeding date in Southern Alberta. *Can. J. Plant Sci.*, 76: 59-65.

Manupeerapan, T. and Pearson, C.J., 1993. Apex size, flowering date and grain yield of wheat as affected by sowing date. *Field Crops Res.*, 32: 41-57.

Meinke, H., Stone, R.C., and Hammer, G.L., 1996. SOI phases and climatic risk to peanut production: A case study for Northern Australia. *Int. J. Climat.*, 16: 783-789.

Monteith, J.L., 1994. Principles of resource capture by crops. In J.L. Monteith, R.K. Scott, and M.H. Unsworth (Editors), *Resource Capture by Crops.* Nottingham University Press, Nottingham, UK, pp. 1-15.

Monteith, J.L. and Elston, J., 1993. Climatic constraints on crop production. In L. Fowden, T. Mansfield, and J. Stoddart (Editors), *Plant Adaptation to Environmental Stress,* Chapman and Hall, London, pp. 3-18.

Muchow, R.C., Hammer, G.L., and Vanderlip, R.L., 1994. Assessing climatic risk to sorghum production in water-limited subtropical environments: II. Effects of planting date, soil water at planting and cultivar phenology. *Field Crops Res.*, 36: 235-246.

Mulholland, B.J., Craigon, J., Black, C.R., Stokes, D.T., Zhang, P., Colls, J.J., and Atherton, J.J., 1997. Timing of critical developmental stages and leaf production in field-grown spring wheat for use in crop models. *J. Agric. Sci., Camb.*, 129: 155-161.

Murphy, D.P.L., Frost, D.L., and Evans, E.J., 1993. Plant development and grain yield in winter wheat as influenced by sowing date and variety. *Aspects of Applied Biology,* 34, Physiology of varieties: 99-104. Association of Applied Biologists, Warwick, UK.

NIAB, 1997. *NIAB Cereal Variety Handbook.* National Institute of Agricultural Botany, Cambridge, UK.

Nix, H.A., 1975. The Australian climate and its effects on grain yield and quality. In A. Lazenby and E.M. Matheson (Editors), *Australian Field Crops Volume 1: Wheat and Other Temperate Cereals.* Angus and Robertson, Sydney, pp. 183-226.

Ortiz-Monasterio, R.Jl., Dhilon, S.S., and Fischer, R.A., 1994. Date of sowing effects on grain yield and yield components of irrigated spring wheat cultivars and relationships with radiation and temperature in Ludhiana, India. *Field Crops Res.*, 37: 169-184.

Porter, J.R., Jamieson, P.D., and Wilson, D.R., 1993. Comparison of wheat simulation models AFRCWHEAT2, CERES-Wheat and SWHEAT for non-limiting conditions of crop growth. *Field Crops Res.*, 33: 131-157.

Prew, R.D., Beane, J., Carter, N., Church, B.M., Dewar, A.M., Lacey, J., Penny, A., Plumb, R.T., Thorne, G.N., and Todd, A.D., 1986. Some factors affecting the growth and yield of winter wheat grown as a third cereal with much or negligible take-all. *J. Agric. Sci., Camb.*, 107: 639-671.

Prew, R.D., Carter, N., Church, B.M., Dewar, A.M., Lacey, J., Magan, N., Penny, A., Plumb, R.T., Thorne, G.N., Todd, A.D., and Williams, T.D., 1985. Some factors limiting the growth and yield of winter wheat and their variation in two seasons. *J. Agric. Sci., Camb.*, 104: 135-162.

Rickman, R.W. and Klepper, B., 1991. Environmentally driven cereal crop growth models. *Annu. Rev. Phytopathol.*, 29: 361-380.

Russell, G., van Gardingen, P., and Wilson, G.W., 1993. Using physiological information about varieties: The way forward? *Aspects of Applied Biology,* 34, Physiology of varieties: 47-56. Association of Applied Biologists, Warwick, UK.

Shackley, B.J. and Anderson, W.K., 1995. Responses of wheat cultivars to time of sowing in the southern wheatbelt of Western Australia. *Aust. J. Exp. Agric.*, 35: 579-587.

Stewart, J.I., 1988. Risk analysis and response farming. In *Challenges in Dryland Agriculture: A Global Perspective.* Texas Agricultural Experimental Station, Bushland, TX, pp. 322-324.

Stewart, J.I., 1991. Principles and performance of response farming. In R.C. Muchow and J.A. Bellamy (Editors). *Climatic Risk in Crop Production: Models and Management for the Semiarid Tropics and Subtropics,* CAB, Wallingford, UK, pp. 361-382.

Stone, R.C., Hammer, G.L., and Woodruff, D., 1993. Assessment of risk associated with climate prediction in management of wheat in north-eastern Australia. In *Proc. 7th Australian Agronomy Conference,* Australian Society of Agronomy, Adelaide, pp. 174-177.

Subedi, K.D., Floyd, C.N., and Budhathoki, C.B., 1998. Cool temperature induced sterility in spring wheat (*Triticum aestivum*) at high altitudes in Nepal: Variation among cultivars in response to sowing date. *Field Crops Res.,* 55: 141-151.

Sylvester-Bradley, R., Stokes, D.T., Scott, R.K., and Willington, V.B.A., 1990. A physiological analysis of the diminishing responses of winter wheat to applied nitrogen. 2. Evidence. *Aspects of Applied Biology,* 25, Cereal Quality II: 289-300. Association of Applied Biologists, Warwick, UK.

Weightman, R.M., Kirby, E.J.M., Sylvester-Bradley, R., Scott, R.K., Clarke, R.W., and Gillett, A., 1997. Prediction of leaf and internode development in wheat. *J. Agric. Sci., Camb.,* 129: 385-396.

Weir, A.H., 1988. Estimating losses in the yield of winter wheat as a result of drought, in Engand and Wales. *Soil Use and Management,* 4: 33-40.

Wibberley, E.J., 1989. *Cereal Husbandry.* Farming Press, Ipswich, UK.

Chapter 7

Plant Density and Distribution As Modifiers of Growth and Yield

Emilio H. Satorre

INTRODUCTION

Plant density is one of the major factors determining the ability of the crop to capture resources; it is of particular importance because it is under fairly close control by the farmer in most wheat-producing systems. It may be strongly difficult to disentangle the effects of crop density from those of other factors, under extensive grain production (Snaydon, 1984). However, there has been interest in defining the relationships between density and crop yield quantitatively in order to establish optimum populations and maximum attainable yields under various situations. As a result, the effect of density on wheat plant size and crop productivity has received attention (Donald, 1963; Harper, 1977; Willey and Heath, 1969).

Wheat is an inbreeding species, heavily selected for uniformity so that most individuals are genetically identical and phenotypically similar, because of the uniformity of seed size and the fact that sown seeds tend to germinate synchronously. Since wheat crops are normally grown in single-species stands, intraspecific competition is intense. Competition is therefore the ecological process largely determining the response of wheat plants to density and plant arrangement. The term competition refers to the process whereby plants share resources (e.g., mineral nutrients, water, and light) that are in insufficient supply for their joint requirements (Satorre, 1988). Competition causes a reduction in survival, dry matter growth, and grain yield of individual wheat plants. However, the management of crop competition through density selection may allow maximum yields per unit area to be achieved.

Identifying and understanding the ecophysiological basis of wheat yield-density response could allow researchers to predict the effect of crop management practices on wheat production and help farmers and technicians to properly design crop production systems under various ecological conditions. In this chapter, I will discuss how wheat density affects crop functioning and how environmental conditions or crop management decisions may affect crop density response. Finally, a brief discussion on how yield-density responses may be mathematically described will be presented.

ECOPHYSIOLOGICAL BASIS OF DENSITY RESPONSE

Wheat dry matter production under potential conditions is determined by the intercepted solar radiation and the radiation use efficiency of the crop canopy. Crop density would mainly affect the ability of the crop to intercept radiation (or to capture the light resource, from an ecological perspective), since there is little evidence on the effects of density on resource use efficiency.

It is known that much of the incident radiation is not available for crop growth during the early stages, due to the fact that low leaf area expansion determines low light interception. Therefore, increasing density may increase the leaf area index of the crop and the proportion of the incident light that is intercepted. Improving plants' spatial distribution may also help them intercept more light at these early stages of crop growth, when leaf area index values are low. Early results from Puckridge and Donald (1967) clearly showed that density manipulation in the range 1.4 to 1078 pl/m^2 (plants per square meter) could successfully increase light interception of wheat crops under western Australian conditions. The greater proportion of intercepted light in high-density crops explained most of the differences in early crop growth rates calculated from the results of that research; that is, growth rates were greater in crops sown at the higher densities.

During the early stages of development, competition among small wheat plants may be evident only at very high densities, and all individual plants tend to have similar dry matter production while crop productivity per unit area tends to increase linearly with density. Plotting the logarithm of plant dry matter against the logarithm of density as suggested by Kira, Ogawa, and Shinozaki (1953) helps to show the way individual plants and crops would tend to respond (see Figure 7.1). As crop development progresses, plants increase in size and leaf area and the onset of competition is evident at lower densities; that is, individual plant size is reduced as density increases above a density threshold (Figure 7.1).

This density threshold may vary at any time according to environmental conditions and wheat variety characteristics. In crop stands above the

FIGURE 7.1. The Linear Relationship of Log Plant Weight to Log Density for Various Crop Growth Stages in a Kira Plot

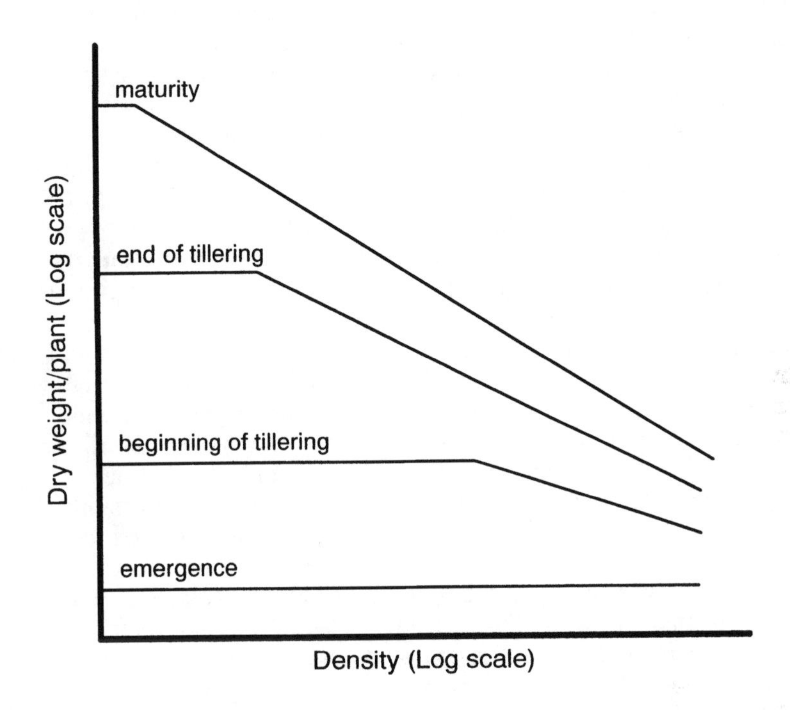

threshold density wheat leaf area index allows full light interception and use of the available resources; therefore, crop growth rate will be maximum for the set of environmental conditions experienced by the crop. At this stage, various crop densities (all those above the competition threshold) would allow maximum growth rates per unit area. Sustaining maximum crop growth rates is important, especially during the critical period for yield determination; that is, from the beginning of ear growth to the beginning of grain growth (Fischer, 1984; Savin and Slafer, 1991). When density and/or environmental conditions have been suitable for the crop to reach 95 percent light interception before the beginning of ear growth (i.e., twenty days before anthesis), crop performance would tend to be independent of the number of plants established. Therefore, a wide range of crop densities would allow ear dry weight at anthesis and grain number to be high enough to maximize grain yield per unit area.

Wheat planting densities therefore appear to strongly influence early dry matter accumulation. There is experimental evidence that increasing density not only improves light capture early in the season but also the uptake of soil resources such as water. In a Mediterranean environment, high sowing densities contributed to reduce soil evaporation and to increase biomass production and water use in the early phases of crop growth (Boogaard et al., 1996). As plant size increases, crop growth rate may depend more on resource availability than on plant density; therefore, growth rates during the critical periods of yield determination may be only influenced by density if plant number is below the competition threshold at that time. Environmental and management practices could alter the crop yield-density response, altering the density above which crop growth rate is maximum.

In general, planting densities chosen for field wheat crops are aimed to produce a crop able to use all above and below-ground resources, allowing the crop to maximize growth rates during critical stages. Increasing early resource capture through high plant densities may not necessarily maximize growth rate at critical crop stages. For example, large water use by a crop in the early phases of growth may reduce soil water availability late in the season, leading to yields similar to those obtained with low plant densities. In temperate subhumid areas, this pattern of water use has led to reduced sowing rates, as in the southern pampas of Argentina (Gallez et al., 1986) and to the finding that there is rarely any difference due to sowing rate in grain yield of crops established in the semiarid Brown soil zone of Canada during dry years (McLeod et al., 1996).

Dry Matter Production

Wheat plants have the ability to compensate for low plant populations by producing more tillers. The total shoot weight of wheat per unit area of land usually increases asymptotically as density increases. The asymptote extends over a wide range of densities, due mainly to the large plasticity of individual plant size, which determines that mean plant weight declines to exactly compensate for increases in density; that is, proportionate reductions in plant weight occur as densities increase above the normal sown density (Shinozaki and Kira, 1956; Holliday, 1960; Donald, 1963; Lerner and Satorre, 1990). Fortyfold differences in plant weight have been reported to occur between widely spaced plants and plants growing in stands at normal seed rates (Puckridge and Donald, 1967; Arias, Satorre, and Guglielmini, 1994), though phenotypic plasticity is highly variable among cultivars. For example, Arias, Satorre, and Guglielmini (1994) reported that widely spaced plants of cultivar Leones, with low tillering ability, weighed 14.3 g at maturity while plants of cultivar Marcos Juarez, with high tillering potential, weighed 24.5 g. Plants of both cultivars sown in 400 pl/m^2 stands were not significantly different and weighed 1.6 g on average. It is important to note that the performance of wheat cultivars under noncompetitive conditions was found to be unrelated to their performance under dense stand conditions (Syme, 1972; Fischer and Kertesz, 1976).

It is expected that, in the range of densities at which maximum yields occur, the crop achieves its maximum possible utilization of limiting resources, such as mineral nutrients, water, and light. The supply of limiting resources and the genetic ability to capture and use them largely determine the maximum shoot yield per unit area of any particular wheat crop in a particular environment (Willey and Heath, 1969). Environmental factors such as temperature or daylength, which affect the length of the growing season, can also affect maximum shoot weight by modifying the ability of the crop to capture resources. For example, long days at high latitudes inhibit wheat tillering, which determines that sowing rates of 500 to 700 pl/m^2 are commonly used for spring wheat crops in places such as Finland (Peltonen and Peltonen-Sainio, 1997).

In some cases, the relationship between density and total shoot weight per unit area is better described by a parabolic model; that is, there is a distinct maximum yield at a particular density, and shoot yield declines as density increases above this point. In these cases, crowding reduces the efficiency of the crop to capture or use environmental resources (Donald, 1963).

Grain Yield

Following Holliday (1960) it has generally been accepted that the response to density of storage organs (e.g., grains) is better described by a parabolic model; that is, grain yield and most other yield components decrease at higher plant densities. This has been generally attributed to the fact that the allocation of resources to storage organs is greatly altered by competition (Harper, 1977). In general, harvest index tends to decline progressively with increasing density (Donald and Hamblin, 1976). Despite the fact that parabolic responses are generally accepted for grain yield-density relationships, some recent studies indicate that yield-density relationships for grain may sometimes be better described by asymptotic models (Martin and Field, 1987; Willey and Holliday, 1971).

The grain yield-density relationship largely depends on the performance of individual yield components. As mentioned before, wheat crop genotypes usually have great plasticity. Evidence of this may be obtained by comparing the performance of plants at wide spacing with that of plants at densities which give the maximum crop yield. Wheat plants under density stress are smaller and have fewer tillers and grains than widely spaced plants (Donald, 1963; Puckridge and Donald, 1967; Faris and De Pauw, 1981; Lerner and Satorre, 1990; Rana, Ganga, and Pachauri, 1995). The analysis of numerical yield components from the work of Puckridge and Donald (1967) showed that, in the range 1.4 to 1078 pl/m^2, the weight per grain was only reduced by 4 percent while the number of grains per plant was reduced by 99 percent. The number of ears per plant was the most affected component of grain number, while the number of grains per spikelet was least affected. Lerner and Mac Maney (1986) and McLeod et al. (1996), among others, also found that grain weight was not affected by crop density, but Scott, Dougherty, and Langer (1977) and Darwinkel (1978) found that increasing sowing rate reduced grain weight. However, the number of grains per unit area they harvested was very high (19,000 to 22,000 grains/m^2) and larger than those harvested in the previously mentioned experiments. Therefore, greater grain yields obtained with higher densities appear to be related to a greater number of spikes per unit area. Due to low availability of crop growth resources, plants under severe competition show a progressive reduction in growth rate, which will markedly affect grain number determination during the critical period of 20 days prior to anthesis to 10 days postanthesis. Therefore, shoot growth may become less affected than grain yield, causing a reduction in harvest index. It is clear from this data that the ear population is the main factor controlling the grain yield-density response of wheat crops. Surveys of commercial crops have demonstrated that grain yields are positively correlated with ear populations (Harper, 1983; Calderini et al., 1995). In high-yielding systems, it appears to be essential to have large ear populations, which under good conditions may be obtained using low seeding rates, due to tillering contribution, or high seeding rates (280-500 pl/m^2) with a uniform kernel set mostly on main stem spikes. Therefore, in low-density crops tillers are an essential component in determining the number of ears at harvest. Tillering has been recognized as a complex phenomenon, controlled by endogenous and environmental factors (Sánchez et al., 1993). The reduction in the number of tillers produced by plants at high density has been for a long time almost exclusively associated with the low availability of per-plant resources. However, recent studies have found that the effect of resources, such as photosynthetically active radiation, were related to tiller and dry matter production, but with the same resource availability tillering was lower when the red/far red ratio of light composition was also lower (Barnes and Bugbee, 1991). These results indicate that, within a wide range of resource availability, wheat tiller production may be influenced by photomorphogenic reactions mediated by light quality. Therefore, tiller number will be the result of the interaction between resource-based and non-resource-based environmental signals.

Although we may consider that within a normal range of sowing densities the rate and duration of crop developmental phases are not modified, there is some evidence suggesting that some developmental processes in yield formation may be early altered by density (Kirby and Faris, 1972; Yu, Van Sandford, and Egli, 1988; Lerner and Cerri, 1990). The increase of plant density in the range 180 to 930 pl/m^2 reduced the number of leaves, spikelets, and initiated florets of two Argentinian wheat cultivars (Lerner and Cerri, 1990).

Darwinkel (1980) has suggested that high-density crops would produce a large number of spikes and have a greater risk of lodging and insect and disease damage. Although different plant densities (325 to 650 pl/m^2) had little effect on carbon photoassimilation in top plant structures, including spike structures, high densities tend to affect partitioning to the stem internodes and leaf senescence (Verona, Loffer, and Fernández, 1980; Sarandon and Chidichimo, 1985; Wang et al., 1997). The stem tends to export photoassimilates during grain filling at high plant density and the rate of leaf senescence is increased. Both processes could also contribute to low grain yields under high sowing rates.

THE EFFECT OF RESOURCES ON CROP DENSITY RESPONSE

The supply of limiting resources such as water, light, and nutrients affect the form and parameters of the yield-density relationships, largely through their effect on the maximum yield per unit area that can be achieved at very high densities (see Figure 7.2). They have little effect on the yield of widely spaced plants; that is, yield per plant at very low densities (see Table 7.1). For example, nitrogen fertilizer applications tended to increase maximum wheat yields per unit area under field conditions, but only slightly affect the maximum yield when plants were sown widely spaced (Arias, Satorre, and Guglielmini, 1994; Table 7.1).

The response to density depends upon the supply of limiting resources, since competition for limiting resources is the driving process in crop stands. Nitrogen, a major limiting factor, interacts strongly with density (Snaydon, 1984). The optimum density and, therefore, the maximum grain yield tend to be greater with increasing applications of nitrogen, if nitrogen is the yield-limiting factor (Figure 7.2). When the major limiting resource is supplied, the crop system might be enabled to sustain a higher

FIGURE 7.2. Relationship Between Density and Shoot Yield at Two Levels of Nitrogen Fertilization

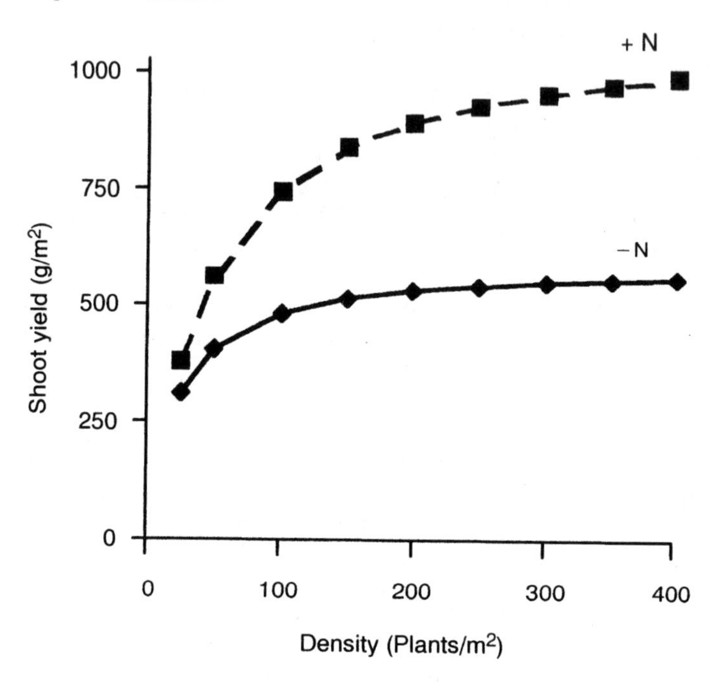

Note: +N: plus eq. 100 kg N/ha; − N: no fertilizer was applied. Adapted from Arias, Satorre, and Guglielmini, 1994.

density and hence a greater grain yield. Within the range of crop densities that allow maximum grain yields, the demand of growth resources will tend to be similar; therefore the nitrogen application rate or water provision required will tend to be similar between the extreme low and high densities that allow maximum yields (Salazar et al., 1996).

Nitrogen has a great impact on early vegetative growth; while this increases the shoot weight it generally reduces the harvest index at any crop density (Donald and Hamblin, 1976). Water provision interacts markedly with density and the response to nitrogen fertilizer application; where water supply is inadequate during the late developmental stages high densities and increase nitrogen applications may severely reduced grain yield.

TABLE 7.1. Shoot and Spike Yield of Two Wheat Cultivars Sown at Very Low and High Densities at Two Levels of Fertilizer Application (0 and 100 kg N/ha)

	Shoot yield (g/m^2)		Ear yield (g/m^2)	
	0 kg N/ha	100 kg N/ha	0 kg N/ha	100 kg N/ha
Leones INTA				
1 pl/m^2	14	15	7	8
400 pl/m^2	532	592	270	384
Marcos Juarez				
1 pl/m^2	26	23	14	13
400 pl/m^2	536	912	345	454

Source: Data was adapted from Arias, Satorre, and Guglielmini, 1994.

THE EFFECT OF SOWING DATE AND PLANT ARRANGEMENT

Among the management factors that may affect wheat yield-density relationships, the date of sowing and the planting arrangement of the crop are the most important; both factors are under close control by farmers. Date of sowing may affect plant weight under both competition-free conditions and high-density conditions (Harper, 1977), though its effect on yield-density response has not been extensively studied. The effect of sowing date on maximum yield per unit area can be inferred from the increase or decrease in the maximum plant yield as the length of the growth period increases or decreases (Donald, 1951; Shinozaki and Kira, 1956; Bleasdale, 1966). This effect has been confirmed with barley cultivars sown at a wide range of densities on various sowing dates from autumn to spring (Kirby, 1969). Under extensive crop production conditions, a delay in sowing from an optimum sowing date may require a density increment to reach maximum attainable yields, since individual plant growth is reduced (Senigagliesi et al., 1986; Popa, 1995). However, the interaction between sowing date and density response may also be interpreted from the way sowing date modified the maximum yield of the crop per unit area, by modifying the total available resources, in general, or environmental conditions during the critical periods of yield determination, in particular. It is convenient to keep in mind that by increasing crop density wheat crops can hardly compensate yield loss caused by delays in sowing date, though this practice may allow maximum achievable yields under late-sown conditions. Low yields associated with delays in sowing usually reflect poor environmental conditions during the critical period of yield component determination. Late-sown crops may therefore tend to produce fewer grains per unit area and/or smaller grains; it is possible to partially compensate grain number production through density management but any positive effect on grain weight determination is hardly expected by increasing density of late-sown crops.

At a given crop density sowing patterns can be random, clumped, or regular; wheat crops are normally sown in rows, that is, in a clumped arrangement. The planting arrangement of crops is often described by their "rectangularity," that is, the ratio of the distance between rows to the distance between plants within a row. In general, wheat yield is greatest, at any density, if the plants are arranged regularly, that is, the rectangularity is 1:1 (Holliday, 1963; Fischer and Miles, 1973; Pant, 1979; Auld, Kemp, and Medd, 1983), though the effect of planting arrangement is often not significant if densities are at or above those required to achieve maximum yield (Auld, Kemp, and Medd, 1983; Nerson, 1980). The extent to which rectangularity affects the yield of the crop depends on the plasticity of the individual plants of any particular wheat variety and the environmental conditions. Holliday (1963) reported yield increases between 8 to 33 percent due to reducing row spacing from 20 to 10 cm. Similarly, most research worldwide has shown that closer row spacing (15 to 18 cm) gave higher grain yields than wider row spacings (usually greater than 23 cm; Gallez et al., 1986; Panwar et al., 1995; Thakur et al., 1996; Malik, Haroon-ur-rasheed, and Razzaq, 1996; Panda et al., 1996) though a few experiments showed that wider row spacing did not result in wheat yield losses (Lafond et al., 1996).

Patterns of effect are not consistent through literature when rectangularity is analyzed. Several experiments reported that there was a consistent yield depression at low sowing rates as rectangularity was increased (Fawcett, 1964; Auld, Kemp, and Medd, 1983) (see Figure 7.3a). In some other cases crop yields per unit area gradually declined as rectangularity increased either by increasing plant density or increasing row width; though yield declines at high density were greater than at low density (Willey and Heath, 1969) (see Figure 7.3b). However, results consistently show that there are better possibilities to equal high-density crop yields if low-density crops are sown in a square pattern; grain yield advantages from a better plant spatial distribution are expected when crops are sown at low density rather than at high density. The reported evidence suggests that any advantage derived from spatial arrangement is brought about by improving a crop's ability to exploit available resources (Kemp, Auld, and Medd,

FIGURE 7.3. The Effects of Rectangularity and Density on Grain Yield of Wheat

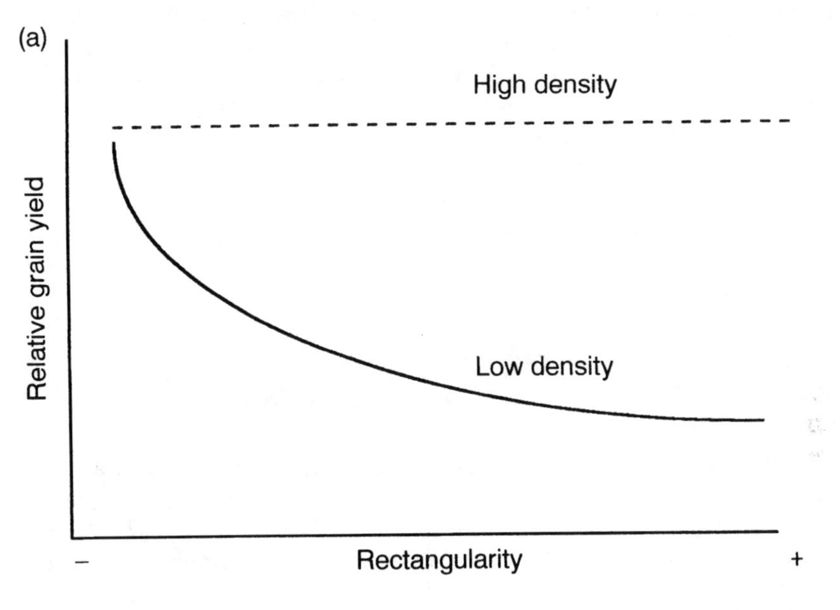

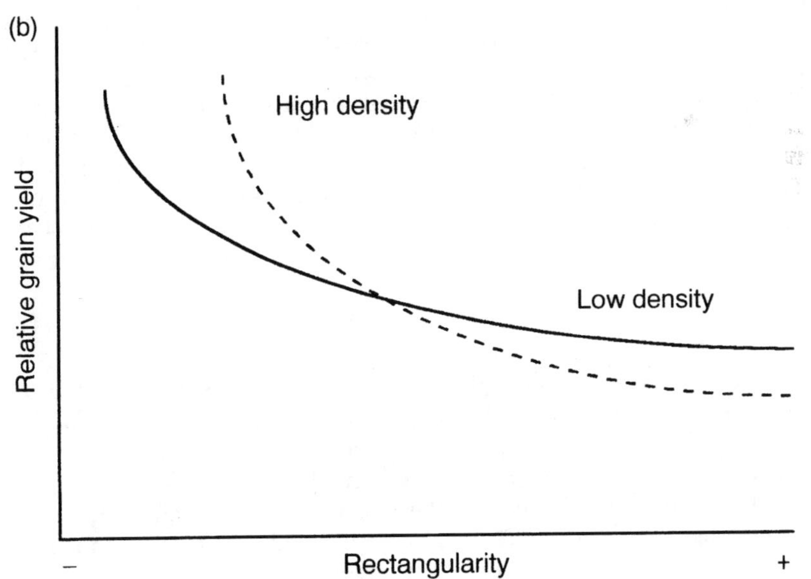

1983). Rectangular arrangements possibly cause an early reduction of crop growth rate, which delays or even avoids use of distant resources.

In addition, reducing the distance between rows and rectangularity and increasing plant density to rapidly exploit resources can enhance the competitiveness of wheat against weeds (Ross and Harper, 1972; Medd et al., 1985; Flemming, Young, and Ogg, 1988; Satorre and Arias, 1990; Arias and Satorre, 1991; Solie et al., 1991). For example, Solie et al. (1991) concluded that yield increases of 18 percent could be obtained when row spacing was reduced from 23 to 7.5 cm in both cheat- (*Bromus secalinus*) free and cheat-infested fields.

MATHEMATICAL RELATIONSHIPS BETWEEN PLANT YIELD AND DENSITY

Mathematical models are useful for interpreting and predicting the effect of density on yield. Interest in quantitative relationships between crop yield and density was largely stimulated by the need to clearly define optimum sowing densities for crop production. Such relationships should, ideally, take account of environmental factors, such as limiting resources, which affect the response to density. Willey and Heath (1969), Mead (1979), and Ratkowsky (1983), among others, made comprehensive reviews of the various models used to describe yield-density relationships, pointing out the advantages and disadvantages of the various mathematical functions proposed. Several types of equations have been described, but only the so-called reciprocal equations for yield per plant have proved to be satisfactory, and to provide a simple biological interpretation for the equation parameters. Four main forms of the reciprocal equation have been proposed:

$$w^{-1} = a + b \cdot z \qquad (1)$$

where w is the shoot or grain yield per plant, z is the crop density (number of plants per unit area) and a and b are parameters of the equation. This is the simplest equation and describes only the asymptotic model. Equation (1) was first proposed by Shinozaki and Kira (1956) and its parameters (a and b) have biological meaning. In this equation, a is the reciprocal of the yield per plant (w^{-1}) at an infinitely low density, that is, it is an estimation of the yield of wide-spaced plants, in competition-free conditions. Willey and Heath (1969) regarded this as a measure of the "genetic potential" of individual plants of the crop. The fitted value of a indeed largely depends on the genetic characteristics of a particular variety. If the yield-density relationship is truly asymptotic, b is inversely related to the asymptote, that is, the maximum yield per unit area at very high densities; this may be considered a measure of the "environmental potential" (Willey and Heath, 1969). This equation has been successfully used to describe results from field experiments using large plots or small plots with systematic experimental designs (Lerner and Satorre, 1990).

Other equations incorporate a third parameter, so that they can also describe parabolic yield-density relationships. For example, Bleasdale (1966) proposed the expression:

$$(w)^{-\emptyset} = a + b \cdot z \tag{2}$$

which is a simplified form of a model proposed by Bleasdale and Nelder (1960). If $\theta = 1.0$ the model describes the asymptotic relationship, but if $\theta < 1.0$ it describes a parabolic relationship.

Holliday (1960b) proposed the expression:

$$w^{-1} = a + b \cdot z + c \cdot z^2 \tag{3}$$

The quadratic term accounts for parabolic relationships; if the parameter $c = 0.0$ then the equation reduces to that of Shinozaki and Kira (1956). Farazdaghi and Harris (1968) proposed the expression:

$$w^{-1} = a + b \cdot z^{\theta} \tag{4}$$

Like the third expression, this becomes identical to the first equation when $\theta = 1$.

Ratkowsky (1983) found that all the expressions (1-4) gave good fit to the various data sets that he analyzed, though equation (3) gave the best linear fit. In spite of the good fit, the reciprocal equations have some limitations. In particular, they all fail to provide a variance structure that is adequate for statistical analysis, and for correct interpretation of parameters. However, novel statistical techniques, using data transformation and nonlinear procedures, have improved the statistical treatment of all four models presented here (Ratkowsky, 1983).

The response of wheat crops to plant density should be also defined in terms of the spatial arrangement of the plants. Few researchers have intended to distinguish between density and plant arrangement when modeling crop response, since response to different densities usually has been studied at a constant row width. Although the extent to which spatial arrangement may affect wheat yield is strongly dependent on the plasticity of the genotype (Willey and Heath, 1969), it appears to be important to consider the effect of rectangularity on the yield-density equations, particularly when crops sown at various densities and distances between rows are compared. In these cases, equations should be able to describe the effects of density as well as those of rectangularity. For this purpose it has been proposed to include intrarow and interrow spacing as variables in some of these equations (Willey and Heath, 1969). It is desirable to found any mathematical description of the yield-density-rectangularity relationship on parameters with some biological meaning and that may be experimentally verifiable. The reciprocal equations discussed here have proved to be robust in both aspects and provide a tool to describe asymptotic or parabolic yield-density relationships. However, there has been insufficient research on the effect of rectangularity and the equations have been only tested on wheat crops in a few cases.

CONCLUDING REMARKS

This chapter has pointed out that highly productive wheat crops depend on individual crop plants experiencing a severe competitive stress. It has concentrated on the effect of density on crop growth and yield; it has deliberately avoided any discussion of the effect of density on the mortality of plants since this hardly occurs, at least as a density-dependent process, in most crop production conditions. It has attempted to examine the response of wheat crops to density from an ecophysiological and ecological point of view. Plant functional responses and competitive interactions are responsible for the effects of density on yield in various environments and crop management situations. A conceptual framework was intended to be built in order to understand the basis of the response and to provide a rationale for predicting and manipulating plant density in experimental or commercial crops. Modifying crop density and plant arrangement may be seen as a way to change crop spatial and temporal structure and, by this means, the use of crop resources. High densities for any particular environment should provide a greater use of early available environmental resources; however, yield-density responses still rely strongly on the way resources are used during the critical periods of yield components determination. Taking this into account, density manipulation should be effective only when it assures that the crop must be able to achieve maximum possible growth rates during those periods.

REFERENCES

Arias, S. and Satorre, E.H. (1991). Competencia entre trigo (*Triticum aestivum*) y *Brassica* sp. El efecto de la densidad del cultivo y la maleza sobre el rendimiento de grano. *Actas de la XII Reunión Argentina sobre la maleza y su control (Mar del Plata, Argentina)* 2:5-10.

Arias, S., Satorre, E.H., and Guglielmini, A. (1994). Competencia entre trigo (*Triticum aestivum* L.) y *Brassica* sp. El efecto de la densidad del cultivo y la maleza en dos niveles de fertilización nitrogenada. *Actas III Congreso Nacional de Trigo (Bahía Blanca, Argentina)* 1:193-194.

Auld, B.A., Kemp, D.R., and Medd, R.W. (1983). The influence of spatial arrangement on grain yield of wheat. *Australian Journal of Agricultural Research* 34:99-108.

Barnes, C. and Bugbee, B. (1991). Morphological responses of wheat to changes in phytochrome photoequilibrium. *Plant Physiology* 97:359-365.

Bleasdale, J.K.A. (1966). Plant growth and crop yield. *Annals of Applied Biology* 57:173-182.

Bleasdale, J.K.A. and Nelder, J.A. (1960). Plant population and crop yield. *Nature* 188:342.

Boogaard, R. van den, Veneklass, E.J., Peacock, J.M., and Lambers, H. (1996). Yield and water use of wheat (*Triticum aestivum*) in a mediterranean environment: Cultivar differences and sowing density effects. *Plant and Soil* 181(2):251-262.

Calderini, D.F., Maddonni, G.A., Miralles, D.J., Ruiz, R.A., and Satorre, E.H. *ex aequo* (1995). Trigo: Modelos de alta producción. *Revista CREA* 177:44-47.

Darwinkel, A. (1978). Patterns of tillering and grain production of winter wheat at a wide range of plant densities. *Netherland Journal of Agricultural Science* 26:383-398.

Darwinkel, A. (1980). Ear development and formation of grain yield in winter wheat. *Netherland Journal of Agricultural Science* 28:156-163.

Donald, C.M. (1951). Competition among pasture plants. I. Intraspecific competition among annual pasture plants. *Australian Journal of Agricultural Research* 2:355-376.

Donald, C.M. (1963). Competition among crop and pasture plants. *Advances in Agronomy* 15:1-118.

Donald, C.M. and Hamblin, J. (1976). The biological yield and harvest index of cereals as agronomic and plant breeding criteria. *Advances in Agronomy* 28: 361-405.

Farazdaghi, H. and Harris, P.M. (1968). Plant competition and crop yield. *Nature* 217:289-290.

Faris, D.G. and De Pauw, R.M. (1981). Effects of seeding rate on growth and yield of three spring wheat cultivars. *Field Crops Research* 3:289-301.

Fawcett, R.G. (1964). Effect of certain conditions on yield of crop plants. *Nature* 204:858-860.

Fischer, R.A. (1984). Wheat. In W.H. Smith and S.J. Barta (eds.). *Symposium on potential productivity of field crops under different environments.* IRRI, Los Baños, Philippines, 129-153.

Fischer, R.A. and Kertesz, Z. (1976). Harvest index in spaced populations and grain weight in microplots as indicators of yielding ability in spring wheat. *Crop Science* 16:56-59.

Fischer, R.A. and Miles, R.E. (1973). The role of spatial pattern in the competition between crop plants and weeds. A theoretical analysis. *Mathematical Biosciences* 18:335-350.

Flemming, G.F., Young, F.L., and Ogg, A.C. (1988). Competitive relationships among winter wheat (*Triticum aestivum*), jointed goatgrass (*Aegilop scylindrica*), and downy brome (*Bromus tectorum*). *Weed Science* 36:479-486.

Gallez, L.M., Mockel, F.E., Cantamutto, M.A., Gullace, G.D., and Vallati, A.R. (1986). Densidad de siembra y separación entre hileras: Su influencia sobre el rendimiento de trigo en la región pampeana semiárida. *Actas I Congreso Nacional de Trigo (Pergamino, Argentina)* 3:167-177.

Harper, F. (1983). *Principles of arable crop production.* Granada Publishing, London.

Harper, J.L. (1977). *Population biology of plants.* Academic Press, London.

Holliday, R. (1960). Plant population and crop yield. *Field Crop Abstracts* 13(3):159-167.

Holliday, R. (1963). The effect of row width on the yield of cereals. *Field Crop Abstracts* 16:71-81.

Kemp, D.R., Auld, B.A., and Medd, R.W. (1983). Does optimizing plant arrangements reduce interference or improve the utilization of space? *Agricultural Systems* 12:31-36.

Kira, T., Ogawa, H., and Shinozaki, K. (1953). Intraspecific competition among higher plants. 1. Competition-density-yield inter-relationships in regularly dispersed populations. *Journal of the Institute Polytech, Osaka City University* 4:1-16.

Kirby, E.J.M. (1969). The effect of sowing date and plant density on barley. *Annals of Applied Biology* 63:513-521.

Kirby, E.J.M. and Faris, D.G. (1972). The effects of plant density on tiller growth and morphology in barley. *Journal of Agricultural Science, Cambridge* 78:281-288.

Lafond, G.P., Domitruk, D., Bailey, K.L., and Derksen, D. (1996). Effects of row spacing, seeding rate and seed-placed phosphorus on wheat and barley in the Canadian prairies. *Better Crops with Plant Food* 80(4):20-22.

Lerner, S.E. and Cerri, A.M. (1990). Generación de macollos, espiguillas y flores en trigo (*Triticum aestivum* L.). Efectos de la densidad de siembra. *Actas II Congreso Nacional de Trigo (Pergamino, Argentina)* 1:59-69.

Lerner, S.E and Mac Maney, M. (1986). Densidad de siembra y fertilización nitrogenada en trigo. III Cultivares Leones INTA, Norkinpan 70 y Victoria INTA. *Actas I Congreso Nacional de Trigo (Pergamino, Argentina)* 3:47-65.

Lerner, S.E. and Satorre, E.H. (1990). Aplicación de un diseño experimental sistemático al estudio de la respuesta a la densidad de cultivares de trigo. *Actas II Congreso Nacional de Trigo (Pergamino, Argentina)* 1:44-50.

Malik, M.A., Haroon-ur-rasheed, M., and Razzaq, A. (1996). Row spacing study in two wheat varieties under rainfed conditions. *Sarhad Journal of Agriculture* 12(1):31-36.

Martin, M.P.L.D. and Field, R.J. (1987). Competition between vegetative plants of wild oat (*A. fatua*) and wheat (*Triticum aestivum* L.). *Weed Research* 27: 119-124.

McLeod, J.G., Campbell, C.A., Gan, Y., Dyck, F.B., and Vera, C.L. (1996). Seeding depth, rate and row spacing for winter wheat grown on stubble and chemical fallow in the semiarid prairies. *Canadian Journal of Plant Science* 76(2): 207-214.

Mead, R. (1979). Competition experiments. *Biometrics* 35:41-54.

Medd, R. W., Auld, B.A., Kemp, D.R., and Murison, R.D. (1985). The influence of wheat density and spatial arrangement on annual ryegrass, *Lolium rigidum,* competition. *Australian Journal of Agricultural Research* 36:361-371.

Nerson, H. (1980). Effects of population density and number of ears on wheat yield and its components. *Field Crops Research* 3:225-234.

Panda, S.C., Pattanaik, A., Rath, B.S., Tripathy, R.K., and Behera, B. (1996). Response of wheat (*Triticum aestivum*) to crop geometry and weed management. *Indian Journal of Agronomy* 41(4):553-557.

Pant, M.M. (1979). Dependence of plant yield on density and planting pattern. *Annals of Botany* 44:513-516.

Panwar, R.S., Malik, R.K., Balyan, R.S., and Singh, D.P. (1995). Effect of isoproturon, sowing method and seed rate on weeds and yield of wheat (*Triticum aestivum*). *Indian Journal of Agricultural Sciences* 65(2):109-111.

Peltonen, J. and Peltonen-Sainio, P. (1997). Breaking uniculm growth habit of spring cereals at high latitudes by crop management. II. Tillering, grain yield and yield components. *Journal of Agronomy and Crop Science* 178(2):87-95.

Popa, M. (1995). Results on the formation of grain yields under the influence of agronomic factors. *Probleme de Agrofitotehnie Teoretica si Aplicata* 17(1): 57-67.

Puckridge, D.W. and Donald, C.M. (1967). Competition among wheat plants sown at a wide range of densities. *Australian Journal of Agricultural Research* 18:193-211.

Rana, D.S., Ganga, S., and Pachauri, D.K. (1995). Response of wheat to seeding rates and row spacing under dryland conditions. *Annals of Agricultural Research* 16:339-342.

Ratkowsky, D.A. (1983). *Non linear regression modelling. A unified practical approach.* Marcel Dekker, New York and Basel.

Ross, M.A. and Harper, J.L. (1972). Occupation of biological space during seedling establishment. *Journal of Ecology* 60:77-88.

Salazar, G.M., Moreno, R.O., Salazar, G.R., and Carrillo, M.L. (1996). Wheat production as affected by seeding rate × fertilization interaction. *Cereal Research Communications* 24(2):231-237.

Sánchez, R.A., Casal, J.J., Ballaré, C.L., and Scopel, A.L. (1993). Plant responses to canopy density mediated by photomorphogenic processes. In Buxton, D.R., Shibles, R., Forsberg, R.A., Blad, B.L., Asay, K.H., Paulsen, G.M., and Wilson, R.F. (Editors). *International Crop Science I,* 779-786. Crop Science Society of America, Inc., Madison, WI.

Sarandon, S.J. and Chidichimo, H.O. (1985). Efecto de la densidad de siembra sobre la acumulación y redistribución del nitrógeno en tres cultivares de *T. aestivum* L. *Revista de la Facultad de Agronomía (La Plata)* 61-62:105-122.

Satorre, E.H. (1988). The competitive ability of spring cereals. PhD thesis, University of Reading, Reading, UK.

Satorre, E.H. and Arias, S.P. (1990). Competencia entre trigo (*Triticum aestivum*) y malezas. III. El efecto de la densidad del cultivo y la maleza. *Actas II Congreso Nacional de Trigo (Pergamino, Argentina)* 4:1-10.

Satorre, E.H. and Ghersa, C.M. (1987). Relationship between canopy structure and weed biomass. *Field Crops Research* 17:37-43.

Savin, R. and Slafer, G.A. (1991). Shading effects on yield of an Argentinian wheat cultivar. *Journal of Agricultural Sciences, Cambridge* 116:1-7.

Scott, W.R., Dougherty, C.T., and Langer, H.M. (1977). Development and yield components of high yielding wheat crops. *New Zealand Journal of Agricultural Research* 20:205-212.

Senigagliesi, C.A., García, R., Meira, S., Rivero de Galetto, M., and Stornini, M.T. (1986). Producción de trigo en el área maicera agrícola-ganadera de influencia de la EEA INTA Pergamino. *Actas I Congreso Nacional de Trigo (Pergamino, Argentina)* 3:81-105.

Shinozaki, K. and Kira, T. (1956). Intraspecific competition among higher plants. VII. Logistic Theory of the C-D effect. *Journal of the Institute Politech. Osaka City University* 7:35-72.

Snaydon, R.W. (1984). Plant demography in an agricultural context. In R. Dirzo and J. Sarukhan (Eds.). *Perspectives on plant population biology,* 389-408. Sinauer Associates Inc., Sunderland, MA.

Solie, J.B., Solomon Jr., S.G., Self, K.P., Peeper, T.F., and Koscelny, J.A. (1991). Reduced row spacing for improved wheat yields in weed-free and weed-infested fields. *Transactions of the ASAE* 34(4):1654-1660.

Syme, J.R. (1972). Single plant characters as a measure of field plot performance of wheat cultivars. *Australian Journal of Agricultural Research* 23:753-760.

Thakur, S.S., Pandey, I.B., Singh, S.J., and Mishra, S.S. (1996). Effect of seed rate and row spacing on late sown wheat in alluvial calcareous soil. *Journal of Research, Birsa Agricultural University* 8(2):123-125.

Verona, C.A., Loffer, C.M., and Fernández, O.N. (1980). Efecto de la densidad de plantas sobre el rendimiento y la distribución de N en *T. durum* Def. *Revista de Investigaciones Agropecuarias, INTA* 15(1):75-95.

Wang Zhen Lin, Fu Jin Min, He Ming Rong, Yin Yan Ping, and Cao Hong Min (1997). Planting density effects on assimilation and partitioning of photosynthates during grain filling in the late sown wheat. *Photosynthetica* 33(2): 199-204.

Willey, R.W. and Heath, S.B. (1969). The quantitative relationships between plant population and crop yield. *Advances in Agronomy* 21:281-321.

Willey, R.W. and Holliday, R. (1971). Plant population, shading and thinning studies in wheat. *Journal of Agricultural Science, Cambridge* 77:453-461.

Yu, Z., Van Sandford, D.A., and Egli, D.B. (1988). The effects of population density on floret initiation, development and abortion in winter wheat. *Annals of Botany* 62:295-302.

Chapter 8

Wheat Yield As Affected by Weeds

Robert J. Froud-Williams

INTRODUCTION

In terms of the area occupied and grain yield produced, on a global basis wheat is the most important field crop, accounting for 222 million ha and the production of 564 million tons. Although second to maize in area cropped in the United States, it is the most important field crop in Europe, Canada, and Australia (Oerke et al., 1994). In the United Kingdom wheat occupies the greatest cultivated area (2.0 million ha) despite recent reductions following the introduction of set-aside.

More herbicides are registered for use in cereals than any other crop and world crop usage is dominated by small-grain cereals. In the United Kingdom a greater amount of active ingredient is applied to winter wheat than any other field crop, with an average of 2.32 products applied per ha (Orson, 1987).

WEEDS OF WHEAT

Weed flora composition of wheat is a reflection of season of planting, cultural and chemical control, geographic location, and soil type. Thus winter wheat will be dominated by autumn-germinating species whereas spring wheat will favor summer annuals. Undoubtedly the characteristic broad-leaved flora of cereals has declined markedly following the discovery of selective growth regulator type herbicides (aryloxy alkanoic acids), as indicated by Andreasen, Stryhn, and Streibig (1996). In Europe the greater emphasis on early-sown winter wheat established by minimal cultivation techniques has favored infestation by annual and perennial grass weeds (Froud-Williams, 1988). In North America the development of *Setaria viridis* in continuous wheat and *Hordeum jubatum* in the absence of tillage have been documented by Hume, Tessier, and Dyck (1991). Infestation of continuous winter wheat by *Bromus tectorum* has been reported by Blackshaw et al. (1994), who also noted fewer weeds in rotational cropping than continuous winter wheat. Conversely, greater weed infestations have been observed in zero-till than either minimum or conventional tillage (Blackshaw et al., 1994). These observations have been confirmed by Gill and Arshad (1995), although Derksen et al. (1993) failed to detect an increased association between annual grass weeds and zero-tillage. Tillage factors affecting weed flora composition are reviewed by Moyer et al. (1994).

Increased reliance on the substituted phenylureas isoproturon and chlorotoluron in the United Kingdom has led to increased herbicide resistance in *Alopecurus myosuroides,* while resistance to aryloxy-phenoxypropionates and cyclohexanediones has been reported for *Lolium* spp. and *Avena* spp. in Australia, North America, and the United Kingdom (Heap, 1997).

Major weeds of wheat reported by Oerke et al. (1994) include *Avena* spp., *Elytrigia repens, Cirsium arvense,* and *Convolvulus arvensis.* Although *Avena* spp. are of widespread distribution, *Setaria* spp. and *Bromus tectorum* are particularly problematic in North America while *Alopecurus myosuroides* is largely confined to northwest Europe. *Apera spica-venti* occurs mainly in eastern Europe and *Bromus diandrus* is especially prevalent in southern Europe. Major weeds of wheat in Australia are *Bromus diandrus, Lolium rigidum,* and *Phalaris paradoxa. Phalaris minor* is a major problem of wheat in India. Other weeds associated with spring-sown crops include *Polygonum* spp., *Chenopodium album, Stellaria media*, and *Galeopsis tetrahit* (Salonen, 1992a).

Surveys of weeds of wheat in the United Kingdom have identified *Avena* spp., *Elytrigia repens, Alopecurus myosuroides, Poa* spp., and *Bromus sterilis* as the predominant grass weeds and *Galium aparine, Viola arvensis, Stellaria media, Myosotis arvensis, Lamium purpureum, Veronica persica*, and *Matricaria* spp. as the dominant annual broad-leaved weeds of winter wheat (Froud-Williams and Chancellor, 1982; Chancellor and Froud-Williams, 1984; Whitehead and Wright, 1989). In southern Spain Saavedra et al. (1989) report that the most frequent grass weeds are *Avena sterilis, Lolium rigidum, Phalaris* spp., *Cynodon dactylon,* and *Bromus diandrus.*

YIELD LOSSES

Globally, weeds account for substantial yield losses in wheat (9.8 percent) based on estimates reported by Cramer (1967), but more recent estimates suggest even greater yield reductions (13.1 percent) due to the introduction of less competitive varieties and greater use of fertilizers and plant growth regulators (Oerke et al., 1994).

In Canada yield loss of 10 percent is attributed to competition with *Avena fatua* equivalent to 360,000 t in Alberta alone (Nalewaja, 1977). In Saskatchewan losses attributed to *Setaria viridis* were 7.8 percent. Nalewaja (1977) estimated loss of wheat and barley at 3.5 million t for Canada as a whole, indicating losses in excess of 20 percent in the prairie provinces.

In the United States *Bromus* spp. are the major cause of yield loss and in trials conducted between 1983 and 1986 in Oklahoma, yield losses ranged from 46.3 to 48.7 percent (Ratliff and Peeper, 1987). Densities of 24, 40 and 65 *B. tectorum* plants m^{-2} reduced winter wheat yields by 10, 15 and 20 percent respectively (Stahlman and Miller, 1990). Losses incurred by *Avena* spp. were estimated at 2.7 million t (Nalewaja, 1977), while for South America and Australia losses due to wild oat competition were estimated at 15 percent. Similarly, Martin, McMillan, and Cook (1988) indicate that based on a yield of 2.47 t ha^{-1}, losses due to weeds in New South Wales (1983-1985) were calculated as 0.37 t ha^{-1} (equivalent to 15 percent).

In Northern Europe average losses due to *Avena* were estimated at 3 percent but reductions as great as 12 percent were observed. Generally, control of broad-leaved weeds (Aamisepp, 1985; Davies, 1988) has resulted in lower yield responses than removal of grass weeds (c. 10 percent). Nonetheless, in the United Kingdom Wilson and Cussans (1983) report yield reductions of 11.3 percent from broad-leaved weeds and 24.9 percent for grass weeds, while yield responses of 45 to 120 percent were observed following control of *Alopecurus myosuroides* (Clarke, 1987). Likewise yield losses of 67 to 85 percent and 34 to 71 percent have been reported for low- and high-density winter wheat crops over a range of *B. sterilis* infestations (Cousens, Pollard, and Danner, 1985). Unfortunately those weeds that are considered most competitive (namely annual grasses) are also the most expensive to control (Keen, 1991).

Infestation of winter wheat by *Apera spica-venti* in Poland incurred grain yield losses of 60 percent as a consequence of reduced ear numbers, ear length, and grain weight (Rola and Rola, 1983). Regression analysis indicated a linear relationship between panicle number and yield penalty, such that for every increase of 5 panicles m^{-2} a 0.4 percent yield loss was incurred, enabling economic thresholds of 10 to 20 m^{-2} to be calculated.

Afentouli and Eleftherohorinos (1996) indicated that although *Phalaris minor* and *Phalaris brachystachys* at a density of 76 plants m^{-2} did not affect yield, densities of 150 and 304 plants m^{-2} reduced yield by 23 to 28 percent and 36 to 39 percent respectively when present until harvest.

In India, Mehra and Gill (1988) reported that *Phalaris minor* at 50 and 250 plants m^{-2} reduced yield by 8 and 44 percent respectively. Likewise Balyan and Malik (1989) report that 150 plants m^{-2} reduced yield by 30 percent.

COMPETITION

Surprisingly little information has been published concerning the competitive effects of weeds in wheat relative to the other major crops, maize, soybean, cotton, rice, and barley. Selected references published prior to 1978 on the effects of weeds on wheat are included in a review of weed-crop competition by Zimdahl (1980). Unlike other field crops, information on the critical period of competition (weed-free maintenance period) is generally lacking with but few exceptions (Rydrych, 1974; Welsh et al., in press). Nonetheless Wilson and Cussans (1978) have demonstrated that yields were influenced to a greater extent by date of weed removal than herbicide efficacy. Thus, for example, control of *Avena fatua* in winter wheat with barban at the three-leaf stage provided a greater yield advantage than application of benzoylprop-ethyl in the spring. Greatest yield response was attained from a sequential application of each herbicide applied at half recommended rate. Similar results were obtained following autumn removal of *A. fatua* and *Alopecurus myosuroides* with difenzoquat and clofop-isobutyl respectively despite incomplete weed emergence (Wilson, 1979). No further yield benefit resulted from subsequent spring treatment and yields were not significantly different following half-rate application.

Likewise, although overall weed control was less effective following autumn application of herbicides, yield response from broad-leaved weed control was greater for autumn than spring treatment (Wilson, 1980). Conversely, in a further series of trials there was no yield advantage in herbicide application at the three-leaf stage (GS 1.3) in November compared with the pseudostem stage (GS 3.0) in late April, partly attributed to severe winter weather (Wilson, 1982). Thus it was concluded that autumn removal of broad-leaved weeds is not always necessary for maximum yield response provided that they are removed prior to crop jointing.

In a comparison of twelve winter wheat trials conducted during harvest years 1980 and 1981 significant yield benefits occurred in five of the trials, but in no instance did autumn treatments result in greater yield than spring treatments (Orson, 1982). This led Orson to conclude that spring removal prior to pseudostem erect was at least as effective as autumn control in the majority of trials evaluated.

Subsequently Wilson, Thornton, and Lutman (1985) indicated that the response to autumn removal of broad-leaved weeds was limited (<1 t ha^{-1}) with the exception of one trial where control of *Galium aparine* resulted in a yield increase of 4 t ha^{-1}. However, in contrast to broad-leaved weed control, failure to remove *A. myosuroides* in autumn led to substantial yield penalties often in excess of 4 t ha^{-1}. Previously Wilson (1980) had demonstrated that for mixed infestations of broad-leaved weeds and *A. myos-*

uroides, autumn treatment resulted in greatest yield benefit provided that *A. myosuroides* was adequately controlled. Likewise, Orson (1980), in a comparison of eleven winter wheat trials, concluded that significant yield responses were obtained from autumn removal of *Stellaria media,* although significant responses only occurred in three of the trials.

Simulation models of competition between annual broad-leaved weeds and winter wheat in the Netherlands indicate that spring-emerging weeds have negligible effects on yield whereas autumn-emerging cohorts could exact yield losses as great as 20 percent (Lotz, Kropff, and Groeneveld, 1990).

Competition Indices

The relationship between crop yield loss and weed density commonly conforms to a rectangular hyperbola such that yield loss per individual weed plant decreases as weed density increases and approaches an upper asymptote at high weed density. Examples of this relationship have been demonstrated between wheat and *Avena fatua* (Cousens et al., 1987; Martin, Cullis, and McNamara, 1987), *Bromus diandrus* (Gill, Poole, and Holmes, 1987), *Bromus sterilis* (Cousens et al., 1988), *Bromus tectorum* (Stahlman and Miller, 1990; Blackshaw, 1993), *Matricaria perforata* (Douglas et al., 1991), *Galium aparine* (Zanin, Berti, and Toniolo, 1993), *Phalaris* spp. (Afentouli and Eleftherohorinos, 1996), and *Veronica hederifolia* (Angonin, Caussanel, and Meynard, 1996).

The decision to control weeds will depend in the short term on whether a financial benefit will accrue. Below a specific density a yield benefit from weed removal may not be detectable. The economic threshold is the density at which the cost of an herbicide and its application is just balanced by the financial benefits of yield response. In practice a longer-term approach is required to account for the implications of further seed replenishment on future weed infestations and is referred to as the economic optimum threshold. Models enabling the determination of various thresholds have been constructed for the annual grass weeds *Avena fatua* (Cousens et al., 1986) and *Alopecurus myosuroides* (Doyle, Cousens, and Moss, 1986). For *A. fatua* an economic optimum threshold of two to three seedlings m^{-2} was calculated for winter wheat treated with difenzoquat and for *A. myosuroides* 7.5 seedlings m^{-2} if treated early postemergence with chlorotoluron. The values are considerably lower than the respective single-year economic thresholds of 8 to 12 and 30 to 50 m^{-2}. However, in practice an element of risk should be accepted to allow for impaired herbicide efficacy or atypical emergence resulting in thresholds of 1 and 5 m^{-2} respectively. Similarly, assuming a grain price of £100^{-1}, an herbicide cost including application of £40 ha^{-1}, and 100 percent herbicide efficacy, an economic threshold of 7.1 plants m^{-2} was derived for *Bromus sterilis* in winter wheat (Cousens et al., 1988). Such a threshold resulted in predicted yield reductions of 0.32 to 0.45 t ha^{-1}.

In Germany Gerowitt and Heitefuss (1990) observed fixed threshold values of 20 to 30m^{-2} for grass weeds including *A. myosuroides* but excluding *A. fatua.* For broad-leaved weeds values were higher (40-50 m^{-2}) with the exceptions of *Galium aparine* (0.1-0.5 m^{-2}) and *Fallopia convolvulus* (2.0 m^{-2}). Likewise, in Italy, Zanin, Berti, and Toniolo (1993) estimated economic thresholds for *Avena sterilis* ssp. *ludoviciana* in winter wheat of 7 to 12 m^{-2} depending on choice of herbicide. For *Alopecurus myosuroides* and *Lolium multiflorum* thresholds varied between 25 and 35 m^{-2}, for *Bromus sterilis* <40 m^{-2}, whereas for *Galium aparine* it was only 2m^{-2}. However, thresholds were greatly influenced by nitrogen availability, such that in its absence thresholds for both *B. sterilis* and *G. aparine* were negligible.

In reality no single threshold value can be applied, for its determination will depend on choice of herbicide, differing in cost and efficacy (Cousens et al., 1988; Pannell, 1994) while thresholds will vary depending on the potential yield of the crop (Streibig et al., 1989; Stahlman and Miller, 1990). Nor do thresholds take account of the fact that weeds are often aggregated and not uniformly distributed (Thornton et al., 1990). In addition they tend to occur as mixed infestations (Hume, 1989; Pannell and Gill, 1994). Furthermore, allowance needs to be made for the fact that individual populations will differ in vigor and environmental factors will vary between years (Firbank et al., 1990; Melander, 1994; Angonin, Caussanel, and Meynard, 1996). Nonetheless, weed thresholds may be impracticable, for yield loss may occur at relatively low density, with, for example, infestations of *A. fatua* resulting in 1 percent yield loss per each individual plant m^{-2} (Wilson, Cousens, and Wright, 1990).

Wilson (1986) demonstrated that in 42 winter wheat trials yield responses to broad-leaved weed control were influenced more by weed species present than by density. In practice, broad-leaved weeds tend to occur in mixed infestations, each species of differing competitive ability, thus precluding a direct relationship between yield response and removal of individual species. A system of crop equivalent values or competitive index would enable competition from mixed infestations to be integrated, yield loss predicted, and thresholds applied. However, as weed density increases, so too intraspecific competition increases, reducing the inherent interspecific competition between crop and weed. Thus as weed density increases, yield loss per plant m^{-2} reduces to a greater extent for the more competitive species.

Wilson and Wright (1990) investigated the growth and competitiveness of twelve annual weed species at a wide range of densities in crops of winter wheat at two sites of similar soil type and location in subsequent years. Using the hyperbolic yield density relationship they were able to derive the percentage yield loss per weed m^{-2} at which it would be uneconomic to apply herbicides. Thus assuming an arbitrary yield loss of 2 percent as acceptable it was demonstrated uneconomic to control *A. fatua* infestations $<0.5\ m^{-2}$ and *Veronica hederifolia* $<39\ m^{-2}$. Observed densities at which a 2 percent yield loss would have been incurred were greater than predicted for those species that senesce early in the life of the crop, e.g., *Viola arvensis* spp. ($109\ m^{-2}$) but lower than those which persist, e.g., *Matricaria* spp. ($1.7\ m^{-2}$).

Black and Dyson (1993) developed an economic threshold model to calculate the benefits of early weed removal based on converting weed densities to weed units as a function of weed competitiveness. They provided evidence for a linear relationship between weed density after herbicide application and grain yield in contrast to the curvilinear relationship between unsprayed weed density and crop yield loss, attributable to the reduced competitiveness of survivors.

Hume (1993) developed multiple regression equations to describe the relationship between yield loss of spring wheat and density and shoot weight of multispecies weed infestations dominated by *Setaria viridis*. The inclusion of crop density as a variable improved the prediction of yield loss, while rainfall and thermal time were significantly related to crop yield loss. Hyperbolic and sigmoidal equations were less efficient at describing the data than were multiple regression equations. Likewise Vitta and Fernandez-Quintanilla (1996) observed that simple regression models of crop yield loss based on weed leaf area were as accurate as those based on weed density. Subsequently, Debaeke et al. (1997) have developed a simulation model based on a limited number of parameters including light interception and competition for water and soil nitrogen of two or more species in mixture and have demonstrated a reasonably good prediction between wheat and cultivated oats.

Onset of Competition

Time of weed emergence relative to that of the crop will have important implications for competitiveness and yield reduction. Thus, for example, early-emerging cohorts of *Avena fatua* (Kirkland, 1993) and *Avena sterilis* ssp. *ludoviciana* (Gonzalez-Andujar and Fernandez-Quintanilla, 1991) were demonstrated to be most competitive and most fecund respectively. Likewise O'Donovan et al. (1985) have demonstrated a significant relationship between yield loss of spring wheat and relative time of emergence of *A. fatua*. For a given density, percentage yield loss increased the earlier that wild oats emerged relative to the crop and gradually diminished the later emergence occurred. Yield losses derived from regression analyses of pooled data indicate that for every day *A. fatua* emerged before the crop, yield loss increased by 3 percent. Similarly, yield loss decreased by a comparable amount for every day emergence occurred subsequent to the crop for periods of six days either side of crop emergence.

The influence of time of emergence of *A. fatua* relative to wheat was investigated by Martin and Field (1987, 1988), using boxes that allowed separation of root and shoot competition. Relative yield total for mixtures of wild oat and wheat under different forms of competition and different sowing times was close to unity, indicating that the two species competed fully for limiting resources. Wild oat was more competitive than wheat when the two species were sown simultaneously, largely as a consequence of its greater root competitive ability, shoot competitive ability being similar. When *A. fatua* was sown three or six weeks later than wheat, wheat was more competitive and panicle production was prevented. Hence, to prevent replenishment of the seedbank it was necessary to control cohorts emerging within the first three weeks of drilling. Satorre and Snaydon (1992) indicated that competition for below-ground resources, particularly nitrogen, was more severe than aerial competition. While cultivars of wheat, barley, and oats differed little in their below-ground competitive ability, differences in shoot-competitive ability differed markedly. Previous studies have suggested a greater rate of root development by *A. fatua* relative to spring wheat, but this is refuted by Bingham (1995).

Resource Limitation

Henson and Jordan (1982) observed that competition between *A. fatua* and wheat reduced the efficacy of nitrate uptake by the crop and reduced grain N content. Similar observations are reported by Gonzalez Ponce (1988). That weeds may compete more effectively for nitrogen is demonstrated by Carlson and Hill (1985a), who showed a reduction of wheat yield with increased fertilization, whereas panicle density of wild oat actually increased. Likewise, Anderson (1991) indicated that *Bromus tectorum* was more responsive to nitrogen than wheat, but that time of application affected responsivity. *Galium aparine* was less competitive than the crop at low levels of nitrogen but became more competitive as nitrogen level was increased (Baylis and Watkinson, 1991). Conversely, Valenti and

Wicks (1992) reported that increased nitrogen use in winter wheat decreased growth of *Echinochloa crus-galli* and *Setaria viridis* as a consequence of increased light interception by the crop.

While nitrogen did not significantly change the competitive ranking of individual cereal cultivars, its application alleviated the effects of competition on *A. fatua* to some extent (Satorre and Snaydon, 1992). Likewise application of nitrogen partially alleviated the effects of root competition with *Alopecurus myosuroides,* but did not affect shoot competition (Exley and Snaydon, 1992).

Grundy, Froud-Williams, and Boatman (1993) reported that increased nitrogen up to 160 kg N ha^{-1} reduced total aboveground weed biomass, albeit some species, e.g., *Viola arvensis,* responded positively. Similarly, Jørnsgård et al. (1996) reported that *Stellaria media, Lamium* spp., and *Veronica* spp. actually had lower nitrogen optima than the crop, suggesting that reduced nitrogen rates could actually increase weed competitiveness for those species. Differences in responsivity to nitrogen on net photosynthetic rate of *Phalaris minor, Chenopodium album*, and *Sinapis arvensis* indicated that at high nitrogen (120 kg N ha^{-1}) *P. minor* was less competitive than winter wheat, whereas *S. arvensis* was more competitive. *C. album* was the better competitor at both low (20 kg N ha^{-1}) and high nitrogen levels (Iqbal and Wright, 1997).

Evidence that competition between wheat and *Veronica hederifolia* was primarily for nitrogen is apparent from the fact that crop nitrogen content was reduced in competition with *V. hederifolia* but that delayed application of nitrogen until the onset of weed senescence alleviated this effect (Angonin, Caussanel, and Meynard, 1996).

Alleviation of Competition

Increased seed rate may be used to enhance crop competitiveness (Moss, 1985a; Carlson and Hill, 1985b; Martin, Cullis, and McNamara, 1987; Cudney et al., 1989; Grundy, Froud-Williams, and Boatman, 1993). Thus, for example, Moss (1985a) demonstrated that increased crop density resulted in an increased yield response as a consequence of suppression of *Alopecurus myosuroides* as indicated by reduced blackgrass inflorescence number. Lemerle et al. (1996) demonstrated that a doubling of seed rate reduced infestation by *Lolium rigidum.* However, no such benefits were observed by Appleby and Brewster (1992) as a consequence of increased crop lodging. Nonetheless, other studies have shown benefits from increased crop seed rate (Samuel and Guest, 1990; Whiting and Richards, 1990; Koscelny et al., 1990, 1991; Doll, Holm, and Sogard, 1995; Wilson et al., 1995).

Thus Wilson et al. (1995) demonstrated that reducing crop density by half (from 200 to 100 m^{-2}) increased June biomass of *Viola arvensis* by 74 percent and that of *Papaver rhoeas* by 63 percent. Weed biomass was more than doubled when the crop was reduced from 200 to 40 plants m^{-2}. Crop yield losses were greater at low crop populations.

Likewise, manipulation of spatial arrangement may also be advantageous (Kemp, Auld, and Medd, 1983; Medd et al., 1985). Thus Koscelny et al. (1990) report that decreased row spacing increased yield of *Bromus secalinus* infested wheat in six out of ten experiments. Greater benefits were achieved for early-sown than later-sown crops (Koscelny et al., 1991).

A number of authors have reported greater yield loss from earlier drilling in the presence of *Alopecurus myosuroides* (Moss, 1985b; Cosser et al., 1997), *Alopecurus myosuroides*, and *Apera spica-venti* (Melander, 1995), *Phalaris minor* (Singh et al., 1995).

Choice of cultivar too may offset weed competitiveness (Challaiah et al., 1986; Gonzalez Ponce, 1988; Ramsel and Wicks, 1988; Richards and Davies, 1991; Balyan et al., 1991; Satorre and Snaydon, 1992; Valenti and Wicks, 1992; Blackshaw, 1994; Lemerle et al., 1996; Froud-Williams, 1997).

Balyan et al. (1991) observed that differences in competitive ability between wheat cultivars were related more to height and dry matter than tillering capacity. Other authors have reported greater competitive ability of traditional long-strawed than semidwarf cultivars (Challaiah et al., 1986; Wicks et al., 1986, 1994; De Lucas Bueno and Froud-Williams, 1994; Cosser et al., 1997). Thus greater competition from *Setaria viridis* (Blackshaw, Stobbe, and Sturko, 1981) and *Phalaris minor* (Gill and Mehra, 1981) occurred in semidwarf than conventional wheats.

JUSTIFICATION FOR WEED REMOVAL

Studies of the effects of weed removal on wheat yield have on occasion indicated noneconomic benefits from chemical control, although greater responses (>10 percent) have often been reported for grass-weed control, but then only at high levels of infestation (Snaydon, 1982). However, Zanin, Berti, and Giannini (1992) indicated that yield reductions in winter wheat resulting from weeds ranged from 23 to 30 percent, with a "break-even" yield loss calculated as between 1.4 and 15 percent and a probability of a positive net return from weed control always greater than 80 percent.

In an economic appraisal of the benefits of prophylactic weed control in wheat in New Zealand it was concluded that herbicide application would have been uneconomic in 24 percent of the instances analyzed (Bourdôt et al., 1996). Nonetheless, a comparison of 17 on-farm and two experiment station trials conducted in Oklahoma indicated that farmers neither expected a yield response from broad-leaved weed control in hard red winter wheat nor obtained a yield increase, with the exception of two sites (Scott and Peeper, 1994). Their primary objective was to have weed-free crops at harvest. Thus the objectives of weed control are not solely of increased yield, but of ease of harvesting, avoidance of grain contamination, and freedom from pest, pathogen, and toxicological problems. In addition the desire to avoid accumulation within the soil seedbank also deserves attention. Contamination of the harvested grain may adversely affect quality, possibly leading to crop rejection such that minimum purity standards are required, e.g., within the European Union less than one seed of *Avena fatua* per 500 g of grain.

A consequence of weeds present at crop maturity is to delay harvesting operations as a result of the greater throughput of extraneous plant material (matter other than grain), and if the weeds are still green they may also incur drying costs (Elliott, 1980). Thus Elliott has shown that competition with *Alopecurus myosuroides* has little effect on harvesting or grain drying as it senesces prior to crop maturity, but may account for 50 percent crop yield reduction. Conversely, species that are still photosynthetic at harvest, e.g. *Elytrigia repens* and *Galium aparine,* may not only reduce yield but also contribute significantly to harvesting costs through the increase of matter other than grain.

EFFECTS ON YIELD COMPONENTS AND GRAIN QUALITY

Relatively few investigations have considered the effects of weed competition on yield components and grain quality (Angonin, Caussanel, and Meynard, 1996; Cosser et al., 1997). Likewise, the phytotoxic effects of herbicides on quality has received scant attention (Grundy, Boatman, and Froud-Williams, 1996). Early competition reduces tiller number and hence ear number and possibly grains per ear, whereas late competition will affect grain filling and thousand-grain weight. Thus for example yield reductions from competition between wheat and *Lolium multiflorum* were primarily attributed to reduced tillering capacity rather than individual grain weight (Liebl and Worsham, 1987). Similar observations are reported for *Avena fatua* (Medd et al., 1985). Unusually, Cousens et al. (1988) report that for low densities of *Bromus sterilis* grain number per ear was more affected than ear number whereas the converse was observed at high weed densities. Grain weight increased in the presence of *B. sterilis.* The authors were unable to explain these results, but suggested that one explanation could be that competition occurred early in the life of the crop, such that following senescence of the weed, the only compensatory response was that of increased grain size. Likewise, Satorre and Snaydon (1992) observed that the greatest effect of root competition was on grains per ear, indicating that competition occurred between GS30 and GS70. Little effect on grain weight indicated little evidence of competition during later stages of development.

Further evidence that individual species differ in the onset and duration of their competitive effects is provided by Wright and Wilson (1992). Thus whereas early competition from *Alopecurus myosuroides* affects ear number and/or grain number, later competition from *Galium aparine* affects thousand-grain weight. Angonin, Caussanel, and Meynard (1996) demonstrated that competition from *Veronica hederifolia* occurs early in the crop cycle, resulting in increased tiller mortality with the consequence of reduced ear and grain number.

PHYTOTOXIC EFFECTS OF HERBICIDES

If removal of weed competition results in negligible yield benefit, any yield penalty arising from herbicide phytotoxicity is unlikely to be acceptable (Salonen, 1992b). Specific spray intervals (spray windows) are available during which individual herbicides may be applied with minimal risk of crop damage. Penalties from incorrect herbicide timing are difficult to quantify but experimental evidence suggests that incorrect application of broad-leaved herbicides may incur yield penalties of 5 to 10 percent or more (Tottman, 1977a, 1978). Comparatively few days are available for herbicide application during the "safe stage" in spring, while the introduction of new varieties and the trend toward earlier autumn drilling have contributed to the appearance of the first node detectable stage (GS31) somewhat earlier, application of herbicides beyond which may incur damage (Tottman, Appleyard, and Sylvester-Bradley, 1985). Too early an application of growth regulator-type herbicides such as mecoprop may result in ear deformities, whereas late applications of benzoic acids may actually cause yield reduction through incomplete grain filling (Tottman, 1977b, 1978; Ivany and Nass, 1984).

Variation in susceptibility between cultivars of winter wheat and the wild oat herbicides difenzoquat and diclofop-methyl have been reported by Tottman et al. (1982) and Tottman, Lupton, and Oliver (1984). High rates of difenzoquat damaged cultivars Bouquet, Sportsman, Hobbit, and Maris Huntsman but not Hustler (Tottman, Lupton, and Oliver, 1984). Treatment of the sensitive varieties during tillering stimulated excessive tiller production, resulting in more ears with fewer and smaller grains, whereas later applications reduced grain number and depressed yield. In contrast, early applications of diclofop-methyl were more damaging than later treatment and cv. Bouquet was the most tolerant cultivar evaluated.

Significant yield reductions were caused by treatment of cv. Mardler with diclofop-methyl alone or in sequence with ioxynil + bromoxynil + mecoprop (Askew and Scourey, 1982). Likewise Grundy, Boatman, and Froud-Williams (1996) report herbicide phytoxic effects of fluroxypyr + ioxynil + clopyralid on yield of cv. Tonic. Elsewhere, in Nebraska 2,4-D ester, 2,4-D amine + dicamba + metsulfuron-methyl and metsulfuron-methyl + 2,4-D reduced grain yields of winter wheat compared with hand-weeded treatments (Wicks, Martin, and Mahnken, 1995).

Soil-acting herbicides may also incur yield penalty if timed incorrectly. Thus O'Sullivan, Weiss, and Friesen (1985) observed that spring application of trifluralin was more damaging to spring wheat than autumn application, possibly due to elevated temperatures (Heath, McKercher, and Ashford, 1985). Likewise Olson and McKercher (1985) report greater trifluralin injury to spring wheat than either durum wheat or rye.

Blair et al. (1991) demonstrated an interaction between sowing depth and winter wheat cultivar in susceptibility to postemergence applications of isoproturon and chlorotoluron. For chlorotoluron-treated plants phytotoxicity was unaffected by depth of sowing, but cultivars Slejpner and Galahad were more sensitive than Avalon and Fenman. In contrast the ED_{50} for IPU was strongly correlated with sowing depth for both cv. Avalon and Slejpner. The greater the depth of sowing, the greater protection from injury.

Whiting and Davies (1990) reported that autumn application of diflufenican/isoproturon resulted in crop yield exceeding that of untreated winter wheat, whereas spring application of metsulfuron-methyl + mecoprop resulted in crop damage and yield loss. Conversely, Ivany, Nass, and Sanderson (1990) reported that autumn application of either 2,4-D or a 2,4-D mecoprop + dicamba mixture significantly reduced yields of winter wheat cv. Lennox compared with spring application. Likewise, yield losses were incurred by spring applications of dicamba but not MCPA/dicamba. Nonetheless, mechanical weed control may also reduce ear numbers, although individual grain weight may compensate for this (Wilson, Wright, and Butler, 1993).

REFERENCES

Aamisepp, A. (1985) Weed control in winter cereals according to need. Final report weeds and weed control. *26th Swedish Weed Conference,* Uppsala, Sweden, 1-18.

Afentouli, C.G. and Eleftherohorinos, I.G. (1996) Little seed canary grass (*Phalaris minor*) and short-spiked canary grass (*Phalaris brachystachys*) interference in wheat and barley. *Weed Sci.*, 44, 560-565.

Anderson, R.L. (1991) Timing of nitrogen application affects downy brome (*Bromus tectorum*) growth in winter wheat. *Weed Technol*, 5, 582-585.

Andreasen, C., Stryhn, H., and Streibig, J.C. (1996) Decline of the flora in Danish arable fields. *J. Appl. Ecol.*, 33, 619-626.

Angonin, C., Caussanel, J-P., and Meynard, J-M. (1996) Competition between winter wheat and *Veronica hederifolia:* Influence of weed density and the amount and timing of nitrogen application. *Weed Res.*, 36, 175-187.

Appleby, A.P. and Brewster, B.D. (1992) Seedling arrangement on winter wheat (*Triticum aestivum*) grain yield and interaction with Italian ryegrass (*Lolium multiflorum*). *Weed Technol.*, 6, 820-823.

Askew, M.F. and Scourey, L.R.K. (1982) The effect of some herbicides and herbicide sequences on the yield of two winter wheat cultivars in the absence of weeds. *Aspects of Applied Biology 1*, Broad-leaved weeds and their control in cereals, 211-217. Association of Applied Biologists, Warwick, UK.

Balyan, R.S. and Malik, R.K. (1989) Influence of nitrogen on competition of wild canary grass. *Pestology*, 13, 5-6.

Balyan, R.S., Malik, R.K., Panwar, R.S., and Singh, S. (1991) Competitive ability of winter wheat cultivars with wild oat. *Weed Sci.*, 39, 154-158.

Baylis, J.M. and Watkinson, A.R. (1991) The effect of reduced nitrogen fertilizer inputs on the competitive effect of cleavers (*Galium aparine*) on wheat (*Triticum aestivum*). *Proc. 1991 Brighton Crop Prot. Conf.—Weeds*, 129-134. British Crop Protection Council, Surrey, UK.

Bingham, I.J. (1995) A comparison of the dynamics of root growth and biomass partitioning in wild oat (*Avena fatua* L.) and spring wheat. *Weed Res.*, 35, 57-66.

Black, I.D. and Dyson, C.B. (1993) An economic threshold model for spraying herbicides in cereals. *Weed Res.*, 33, 279-290.

Blackshaw, R.E. (1993) Downy brome (*Bromus tectorum*) density and relative time of emergence affect interference in winter wheat (*Triticum aestivum*). *Weed Sci.*, 41, 555-556.

Blackshaw, R.E. (1994) Differential competitive ability of winter wheat cultivars against downy brome. *Agron. J.*, 86, 649-654.

Blackshaw, R.E., Larney, F.O., Lindwall, C.W., and Kozub, G.C. (1994) Crop rotation and tillage effects on weed populations on the semi-arid Canadian Prairies. *Weed Technol.*, 8, 231-237.

Blackshaw, R.E., Stobbe, E.H., and Sturko, A.R.W. (1981) Effect of seeding dates and densities of green foxtail (*Setaria viridis*) on the growth and productivity of spring wheat (*Triticum aestivum*), *Weed Sci.*, 29, 212-217.

Blair, A.M., Martin, T.D., Brain, P., and Cotterill, E.G. (1991) The interaction between planting depth of four winter wheat cultivars, *Alopecurus myosuroides* Huds. and *Bromus sterilis* L. and their susceptibility to post-emergence applications of isoproturon and chlorotoluron. *Weed Res.*, 31, 285-294.

Bourdôt, G.W., Saville, D.J., Hurrell, G.A., and Daly, M.J. (1996) Modelling the economics of herbicide treatment in wheat and barley using data on prevented grain yield losses. *Weed Res.*, 36, 449-460.

Carlson, H.L. and Hill, J.E. (1985a) Wild oat (*Avena fatua*) competition with spring wheat: Effects of nitrogen fertilization. *Weed Sci.*, 34, 29-33.

Carlson, H.L. and Hill, J.E. (1985b) Wild oat (*Avena fatua*) competition with spring wheat: Plant density effects. *Weed Sci.*, 33, 176-181.

Challaiah, Burnside, O.C., Wicks, G.A., and Johnson, V.A. (1986) Competition between winter wheat (*Triticum aestivum*) cultivars and downy brome (*Bromus tectorum*). *Weed Sci.*, 34, 689-693.

Chancellor, R.J. and Froud-Williams, R.J. (1984) A second survey of cereal weeds in central southern England. *Weed Res.*, 24, 29-36.

Clarke, J.H. (1987) Evaluation of herbicides for the control of *Alopecurus myosuroides* (blackgrass) in winter cereals. Summary of results of ADAS trials 1985 and 1986 harvest years. *Proc. 1987 Brighton Crop Prot. Conf.—Weeds,* 375-382. British Crop Protection Council, Surrey, UK.

Cosser, N.D., Gooding, M.J., Thompson, A.J., and Froud-Williams, R.J. (1997) Competitive ability and tolerance of organically grown wheat cultivars to natural weed infestations. *Ann. Appl. Biol.*, 130, 523-535.

Cousens, R., Brain, P., O'Donovan, J.T., and O'Sullivan, P.A. (1987) The use of biologically realistic equations to describe the effects of weed density and relative time of emergence on crop yield. *Weed Sci.*, 35, 720-725.

Cousens, R., Doyle, C.J., Wilson, B.J., and Cussans, G.W. (1986) Modelling the economics of controlling *Avena fatua* in winter wheat. *Pestic. Sci.*, 17, 1-12.

Cousens, R., Firbank, L.G., Mortimer, A.M., and Smith, R.G.R. (1988) Variability in the relationship between crop yield and weed density for winter wheat and *Bromus sterilis*. *J. Appl. Ecol.*, 25, 1033-1044.

Cousens, R., Pollard, F., and Denner, R.A.P. (1985) Competition between (*Bromus sterilis*) and winter cereals. *Aspects of Applied Biology, 9.* The biology and control of weeds in cereals, 67-74. Association of Applied Biologists, Warwick, UK.

Cramer, H.H. (1967) *Plant Protection and World Crop Production.* Bayer, AG, Leverkusen, Germany.

Cudney, D.W., Jordan, L.S., Holt, J.S., and Reints, J.S. (1989) Competitive interactions of wheat (*Triticum aestivum*) and wild oats (*Avena fatua*) grown at different densities. *Weed Sci.*, 37, 538-543.

Davies, D.H.K. (1988) Yield responses to the use of herbicides in cereal crops, East of Scotland trials 1979-1988. *Aspects of Applied Biology, 18,* Weed control in cereals and the impact of legislation on pesticide application, 47-56. Association of Applied Biologists, Warwick, UK.

De Lucas Bueno, C. and Froud-Williams, R.J. (1994) The role of varietal selection for enhanced crop competitiveness in winter wheat. *Aspects of Applied Biology* 40, 343-346. Association of Applied Biologists, Warwick, UK.

Debaeke, P., Caussanel, J.-P., Kiniry, J.R., Kafiz, B., and Mondragon, G. (1997) Modelling crop:weed interactions in wheat with ALMANAC. *Weed Res.,* 37, 325-341.

Derksen, D.A., Lafond, G.P., Thomas, A.G., Loeppky, H.A., and Swanton, C.J. (1993) Impact of agronomic practices on weed communities: Tillage systems. *Weed Sci.,* 41, 409-417.

Doll, H., Holm, V., and Sogaard, B. (1995) Effect of crop density on competition by wheat and barley with *Agrostemma githago* and other weeds. *Weed Res.,* 35, 391-396.

Douglas, D.W., Thomas, A.G., Peschken, D.P., Bowes, G.G., and Derksen, D.A. (1991) Effects of summer and winter annual scentless chamomile (*Matricaria perforata Mérat*) interference on spring wheat yield. *Can. J. Plant Sci.,* 71, 841-850.

Doyle, C.J., Cousens, R., and Moss, S.R. (1986) A model of the economics of controlling *Alopecurus myosuroides* Huds. in winter wheat. *Crop Prot.,* 5, 143-150.

Elliott, J.G. (1980) The economic significance of weeds in the harvesting of grain. *Proc. 1980 Brighton Crop Prot. Conf,—Weeds,* 787-797. British Crop Protection Council, Surrey, UK.

Exley, D.M. and Snaydon, R.W. (1992) Effects of nitrogen fertilizer and emergence date on root and shoot competition between wheat and blackgrass. *Weed Res.,* 32, 175-182.

Firbank, L.G., Cousens, R., Mortimer, A.M., and Smith, R.G.R. (1990) Effects of soil type on crop yield-weed density relationships between winter wheat and *Bromus sterilis. J. Appl. Ecol.,* 27, 308-318.

Froud-Williams, R.J. (1988) Changes in weed flora with different tillage and agronomic management systems. In *Weed Management: Ecological Approaches.* Eds. Altieri, M.A. and Liebman, M. CRC Press, Boca Raton, FL, 213-236.

Froud-Williams, R.J. (1997) Varietal selection for weed suppression. *Aspects of Applied Biology,* 50, 355-360.

Froud-Williams, R.J. and Chancellor, R.J. (1982) A survey of grass weeds in cereals in central southern England. *Weed Res.,* 22, 163-171.

Gerowitt, B. and Heitefuss, R. (1990) Weed economic thresholds in cereals in the Federal Republic of Germany. *Crop Prot.,* 9, 323-331.

Gill, G.S., Poole, M.S., and Holmes, J.E. (1987) Competition between wheat and brome grass in Western Australia. *Aus. J. Exp. Agric.,* 27, 291-294.

Gill, H.S. and Mehra, S.P. (1981) Growth and development of *Phalaris minor* Retz: *Chenopodium album* L. and *Melilotus indica* L. in wheat crop ecosystems. In *Proc. 8th Asian-Pacific Weed Sci. Soc. Conf.* 175. Punjab Agricultural University, Ludhiana, India.

Gill, K.S. and Arshad, M.A. (1995) Weed flora in the early growth period of spring crops under conventional, reduced and zero tillage systems on a clay soil in northern Alberta, Canada. *Soil and Till. Res.,* 33, 65-79.

González-Andujar, J.L. and Fernandez-Quintanilla, C. (1991) Modelling the population dynamics of *Avena sterilis* under dry-land cereal cropping systems. *J. Appl. Ecol.*, 28, 16-27.

González Ponce, R. (1988) Competition between *Avena sterilis* ssp. *macrocarpa* Mo. and cultivars of wheat. *Weed Res.*, 28, 303-307.

Grundy, A.C., Boatman, N.D., and Froud-Williams, R.J. (1996) Effects of herbicide and nitrogen fertilizer application on grain yield and quality of wheat and barley. *J. Agric. Sci.*, 126, 379-385.

Grundy, A.C., Froud-Williams, R.J., and Boatman, N.D. (1993) The use of cultivar, crop seed rate and nitrogen level for the suppression of weeds in winter wheat. *Proc. Brighton Crop Prot. Conf.—Weeds*, 997-1002. British Crop Protection Council, Surrey, UK.

Heap, I.M. (1997) The occurrence of herbicide resistant weeds worldwide. *Pestic. Sci.*, 51, 235-243.

Heath, M.C., McKercher, R.B., and Ashford, R. (1985) Influence of high soil temperature, ammonium ion and rapeseed residue on trifluralin phytotoxicity to wheat. *Can. J. Plant Sci.*, 65, 151-161.

Henson, J.F. and Jordan, L.S. (1982) Wild oat (*Avena fatua*) competition with wheat (*Triticum aestivum* and *T. turgidum durum*) for nitrate. *Weed Sci.*, 30, 297-300.

Hume, L. (1989) Yield losses in wheat due to weed communities dominated by green foxtail (*Setaria viridis* (L.) Beauv.): A multispecies approach. *Can. J. Plant Sci.*, 69, 521-529.

Hume, L. (1993) Development of equations for estimating yield losses caused by multi-species weed communities dominated by green foxtail (*Setaria viridis* (L.) Beauv.) *Can. J. Plant Sci.*, 73, 625-635.

Hume, L., Tessier, S., and Dyck, F.B. (1991) Tillage and rotation influences on weed community composition in wheat (*Triticum aestivum* L.) in south western Saskatchewan. *Can. J. Plant Sci.*, 71, 783-789.

Iqbal, J. and Wright, D. (1997) Effects of nitrogen supply on competition between wheat and three annual weed species. *Weed Res.*, 37, 391-400.

Ivany, J.A. and Nass, H.G. (1984) Effect of herbicides on seedling growth, head deformation and grain yield of spring wheat cultivars. *Can. J. Plant Sci.*, 64, 25-30.

Ivany, J.A., Nass, H.G., and Sanderson, J.B. (1990) Effect of time of application of herbicides on yield of three winter wheat cultivars. *Can. J. Plant Sci.*, 70, 605-609.

Jørnsgård, B., Rasmussen, K., Hill, J., and Christiansen, J.L. (1996) Influence of nitrogen on competition between cereals and their natural weed populations. *Weed Res.*, 36, 461-470.

Keen, B.W. (1991) Weed control in small grain cereals in a low profit situation. The short term view. *Proc. 1991 Brighton Crop Prot. Conf.—Weeds*, 113-120. British Crop Protection Council, Surrey, UK.

Kemp, D.R., Auld, B.A., and Medd, R.W. (1983) Does optimizing plant arrangement reduce interference or improve the utilization of space. *Agric. Sys.*, 12, 31-36.

Kirkland, K.J. (1993). Spring wheat (*Triticum aestivum*) growth and yield as influenced by duration of wild oat (*Avena fatua*) competition. *Weed Technol.*, 7, 890-893.

Koscelny, J.A., Peeper, T.F., Solie, J.B., and Solomon, S.G. (1990) Effect of wheat (*Triticum aestivum*) row spacing, seeding rate and cultivar on yield loss from cheat (*Rbomus secalinus*). *Weed Technol.*, 4, 487-492.

Koscelny, J.A., Peeper, T.F., Solie, J.B., and Solomon, S.G. (1991) Seeding date, seeding rate and row spacing affect wheat (*Triticum aestivum*) and cheat (*Bromus secalinus*). *Weed Technol.*, 5, 707-712.

Lemerle, D., Verbeek, B., Cousens, R.D., and Coombes, N.E. (1996) The potential for selecting wheat varieties strongly competitive against weeds. *Weed Res.*, 36, 505-513.

Liebl, R. and Worsham, A.D. (1987) Interference of Italian ryegrass (*Lolium multiflorum*) in wheat (*Triticum aestivum*). *Weed Sci.*, 35, 819-823.

Lotz, L.A.P., Kropff, M.J., and Groeneveld, R.M.W. (1990) Modelling weed competition and yield losses to study the effect of omission of herbicides in winter wheat. *Neth. J. Agric. Sci.*, 38, 711-717.

Martin, M.P.L.D. and Field, R.J. (1987) Competition between vegetative plants of wild oat (*Avena fatua* L.) and wheat (*Triticum aestivum* L.) *Weed Res.*, 27, 119-124.

Martin, M.P.L.D. and Field, R.J. (1988) Influence of time of emergence of wild oat on competition with wheat. *Weed Res.*, 28, 111-116.

Martin, R.J., Cullis, B.R., and McNamara, D.W. (1987) Prediction of wheat yield loss due to competition by wild oats (*Avena* spp). *Aust. J. Agric. Res.*, 38, 487-499.

Martin, R.J., McMillan, M.G., and Cook, J.B. (1988) Survey of farm management practices of the northern wheat belt of New South Wales. *Aust. J. Exptl. Agric.*, 28, 499-509.

Medd, R.W., Auld, B.A., Kemp, D.R., and Murison, R.D. (1985) The influence of wheat density and spatial arrangement on annual ryegrass (*Lolium rigidum*) competition. *Aust. J. Agric. Res.*, 36, 361-371.

Mehra, S.P. and Gill, H.S. (1988) Effect of temperature on germination of *Phalaris minor* Retz. and its competition in wheat. *Punjab Agric. Univ. J.*, 25, 529-533.

Melander, B. (1994) Modelling the effects of *Elymus repens* (L.) Gould competition on yield of cereals, peas and oilseed rape. *Weed Res.*, 34, 99-108.

Melander, B. (1995) Impact of drilling date on *Apera spica-venti* L. and *Alopecurus myosuroides* Huds. in winter cereals. *Weed Res.*, 35, 157-166.

Moss, S.R. (1985a) The influence of crop variety and seed rate on *Alopecurus myosuroides* competition in winter cereals. *Proc. Brighton Crop Prot. Conf.—Weeds.*, 701-708. British Crop Protection Council, Surrey, UK.

Moss, S.R. (1985b) The effect of drilling date, pre-drilling cultivations and herbicides on *Alopecurus myosuroides* (blackgrass) populations in winter cereals. *Aspects of Applied Biology,* 9, The biology and control of weeds in cereals. 31-39. Association of Applied Biologists, Warwick, UK.

Moyer, J.R., Roman, E.S., Lindwall, C.W., and Blackshaw, R.E. (1994) Weed management in conservation tillage systems for wheat production in North and South America. *Crop Prot.,* 13, 243-259.

Nalewaja, J.D. (1977) Wild oats: Global gloom. *Proc. Western Society of Weed Science,* 30, 21-32.

O'Donovan, J.T., de St. Remy, E.A., O'Sullivan, P.O., Dew, D.A., and Sharma, A.K. (1985) Influence of the relative time of emergence of wild oat (*Avena fatua*) on yield loss of barley (*Hordeum vulgare*) and wheat (*Triticum aestivum*). *Weed Sci.,* 33, 498-503.

Oerke, E.-C., Dehne, H.-W., Schönbeck, F., and Weber, A. (1994) *Crop Production and Crop Protection: Estimated Losses in Major Food and Cash Crops.* Elsevier, Amsterdam.

Olson, B.M. and McKercher, R.B. (1985) Wheat and triticale root development as affected by trifluralin. *Can. J. Plant Sci.,* 65, 723-729.

Orson, J.H. (1980) Annual broad-leaved weed control in winter wheat and winter barley: Autumn and spring treatments compared. *Proc. 1980 Brighton Crop Prot. Conf.—Weeds,* 251-258. British Crop Protection Council, Surrey, UK.

Orson, J.H. (1982) Annual broad-leaved weed control in winter wheat and winter barley, autumn and spring treatments compared. *Aspects of Applied Biology 1,* Broad-leaved weeds and their control in cereals, 43-52. Association of Applied Biologists, Warwick, UK.

Orson, J.H. (1987) Growing practices—an aid or hindrance to weed control in cereals. *Proc. Brighton Crop Prot. Conf.—Weeds,* 87-96. British Crop Protection Council, Surrey, UK.

O'Sullivan, P.A., Weiss, G., and Friesen, D. (1985) Tolerance of spring wheat (*Triticum aestivum* L.) to trifluralin deep-incorporated in the autumn or spring. *Weed Res.,* 25, 275-280.

Pannell, D.J. (1994) Mixtures of wild oats (*Avena fatua*) and ryegrass (*Lolium rigidum*) in wheat: Competition and optimal economic control. *Crop Prot.,* 13, 371-375.

Pannell, D.J. and Gill, G.S. (1994) Mixtures of wild oats (*Avena fatua*) and ryegrass (*Lolium rigidum*) in wheat: Competition and optimal economic control. *Crop Prot.,* 13, 371-375.

Ramsel, R.E. and Wicks, G.A. (1988) Use of winter wheat (*Triticum aestivum*) cultivars and herbicides in aiding weed control in an ecofallow corn (*Zea mays*) rotation. *Weed Sci.,* 36, 394-398.

Ratliff, R.L. and Peeper, T.F. (1987) Bromus control in winter wheat (*Triticum aestivum*) with the ethylthio analog of metribuzin. *Weed Technol.,* 1, 235-241.

Richards, M.C. and Davies, D.H.K. (1991) Potential for reducing herbicide inputs/rates with more competitive cereal cultivars. *Proc. 1991 Brighton Crop Prot. Conf.—Weeds,* 1233-1240. British Crop Protection Council, Surrey, UK.

Rola, H. and Rola, J. (1983) Competition of *Apera spica-venti* in winter wheat. *Proc. 10th Int. Congress Plant Protection,* 122. Brighton, British Crop Protection Council, Surrey, UK.

Rydrych, D.J. (1974) Competition between winter wheat and downy brome. *Weed Sci.,* 22, 211-214.

Saavedra, M., Cuevas, J., Mesa-García, J., and García-Torres, L. (1989) Grassy weeds in winter cereals in southern Spain. *Crop. Prot.,* 8, 181-187.

Salonen, J. (1992a) Efficacy of reduced herbicide doses in spring cereals of different competitive ability. *Weed Res.,* 32, 483-491.

Salonen, J. (1992b) Yield responses of spring cereals to reduced herbicide doses. *Weed Res.,* 32, 493-499.

Samuel, A.M. and Guest, S.J. (1990) Effect of seed rates and within crop cultivations in organic winter wheat. BCPC Monograph No. 43, *Organic and Low Input Agriculture,* 49-54. British Crop Protection Council, Surrey, UK.

Satorre, E.H. and Snaydon, R.W. (1992) A comparison of root and shoot competition between spring cereals and *Avena fatua* L. *Weed Res.,* 32, 45-55.

Scott, R.C. and Peeper, T.F. (1994) Economic returns from broad-leaf weed control in hard red winter wheat (*Triticum aestivum*). *Weed Technol.,* 8, 797-806.

Singh, S., Malik, R.K., Panwar, R.S., and Balyan, R.S. (1995) Influence of sowing time on winter wild oat (*Avena ludoviciana*) control in wheat (*Triticum aestivum*) with isoproturon. *Weed Sci.,* 43, 370-374.

Snaydon, R.W. (1982) Weeds and crop yield. *Proc. 1982 Brighton Crop Prot. Conf.—Weeds,* 729-739. British Crop Protection Council, Surrey, UK.

Stahlman, P.W. and Miller, S.D. (1990) Downy brome (*Bromus tectorum*) interference and economic thresholds in winter wheat (*Triticum aestivum*). *Weed Sci.,* 38, 224-228.

Streibig, J.C., Combellack, J.H., Pritchard, G.H., and Richardson, R.G. (1989) Estimation of thresholds for weed control in Australian cereals. *Weed Res.,* 29, 117-126.

Thornton, P.K., Fawcett, R.H., Dent, J.B., and Perkins, T.J. (1990) Spatial weed distribution and economic thresholds for weed control. *Crop Prot.,* 9, 337-342.

Tottman, D.R. (1977a) The identification of growth stages in winter wheat with reference to the application of growth-regulator herbicides. *Ann. Appl. Biol.,* 87, 213-224.

Tottman, D.R. (1977b) A comparison of the tolerance by winter wheat of herbicide mixtures containing dicamba and 2,3,6-TBA, or ioxynil. *Weed Res.,* 17, 273-282.

Tottman, D.R. (1978) The effects of a dicamba herbicide mixture on the grain yield components of winter wheat. *Weed Res.,* 18, 335-339.

Tottman, D.R., Appleyard, M., and Sylvester-Bradley, R. (1985). Cereal growth stages—stem extension and apical development. *Aspects of Applied Biology 10,* Field trials methods and data handling, 401-413. Association of Applied Biologists, Warwick, UK.

Tottman, D.R., Lupton, F.G.H., and Oliver, R.H. (1984) The tolerance of difenzoquat and diclofop-methyl by winter wheat varieties at different growth stages. *Ann. Appl. Biol.*, 104, 151-159.

Tottman, D.R., Lupton, F.G.H., Oliver, R.H., and Preston, S.R. (1982) Tolerance of several wild oat herbicides by a range of winter wheat varieties. *Ann. Appl. Biol.*, 100, 365-373.

Valenti, S.A. and Wicks, G.A. (1992) Influence of nitrogen rates and wheat (*Triticum aestivum*) cultivars on weed control. *Weed Sci.*, 40, 115-121.

Vitta, J.I. and Fernandez-Quintanilla, C. (1996) Canopy measurements as predictors of weed crop competition. *Weed Sci.*, 44, 511-516.

Welsh, J.P., Bulson, H.A.P., Stopes, C.E., Froud-Williams, R.J., and Murdoch, A.J. (in press) The critical weed free period in organically grown winter wheat. *Ann. Appl. Biol.*

Whitehead, R. and Wright, H.C. (1989) The incidence of weeds in winter cereals in Great Britain. *Proc. Brighton Crop Prot. Conf.—Weeds,* 107-112. British Crop Protection Council, Surrey, UK.

Whiting, A.J. and Davies, D.H.K. (1990) Response of winter wheat to herbicide rate and timing. *Proc. Crop Protection in Northern Britain,* 77-82. The Association for Crop Protection in Northern Britain, Dundee, Scotland.

Whiting, A.J. and Richards, M.C. (1990) Crop competitiveness as an aid to weed control in cereals. BCPC Monograph No. 45, *Organic and Low Input Agriculture,* 197-200. British Crop Protection Council, Surrey, UK.

Wicks, G.A., Martin, D.A., and Mahnken, G.W. (1995) Cultural practices in wheat (*Triticum aestivum*), on weeds in subsequent fallow and sorghum (*Sorghum bicolor*). *Weed Sci.*, 43, 434-444.

Wicks, G.A., Nordquist, P.T., Hanson, G.E., and Schmidt, J.W. (1994) Influence of winter wheat (*Triticum aestivum*) cultivars on weed control in sorghum (*Sorghum bicolor*). *Weed Sci.*, 42, 27-34.

Wicks, G.A., Ramsel, R.E., Nordquist, P.T., Schmidt, J.W., and Challaiah (1986) Impact of wheat cultivars on establishment and suppression of summer annual weeds. *Agron J.*, 78, 59-62.

Wilson, B.J. (1979) The effect of controlling *Alopecurus myosuroides* Huds. and *Avena fatua* L. individually and together, in mixed infestations on the yield of wheat. *Weed Res.*, 19, 193-199.

Wilson, B.J. (1980) The effect on yield of mixtures and sequences of herbicides for the control of *Alopecurus myosuroides* Huds. and broad-leaved weeds in winter cereals. *Weed Res.*, 20, 65-70.

Wilson, B.J. (1982) The yield response of winter cereals to autumn or spring control of broad-leaved weeds. *Aspects of Applied Biology 1,* Broad-leaved weeds and their control in cereals, 53-61. Association of Applied Biologists, Warwick, UK.

Wilson, B.J. (1986) Yield responses of winter cereals to the control of broad-leaved weeds. *Proc. EWRS Symp. Economic Weed Control.*, 75-82. Stuttgart, Germany.

Wilson, B.J. and Cussans, G.W. (1978) The effects of herbicides applied alone and in sequence, on the control of wild-oats (*Avena fatua*) and broad-leaved weeds, and on yield of winter wheat. *Ann. Appl. Biol.*, 89, 459-466.

Wilson, B.J. and Wright, K.J. (1990) Predicting the growth and competitive effects of annual weeds in wheat. *Weed Res.,* 30, 201-211.

Wilson, B.J., Cousens, R., and Wright, K.J. (1990) The response of spring barley and winter wheat to *Avena fatua* population and density. *Ann. Appl. Biol.,* 116, 601-609.

Wilson, B.J., Thornton, M.E., and Lutman, P.J.W. (1985) Yields of winter cereals in relation to the timing of control of black-grass (*Alopecurus myosuroides* Huds.) and broad-leaved weeds. *Aspects of Applied Biology* 9, The biology and control of weeds in cereals, 41-48, AAB, Warwick, UK.

Wilson, B.J., Wright, K.J., Brain, P., Clements, M., and Stephens, E. (1995) Predicting the competitive effects of weed and crop density on weed biomass, weed seed production and crop yield in wheat. *Weed Res.,* 35, 265-278.

Wilson, B.J., Wright, K.J., and Butler, R.C. (1993) The effect of different frequencies of harrowing in the autumn or spring on winter wheat, and on the control of *Stellaria media* (L.) Vill., *Galium aparine* L. and *Brassica napus* L. *Weed Res.,* 33, 501-506.

Wilson, B.W. and Cussans, G.W. (1983) The effect of weeds on yield and quality of winter cereals in the U.K. *Proc. 10th Int. Congress Plant Protection,* 121. Brighton, British Crop Protection Council, Surrey, UK.

Wright, K.J. and Wilson, B.J. (1992) Effects of nitrogen fertiliser on competition and seed production of *Avena fatua* and *Galium aparine* in winter wheat. *Aspects of Applied Biology,* 30, 381-386. Association of Applied Biologists, Warwick, UK.

Zanin, G., Berti, A., and Giannini, M. (1992) Economics of herbicide use on arable crops in north-central Italy. *Crop Prot.,* 11, 174-180.

Zanin, G., Berti, A., and Toniolo, L. (1993) Estimation of economic thresholds for weed control in winter wheat. *Weed Res.*, 33, 459-467.

Zimdahl, R.L. (1980) *Weed-Crop Competition: A Review.* International Plant Protection Center, Corvallis, OR, p. 196.

Chapter 9

Wheat Growth, Yield, and Quality As Affected by Insect Herbivores

Victor O. Sadras
Alberto Fereres
Roger H. Ratcliffe

INTRODUCTION

A pest is any organism that reduces crop yield or quality (Fick and Power, 1992); "key pests" are persistent, occur perennially, and usually reach economically damaging levels (Hearn and Fitt, 1992). Key arthropod pests of wheat have been identified for the main temperate cropping systems in the world (FAO, 1987a; Hatchett, Starks, and Webster, 1987; Pearson, 1992). Aphids (Hemiptera: Aphididae) stand out as major pests as they are found in virtually all temperate cropping systems and have the potential to reduce yield substantially (see Table 9.1). Aphids are also important in the tropics (Hill and Waller, 1988). Hessian fly and blossom midges (Diptera: Cecidomyiidae) and sawflies (Hymenoptera: Cephidae) make up another conspicuous group of pests with significant potential for yield reduction (Table 9.1).

Plant/herbivore relationships can be investigated with an entomological focus whereby plants are considered hosts that provide animals with food and shelter. On the other hand, plant/herbivore relationships can be seen from a "phytocentric" perspective (Baldwin, 1993), which emphasizes physiological and morphological plant adaptations and responses to herbivory. Given the aim of the book, this chapter takes a phytocentric view. The wheat crop/herbivore interface will be considered in relation to the linkage between crop simulation models and pests.

The physiology and ecology of wheat herbivores, the structure and function of the third trophic level, and pest management issues are closely

TABLE 9.1. Estimated Potential Yield Losses Due to Arthropod Pests in Various Cropping Systems

Pest[a]	Cropping system	Yield loss (%)	Reference
Hessian fly	semiarid Morocco	35[b]	Amri et al. (1992)
Hessian fly	Spain (Badajoz)	35[c]	Arias Giralda and Bote Velasco (1992)
Army cutworm	U.S.A. (Kansas)	25[d]	Bauernfeind and Wilde (1993)
Bulb fly	U.K.	22[e]	Bardner (1968)
Leaf beetle	Poland	7[d]	Kaniuczak (1987)
Leaf beetle	U.S.A. (Michigan)	55[d]	Webster, Smith, and Lee (1972)
Midge	U.K.	30[f]	Oakley (1994)
Midge	Finland	29[d]	Kurppa (1988)
Russian aphid	U.S.A. (Texas)	80[d]	Archer and Bynum (1992)
Stem sawfly	Canada (Great Plains)	22[d]	Holmes (1977)

[a] Hessian fly, *Mayetiola destructor;* stem sawfly, *Cephus* spp.; bulb fly, *Delia coarcata;* Russian aphid, *Diuraphis noxia;* leaf beetle, *Oulema* spp.; army cutworm, *Euxoa auxilaris;* midges, *Sitodiplosis mosellana, Contarinia tritici.*

[b] Yield reduction in susceptible varieties relative to resistant varieties.

[c] Maximum yield reduction estimated as a function of damaged plants and the association between damage and yield components.

[d] Yield reduction in unprotected crops where populations of pests were allowed to progress unchecked relative to insecticide-protected control.

[e] Yield reduction in unprotected crops where populations of pests were allowed to progress unchecked relative to control crops in which soil was covered with polythene sheets to prevent egg laying.

[f] Yield reduction in heavily infested crops relative to uninfested fields.

related to, but beyond the scope of this chapter. Nonetheless, we will consider the effects of the wheat plant on its pests when necessary to characterize mechanisms of plant resistance to herbivores. Aspects of the interaction between the crop and organisms at the third trophic level will be briefly discussed.

Plants in the field are frequently exposed to multiple stresses (Chapin et al., 1987; Mooney, Winner, and Pell, 1991) and actual yield responses to herbivores are influenced by a number of environmental and management factors. Hence the responses of wheat to herbivores in combination with other biotic (e.g., pathogens) and abiotic (e.g., water deficit) stresses are also discussed in this chapter.

Yield responses to herbivory are primarily related to (a) physiological and morphological responses of injured plants, and (b) changes in competitive relationships among plants. Physiological and morphological changes in plants damaged by herbivores are often obvious and have received broad attention; these are discussed in the sections "Key Pests" and "Other Pests." Less obvious are the changes in intra- and interspecific competitive relationships among plants derived from differential damage (Crawley, 1989; McNaughton, 1983b), which in turn derives from the nonuniform temporal and spatial distribution of pests in the field (Taylor, 1984). Crop-level responses associated with differential damage are emphasized in the section, "Population-Level Responses to Herbivory."

YIELD RESPONSES TO HERBIVORY

Plant growth and yield can be reduced, unaffected, or increased by herbivory. This continuum of responses has been the subject of reviews dealing with natural (Maschinski and Whitham, 1989; McNaughton, 1983a) and agricultural systems (Bardner and Fletcher, 1974; Harris, 1974; Jameson, 1963). It is of course easier to find examples in which herbivores reduced rather than increased crop yield but reports of no yield loss are common (Wood, 1965) and moderate yield increases of crops associated with insect damage have been analyzed by Harris (1974). The section "Other Pests" illustrates the interaction between leaf loss and temperature that could lead to enhanced yield of defoliated wheat crops.

On average, yield losses due to insects in wheat seem to be lower than in other crops. From data derived by the FAO in 1978, for instance, worldwide yield losses due to insect pests averaged 5 percent in wheat compared to 28 percent in rice (FAO, 1987b). Nonetheless, large yield losses due to herbivory can occur in unprotected crops (Table 9.1) and reduction in quality can further contribute to economic losses.

Actual yield responses to herbivory are difficult to predict because they depend on (a) pest factors, such as spatial insect distribution, intensity, timing, and duration of damage; (b) crop factors, e.g., phenological development; and (c) environmental and management factors such as temperature, rainfall, and sowing date that usually affect growth and development of the crop, its pests, and the pests' natural enemies.

FUNCTIONAL CLASSIFICATION OF HERBIVORES

A guild is a group of species that exploit the same class of environmental resources in a similar way (Hawkins and MacMahon, 1989). On this basis, "feeding guilds" have been defined that group herbivores, irrespective of their taxonomy and biology, in classes based upon the plant organ that is damaged and the feeding mechanisms of the herbivores. The aim of the classifications of herbivores by Boote et al. (1983) and Johnson (1987) was to develop interfaces to link crop simulation models and pests. Boote et al. (1983) defined seven classes: (1) *tissue consumers*, which remove host material after assimilate has been converted into host tissue; (2) *photosynthetic rate reducers,* which reduce the rate of photosynthesis (and normally transpiration) per unit leaf area; (3) *leaf senescence accelerators*, which induce or accelerate the rate of leaf senescence and abscission; (4) *light stealers,* which reduce the amount of light available for the crop irrespective of changes in leaf area; (5) *assimilate sappers*, which remove soluble assimilates; (6) *turgor reducers,* which damage roots or vascular tissue, disrupting water and nutrient uptake and translocation; and (7) *stand reducers,* which destroy seeds or seedlings. Johnson (1987) proposed a classification with two categories: pests that reduce the amount of radiation intercepted by the canopy, RI; and pests that reduce its radiation-use efficiency, RUE. This classification fits crop simulation models that calculate crop biomass as a function of RI and RUE (Rossing et al., 1992; van Emden and Hadley, 1994). The phytocentric approach of this chapter justifies dealing with groups of pests rather than with individual species despite a series of limitations outlined below.

None of these classifications account for the effects of herbivores as carriers or facilitators of diseases. Economically important effects on grain quality and harvestable yield need also to be considered. Stem borers, for example, can reduce yield by damaging vascular tissue but they can also cause significant losses by increasing lodging.

The categories in these classifications of pests are not mutually exclusive. For instance, aphids are both assimilate sappers and light stealers (or RUE reducers). Furthermore, studies at the biochemical and cytological levels show that the effects of aphids go beyond the simple removal of assimilates or the indirect damage associated with honeydew production. Hence, trying to force the many complex and often poorly understood effects of herbivores on plants into a set of fixed categories could be an obstacle, rather than an aid for further understanding.

Two recent reviews discuss the role of simulation models in the research of plant/herbivore issues (Sadras and Felton, 1997) and the opportunities for using system approaches in pest management (Rossing and Heong, 1997).

KEY PESTS

This section focuses on four major insect pests of wheat: aphids, Hessian fly, sawfly, and blossom midges.

Effects on Wheat Development, Growth, Yield, and Quality

Aphids

Wheat crops can be colonized by several species of aphids, although most of the economic damage is produced by the English grain aphid, *Sitobion avenae* (F.); the rose-grain aphid, *Metopolophium dirhodum* (Walker); the bird cherry-oat aphid, *Rhopalosiphum padi* (L.); the greenbug, *Schizaphis graminum* (Rondani); and the Russian wheat aphid, *Diuraphis noxia* (Mordvilko). Although the five species are distributed in most cereal-producing regions of the world, the three first species are most important in Western Europe and North Africa, and the others often reach damaging levels in North and South America, South Africa, and Eastern Europe. Since its appearance as a pest in North America in 1986 (Fereres, Araya, and Foster, 1986), special attention has been paid to the ecology, damage, and control of *D. noxia*. In addition to the diversity of aphid species feeding on wheat, several biotypes of *S. graminum* have been described according to their ability to overcome plant resistance. Details of aphid evolution and genetics can be found in Hales et al. (1997).

Aphids search for a suitable host plant by making very brief superficial intracellular probes (López-Abella, Bradley, and Harris, 1988). Once they receive a positive stimulus, they search for the phloem, where they keep feeding most of the time (Pollard, 1973). Direct and indirect effects of aphids on wheat span from the biochemical to the crop level of organization.

Direct damage involves depletion of carbohydrates and amino acids, or even plant cell organelles (Saxena and Chada, 1971). Direct damage also includes toxicosis induced by injection of active salivary secretions (Al Mousawi, Richardson, and Burton, 1983; Chatters and Schlehuber, 1951). *S. graminum* feeding on wheat may cause ultrastructural alterations including disruption of chloroplasts and cellular membranes and enlargement of the plastoglobuli within the chloroplast, finally leading to a macroscopic necrotic lesion surrounded by a chlorotic halo (Morgham et al., 1994). Ciepiela (1989a, 1989b) found that *S. avenae* can induce an increase in the content of some precursors of plant phenolic compounds (phenylalanine and tyrosine), which seem to be involved in the protection of wheat against aphids (Leszczynski, 1985). Ciepiela (1989b) also found that the increase in the activity of two key enzymes in phenol biosynthesis, phenylalanine ammonia-lyase and tyrosine ammonia-lyase, was greater in ears of resistant wheat 'Atlas 66' than in susceptible 'Bezostaya 1.' The activity of these lyases was correlated with an increase in phenylalanine and tyrosine after infestation. An increase in ethylene production by wheat has been reported after *S. graminum* infestation (Anderson and Peters, 1994). Ethylene production seems to be only a symptom of aphid-infested plants and is not related to the induction of chlorotic and necrotic lesions observed after feeding by *S. graminum* (Anderson and Peters, 1995), but it could influence the rate of leaf senescence. It has been suggested that *S. graminum* can accelerate senescence of wheat leaves by (a) reducing chlorophyll content (Ryan et al., 1987) and (b) accelerating the breakdown of wheat proteins and mobilization of amino acids that stimulate ingestion of phloem constituents by the aphid (Morgham et al., 1994). *S. graminum* can reduce fructan content of wheat stems (Holmes et al., 1991) and consequently may affect wheat responses to cold stress.

Indirect damage caused by aphids includes (a) transmission of viruses and (b) excretion of honeydew, which reduces photosynthesis and promotes leaf senescence and the growth of saprophytic fungi. In a series of laboratory and field experiments, Rabbinge et al. (1981) found that the excretion of honeydew by *S. avenae* and *M. dirhodum* blocked the stomata and reduced radiation use efficiency and the maximum rate of photosynthesis of wheat leaves. More recently, Rossing and van de Wiel (1990) showed that the application of artificial honeydew to the flag leaf of wheat reduced CO_2 assimilation, mesophyll conductance, and leaf nitrogen content 15 days after treatment under high irradiance and hot, dry conditions. The same experiments showed that the rate of dark respiration was increased one day after the application of honeydew, and that the initial radiation use efficiency was significantly lower 15 days after treatment with honeydew substitute at moderate temperature and humidity. Rossing (1991) found that indirect damage due to honeydew increased at higher attainable yield.

Damage of aphids to wheat is thus a complex combination of direct and indirect effects that can lead to substantial yield reductions (Table 9.1). At the plant and crop levels, aphid injury can translate into reductions in leaf area, photosynthesis per unit leaf area, carbohydrate reserves, shoot and root growth, and kernel number and mass (Aalbersberg, van der Westhuizen, and Hewitt, 1989; Gray et al., 1990; Holmes et al., 1991; Rabbinge et al., 1981; Riedell and Kieckhefer, 1995; Rossing and van de Wiel, 1990). Depending on whether maximum aphid damage concentrates around anthesis and/or during grain filling, kernel number or mass are more likely to be affected (Aalbersberg, van der Westhuizen, and Hewitt, 1989; Wratten and Redhead, 1976). The effect of timing of damage on yield was further investigated by several authors. Rossing (1991) showed that *S. avenae* damage per aphid-day decreased from flowering to ripeness because damage caused by honeydew decreased with crop age while direct feeding damage was about constant. According to Kieckhefer and Gellner (1992), yield losses caused by *R. padi, S. graminum*, or *D. noxia* are in the range of 35 to 40 percent at a constant density of 15 aphids per plant during a 30-day period at 10°C. It appears, however, that the duration of the infestation can be more important than its intensity (Burd and Burton, 1992; Kieckhefer, Gallner, and Riedell, 1995). For instance, a combination of 25 aphids × 12 days produced higher damage to wheat than other 300 aphid-day combinations (150 aphids × 2 days; 75 aphids × 4 days; 50 aphids × 6 days) (Kieckhefer, Gellner, and Riedell, 1995).

Aphids (e.g., *S. avenae* and *M. dirhodum*) can also reduce grain quality of wheat, but only at high density, causing large yield reductions (Oakley and Walters, 1994).

Hessian Fly

Hessian fly, *Mayetiola destructor* (Say) is a destructive pest of wheat and is widely distributed throughout most wheat-growing regions of Europe, North Africa, Asia, and North America. A recent review of Hessian fly distribution, biology and ecology, host range, crop damage, losses, and control methods was published by Ratcliffe and Hatchett (1997).

In winter wheat areas, the typical life cycle of the Hessian fly begins with fall emergence of adults from infested wheat stubble or volunteer wheat. Soon after mating, the female begins ovipositing on the upper leaf surface of young plants and lays between 200 and 300 glossy red eggs. Adults do not feed, are short lived and die within a few days after emergence. Newly hatched larvae crawl behind the leaf sheaths and migrate to the crown of the plant, where they begin to feed. When the larva is full grown, the outer skin hardens and forms a protective puparium in which it overwinters; the puparium is commonly called "flaxseed" because of its resemblance to the seed of flax. The following spring, these larvae transform to pupae, and adults emerge and infest wheat about the time the plants begin to joint. Most of the larvae of the spring generation are found just above the nodes under the leaf sheaths. The larvae pass the summer inside the puparia in the dry stubble. In late summer or fall, the larvae pupate, and adults emerge and infest volunteer or early-seeded wheat.

Gagne and Hatchett (1989) provided a detailed description of the larval instars.

Wheat injury by the Hessian fly is caused entirely by the larva. Its feeding mechanism and how it obtains food from the wheat plant are not fully understood. No visible injury to plant tissue has been observed at feeding sites, although infested plants show a characteristic stunted appearance. A study of the larval mouthparts has revealed highly specialized mandibles that probably are used to inject salivary fluids into plants (Hatchett, Kreitner, and Elzinga, 1990). These secretions are believed to contain enzymatic substances that inhibit plant growth and increase cell wall permeability, which allows the larva to suck the juices from the plant (Byers and Gallun, 1972b, 1972c).

Studies of the effects of Hessian fly on wheat range from the biochemical to the crop level. Refai, Jones, and Miller (1955) recorded twice the amount of total sugar accumulation in wheat leaves after Hessian fly infestation compared with control plants and concluded that plant phosphorylases that are involved in the splitting of glycosidic bonds were inactivated or inhibited, thus causing the amount of sugars to increase. Miller et al. (1958) found that the center leaves of Hessian fly-infested wheat plants contained greater percentages of the lipid-soluble pigments chlorophyll, carotene, and xanthophyll than leaves of uninfested plants. They suggested that the unnatural dark blue-green color of the center leaves of infested plants may be due to the higher concentrations of chlorophyll a and b. Wellso et al. (1986) reported changes in the concentration of carbohydrates that affect wheat responses to low temperature (see the section on Hessian fly under "Interactions Between Herbivory and Other Factors"). McMullen and Walgenbach (1986) described wheat cytological changes induced by Hessian fly feeding. Distinctive features of affected cells include disruption of cellular membranes and plastids.

Leaves of infested plants appear more erect and are shorter and darker green than those of uninfested plants (Gallun, Ruppel, and Everson, 1966). Cartwright, Caldwell, and Compton (1959) reported that larval feeding on susceptible plants reduced the growth of the second leaf by 59 percent, completely stunted the third leaf, and prevented fourth-leaf development. In the spring when plants are in the jointing stage, larval feeding causes injury that prevents normal elongation of internodes and transport of nutrients to the developing spike. Byers and Gallun (1972a) showed that stunting occurred earlier when more larvae were feeding on the plant. The number of stems and leaves, plant weight and length, crown weight, leaf weight and length, and root weight are reduced in plants infested by Hessian fly (Wellso, Hoxie, and Olien, 1989). When infestations are severe, many of the young plants die after larvae mature. Because successful larvae typically kill the stems where they feed, tiller development of fly-infested wheat is an important tolerance trait for which intraspecific differences have been reported (Figure 9.1A). Reductions in plant growth due to larval feeding injury may lead to reductions in both yield (Table 9.1) and grain quality. Weakened stalks, which often break and lodge as the plant matures, contribute further to economic losses (Buntin and Chapin, 1990; Painter, 1960).

Stem Sawfly

The most common species of sawfly are *Cephus cinctus* (Norton) in North America (Weiss and Morrill, 1992) and *Cephus pygmeus* (L.) and *Trachelus* spp. in Europe, Asia, and Mediterranean Africa (Alvarado et al., 1992; Banita et al., 1992; Miller, 1992). Weiss and Morrill (1992) reviewed wheat stem sawfly biology and ecology, crop damage and associated yield losses, and methods for its control. Briefly, adult females deposit eggs in wheat stems, showing preference for elongating internodes of succulent, thick (about 3 mm diameter) stems. Adult females are very selective in their choice of stems for egg laying; they prefer stems with three elongating internodes to those with only two or five (Holmes and Peterson, 1960), an important factor that can be exploited in developing resistant varieties. Shortly after hatching, larvae move down the stem, where they feed on parenchyma and vascular tissue. Accordingly, stem sawfly is primarily a "turgor reducer" in Boote et al.'s (1983) classification. Larval damage to vascular tissue, however, affects not only water but also nutrient and probably hormonal fluxes. This damage normally translates into reduced number and mass of wheat kernels. Sawfly damage can also reduce the protein content of kernels. After completion of larval development (4th-5th instar), the larva girdles the stem with a notch. Thus, in addition to yield reductions associated with its feeding, sawfly-weakened stems are more susceptible to lodging, and yield losses due to reduced harvestability can be significant.

Blossom Midges

Common species of blossom midges are the orange midge *Sitodiplosis mosellana* (Géhin), widespread in the Northern Hemisphere, and the yellow or lemon midge *Contarinia tritici* (Kirby), restricted to Europe

FIGURE 9.1. Wheat Responses to Insect Herbivores As Affected by Cultivar and Environmental Conditions

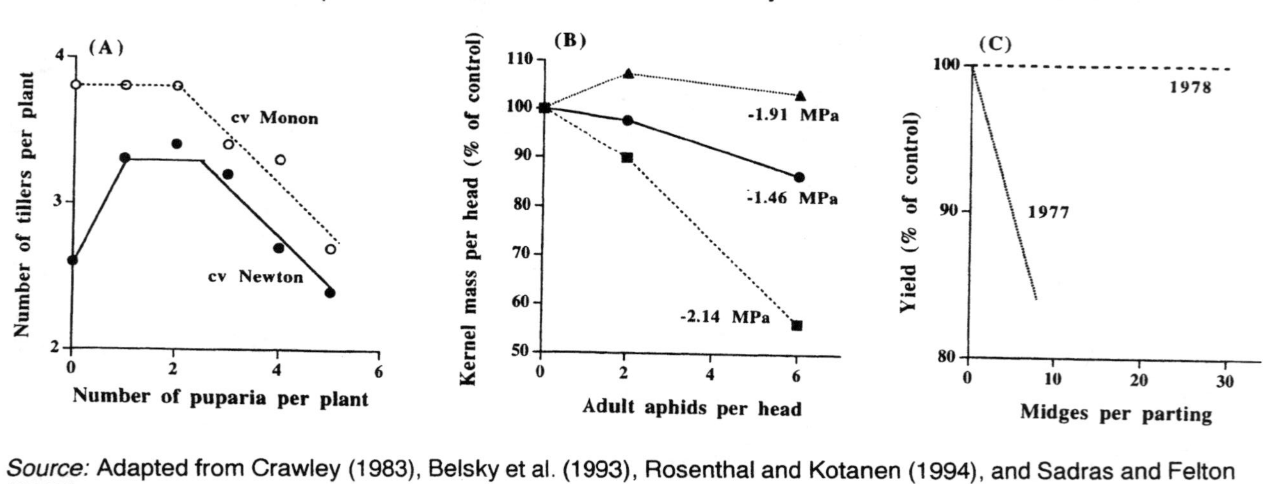

Source: Adapted from Crawley (1983), Belsky et al. (1993), Rosenthal and Kotanen (1994), and Sadras and Felton (1997).

(A) Tillering of 'Newton' and 'Monon' in response to intensity of Hessian fly (Biotype L) infestation. Number of puparia (included third instars) is taken as indicative of previous larval infestation. Note the stimulation of tillering in the low-tillering cultivar caused by 1-3 puparia. Data from Wellso and Hoxie (1994) (their Table 1).

(B) Reduction in kernel mass per head of 'Talento' as affected by *Sitobion avenae* and water deficit between head emergence (GS 58) and milky ripe stage (GS 78). Average leaf water potential is shown for the well-watered (circles), moderately (triangles), and severely (squares) stressed treatments (adapted from Fereres et al., 1988, their Fig. 3 and Table 3).

(C) Relationship between yield and midge density in two seasons in Britain. In 1977, midges accounted for 52 percent of the variance in yield in contrast to the nonsignificant relationship in 1978 when cooler weather slowed the rate of development of midges, causing 90 percent mortality. Adapted from Oakley (1994) (Figs. 3 and 4).

(Oakley, 1994). Oakley (1994), Kurppa and colleagues (Kurppa, 1989a, 1989b; Kurppa and Husberg, 1989) and Hjorth (1992) discussed aspects of the biology of blossom midges, crop damage, insect control, and varietal resistance. Oakley (1994) and Hjorth (1992) both emphasized the large spatial and temporal fluctuations of midge populations and the consequent need for forecasting methods to aid pest control (Oakley et al., 1994).

The following summary is based on Oakley's (1994) review. Midge larvae overwinter in cocoons in the soil where they can remain up to 3 years (*C. tritici*) or up to 13 years (*S. mosellana*). Adults emerge after a cold temperature vernalization period followed by warmer temperature, mate at the site of emergence, and females lay their eggs on the glume, lemma, or palea of florets before anthers dehisce (*S. mosellana*) or between the lemma and palea (*C. tritici*). Larvae feed on developing kernels by exuding enzymes through their cuticles that break down kernel tissues, and absorbing the resulting mush (*S. mosellana*) or on stigmas and anthers, hence preventing pollination (*C. tritici*). *C. tritici* therefore reduces yield by reducing grain number; the relationship between yield reduction and kernel number is linear at high intensity of damage, but some compensation associated with increase in healthy kernel mass may occur at low intensity of attack. *S. mosellana* reduces kernel number when damage occurs early (i.e., before the kernel starts to grow) and/or when damage is intense, i.e., ≥ 3 larvae per kernel; otherwise it reduces kernel mass.

Besides the reduction in yield, *S. mosellana* can reduce wheat quality by inducing prematurity sprouting, reducing Hagberg falling number, and affecting other flour properties (Oakley, 1992). Lunn et al. (1995) demonstrated secretion of a midge α-amylase during feeding at the dough stage while absence of amylase activity from the mature kernel indicated degradation of the enzyme after cessation of feeding. Lunn et al. (1995) also detected high levels of sprouting-specific isozymes α-AMY1+2 in damaged kernels but sprouting was not necessarily associated with the activity of these enzymes. The conclusion of Lunn et al. (1995) that germination of infested grains results from the interaction between midge damage and weather conditions agrees with the hypothesis of Oakley (1994), that midge damage loosens the kernel's pericarp, facilitating water uptake and hence sprouting in poor weather.

Wheat Resistance to Herbivores

Plant responses to a given pest grade continuously from full resistance to the extreme sensitivity of those plants that are unprotected and unable to regrow after damage (Belsky et al., 1993; Hooker, 1984; Painter, 1951).

Plant resistance to herbivory has two components: avoidance and tolerance (Figure 9.2A). Avoidance includes phenological escape and defense. Selection of varieties with phenological cycles adjusted to escape injury, combined with appropriate sowing date, plays an important role in crop protection at the farm level. The degree of synchrony in planting and crop development also influences the relationships between crops, herbivores, and predators at the regional level (Ives and Settle, 1997). It is worth noting that, from an entomological perspective, phenological escape has not been regarded as a true mechanism of plant resistance to herbivory (Kogan and Ortman, 1978; Painter, 1951).

Defenses can be classified according to several criteria: (a) according to the type of plant trait, i.e., *morphological* versus *chemical* (Figure 9.2A); (b) according to the evolution of the plant/herbivore system, *induced* defense is viewed as a strategy that adjusts the level of defenses to the prevailing risk of herbivory in contrast to *constitutive*, invariant defenses (Åström and Lundberg, 1994); (c) according to their effects on herbivores,

FIGURE 9.2. (A) Plant- and (B) Population-Level Mechanisms of Resistance to Herbivory

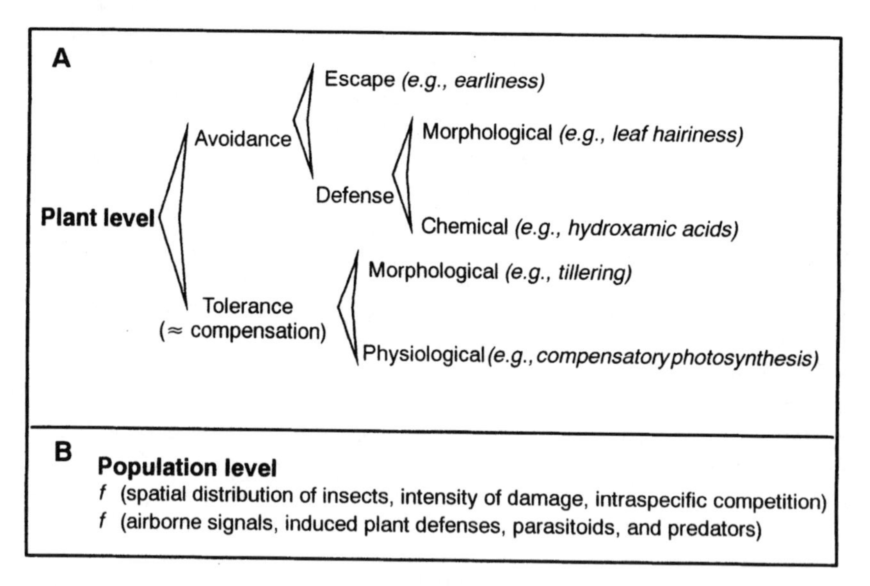

Source: Adapted from Crawley (1983), Belsky et al. (1993), Rosenthal and Kotanen (1994), and Sadras and Felton (1997).

antibiosis involves plant properties that adversely affect the metabolism of an herbivore feeding on a plant, while *antixenosis* or *nonpreference* involves behavioral responses of the herbivore that result in avoidance of the plant as food or as an oviposition substrate (Kogan and Ortman, 1978; Painter, 1951).

Tolerance depends on morphological and physiological traits that, rather than protect the plants from damage, allow them to recover after damage has occurred (Belsky et al., 1993). "Compensation" or "compensatory growth" are terms that are usually used as synonyms for "tolerance" (Rosenthal and Kotanen, 1994). General aspects of plant defenses against herbivory and tolerance mechanisms have been reviewed recently (Baldwin, 1993; Bennett and Wallsgrove, 1994; Rosenthal and Kotanen, 1994; Trumble, Kolodny-Hirsch, and Ting, 1993). The rest of this section deals with wheat resistance to key pests in the context of the previous definitions.

Aphids

Morphological and chemical defenses may contribute to the resistance of wheat to aphids. Leaf pubescence is associated with resistance against the yellow sugarcane aphid, *Sipha flava* (Forbes) in some wheat lines (Webster et al., 1994). Although pubescence can potentially improve wheat resistance to some aphids and beetles it may also generate problems with other pests such as the wheat curl mite *Eriophyes tulipae* (Keifer), whose landing efficiency increases on cultivars with high density of leaf trichomes (Harvey, Martin, and Seifers, 1990).

Chemical defenses include hydroxamic acids, phenols, and indole alkaloids (Ciepiela, 1989a; Leszczynski, 1992; Leszczynski, Wright, and Bakowski, 1989). Hydroxamic acids as well as phenols are present in undamaged host plants as glucosides (Wahlross and Virtanen, 1959). The glucosides are enzymatically hydrolized into aglucones when the plant tissue is injured and intracellular compartmentalization is destroyed (Hoffman and Hofmanova, 1970; Katagiri, 1979). As the aglucones are more toxic to aphids than glucosides, this results in an increase in the antibiotic effect, i.e., aphid performance is negatively affected. Some aphid species such as the corn leaf aphid, *Rhopalosiphum maidis* (Fitch), have developed an intracellular puncture avoidance strategy when feeding on wheat that allows the insect to avoid contact with hydroxamic acid compounds (Givovich and Niemeyer, 1995). Hydroxamic acid concentration in wheat is usually (Corcuera et al., 1992; Givovich et al., 1994; Leszczynski, Wright, and Bakowski, 1989) but not always (Kazemi and Van Emden, 1992) negatively correlated with aphid performance. Hydroxamic acids may act as feeding deterrents (antixenotics) for cereal aphids, which lead to selection of plants with lower content of these allelochemicals by both wingless (Givovich and Niemeyer, 1991) and winged aphids (Nicol et al., 1992). At the sieve elements, hydroxamic acids may act as a double barrier as they have both antixenotic and antibiotic effects, decreasing aphid performance in terms of mean relative growth rate (Thackray et al., 1990), survival (Argandoña, 1994), reproductive rate (Corcuera, Argandoña, and Peña, 1982), or intrinsic rate of population increase, r_m (Bohidar, Wratten, and Niemeyer, 1986). However, some aphid species (e.g., *S. avenae*) have been able to overcome these plant defenses by detoxification of the DIMBOA aglucone ingested from wheat plants to MBOA, a less toxic compound (Leszczynski and Dixon, 1990). Some of the detoxifying enzymes (UDP-glucose transferases) that metabolize these toxic aglucones have been found in the cytosolic and microsomal fractions of *S. avenae* (Leszczynski, Matok, and Dixon, 1992).

The concentration of free phenols in wheat is positively correlated with the level of resistance to *S. avenae* (Ciepiela, 1989a; Leszczynski, Wright, and Bakowski, 1989). Ciepela (1989a) found that a "toxicity index" (ratio between the concentrations of free phenols and free amino acids) could be associated with the level of resistance of wheat to *S. avenae*. Also, the concentration in wheat of essential amino acids such as alanine, histidine, and threonine has been positively correlated with the fecundity of *R. padi* (Kazemi and Van Emden, 1992). Furthermore, the presence of the indole alkaloid gramine in wheat has been associated with antibiotic and antixenotic types of resistance to *M. dirhodum* and *R. padi* (Lamb and MacKay, 1995). However, Lesczynski, Wright, and Bakowski (1989) found that the concentration of this allelochemical was not correlated with performance (measured as r_m) of *S. avenae*, although resistant wheat cultivars had greater concentrations of indole alkaloids, hydroxamic acids, and phenols than their susceptible counterparts.

Several sources of resistance to aphids can be used in wheat. Four perennial grasses, *Agropyron* spp. and *Elymus angustus*, and hybrids between these grasses and cultivated wheat have been shown to differ in antixenotic and antibiotic components of resistance against *R. padi* (Tremblay, Cloutier, and Comeau, 1989). High levels of antibiosis to *R. padi* have been found in wheat cvs. Ommid (Kazemi and Van Emden, 1992) and GK Zombor (Papp and Mesterházy, 1993). Some lines of *Triticum monococcum* L. showed resistance to *S. avenae*, although some aphid clones were more adversely affected than others by the resistant germplasm (Caillaud et al., 1995). Various resistant wheat cultivars have been released to control *S. graminum* biotypes (Joppa, Timian, and Williams, 1980; Starks, Burton, and Merkle, 1983). Sources of resistance to this species include Largo wheat (Tyler, Webster, and Merkle, 1987), *Hordeum chilense* (Castro, Martin, and Martin, 1996), and 'Insave F.A.' rye, *Secale cereale* L. (Sebesta et al., 1995). In the past decade, an enormous effort has been made to find sources of resistance to *D. noxia*. Several sources of resistance (mainly antibiotic and tolerance-type) have been found in several plant introductions from Tunisia (Formusoh et al., 1992), Iran, Afghanistan, and the former Soviet Union (Porter, Webster, and Baker, 1993). Also, resistant genes have been identified from wheat plant introductions PI 137739 (gene *Dn1* for antibiosis) and PI 262660 (gene *Dn2* for tolerance) (Quisenberry and Schotzko, 1994). Resistance to *D. noxia* in three triticale lines is controlled by a single dominant gene at a common locus (Nkongolo, Lapitan, and Quick, 1996).

Hessian Fly

During the past 50 years, resistant wheat varieties have provided the most reliable and economical control of Hessian fly. Sixty varieties resistant to Hessian fly were released during the period from 1950 to 1983 (Hatchett, Starks, and Webster, 1987). In areas where resistant varieties have been grown for several years, losses from the Hessian fly have been reduced to <1 percent (Maxwell, Jenkins, and Parrott, 1972). Growing resistant wheat varieties to suppress Hessian fly populations is compatible with biological, cultural, and chemical control methods that have been studied, but of these methods, only cultural control has been widely adapted along with the use of resistant varieties. The most successful cultural methods have been delayed seeding of winter wheat to escape fall infestation and destruction of volunteer wheat, which serves as a summer reservoir for the Hessian fly (Ratcliffe and Hatchett, 1997). Although systemic insecticides applied at seeding provide control of the Hessian fly, they are no more effective than resistant wheat varieties (Buntin and Chapin, 1990).

Numerous sources of resistance to Hessian fly in wheat have been identified and are being utilized in breeding programs. Resistance in these sources is dominant, partially dominant, or recessive, and conditioned by single, duplicate, or multiple genetic factors derived from common and durum wheats, the wild wheat *Triticum tauschii* (Coss) Schmal., and rye. Twenty-seven major genes designated H1 through H27 have been described. Ratcliffe and Hatchett (1997) present information about the source of the 27 genes, chromosome location, if known, and selected literature references. Antibiosis is the primary mechanism of resistance associated with these genes and is expressed as the death of the first-instar larvae within two to three days after establishment. The biochemical nature of antibiosis in wheat to the Hessian fly has not been determined for any gene. Miller and co-workers (Miller et al., 1960; Miller and Swain, 1960) studied the influence of silica in sheaths of wheat plants and free amino acids, organic acids, and sugars in plants on resistance to Hessian fly. None of these factors were linked significantly to resistance. Indirect evidence supports the hypothesis that hypersensitivity involving "recognition" of an avirulence gene product or process is the phenotypic basis of wheat resistance to Hessian fly (Grover, 1995; Shukle, Grover, and Foster, 1990).

Other genetic factors for resistance to the Hessian fly have been reported. Resistance derived from wheats 'Kawvale' and 'Marquillo' may include tolerance to larval feeding as well as antibiosis (Painter, 1951). Both varieties, but especially 'Marquillo,' have been used in hard red winter wheat breeding programs in Kansas and Nebraska since the 1950s, and extensively since the 1980s. Antixenosis associated with leaf pubescence, which influences oviposition by Hessian fly females, has been reported by Roberts et al. (1979). In addition to reduced oviposition, they found that egg hatch and larval establishment was lower on the pubescent-leaved 'Vel' than on glabrous-leaved 'Arthur,' indicating possible antibiosis, as well. Wellso and Araya (1993) and Wellso and Hoxie (1994) showed that infested wheat plants may compensate for Hessian fly injury by normal production of tillers if the infested main stem is not killed, and also demonstrated the existence of intraspecific variation in tillering responses to Hessian fly damage (Figure 9.1A).

The development of virulent biotypes of the Hessian fly poses the greatest threat to the permanence of resistance in wheat varieties (Ratcliffe and Hatchett, 1997). Ratcliffe and colleagues (Ratcliffe and Hatchett, 1997; Ratcliffe et al., 1996; Ratcliffe et al., 1994) reported that major shifts in biotype composition and virulence to resistance genes in wheat occurred in Hessian fly populations throughout the eastern United States from the mid-1980s to mid-1990s. The increased frequency of virulent biotypes in Hessian fly populations resulted in greatly reduced effectiveness of some resistance genes. Ratcliffe and Hatchett (1997) emphasized that the optimum use of present and future resistance genes in wheat breeding programs depends on the development of gene deployment strategies that improve the durability of resistance genes in wheat varieties exposed to Hessian fly populations. Strategies suggested included gene stacking, use of moderately resistant genes or combinations of genes with different levels of resistance or temperature-sensitive or insensitive traits, and selective use of genes depending upon the biotype composition of Hessian fly populations as determined by field surveys. Such strategies should reduce selection pressure on Hessian fly populations for virulent genotypes, and increase stability of resistance in wheat cultivars.

Stem Sawfly

Weiss and Morrill (1992) summarized the development of varieties resistant to sawfly from the discovery in the 1920s of stem solidness (i.e., pith growth inside the stem) as a key trait to the characteristics of resistant varieties currently used in North America. They emphasized the main drawback of resistant varieties, i.e., their low yield potential. In the absence of sawfly, the yield reduction associated with resistance calculated by Weiss and Morrill was up to 1.4 t ha^{-1}. They also calculated, for different localities, the probability of this yield penalty associated with resistance being offset by losses due to sawfly. The negative association between yield and solid stems, however, has been attributed to undesirable genes remaining from the original solid-stemmed source, S-615 (Hayat et al., 1995).

Another limitation related to the use of plant resistance based on stem solidness is the strong environmental variance of this trait. High stand density, for instance, can significantly reduce stem solidness of both durum and bread wheats selected for resistance to sawfly (Miller, El Masri, and Al Jundi, 1993). Recent surveys of durum wheat landraces in Turkey, nonetheless, have identified germplasm combining both stem solidness and general stress tolerance (Damania et al., 1997).

Avoidance traits play an important role in resistance to sawfly, mostly in relation to the marked selectivity of adult females in their choice of stems for egg laying. This is best illustrated by the comparison between winter Canadian cultivars, which often escape damage because they are too mature when the sawfly flight occurs, and the more susceptible spring varieties (Holmes and Peterson, 1960). Holmes and Peterson (1960) showed that phenological escape also contributes to the resistance of spring varieties.

Blossom Midges

Phenological escape (Figure 9.2A) is the main trait involved in wheat resistance to midges, as most of the variation between spring and winter wheats and among varieties within each group relates to the degree of coincidence between the rather narrow susceptible crop stages and the flight of the midges (Kurppa, 1989a; Oakley, 1994). Other traits that have been considered in wheat resistance to midges are hypersensitivity and tightness of the ear, a function of the gap between lemma and palea (Oakley, 1994). Hypersensitive varieties identified among Finnish spring wheats tend to lose damaged kernels, preserving the quality of harvested kernels and allowing for compensation by increase in the mass of undamaged kernels. Further details can be found in the review by Oakley (1994) including an interesting discussion on the variable ranking of varieties screened for resistance to midges depending on (a) the methods used in screening for resistance, e.g., small versus large plots, hand versus machine harvest, cleaned versus uncleaned samples and (b) large genotype × environment interactions.

Interactions Between Herbivores and Other Factors

Aphids

Growth and yield responses of wheat to aphid damage are influenced by factors such as plant viruses, plant nutrition, water deficit, oxidative stress, and extreme temperature.

Aphids are the vectors of barley yellow dwarf virus (BYDV), probably the most harmful and widespread disease in wheat (Conti et al., 1990). BYDV infection, in turn, can improve the nutritional quality of wheat to aphids by increasing amino acid content (Ajayi, 1986) and/or soluble carbohydrates (Fereres et al., 1990; Jensen, 1972). These physiological changes induced by BYDV infection probably contribute to the increased reproduction rate and shorter developmental time observed for *R. padi* (Araya and Foster, 1987), *S. avenae* (Fereres et al., 1989), and *S. graminum* (Montllor and Gildow, 1986) feeding on BYDV-infected plants. The increased fecundity observed for *S. avenae* on BYDV-infected plants was greater on the BYDV-sensitive 'Abe' and 'Talento' wheats than with the symptomatically resistant 'Caldwell' (Fereres et al., 1989). This indicates that cereal aphids and BYDV interact for their mutual benefit, resulting in a higher potential for population growth of both vector and virus. In addition, the virus may spread faster, because feeding on infected hosts induces a higher proportion of alates in the vector population (Gildow, 1983). Interestingly, the putative mutualistic relationship between *S. avenae* and BYDV was not evident for the nonvector *D. noxia* (Mowry, 1994). McKirdy and Jones (1997) investigated the interactions among sowing date, aphid control, and BYDV incidence in cropping systems of Western Australia. They (1) showed that the earlier the sowing date, the longer the exposure to migrating aphids and the greater the threat of early BYDV infection; (2) compared their findings with those of wheat crops in New Zealand and United Kingdom; (3) discussed the trade-off between yield losses associated with BYDV incidence and yield losses due to delaying sowing date; and (4) assessed the role of stubble retention to reduce aphid landings on wheat plants.

The water and nutritional status of the wheat crop can modify cereal aphid biology and life cycle. Analysis of insect responses to plant growing conditions is, however, beyond the scope of this chapter. A review by Waring and Cobb (1992) indicates that, in general, sucking insects grow better in plants with adequate supply of water, N, and P, but negative and nonlinear responses have also been reported. In contrast to N and P, K fertilization often affects sucking insects adversely. The work by Duffield et al. (1997) illustrates the complexity of the interactions between aphid species (*S. avenae, M. dirhodum*), rate of nitrogen fertilizer, rainfall, and entomopathogenic fungi in wheat crops. High doses of N fertilizer increased populations of *M. dirhodum* irrespective of other environmental conditions while populations of *S. avenae* increased with increasing supply of N in years favorable for aphid growth. When high rainfall created unfavorable conditions for aphid growth, high N supply reduced populations of *S. avenae* and this effect was partially attributed to an increase in the incidence of entomopathogenic fungi (Duffield et al., 1997).

Early work by Kloft (1954) proved that aphids can affect water uptake and transpiration of plants. In the study by Dorschner et al. (1986), *S. graminum* disrupted two potentially adaptive responses of wheat to drought: (a) the aphid eliminated the potential increase in cell membrane stability that is associated with wheat conditioned to drought stress, and (b) osmotic potential was reduced less in the greenbug plus drought stress treatment than in the drought stress-only treatment. This indicates that aphids are potentially more damaging when wheat is grown under water stress, a prediction further supported by direct measurements made by Fereres et al. (1988) (Figure 9.1B).

Damage by *S. graminum* (Holmes et al., 1991) or *R. padi* (Wellso, Olien, and Hoxie, 1985) may reduce cold hardiness of winter wheat by reducing its fructan reserves. Storlie et al. (1993) showed that *D. noxia* (147 aphids per plant) not only reduced fructan content but also increased osmotic potential of ‘Froid’ (more cold hardy) and ‘Brawny’ (less cold hardy) by 0.24 and 0.35 MPa, respectively. Because cold hardiness and osmotic potential are negatively related (De Noma, Taylor, and Ferguson, 1989) increased osmotic potential may contribute to the reduced cold tolerance of aphid-infested wheat crops. Furthermore, Storlie et al. (1993) found that only the less cold hardy ‘Brawny’ had a significantly lower winter survival and grain yield under *D. noxia* autumn infestation.

Hessian Fly

Many biotic and abiotic factors affect the abundance and destructiveness of the Hessian fly (Barnes, Miller, and Arnold, 1959; Buntin and Chapin, 1990; Cartwright, 1923; Drake and Decker, 1932; Emery, 1937; Larrimer, 1922; Webster, 1906; Webster and Kelly, 1915). When these factors favor the insect's survival and development, populations may increase rapidly from one generation to the next. Adult females tend to lay more eggs at higher humidity and survival of small larvae is higher under these conditions, resulting in higher levels of infestation (Painter, 1954). Sufficient moisture to cause germination and growth of volunteer wheat during the summer may cause the development of a late-summer brood of flies for which volunteer wheat serves as a breeding place (Foster, Taylor, and Araya, 1986). For this reason the destruction of volunteer wheat can be an important method of reducing fall infestation pressure. In contrast, hot, dry summer and dry fall weather retards the development of the Hessian fly (Webster and Kelly, 1915). High summer temperatures induce a more intense aestivation and delayed adult emergence, which may render delayed fall planting of wheat less effective (Wellso, 1991). Foster and Taylor (1975) reported poor emergence of Hessian flies at 26.7°C in laboratory tests, and attributed this condition to higher levels of aestivation in insects reared at this temperature. If fall conditions that induce late Hessian fly emergence are followed by mild winter conditions, higher Hessian fly winter survival may result in heavy spring infestations of wheat (Wellso, 1991).

In most northern areas of North America, low fall temperatures usually stop emergence of adults in any great numbers by late September (Foster, Taylor, and Araya, 1986). Many adults appear and disappear within a period of several weeks, often early enough in the fall for wheat planting to be safely delayed until serious danger of infestation is past. These later dates for fall planting are referred to as "fly-free" dates and vary from north to south with temperature. Delayed planting of winter wheat is not effective in southern areas, however, because of the extended developmental period of the Hessian fly (Buntin and Chapin, 1990).

Similar to the effect of aphid feeding, Hessian fly feeding can reduce cold hardiness of winter wheat. Wellso et al. (1986) and Wellso, Hoxie, and Olien (1987) attributed the reduction in cold hardiness of 'Winoka' and 'Genesee' wheats to depletion of fructan and other sugars by larval feeding. They found that larval feeding reduced fructose, fructan, and glucose, but did not influence the level of sucrose. In further studies, Wellso, Coolbaugh, and Hoxie (1991), Wellso, Hoxie, and Olien (1989), and Wellso, Hoxie, and Taylor (1990) reported that wheat plants on which

Hessian fly larvae are feeding maintain a steady state in soluble sugars independent of larval numbers, although fructans appeared to be depleted by feeding larvae.

There is little published information about the interaction between Hessian fly injury and water or nutritional deficit. Extensive loss of yield may result if moisture stress during grain maturation coincides with Hessian fly damage (Reed, 1986). As in other plant/herbivore systems (Waring and Cobb, 1992), increasing rates of fertilizer applied to wheat crops may increase (Okigbo and Gyrisco, 1962), decrease (Ellet and Wolfe, 1921), or have no effect (Foster and Jeffrey, 1937; Hudson, 1920) on the incidence of Hessian fly injury. Okigbo and Gyrisco (1962) found no definite relationship between Hessian fly infestation and soil reaction or drainage. Despite the lack of consistent responses, the application of N-P-K fertilizers to wheat were recommended to quicken growth, encourage rapid stooling, and develop strong straw (Gossard and Houser, 1906) and hence improve Hessian fly management. Similarly, Davis (1918) recommended application of acid phosphate and lime to southern Indiana wheat fields to hasten maturity of the plants and reduce Hessian fly damage.

Boosalis (1954) reported an interaction between Hessian fly and *Helminthosporium* and *Fusarium* fungi and several nonpathogenic bacteria associated with crown and basal stem rot of wheat. Hessian fly-infested plants had a substantially higher percentage of pathogens than uninfested plants, but the composition of microflora was very similar. Stunting, delay of maturation, and grain injury were more pronounced on Hessian fly-infested wheat with deteriorated crowns.

Stem Sawfly

Weiss and Morrill (1992) provide an interesting account of the interaction among environmental factors, insect biology, and changes in cultural practices that led to the evolution of stem sawfly as a key pest in North America. It all started in the early 1900s when growers adopted a fallow/wheat rotation that enhanced soil moisture and wheat productivity in a region where wheat acreage was rapidly increasing. Two consequences of increased soil moisture were an increase in stem diameter and a longer leaf area duration. The former increased the attractiveness of wheat to ovipositing females and the latter allowed sawfly to complete its cycle in wheat. Furthermore, in an attempt to reduce wind erosion in fallowed fields, farmers adopted a cropping system with alternating, relatively narrow, cropped and uncropped strips. Because adult sawflies are weak fliers, the spatial distribution of host plants (i.e., cropped areas) and sources of adult flies (i.e., fallow) were much more favorable for the dispersion of the insects than in cropping systems using large fallow/crop areas in which infestations only reached the edges of the crop.

An obviously important interaction between sawfly and weather factors is that losses due to lodging and reduced crop harvestability will increase with high winds and heavy rainfall (Weiss and Morrill, 1992). Because phenological escape is a key trait for resistance to sawfly, interactions among variety, sowing date, and environmental conditions (chiefly temperature), which affect the rate of development of both wheat crops and sawfly populations are clearly important. Reductions in stem solidness in response to plant density discussed above could be accounted for, in part, by reduced light intensity and changes in light spectral composition (Holmes, 1984). Reduced light intensity associated with cloudiness can also reduce stem solidness and increase susceptibility to stem sawfly (Holmes, 1984).

Blossom Midges

Wellso and Freed (1982) reported a positive association between the blotch fungus *Septoria nodorum* and the midge *Sitodiplosis mosselana.* It is unclear whether the midge is attracted to plants infected with glume blotch or midge damage facilitates the fungal infection (Wellso and Freed, 1982). Infections by *Fusarium* spp. have also been found in association with *S. mosselana* (Oakley, 1994). Both fungal infections and sprouting of midge-damaged kernels are more likely to occur in damp (e.g., U.K.) than in dry (e.g., Canada) conditions (Oakley, 1994). In Finland, Kurppa (1989a) found no significant interactions between fungicide treatments and damage caused by blossom midges.

Given that damage by midges is strongly dependent on the degree of synchrony between the susceptible stage of the crop and midge flight, interactions among genotype, temperature, and photoperiod that affect the rate of reproductive development of wheat are likely to be important. Oakley (1994) pointed out how the susceptibility of certain varieties can be increased by fluctuating temperatures that extend the duration of the preflowering period.

Oakley's (1994) comparison of the yield/midge relationships between crops in two seasons further illustrates the influence of weather and parasitoids on the crop/insect system (see Figure 9.1C). In 1977, when a warm spell against the run of prevailing weather accelerated egg development, wheat yield and adult midge density were inversely correlated. No correlation was found in 1978, when cool temperature slowed down egg development, and egg mortality increased because the anthers emerging from the floret swept away eggs that had been too slow to develop, and eggs were exposed to natural enemies for a longer period.

OTHER PESTS

Before internode elongation, when the apical and axillary meristems of wheat are close to the ground, soil-dwelling insects may affect the crop differently depending on whether they are surface-feeding or subterranean species. Army cutworms (*Euxoa auxiliaris* Grote), for instance, are surface-feeding insects whose foraging activity, likened to grazing cattle, damages but rarely kills wheat plants when conditions for regrowth are favorable (Bauernfeind and Wilde, 1993). In contrast, subterranean herbivores such as the pale western cutworm (*Agrotis orthogonia* Morrison) can damage the plant's meristems and are therefore more likely to be "stand reducers" (Hill, Byers, and Schaalje, 1992). Crop recovery in the first case depends on both the capacity of individual plants to regrow after damage, and on population-level mechanisms. In the case of subterranean insects that kill seeds or plants, crop recovery obviously depends on population-level mechanisms alone.

A number of studies explored the interactions between cutworms and abiotic factors. In a controlled environment, the rate and amount of damage caused to wheat plants by *A. orthogonia* increased with temperature above 15°C and with soil moisture (Jacobson and Peterson, 1965).

In a field trial including two sites with contrasting moisture conditions, Bauernfeind and Wilde (1993) measured the effects of insecticides on *E. auxiliaris* and wheat yield. Under dry conditions, yield was reduced from 2.6 t ha^{-1} in insecticide-protected crops to 1.9 t ha^{-1} in untreated plots where populations of armyworm were around 80 larvae m^{-2}. In the wetter site, crops tolerated 88 to 108 larvae m^{-2} with no yield losses. Experiments by Buntin (1994) in which wheat plants were mechanically clipped at the soil surface at the one-, two- and four-leaf stages showed delayed spike emergence but no substantial yield reductions, further highlighting the ability of wheat to recover after this type of damage. In similar experiments with rye and triticale, plants defoliated at the two- and four-leaf stages were less damaged by cold temperatures and yielded as much as or more than nondefoliated plants in years with late frosts (Buntin, 1994). Messina, Jones, and Nielson (1993) showed that *D. noxia* established better in stands of grasses that were clipped than in untreated controls. Altogether, these examples highlight the importance of wheat tolerance to early-season defoliation, and the influence of environmental factors (e.g., water availability) on tolerance. Importantly, these examples also indicate how the effects of early pests may interact with (a) later stresses such as frost, by affecting crop phenological development; and (b) further herbivory, by affecting the suitability of the crop to other pests.

In contrast to the recovery of crops defoliated by insects, wheat crops did not recover well and suffered severe yield losses when defoliated by rabbits (Crawley and Putman, 1989). Although the nature of damage caused by rabbits is not strictly comparable to that of insects, comparison between rabbit-grazed and ungrazed wheat plots highlights how defoliation can interact with other biotic factors. Herbicides increased yield only on grazed plots, while aphicide increased yield only on plots that were both grazed and treated with herbicide (Crawley and Putman, 1989). Reductions in the competitive ability of plants following herbivory and the consequent interaction between weed interference and insect damage can affect yield through mechanisms discussed in Sadras (1997).

Leaf beetles (*Oulema* spp.) are an increasing problem in Central Europe and in parts of the United States with yield losses that may justify chemical control in some cases (Table 9.1). Defoliation by direct feeding activity and probably also accelerated leaf senescence are the main known effects of these insects which, at the crop level, can reduce growth by reducing light interception. Leaf hairiness is an important trait for resistance to leaf beetles (Gallun, Ruppel, and Everson, 1966; Haynes and Gage, 1981; Papp and Mesterházy, 1996). In experiments including 26 winter wheat genotypes, Papp, and Mesterházy (1996) found a negative correlation between feeding damage by *O. melanopus* (L.) and trichome length, but not with trichome density. Leaf damage, as a percentage of the flag leaf surface, ranged from 7 percent in the more resistant variety to 69 percent in the less resistant one. Despite this, leaf-feeding damage did not affect yield. This is consistent with McNaughton's (1983a) statement that "growth reduction and tissue destruction are rarely, if ever, translated monotonically into a proportional reduction of final yield" (p. 329) and further illustrates the importance of tolerance traits in wheat resistance to herbivory.

Lepidoptera stem miners can reduce wheat yield in some cropping systems. In part because of their preference for feeding on peduncles, *Ostrinia nubilalis* Hb. larvae feeding on inner stem tissue can produce a "white head" symptom. In the study by Buntin (1992), *O. nubilalis* larval injury reduced both kernel mass and to a lesser extent kernel number; it also reduced kernel test weight and quality. Similarly, Hennig (1987) reported reductions in yield caused by *Cnephasia pumicana* (Zeller) that were mostly associated with lighter kernels. Hennig also highlighted the effect of larval mining on stem breakage. Compensatory responses of wheat to stem borers have not been investigated in great detail, but rice injured by the stem borer *Scirpophaga incertulas* (Walker) showed a series of compensatory mechanisms including (a) increased translocation of carbon assimilate from injured main shoots to primary tillers; (b) increased tillering; and (c) compensatory photosynthesis, i.e. green leaves adjacent to dying leaves had greater photosynthetic rate than green leaves adjacent to green leaves (Rubia et al., 1996).

The biology of the wheat bulb fly (*Delia coarctata* Fallen) and its effects on wheat growth and yield have been widely investigated (Bardner, 1968; Bardner, Fletcher, and Huston, 1969; Gough, 1947a, 1947b; Roloff and Wetzel, 1989). Adult females lay eggs in summer, mainly in bare soil, and neonate larvae make their way from the soil to suitable host plants, where they can hollow and potentially kill one or more shoots per plant. Yield is lowered mostly by a reduction in the density of ear-bearing shoots which depends, in turn, on the proportion of plants that are killed, and the fertility of both damaged and undamaged plants (Bardner, Fletcher, and Huston, 1969). Analysis of *D. coarctata* population dynamics from 1953 to 1990 in the United Kingdom shows a decline in the incidence of this pest associated with an increase in the use of insecticides, elimination of fallow in the rotation, and with apparent changes in temperature and rainfall (Young and Cochrane, 1993). Predicted changes in the geographical distribution of bulb fly with further climatic change highlight the key relationships between the pest and major environmental factors (Evans and Hughes, 1996). Owing to the work by Bardner and colleagues, the wheat crop/bulb fly system is an ideal model to discuss the mechanisms of compensation at the population level, the focus of the following section.

Other arthropod pests that may affect wheat yield and/or quality include sucking bugs (Hemiptera: Pentatomidae) (Jacobson, 1965; Viator, Pantoja, and Smith, 1983); mirids (Heteroptera: Miridae) (Varis, 1991); saddle gall midge (*Haplodiplosis marginata* von Roser) (Petcu, Popov, and Barbulescu, 1987; Skuhravy and Skuhrava, 1986); thrips (Thysanoptera) (Banita, 1987); termites (Isopera) and millipedes (Diplopoda) (Wood and Cowie, 1988); leaf-mining Diptera (Hågvar, Hofsvang, and Trandem, 1994; Massor, Matthes, and Wetzel, 1989); wheat curl mite, the vector of wheat streak mosaic virus (Wood et al., 1995); and other mites (e.g., *Penthaleus major* Dugès, *Petrobia latens* Müller) (Swaine and Ironside, 1983).

POPULATION-LEVEL RESPONSES TO HERBIVORY

Population-level compensation, according to Crawley (1983), occurs when herbivore attack on one individual allows another individual to grow faster. In the following section we also refer, very briefly, to the putative defensive role of volatile chemicals released by injured plants.

Population-Level Compensation

Stand reductions can result from the activity of soil-dwelling arthropods but also from seedling disease, severe defoliation by early-season pests followed by unfavorable conditions for regrowth, and damage caused by stem borers. Tolerance to stand reduction depends on the capacity of the surviving plants to fill the gaps left by dead neighbors (Bardner and Fletcher, 1974). This response corresponds, in a broad sense, to Crawley's definition of population-level compensation and depends on both the intensity and spatial distribution of damage (Figure 9.2B, Figure 9.3). Compensation at the population level can also be expected when differential damage does not kill but delays the growth of injured plants relative to undamaged neighbors. This differential damage "relaxes competition" (*sensu* Crawley, 1983) and undamaged plants grown alongside damaged neighbors can thus compensate, at least partially, for the yield loss of damaged plants.

The importance of population-level compensation is illustrated in Figure 9.3A, where wheat yield is plotted against plant stand reduction caused by various insect pests. All data points in this graph are well above the response expected in the absence of compensation, highlighting the way in which yield losses due to stand reduction can be attenuated by the increase in yield of surviving plants. Crops in Romania, for instance, tolerated stand reductions close to 50 percent with almost no yield loss (Figure 9.3A). Bardner's (1968) analysis of yield components demonstrated how yield losses caused by *D. coarctata* were less than proportional to stand losses because surviving plants produced more fertile shoots, heavier ears, and slightly heavier grain (cf. Figure 9.3A). Yield responses to stand density and uniformity depend on growing conditions, as illustrated by the work of Turner, Prasertsak, and Setter (1994), who showed that unevenness in machine-planted crops did not affect yield in comparison with uniform, hand-sown crops under irrigation in contrast to 20 percent yield reductions attributed to uneven stands in crops water stressed after anthesis. The physiological and morphological mechanisms underlying population-level compensation are the same as those involved in individual plant responses to neighbor interference; Chapter 8 (this volume) discusses these mechanisms in the context of wheat responses to plant population density.

Crop responses to stand reduction also depend on the spatial distribution of damage (Figure 9.3B). Compensation is highest and yield loss is

FIGURE 9.3. Population-Level Compensation After Stand Reduction Caused by Herbivory Depending on the Intensity and Spatial Distribution of Damage

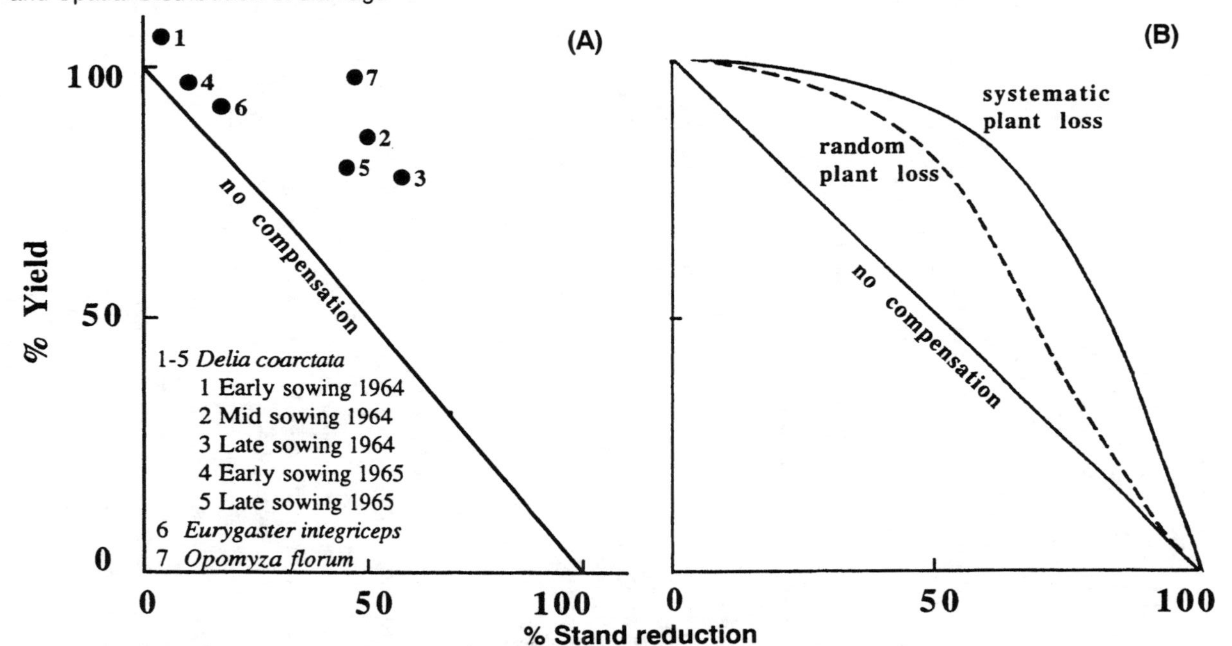

Note: In (A) the distance between data points and the solid line indicates the magnitude of compensation.

Sources: (A) 1-5 Bardner (1968); 6-7 Banita (1987); (B) Jones, Dunning, and Humpries (1955). Bardner's work further illustrates how sowing date can affect yield by shifting critical crop stages away from peak insect activity: early-sown crops (Fig. 9.3A, #1, 4) had more than one shoot when the attack began and plants with only one of their shoots injured survived, while plants injured at this time in late-sown crops (Fig. 9.3A, #3, 5) had a single shoot and died.

lowest when plant loss has a spatially uniform distribution. In comparison, for a given reduction in plant stand, random damage causes a greater reduction in yield (Figure 9.3B). Random distribution of insects in wheat fields have been reported, for instance, for blossom midges in Finland (Kurppa, 1989b), and for bulb fly in Britain (Bardner and Lofty, 1971). Further studies by Bardner and co-workers using artificial gapping to simulate the effects of the wheat bulb fly compared crop yield responses to stand reductions similar in magnitude but varying in spatial arrangements (Bardner and Fletcher, 1969; Bardner and Huston, 1968). While all the plots with gaps showed compensation, the amount of compensatory growth depended greatly on the arrangement of the gaps, being smaller when gaps extended across a larger number of adjacent rows. Further details on population-level compensation can be found in Hughes (1996).

Interactions Among Plants, and Between Plants and the Third Trophic Level Mediated by Airborne Plant Signals

Plants release volatile chemicals that are in many cases by-products of normal metabolism; others may have specific physiological functions (Malone, 1996; Sharkey, 1996). In the study by Buttery, Xu, and Ling (1985) intact wheat plants released 25 volatile compounds, 13 of which were terpenoids.

Volatiles released by injured plants may play a defensive role as elicitors of plant defenses or by attracting natural enemies of herbivores (Bruin, Sabellis, and Dicke, 1995; Dicke and Sabelis, 1989; Paré and Tumlinson, 1996). Owing to the volatile nature of these substances, they could be regarded as defenses at the population level because their supposed benefits are not restricted to the injured plant that has produced them, but could also be extended to undamaged neighbors (Sadras and Felton, 1997). The hypothesis that airborne signals can induce plant defenses or attract "bodyguards" has some empirical support but its relevance in the field needs to be established (Firn and Jones, 1995) notwithstanding the question whether these same volatiles may attract herbivores (Dicke and Sabelis, 1989; Dowdy et al., 1993). Furthermore, physiological considerations indicate that important metabolic costs can be involved in the production, transport, storage, prevention of autotoxication, and release of defensive chemicals (Dicke and Sabelis, 1989). This agrees with ecological studies of communication between the first and third trophic levels by Godfray (1995), who pointed out that any signaling system in which there exists the possibility of a conflict of interest between signaler and receiver will require significant costs for evolutionary stability. Certainly more research is needed in this area. Further discussion of costs or trade-offs related to defenses can be found in Zangerl and Bazzaz (1992) and Gershenzon (1994), who reviewed chemical defenses in general. Baldwin and colleagues emphasized the costs associated with inducible defenses (Baldwin, Karb, and Ohnmeiss, 1994; Baldwin and Ohnmeiss, 1994a, 1994b; Baldwin and Schmelz, 1994; Ohnmeiss and Baldwin, 1994).

In addition to chemicals, quality and intensity of light are important signals involved in plant-plant interactions (Aphalo and Ballaré, 1995) and it has been speculated that changes in the light environment of crops could be involved in the altered relationships between neighboring plants following differential damage (Sadras, 1996).

CONCLUDING REMARKS

The wheat crop is host to a wide range of arthropod herbivores that have the potential to reduce grain yield and quality. Aphids, Hessian fly, stem sawfly, and to some extent blossom midges and bulb fly have the attributes of key pests in many cropping systems, i.e., they are persistent, occur perennially, and usually reach economically damaging levels. Leaf beetles, lepidopterous stem borers, mites, thrips, cutworms, and other soil-dwelling insects can also cause problems in some wheat crops.

The survival of plants depends on traits of resistance to environmental stresses, including herbivory (Trewavas, 1981). In this chapter, we characterized the main mechanisms of wheat resistance to herbivory. Of them, phenological escape is probably the most important one, particularly for pests such as blossom midges and stem sawflies whose damaging activity is restricted to narrow phenological windows. From a plant-centered perspective (Belsky et al., 1993) and even more so from an agronomic perspective, there is justification for regarding the pattern of phenological development as a major trait in herbivory resistance, as much as it is considered the key trait for drought tolerance (Passioura, 1996).

We also discussed the importance of chemical and morphological defenses in resistance to herbivores and emphasized that, because no defense is perfect, yield of insect-damaged crops also depends on their capacity to recover after damage has occurred. We outlined the ecological principles underlying compensation resulting from altered competitive relationships among differentially damaged neighboring plants. The agronomic importance of airborne signals that trigger defensive responses of plants needs to be established.

Control of insect pests in wheat cropping systems is currently achieved by a variable combination of chemical, cultural, and biological methods.

In comparison with other industries (e.g., fruit, vegetables, cotton, rice), wheat production does not rely excessively on the use of environmentally disruptive chemicals for the control of key pests. The role of cultural and biological strategies for pest control is likely to be even more dominant in the future owing to environmental, health, and economic concerns leading to further reductions in the use of chemical pesticides in both wealthy (Pettersson, 1994) and poor countries (Deedat, 1994). Breeding and selection of wheat varieties with enhanced resistance to herbivores is a major avenue to further reduce the need for chemical control of pests. Biotechnological developments, including the continued identification of genes encoding insecticidal proteins and manipulation of induced defenses, will certainly play a major role in the breeding of resistant varieties. This approach is particularly important for insects such as Hessian fly and greenbug that have pathogen-like features in their interaction with the wheat plant. For this approach to be successful, however, a multidisciplinary team effort is required to balance the reductionist focus inherent to research in molecular biology.

REFERENCES

Aalbersberg, Y.K., M.C. van der Westhuizen, and P.H. Hewitt. 1989. Characteristics of the population build-up of the Russian wheat aphid *Diuraphis noxia* and the effect on wheat yield in the eastern Orange Free State. *Annals of Applied Biology.* 114: 231-242.

Ajayi, O. 1986. The effect of barley yellow dwarf virus on the amino acid composition of spring wheat. *Annals of Applied Biology.* 108: 145-149.

Al Mousawi, A.H., P.E. Richardson, and R.L. Burton. 1983. Ultrastructural studies of greenbug (Hemiptera: Aphididae) feeding damage to susceptible and resistant wheat cultivars. *Annals of Entomological Society of America.* 76: 964-971.

Alvarado, M., A. Serrano, A. de la Rosa, and J.M. Duran. 1992. Contribution to the knowledge of the cephids (Hymenoptera; Cephidae) on winter cereals of Western Andalucia. *Boletin de Sanidad Vegetal, Plagas.* 18: 807-816.

Amri, A., M.E. Bouhssini, S. Lhaloui, T.S. Cox, and J.H. Hatchett. 1992. Estimates of yield loss due to the Hessian fly (Diptera: Cecidomyiidae) on bread wheat using near-isogenic lines. *Al Awamia.* No. 77: 75-88.

Anderson, J.A. and D.C. Peters. 1994. Ethylene production from wheat seedlings infested with biotypes of *Schizaphis graminum* (Homoptera: Aphididae). *Environmental Entomology.* 23: 992-998.

Anderson, J.A. and D.C. Peters. 1995. Inhibitors of ethylene biosynthesis and action do not prevent injury to wheat seedlings infested with *Schizaphis graminum* (Homoptera: Aphididae). *Environmental Entomology.* 24: 1644-1649.

Aphalo, P.J. and C.L. Ballaré. 1995. On the importance of information-acquiring systems in plant-plant interactions. *Functional Ecology.* 9: 5-14.

Araya, J.E. and J.E. Foster. 1987. Laboratory study on the effects of barley yellow dwarf virus on the life cycle of *Rhopalosiphum padi* (L.). *Zeitschrift für Pflanzenkrankheiten und Pflanzenschutz.* 94: 578-583.

Archer, T.L. and E.D. Bynum, Jr. 1992. Economic injury level for the Russian wheat aphid (Homoptera: Aphididae) on dryland winter wheat. *Journal of Economic Entomology.* 85: 987-992.

Argandoña, V.H. 1994. Effect of aphid infestation on enzyme activities in barley and wheat. *Phytochemistry.* 35: 313-315.

Arias Giralda, A. and M. Bote Velasco. 1992. Estimate of attack and losses produced by the Hessian fly Mayetiola destructor Say in the southeast of Badajoz. *Boletin de Sanidad Vegetal, Plagas.* 18: 161-173.

Åström, M. and P. Lundberg. 1994. Plant defense and stochastic risk of herbivory. *Evolutionary Ecology.* 8: 288-298.

Baldwin, I.T. 1993. Chemical changes rapidly induced by folivory. In E.A. Bernays (ed.) *Insect-plant interactions.* CRC Press, Boca Raton, FL, pp. 1-23.

Baldwin, I.T., M.J. Karb, and T.E. Ohnmeiss. 1994. Allocation of N-15 from nitrate to nicotine: Production and turnover of a damage-induced mobile defense. *Ecology.* 75: 1703-1713.

Baldwin, I.T. and T.E. Ohnmeiss. 1994a. Coordination of photosynthetic and alkaloidal responses to damage in uninducible and inducible *Nicotiana sylvestris. Ecology.* 75: 1003-1014.

Baldwin, I.T. and T.E. Ohnmeiss. 1994b. Swords into plowshares? *Nicotiana sylvestris* does not use nicotine as a nitrogen source under nitrogen-limited growth. *Oecologia.* 98: 385-392.

Baldwin, I.T. and E.A. Schmelz. 1994. Constraints on an induced defense—The role of leaf area. *Oecologia.* 97: 424-430.

Banita, E. 1987. Capability for attack in the principal wheat pests. *Probleme de Protectia Plantelor.* 15: 201-216.

Banita, E., C. Popov, E. Luca, D. Cojocaru, G. Paunescu, and F. Vilau. 1992. Elements of integrated control of wheat stem sawflies (*Cephus pygmaeus* Latr. and *Trachelus tabidus* L.). *Probleme de Protectia Plantelor.* 20: 169-185.

Bardner, R. 1968. Wheat bulb fly, *Leptohylemyia coarctata* Fall., and its effects on the growth and yield of wheat. *Annals of Applied Biology.* 61: 1-11.

Bardner, R. and K.E. Fletcher. 1969. The distribution of attacked plants. *Rothamsted Experimental Station report for 1968:* Part 1: 199-200. Rothamsted, UK.

Bardner, R. and K.E. Fletcher. 1974. Insect infestations and their effects on the growth and yield of field crops: A review. *Bulletin of Entomology Research* 64: 141-160.

Bardner, R., K.E. Fletcher, and P. Huston. 1969. Recent work on wheat bulb fly. *Proc. 5th Br. Insectic. Fungic. Conf.:* 500-504.

Bardner, R. and P.J. Huston. 1968. Effects of gaps on yield. *Rothamsted Experimental Station Report for 1967:* 207-208. Rothamsted, UK.

Bardner, R. and J.R. Lofty. 1971. The distribution of eggs, larvae, and plants within crops attacked by wheat bulb fly *Leptohylemya coarctata* (Fall.). *Journal of Applied Ecology.* 8: 683-683.

Barnes, H.F., B.S. Miller, and M.K. Arnold. 1959. Some factors influencing the emergence of overwintering Hessian fly larvae. *Entomolgia Experimentalis et Applicata.* 2: 224-239.

Bauernfeind, R.J. and G.E. Wilde. 1993. Control of army cutworm (Lepidoptera: Noctuidae) affects wheat yields. *Journal of Economic Entomology.* 86: 159-163.

Belsky, A.J., W.P. Carson, C.L. Jensen, and G.A. Fox. 1993. Overcompensation by plants: Herbivore optimization or red herring? *Evolutionary Ecology.* 7: 109-121.

Bennett, R.N. and R.M. Wallsgrove. 1994. Secondary metabolites in plant defence mechanisms. *New Phytologist.* 127: 617-633.

Bohidar, K., S.D. Wratten, and H.M. Niemeyer. 1986. Effects of hydroxamic acids on the resistance of wheat to the aphid *Sitobion avenae. Annals of Applied Biology.* 109: 193-198.

Boosalis, M.G. 1954. Hessian fly in relation to the development of crown and basal stem rot of wheat. *Phytopathology.* 44: 224-229.

Boote, K.J., J.W. Jones, J.W. Mishoe, and R.D. Berger. 1983. Coupling pests to crop growth simulators to predict yield reductions. *Phytopathology.* 73: 1581-1587.

Bruin, J., M.W. Sabellis, and M. Dicke. 1995. Do plants tap SOS signals from their infested neighbours? *Trends in Ecology and Evolution.* 10: 167-170.

Buntin, G.D. 1992. Damage by the European corn borer (Lepidoptera: Pyralidae) to winter wheat. *Journal of Entomological Science.* 27: 361-365.

Buntin, G.D. 1994. Simulated insect defoliation of seedlings and productivity of winter small-grain crops. *Journal of Entomological Science.* 29: 534-542.

Buntin, G.D. and J.W. Chapin. 1990. Biology of Hessian fly (Diptera: Cecidomyiidae) in the southeastern United States: Geographic variation and temperature-dependent phenology. *Journal of Economic Entomology.* 83: 1015-1024.

Burd, J.D. and R.L. Burton. 1992. Characterization of plant damage caused by Russian wheat aphid (Homoptera: Aphididae). *Journal of Economic Entomology.* 85: 2017-2022.

Buttery, R.G., C. Xu, and L.C. Ling. 1985. Volatile compounds of wheat leaves (and stems): Possible insect attractants. *Journal of Agriculture and Food Chemistry.* 33: 115-117.

Byers, R.A. and R.L. Gallun. 1972a. Ability of the Hessian fly to stunt winter wheat. 1. Effect of larval feeding on elongation of leaves. *Journal of Economic Entomology.* 65: 955-958.

Byers, R.A. and R.L. Gallun. 1972b. Ability of the Hessian fly to stunt winter wheat. 2. Paper chromatography of extracts of freeze-dried larvae and wheat plants. *Journal of Economic Entomology.* 65: 997-1001.

Byers, R.A. and R.L. Gallun. 1972c. Ability of the Hessian fly to stunt winter wheat. 3. Gibberellic acid and elongation of second and third leaves of infested plants. *Journal of Economic Entomology.* 65: 1001-1004.

Caillaud, C.M., C.A. Dedryver, J.P. Di Pietro, J.C. Simon, F. Fima, and B. Chaubet. 1995. Clonal variability in the response of *Sitobion avenae* (Homoptera: Aphididae) to resistant and susceptible wheat. *Bulletin of Entomological Research.* 85: 189-195.

Cartwright, W.B. 1923. Delayed emergence of Hessian fly for the fall of 1922. *Journal of Economic Entomology.* 16: 432-435.

Cartwright, W.B., R.M. Caldwell, and L.E. Compton. 1959. Response of resistant and susceptible wheats to Hessian fly attack. *Agronomy Journal.* 51: 529-531.

Castro, A.M., A. Martin, and L.M. Martin. 1996. Location of genes controlling resistance to greenbug (*Schizaphis graminum* Rond.) in *Hordeum chilense. Plant Breeding.* 115: 335-338.

Chapin, F.S., A.J. Bloom, C.B. Field, and R.H. Waring. 1987. Plant responses to multiple environmental factors. *BioScience.* 37: 49-57.

Chatters, R.M. and A.M. Schlehuber. 1951. Mechanics of feeding of the greenbug (*Toxoptera graminum,* Rond.) on Hordeum, Avena, and Triticum. *Oklahoma Station Technical Bulletin.* T-41.

Ciepiela, A. 1989a. Biochemical basis of winter wheat resistance to the grain aphid, *Sitobion avenae. Entomologia Experimentalis et Applicata.* 51: 269-275.

Ciepiela, A. 1989b. Changes in phenylalanine and tyrosine content and metabolism in ears of susceptible and aphid resistant winter wheat cultivars upon infestation by *Sitobion avenae. Entomologia Experimentalis et Applicata.* 51: 277-281.

Conti, M., C.J. D'Arcy, H. Jedlinski, and P.A. Burnett. 1990. The "Yellow Plague" of cereals, barley yellow dwarf virus. In Burnett, P.A. (ed.) *World perspectives on barley yellow dwarf virus.* CIMMYT, Mexico City. 1-6.

Corcuera, L.J., V.H. Argandoña, and G.F. Peña. 1982. Effect of benzoxazinone from wheat on aphids. *Proceedings Fifth International Symposium Insect-Plant Relationships.* Pudoc, Wageningen. 33-39.

Corcuera, L.J., V.H. Argandoña, G.E. Zuniga, S.J.H. Rizvi, and V. Rizvi. 1992. Allelochemicals in wheat and barley: Role in plant-insect interactions. S.J.H. Rizvi and V. Rizvi (eds.) *Allelopathy: Basic and applied aspects.* Chapman and Hall, London, pp. 119-127.

Crawley, M. 1983. *Herbivory.* The dynamics of animal-plant interactions., vol. 10. Blackwell Scientific Publications, London, p. 437.

Crawley, M.J. 1989. Insect herbivores and plant population dynamics. *Annual Review of Entomology.* 34: 531-564.

Crawley, M.J. and R.J. Putman. 1989. Rabbits as pests of winter wheat. In R.J. Putman (ed.) *Mammals as pests.* Chapman and Hall, London, pp. 168-177.

Damania, A.B., L. Pecetti, C.O. Qualset, and B.O. Humeid. 1997. Diversity and geographic distribution of stem solidness and environmental stress tolerance in a collection of durum wheat landraces from Turkey. *Genetic Ressources and Crop Evolution.* 44: 101-108.

Davis, J.J. 1918. The control of three important wheat pests in Indiana. *Purdue Agricultural Experiment Station Circular.* 82: 11 pp.

De Noma, J.T., G.A. Taylor, and H. Ferguson. 1989. Osmotic potential of winter wheat crowns for comparing cultivars varying in winterhardiness. *Agronomy Journal.* 81: 159-163.

Deedat, Y.D. 1994. Problems associated with the use of pesticides: An overview. *Insect Science and Its Application.* 15: 247-251.

Dicke, M. and M.W. Sabelis. 1989. Does it pay plants to advertize for bodyguards? Towards a cost-benefit analysis of induced synomone production. In H. Lambers, M. L. Cambridge, H. Konings and T. L. Pons (eds.) *Causes and consequences of variation in growth rate and productivity of higher plants.* SPB Academic Publishing, The Hague, pp. 341-358.

Dorschner, K.W., R.C. Johnson, R.D. Eikenbary, and J.D. Ryan. 1986. Insect-plant interactions: Greenbugs (Homoptera: Aphididae) disrupt acclimation of winter wheat to drought stress. *Environmental Entomology.* 15: 118-121.

Dowdy, A.K., R.W. Howard, L.M. Seitz, and W.H. McGaughey. 1993. Response of *Rhyzopertha dominica* (Coleoptera: Bostrichidae) to its aggregation pheromone and wheat volatiles. *Environmental Entomology.* 22: 965-970.

Drake, C.J., and G.C. Decker. 1932. Late fall activity and spring emergence of the Hessian fly in Iowa. *Annals of the Entomological Society of America.* 25: 345-349.

Duffield, S.J., R.J. Bryson, J.E.B. Young, R. Sylvester-Bradley, and R.K. Scott. 1997. The influence of nitrogen fertiliser on the population development of the cereal aphids *Sitobion avenae* (F.) and *Metopopophium dirhodum* (Wlk.) on field grown winter wheat. *Annals of Applied Biology.* 130: 13-26.

Ellet, W.B. and T.K. Wolfe. 1921. Relation of fertilizer to Hessian fly injury and winter killing. *Journal of the American Society of Agronomy.* 13: 12-14.

Emery, W.T. 1937. Hessian fly eggs and freezing temperatures. *Kansas Entomology Society.* 10: 28-29.

Evans, K.A. and J.M. Hughes. 1996. Methods for predicting changes in pest distribution due to climate change: Wheat bulb fly. Implications of "Global environmental change" for crops in Europe, 13 April 1996, Churchill College, Cambridge, UK. No. 45: 285-292.

FAO. 1987a. *Insect pests of economic significance affecting major crops of the countries in Asia and the Pacific region.* Technical Document, Asia and Pacific Plant Protection Commission, FAO, Thailand. No. 135: 56 pp.

FAO. 1987b. Insecticides: some economic findings. *Phytoma.* 393: 6.

Fereres, A., J.E. Araya, and J.E. Foster. 1986. The Russian wheat aphid: A new pest on cereal crops in the United States and a potential threat to soft red winter wheats. *Purdue University Agricultural Experiment Station Bulletin.* No. 510: 27 pp.

Fereres, A., J.E. Araya, T.L. Housley, and J.E. Foster. 1990. Carbohydrate composition of wheat infected with barley yellow dwarf virus. *Zeitschrift für Pflanzenkrankheiten und Pflanzenschutz.* 97: 600-607.

Fereres, A., C. Gutierrez, P. Del Estal, and P. Castañera. 1988. Impact of the English grain aphid, *Sitobion avenae* (F), (Homoptera : Aphididae) on the yield of wheat plants subjected to water deficits. *Environmental Entomology.* 17: 596-602.

Fereres, A., R.M. Lister, J.E. Araya, and J.E. Foster. 1989. Development and reproduction of the English grain aphid (Homoptera: Aphididae) on wheat cultivars infected with barley yellow dwarf virus. *Environmental Entomology.* 18: 388-393.

Fick, G.W. and A.G. Power. 1992. Pests and integrated control. In C.J. Pearson (ed.) *Field Crops Ecosystems,* vol. 18. Elsevier, Amsterdam, pp. 59-83.

Firn, R.D. and C.G. Jones. 1995. Plants may talk, but can they hear? *Trends in Ecology and Evolution.* 10: 371.

Formusoh, E.S., G.E. Wilde, J.H. Hatchett, and R.D. Collins. 1992. Resistance to Russian wheat aphid (Homoptera: Aphididae) in Tunisian wheats. *Journal of Economic Entomology.* 85: 2505-2509.

Foster, J.E. and P.L. Taylor. 1975. Thermal-unit requirements for development of the Hessian fly under controlled environments. *Environmental Entomology.* 4: 195-202.

Foster, J.E., P.L. Taylor, and J.E. Araya. 1986. The Hessian fly. *Purdue Agricultural Experiment Station Bulletin.* No. 502.

Foster, W.R. and C.E. Jeffrey. 1937. Resistance of winter wheats to Hessian fly. *Canadian Journal of Agricultural Research.* 15: 135-140.

Gagne, R.J. and J.H. Hatchett. 1989. Instars of the Hessian fly (Diptera: Cecidomyiidae). *Annals of the Entomological Society of America.* 82: 73-79.

Gallun, R.L., R. Ruppel, and E.H. Everson. 1966. Resistance of small grains to the cereal leaf beetle. *Journal of Economic Entomology.* 59: 827-829.

Gershenzon, J. 1994. The cost of plant chemical defense against herbivory—a biochemical perspective. In E.A. Bernays (ed.), *Insect-plant interactions.* 5: 105-173.

Gildow, F.E. 1983. Influence of barley yellow dwarf virus-infected oats and barley on morphology of aphid vectors. *Phytopathology.* 73: 1196-1199.

Givovich, A. and H.M. Niemeyer. 1991. Hydroxamic acids affecting barley yellow dwarf virus transmission by the aphid *Rhopalosiphum padi. Entomologia Experimentalis et Applicata.* 59: 79-85.

Givovich, A. and H.M. Niemeyer. 1995. Comparison of the effect of hydroxamic acids from wheat on five species of cereal aphids. *Entomologia Experimentalis et Applicata.* 74: 115-119.

Givovich, A., J. Sandstrom, H.M. Niemeyer, and J. Pettersson. 1994. Presence of a hydroxamic acid glucoside in wheat phloem sap, and its consequences for performance of *Rhopalosiphum padi* (L.) (Homoptera: Aphididae). *Journal of Chemical Ecology.* 20: 1923-1930.

Godfray, H.C.J. 1995. Communication between the first and third trophic levels: An analysis using biological signalling theory. *Oikos.* 72: 367-374.

Gossard, H.A. and J.S. Houser. 1906. The Hessian fly. *Ohio Agricultural Experiment Station Bulletin.* 177: 39 pp.

Gough, H.C. 1947a. Studies on wheat bulb fly, *Leptohylemia coarctata,* Fall. I. Biology. *Bulletin of Entomology Research.* 37: 251-271.

Gough, H.C. 1947b. Studies on wheat bulb fly, *Leptohylemia coarctata,* Fall. II. Numbers in relation to crop damage. *Bulletin of Entomology Research.* 37: 439-454.

Gray, M.E., G.L. Hein, D.D. Walgenbach, and N.C. Elliott. 1990. Effects of Russian wheat aphid (Homoptera: Aphididae) on winter and spring wheat infested during different plant growth stages under greenhouse conditions. *Journal of Economic Entomology.* 83: 2434-2442.

Grover, P.B., Jr. 1995. Hypersensitive response of wheat to the Hessian fly. *Entomologia Experimentalis et Applicata.* 74: 283-294.

Hågvar, E.B., T. Hofsvang, and N. Trandem. 1994. The leafminer *Chromatomyia fuscula* (Diptera: Agromyzidae) and its parasitoid complex in Norwegian barley fields. *Norwegian Journal of Agricultural Sciences.* No. Supp. 16: 369-378.

Hales, D.F., J. Tomiuk, K. Wöhrmann, and P. Sunnucks. 1997. Evolutionary and genetic aspects of aphid biology. *European Journal of Entomology.* 94: 1-55.

Harris, P. 1974. A possible explanation of plant yield increases following insect damage. *Agro Ecosystems.* 1: 219-225.

Harvey, T.L., T.J. Martin, and D.L. Seifers. 1990. Wheat curl mite and wheat streak mosaic in moderate trichome density wheat cultivars. *Crop Science.* 30: 534-536.

Hatchett, J.H., G.L. Kreitner, and R.J. Elzinga. 1990. Larval mouthparts and feeding mechanism of the Hessian fly (Diptera: Cecidomyiidae). *Annals of the Entomological Society of America.* 83: 1137-1147.

Hatchett, J.H., K.J. Starks, and J.A. Webster. 1987. Insect and mite pests of wheat. In E.G. Heyne (ed.) *Wheat and wheat improvement.* Agron. Monogr. No. 13. ASA, CSSA, SSSA, Madison, WI, pp. 625-675.

Hawkins, C.P. and MacMahon, J.A. 1989. Guilds: The multiple meanings of a concept. *Annual Review of Entomology.* 34: 423-451.

Hayat, M.A., J.M. Martin, S.P. Lanning, C.F. McGuire, and L.E. Talbert. 1995. Variation for stem solidness and its association with agronomic traits in spring wheat. *Canadian Journal of Plant Science.* 75: 775-780.

Haynes, D.L. and S.H. Gage. 1981. The cereal leaf beetle in North America. *Annual Review of Entomology.* 26: 259-287.

Hearn, A.B. and G.P. Fitt. 1992. Cotton cropping systems. In C.J. Pearson (ed.) *Ecosystems of the world: Field crop ecosystems.* Elsevier, Amsterdam, pp. 85-142.

Hennig, H. 1987. Investigations on the injuriousness of the corn tortricid, *Cnephasia pumicana* (Zeller) (Lepidoptera: Tortricidae). *Pflanzenschutz.* 3: 12-15.

Hill, B.D., J.R. Byers, and G.B. Schaalje. 1992. Crop protection from permethrin applied aerially to control pale western cutworm (Lepidoptera: Noctuidae). *Journal of Economic Entomology.* 85: 1387-1392.

Hill, D.S. and J.M. Waller. 1988. *Pests and diseases of tropical crops,* volume 2. *Field handbook,* vol. 432. Longman Scientific and Technical, Hong Kong.

Hjorth, A. 1992. Attacks by wheat blossom midges—a review. 33rd Swedish Crop Protection Conference, Uppsala, pp. 221-233.

Hoffman, J. and O. Hofmanova. 1970. 1,4-Benzoxazine derivatives in plants. Sephandex fractionation and identification of a new glucoside. *European Journal of Biochemistry.* 8: 109-112.

Holmes, N.D. 1977. The effect of the wheat stem sawfly, *Cephus cinctus* (Himenoptera: Cephidae), on the yield and quality of wheat. *Canadian Entomologist.* 109: 1591-1598.

Holmes, N.D. 1984. The effect of light on the resistance of hard red spring wheats to the stem sawfly, *Cephus cinctus* (Hymenoptera: Cephidae). *Canadian Entomologist.* 116: 677-684.

Holmes, N.D. and L.K. Peterson. 1960. The influence of the host on oviposition by the wheat stem sawfly, *Cephus cintus* Nort. (Hymenoptera: Cephidae). *Canadian Journal of Plant Science.* 40: 29-46.

Holmes, R.S., R.L. Burton, J.D. Burd, and J.D. Ownby. 1991. Effect of greenbug (Homoptera: Aphididae) feeding on carbohydrate levels in wheat. *Journal of Economic Entomology.* 84: 897-901.

Hooker, A.L. 1984. The pathological and entomological framework of plant breeding. In J.P. Gustafson (ed.) *Gene manipulation in plant improvement, 16th Stadler Genetics Symposium.* Plenum, New York, pp. 177-208.

Hudson, H.F.W. 1920. Wheat, clover and the Hessian fly. *Farmers' Advocate and Home Magazine.* 55: 1485.

Hughes, G. 1996. Incorporating spatial pattern of harmful organisms into crop loss models. *Crop Protection.* 15: 407-421.

Ives, A.R. and W.H. Settle. 1997. Metapopulation dynamics and pest control in agricultural systems. *American Naturalist.* 149: 220-246.

Jacobson, L.A. 1965. Damage to wheat by say stink bug, *Chlorochroa sayi. Canadian Journal of Plant Science.* 45: 413-417.

Jacobson, L.A. and L.K. Peterson. 1965. Interrelations of damage to wheat and feeding by the Pale Western Cutworm, *Agrotis orthogonia* Morrison (Lepidoptera: Noctuidae). *Canadian Entomologist.* 97: 153-158.

Jameson, D.A. 1963. Responses of individual plants to harvesting. *Botanical Review.* 29: 532-594.

Jensen, S.G. 1972. Metabolism and carbohydrate composition in barley yellow dwarf virus-infected wheat. *Phytopathology.* 2: 587-592.

Johnson, K.B. 1987. Defoliation, disease and growth: A reply. *Phytopathology.* 7: 1495-1497.

Jones, F.G.W., R.A. Dunning, and K.P. Humpries. 1955. The effects of defoliation and loss of stand upon yield of sugar beet. *Annals of Applied Biology.* 43: 63-70.

Joppa, L.R., R.G. Timian, and N.D. Williams. 1980. Inheritance of resistance to greenbug toxicity in an amphiploid *Triticum turgidum/Triticum tauschii. Crop Science.* 20: 343-344.

Kaniuczak, Z. 1987. Cereal leaf beetles (*Oulema* spp.)—an increasing threat to cereal crops in Poland. *Materialy Sesji Instytutu Ochrony Roslin.* 27: 61-65.

Katagiri, C. 1979. Alpha-glucosidase in the midgut of the American cockroach, *Periplaneta americana. Insect Biochemistry.* 9: 205-209.

Kazemi, M.H. and H.F. Van Emden. 1992. Partial antibiosis to *Rhopalosiphum padi* in wheat and some phytochemical correlations. *Annals of Applied Biology.* 121: 1-9.

Kieckhefer, R.W. and J.L. Gellner. 1992. Yield losses in winter wheat caused by low-density cereal aphid populations. *Agronomy Journal.* 84: 180-183.

Kieckhefer, R.W., J.L. Gellner, and W.E. Riedell. 1995. Evaluation of the aphid-day standard as a predictor of yield loss caused by cereal aphids. *Agronomy Journal.* 87: 785-788.

Kloft, W. 1954. Über Einwirkungen des Saugaktes von *Myzus padellus* HRL. und Rogers (Aphidinae, Mycini CB) auf den Wasserhaushalt von *Prunus padus* L. *Phytopathogische Zeitschrift.* 22: 454-458.

Kogan, M. and E.F. Ortman. 1978. Antixenosis—A new term proposed to define Painter's "Nonpreference" modality of resistance. *ESA Bulletin.* 24: 175-176.

Kurppa, S.L.A. 1988. Control of orange wheat blossom midge, *Sitodiplosis mosellana* in Finland. *Brighton Crop Protection Conference. Pests and Diseases.* British Crop Protection Council, Surrey, UK. 3: 1017-1022.

Kurppa, S. 1989a. Susceptibility and reaction of wheat and barley varieties grown in Finland to damage by the orange wheat blossom midge *Sitodiplosis mosellana* (Gehin). *Annales Agriculturae Fenniae.* 28: 371-383.

Kurppa, S. 1989b. Wheat blossom midges, *Sitodiplosis mosellana* (Gehin) and *Contarinia tritici* (Kirby) in Finland, during 1981-87. *Annales Agriculturae Fenniae.* 28: 87-96.

Kurppa, S. and G.B. Husberg. 1989. Control of orange wheat blossom midge *Sitodiplosis mosellana* (Gehin), with pyrethroids. *Annales Agriculturae Fenniae.* 28: 103-111.

Lamb, R.J. and P.A. MacKay. 1995. Tolerance of antibiotic and susceptible cereal seedlings to the aphids *Metopolophium dirhodum* and *Rhopalosiphum padi. Annals of Applied Biology.* 127: 573-583.

Larrimer, W.H. 1922. An extreme case of delayed fall emergence of Hessian fly. *Annals of the Entomological Society of America.* 15: 177-180.

Leszczynski, B. 1985. Changes in phenols content and metabolism in leaves of susceptible and resistant winter wheat cultivars infested by *Rhopalosiphum padi* (L.), (Homoptera: Aphididae). *Z. zng. Ent.* 100: 343-348.

Leszczynski, B. 1992. Biochemical interaction between cereal aphids and winter triticale. *Materialy Sesji Instytutu Ochrony Roslin.* 32: 249-253.

Leszczynski, B. and A.F.G. Dixon. 1990. Resistance of cereals to aphids: Interaction between hydroxamic acids and the aphid *Sitobion avenae* (Homoptera: Aphididae). *Annals of Applied Biology.* 117: 21-30.

Leszczynski, B., H. Matok, and A.F.G. Dixon. 1992. Resistance of cereals to aphids: The interaction between hydroxamic acids and UDP-glucose transferases in the aphid *Sitobion avenae* (Homoptera: Aphididae). *Journal of Chemical Ecology.* 18: 1189-1200.

Leszczynski, B., L.C. Wright, and T. Bakowski. 1989. Effect of secondary plant substances on winter wheat resistance to grain aphid. *Entomologia Experimentalis et Applicata.* 52: 135-139.

López-Abella, D., R.H.E. Bradley, and K.F. Harris. 1988. Correlation between stylet paths made during superficial probing and the ability of aphids to transmit nonpersistent viruses. In K.F. Harris (ed.) *Advances in disease vector research*, vol. 5. Springer-Verlag, New York, pp. 251-285.

Lunn, G.D., R.K. Scott, P.S. Kettlewell, and B.J. Major. 1995. Effects of orange wheat blossom midge (*Sitodiplosis mosellana*) infection on pre-maturity sprouting and Hagberg falling number of wheat. Physiological responses of plants to pathogens, 11-13 September 1995, University of Dundee, UK. No. 42: 355-358.

Malone, M. 1996. Rapid, long distance signal transmission in higher plants. *Advances in Botanical Research*. 22: 163-227.

Maschinski, J. and T.G. Whitham. 1989. The continuum of plant responses to herbivory: The influence of plant association, nutrient availability, and timing. *American Naturalist*. 134: 1-19.

Massor, A., P. Matthes, and T. Wetzel. 1989. On the occurrence and monitoring of leaf miners in cereal crops. *Nachrichtenblatt für den Pflanzenschutz in der DDR*. 43: 165-167.

Maxwell, F.G., J.N. Jenkins, and W.L. Parrott. 1972. Resistance of plants to insects. *Advances in Agronomy*. 24: 187-251.

McKirdy, S.J. and R.A.C. Jones. 1997. Effect of sowing time on barley yellow dwarf virus infection in wheat: Virus incidence and grain yield losses. *Australian Journal of Agricultural Research*. 48: 199-206.

McMullen, C.R. and D.D. Walgenbach. 1986. Cytological changes in wheat induced by the Hessian fly. *Journal of the Kansas Entomological Society*. 59: 500-507.

McNaughton, S.J. 1983a. Compensatory plant growth as a response to herbivory. *Oikos*. 40: 329-336.

McNaughton, S.J. 1983b. Physiological and ecological implications of herbivory. In O.L. Lange, P.S. Nobel, C.B. Osmond, and H. Zeigler (eds.) *Physiological plant ecology responses to the chemical and biological environment*. Springer-Verlag, New York, pp. 657-677.

Messina, F.J., T.A. Jones, and D.C. Nielson. 1993. Performance of the Russian wheat aphid (Homoptera: Aphididae) on perennial range grasses: Effects of previous defoliation. *Environmental Entomology*. 22: 1349-1354.

Miller, B.S., H.L. Mitchell, J.A. Johnson, and E.T. Jones. 1958. Lipid soluble pigments of wheat plants as related to Hessian fly infestation. *Plant Physiology*. 33: 413-416.

Miller, B.S., R.J. Robinson, J.A. Johnson, E.T. Jones, and B.W.X. Ponnaiya. 1960. Studies on the relation between silica in wheat plants and resistance to Hessian fly attack. *Journal of Economic Entomology*. 53: 995-999.

Miller, B.S. and T. Swain. 1960. Chromatographic analyses of the free amino acids, organic acids and sugars in wheat plant extracts. *Journal of the Science of Food and Agriculture*. 6: 344-348.

Miller, R.H. 1992. Insect pests of wheat and barley of Mediterranean Africa and West Asia. *Al Awamia*. 77: 3-20.

Miller, R.H., S. El Masri, and K. Al Jundi. 1993. Plant density and wheat stem sawfly (Hymenoptera: Cephidae) resistance in Syrian wheats. *Bulletin of Entomological Research*. 83: 95-102.

Montllor, C.B. and F.E. Gildow. 1986. Feeding responses of two grain aphids to barley yellow dwarf virus-infected oats. *Entomologia Experimentalis et Applicata.* 42: 63-69.

Mooney, H.A., W.E. Winner, and E.J. Pell. 1991. *Response of plants to multiple stresses.* Academic Press, New York.

Morgham, A.T., P.E. Richardson, R.K. Campbell, J.D. Burd, R.D. Eikenbary, and L.C. Sumner. 1994. Ultrastructural responses of resistant and susceptible wheat to infestation by greenbug biotype E (Homoptera: Aphididae). *Annals of the Entomological Society of America.* 87: 908-917.

Mowry, T.M. 1994. Russian wheat aphid (Homoptera: Aphididae) survival and fecundity on barley yellow dwarf virus-infected wheat resistant and susceptible to the aphid. *Environmental Entomology.* 23: 326-330.

Nicol, D., S.V. Copaja, S.D. Wratten, and H.M. Niemeyer. 1992. A screen of worldwide wheat cultivars for hydroxamic acid levels and aphid antixenosis. *Annals of Applied Biology.* 121: 11-18.

Nkongolo, K.K., N.L.V. Lapitan, and J.S. Quick. 1996. Genetic and cytogenetic analyses of Russian wheat aphid resistance in triticale X wheat hybrids and progenies. *Crop Science.* 36: 1114-1119.

Oakley, J.N. 1992. Development of an integrated control strategy for summer aphids in winter wheat. *Proceedings Brighton Crop Protection Conference Pests and Diseases* 1992, vol. 3, Brighton. pp. 1009-1014. British Crop Protection Council, Surrey, UK.

Oakley, J.N. 1994. Orange wheat blossom midge: A literature review and survey of the 1993 outbreak. *HGCA Research Review.* No. 28: 51 pp.

Oakley, J.N., D.I. Green, A.E. Jones, J.B. Kilpatrick, and J.E.B. Young. 1994. Forecasting the abundance of orange wheat blossom midge in wheat. *Proceedings Brighton Crop Protection Conference 1994,* vol. 1. pp. 193-198. British Crop Protection Council, Surrey, UK.

Oakley, J.N. and K.F.A. Walters. 1994. A field evaluation of different criteria for determining the need to treat winter wheat against the grain aphid *Sitobion avenae* and the rose-grain aphid *Metopolophium dirhodum. Annals of Applied Biology.* 124: 195-211.

Ohnmeiss, T.E. and I.T. Baldwin. 1994. The allometry of nitrogen allocation to growth and inducible defense under nitrogen-limited growth. *Ecology.* 75: 995-1002.

Okigbo, B.N. and G.G. Gyrisco. 1962. Effects of fertilizers on Hessian fly infestation. *Journal of Economic Entomology.* 55: 753-760.

Painter, R.H. 1951. *Insect resistance in crop plants.* Macmillan. New York.

Painter, R.H. 1954. Some ecological aspects of the resistance of crop plants to insects. *Journal of Economic Entomology.* 47: 1036-1040.

Painter, R.H. 1960. Entomological problems in developing new wheats. *Cereal Science Today.* 5: 2 pp.

Papp, M. and A. Mesterházy. 1993. Resistance to bird cherry-oat aphid (*Rhopalosiphum padi* L.) in winter wheat varieties. *Euphytica.* 67: 49-57.

Papp, M. and A. Mesterházy. 1996. Resistance of winter wheat to cereal leaf beetle (Coleoptera: Chrysomelidae) and bird cherry-oat aphid (Homoptera: Aphididae). *Journal of Economic Entomology.* 89: 1649-1657.

Paré, P.W. and J.H. Tumlinson. 1996. Plant volatile signals in response to herbivore feeding. *Florida Entomology.* 79: 93-103.

Passioura, J.B. 1996. Drought and drought tolerance. *Plant Growth Regulation.* 20: 79-83.

Pearson, C.J. 1992. *Field crop ecosystems.* Elsevier, Amsterdam.

Petcu, L., C. Popov, and A. Barbulescu. 1987. Aspects concerning the biology, ecology and control of the saddle-gall midge (*Haplodiplosis marginata* Von Roser). *Analele Institutului de Cercetari pentru Cereale si Plante Tehnice, Fundulea.* 54: 363-374.

Pettersson, O. 1994. Reduced pesticide use in Scandinavian agriculture. *Critical Reviews in Plant Science.* 13: 43-55.

Pollard, D.G. 1973. Plant penetration by feeding aphids (Hemiptera: Aphidoidea): A review. *Bulletin of Entomology Research.* 62: 631-714.

Porter, D.R., J.A. Webster, and C.A. Baker. 1993. Detection of resistance to the Russian wheat aphid in hexaploid wheat. *Plant Breeding.* 110: 157-160.

Quisenberry, S.S. and D.J. Schotzko. 1994. Russian wheat aphid (Homoptera: Aphididae) population development and plant damage on resistant and susceptible wheat. *Journal of Economic Entomology.* 87: 1761-1768.

Rabbinge, R., E.M. Drees, M. Graaf, F.C.M. van der Verberne, and A. Wesselo. 1981. Damage effects of cereal aphids in wheat. *Netherlands Journal of Plant Pathology.* 87: 217-232.

Ratcliffe, R.H. and J.H. Hatchett. 1997. Biology and genetics of the Hessian fly and resistance in wheat. In K. Bondari (ed.) *New Developments in Entomology.* Research Signpost, Scientific Information Guild, Trivandrum, India.

Ratcliffe, R.H., H.W. Ohm, F.L. Patterson, S.E. Cambron, and G.G. Safranski. 1996. Response of resistance genes H9 to H19 in wheat to Hessian fly (Diptera: Cecidomyiidae) laboratory biotypes and field populations from the eastern United States. *Journal of Economic Entomology.* 89: 1309-1317.

Ratcliffe, R.H., G.G. Safranski, F.L. Patterson, H.W. Ohm, and P.L. Taylor. 1994. Biotype status of Hessian fly (Diptera: Cecidomyiidae) populations from the eastern United States and their response to 14 Hessian fly resistance genes. *Journal of Economic Entomology.* 87: 1113-1121.

Reed, T. 1986. Cold, drought, disease, and Hessian fly are hurting Alabama wheat. *Western Alabama IPM Newsletter.* 7: 2 pp.

Refai, F.Y., L.E. Jones, and B.S. Miller. 1955. Some biochemical factors involved in the resistance of the wheat plant to attack by the Hessian fly. *Cereal Chemistry.* 32: 437-451.

Riedell, W.E. and R.W. Kieckhefer. 1995. Feeding damage effects of three aphid species on wheat root growth. *Journal of Plant Nutrition.* 18: 1881-1891.

Roberts, J.J., R.L. Gallun, F.L. Patterson, and J.E. Foster. 1979. Effects of wheat leaf pubescence on the Hessian fly. *Journal of Economic Entomology.* 72: 211-214.

Roloff, B. and T. Wetzel. 1989. Investigations on the influence of soil temperature on larval hatching in the wheat bulb fly (*Delia coarctata* (Fall.) and on the expression of injury symptoms. *Archiv für Phytopathologie und Pflanzenschutz.* 25: 403-407.

Rosenthal, J.P. and P.M. Kotanen. 1994. Terrestrial plant tolerance to herbivory. *Trends in Ecology and Evolution.* 9: 145-148.

Rossing, W.A.H. 1991. Simulation of damage in winter wheat caused by the grain aphid Sitobion avenae. 3. Calculation of damage at various attainable yield levels. *Netherlands Journal of Plant Pathology.* 97: 87-103.

Rossing, W.A.H. and K.L. Heong. 1997. Opportunities for using systems approaches in pest management. *Field Crops Research.* 51: 83-100.

Rossing, W.A.H. and L.A.J.M. van de Wiel. 1990. Simulation of damage in winter wheat caused by the grain aphid *Sitobion avenae.* 1. Quantification of the effects of honeydew on gas exchange of leaves and aphid populations of different size on crop growth. *Netherlands Journal of Plant Pathology.* 96: 343-364.

Rossing, W.A.H., M. van Oijen, W. van der Werf, L. Bastiaans, and R. Rabbinge. 1992. Modelling the effects of foliar pests and pathogens on light interception, photosynthesis, growth rate and yield of field crops. In P.G. Ayres (ed.) *Pest and pathogens: Plant responses to foliar to foliar attack.* Bios Scientific Publishers, Oxford, pp. 161-180.

Rubia, E.G., K.L. Heong, M. Zalucki, B. Gonzales, and G.A. Norton. 1996. Mechanisms of compensation of rice plants to yellow stem borer *Scirpophaga incertulas* (Walker) injury. *Crop Protection.* 15: 335-340.

Ryan, J.D., R.C. Johnson, R.D. Eikenbary, and K.W. Dorschner. 1987. Drought/greenbug interactions: Photosynthesis of greenbug resistant and susceptible wheat. *Crop Science.* 27: 283-288.

Sadras, V.O. 1996. Population-level compensation after loss of vegetative buds: Interactions among damaged and undamaged cotton neighbours. *Oecologia.* 106: 432-439.

Sadras, V.O. 1997. Effects of simulated insect damage and weed interference on cotton growth and reproduction. *Annals of Applied Biology.* 130: 271-281.

Sadras, V.O. and G.W. Felton. 1997. Mechanisms of cotton resistance to arthropod herbivory. In J.M. Stewart, D. Oosterhuis, and J.J. Heitholt (eds.) *Cotton physiology.* The Cotton Foundation, Memphis, TN.

Saxena, P.N. and H.L. Chada. 1971. The greenbug, *Schyzaphis graminum.*1. Mouth parts and feeding habits. *Annals of the Entomological Society of America.* 64: 897-904.

Sebesta, E.E., E.A. Wood, Jr., D.R. Porter, J.A. Webster, and E.L. Smith. 1995. Registration of Amigo wheat germplasm resistant to greenbug. *Crop Science.* 35: 293.

Sharkey, T.D. 1996. Emission of low molecular mass hydrocarbons from plants. *Trends in Plant Science.* 1: 78-82.

Shukle, R.H., P.B. Grover, Jr., and J.E. Foster. 1990. Feeding of Hessian fly (Diptera: Cecidomyiidae) larvae on resistant and susceptible wheat. *Environmental Entomology.* 19: 494-500.

Skuhravy, V. and M. Skuhrava. 1986. Control of the saddle midge (*Haplodiplosis marginata* (Von Roser)—an example of integrated pest management in cereal crops. *Nachrichtenblatt für den Pflanzenschutz in der DDR.* 40: 160-161.

Starks, R.J., R.L. Burton, and O.G. Merkle. 1983. Greenbugs (Homoptera: Aphididae) plant resistance in small grains and sorghum to biotype E. *Journal of Economic Entomology.* 76: 877-880.

Storlie, E.W., L.E. Talbert, G.A. Taylor, H.A. Ferguson, and J.H. Brown. 1993. Effects of the Russian wheat aphid on osmotic potential and fructan content of winter wheat seedlings. *Euphytica.* 65: 9-14.

Swaine, G. and D.A. Ironside. 1983. Insect pests of field crops. Queensland Department of Primary Industries Information Series Q18 3006.

Taylor, L.R. 1984. Assessing and interpreting the spatial distributions of insect populations. *Annual Review of Entomology.* 29: 321-57.

Thackray, D.J., S.D. Wratten, P.J. Edwards, and H.M. Niemeyer. 1990. Resistance to the aphids *Sitobion avenae* and *Rhopalosiphum padi* in Gramineae in relation to hydroxamic acid levels. *Annals of Applied Biology.* 116: 573-582.

Tremblay, C., C. Cloutier, and A. Comeau. 1989. Resistance to the bird cherry-oat aphid, *Rhopalosiphum padi* L. (Homoptera: Aphididae), in perennial Gramineae and wheat × perennial Gramineae hybrids. *Environmental Entomology.* 18: 921-932.

Trewavas, A. 1981. How do plant growth substances work? *Plant Cell and Environment.* 4: 203-228.

Trumble, J.T., D.M. Kolodny-Hirsh, and I.P. Ting. 1993. Plant compensation for arthropod herbivory. *Annual Review of Entomology.* 38: 93-119.

Turner, N.T., Prasertsak, P., and Setter, T.L. 1994. Plant spacing, density, and yield of wheat subjected to postanthesis water deficits. *Crop science.* 34: 741-748.

Tyler, J.M., J.A. Webster, and O.G. Merkle. 1987. Designations for genes in wheat germplasm conferring greenbug resistance. *Crop Science.* 27: 526-527.

van Emden, H.F. and P. Hadley. 1994. The application of the concepts of resource capture to the effect of pest incidence on crops. In J.L. Monteith, R.K. Scott, and M.H. Unsworth (eds.) *Resource capture by crops.* Proceedings 52nd Easter School, University of Nottingham, School of Agriculture. Nottingham University Press, Nottingham, UK, pp. 149-165.

Varis, A.L. 1991. Effect of Lygus (Heteroptera: Miridae) feeding on wheat grains. *Journal of Economic Entomology.* 84: 1037-1040.

Viator, H.P., A. Pantoja, and C.M. Smith. 1983. Damage to wheat seed quality and yield by the rice stink bug and southern green stink bug (Hemiptera: Pentatomidae). *Journal of Economic Entomology.* 76: 1410-1413.

Wahlross, O. and A.I. Virtanen. 1959. Precursors of 6-methoxy-benzoxazolinone in maize and wheat plants, their isolation and some of their properties. *Acta Chemica Scandinavica.* 13: 1906-1908.

Waring, G.L. and N.S. Cobb. 1992. The impact of plant stress on herbivore population dynamics. In E.A. Bernays (ed.) *Insect-plant interactions,* vol. 4. CRC Press, Boca Raton, FL, pp. 167-226.

Webster, F.M. 1906. *The Hessian fly.* USDA Cir. #70.

Webster, F.M. and E.O.G. Kelly. 1915. *The Hessian fly situation in 1915.* USDA Cir. #51.

Webster, J.A., C. Inayatullah, M. Hamissou, and K.A. Mirkes. 1994. Leaf pubescence effects in wheat on yellow sugarcane aphids and greenbugs (Homoptera: Aphididae). *Journal of Economic Entomology.* 87: 231-240.

Webster, J.A., D.H. Smith, and C. Lee. 1972. Reduction in yield of spring wheat caused by cereal leaf beetles. *Journal of Economic Entomology.* 65: 832-835.

Weiss, M.J. and W.L. Morrill. 1992. Wheat stem sawfly (Hymenoptera: Cephidae) revisited. *American Entomologist.* 38: 241-245.

Wellso, S.G. 1991. Aestivation and phenology of the Hessian fly (Diptera: Cecidomyiidae) in Indiana. *Environmental Entomology.* 20: 795-801.

Wellso, S.G. and J.E. Araya. 1993. Resistance stability of the secondary tiller of 'Caldwell' wheat after the primary culm was infested with virulent Hessian fly (Diptera: Cecidomyiidae) larvae. *The Great Lakes Entomologist.* 26: 71-76.

Wellso, S.G., R.C. Coolbaugh, and R.P. Hoxie. 1991. Effects of ancymidol and gibberellic acid on the response of susceptible 'Newton' and resistant 'Abe' winter wheat infested by biotype E Hessian flies (Diptera: Cecidomyiidae). *Environmental Entomology.* 20: 489-493.

Wellso, S.G. and R.D. Freed. 1982. Positive association of the wheat midge (Diptera: Cecidomyiidae) with glume blotch. *Journal of Economic Entomology.* 75: 885-887.

Wellso, S.G. and R.P. Hoxie. 1994. Tillering response of 'Monon' and 'Newton' winter wheats infested with biotype L Hessian fly (Diptera: Cecidomyiidae) larvae. *The Great Lakes Entomologist.* 27: 235-239.

Wellso, S.G., R.P. Hoxie, and C.R. Olien. 1987. *Hessian fly, Mayetiola destructor (Say) (Diptera: Cecidomyiidae), induced changes in 'Winoka' wheat.* Dr. W. Junk Publishers, Dordrecht, p. 423.

Wellso, S.G., R.P. Hoxie, and C.R. Olien. 1989. Effects of Hessian fly (Diptera: Cecidomyiidae) larvae and plant age on growth and soluble carbohydrates of Winoka winter wheat. *Environmental Entomology.* 18: 1095-1100.

Wellso, S.G., R.P. Hoxie, and P.L. Taylor. 1990. A comparison of plant parameters and soluble carbohydrates of resistant and susceptible wheat infested with biotype E Hessian flies (Diptera: Cecidomyiidae). *Environmental Entomology.* 19: 1698-1701.

Wellso, S.G., C.R. Olien, and R.P. Hoxie. 1985. Winter wheat cold hardiness and fructan reserves affected by *Rhopalosiphum padi* (Homoptera: Aphididae) feeding. *The Great Lakes Entomologist.* 18: 29-32.

Wellso, S.G., C.R. Olien, R.P. Hoxie, and A.S. Kuhna. 1986. Sugar reserves and cold hardiness of winter wheat reduced by larval feeding of the Hessian fly, *Mayetiola destructor* (Say) (Diptera: Cecidomyiidae). *Enviromental Entomology.* 15: 392-395.

Wood, E.A.J. 1965. Effect of foliage infestation of the English grain aphid on yield of Triumph wheat. *Journal of Economic Entomology.* 58: 778-779.

Wood, E.A.J., E.E. Sebesta, J.A. Webster, and D.R. Porter. 1995. Resistance to wheat curl mite (Acari: Eriophyidae) in greenbug-resistant "Gaucho" triticale and "Gaucho" X and wheat crosses. *Journal of Economic Entomology.* 88: 1032-1036.

Wood, T.G. and R.H. Cowie. 1988. Assessment of on-farm losses in cereals in Africa due to soil insects. *Insect Science and Its Application.* 9: 709-716.

Wratten, S.D. and P.C. Redhead. 1976. Effects of cereal aphids on the growth of wheat. *Annals of Applied Biology.* 84: 437-455.

Young, J.E.B. and J. Cochrane. 1993. Changes in wheat bulb fly (*Delia coarctata*) populations in East Anglia in relation to crop rotations, climatic data and damage forecasting. *Annals of Applied Biology.* 123: 485-498.

Zangerl, A.R. and F.A. Bazzaz. 1992. Theory and pattern in plant defense allocation. In R.S. Fritz and E.L. Simms (eds.) *Plant resistance to herbivores and pathogens: Ecology, evolution and genetics.* University of Chicago Press, Chicago, pp. 363-391.

Chapter 10

Wheat Yield As Affected by Diseases

Erlei Melo Reis
Carlos A. Medeiros
Marta M. Casa Blum

INTRODUCTION

According to the disease concept, pathogens interfere in plant physiological processes, leading to yield reductions. The interference of plant pathogens with the vital processes of wheat leads to classes of pathogenesis (McNew, 1960). The infected plant shows alteration in its physiology and morphology, which leads to damage and eventually yield losses. The terminology of crop yield losses has been reviewed by Zadoks (1985), which states that any visible and measurable symptom caused by a harmful organism is collectively called *injury*; *damage* is defined as reduction in the quantity and/or quality of yield. The reduction in financial return per unit area due to harmful organisms is called *loss*. *Economic damage threshold* is the disease intensity at which the cost of control equals the amount of yield loss.

PHYSIOLOGICAL PROCESSES INTERRUPTED BY WHEAT PATHOGENS

Prevention of Seedling Metabolism (Seed Decay)

Seed decay determines the quick death of seedlings soon after germination and before soil emergence. It is caused by several species of fungi belonging to the *Pythium* genus. These pathogens are found in most agricultural soils of the world, attacking a wide range of hosts including small grain crops. The pathogens in this class are seedling blight fungi such as *Pythium arrenomanes* Drechs., *P. graminicola* Subr., *P. aphanidermatum* (Edson) Fitz., *P. volutum* Vant and Tru, and *P. myriotylum* Drechs. This disease is common in wet, phosphorus-deficient, and low-organic-matter soils. As a rule of thumb, they are aggressive in juvenile tissue but may be restricted as soon as the adjacent tissue begins to achieve more mature differentiation. In some soils where a *Pythium*-specific fungicide, metalaxyl (Apron 5G, 5 kg/ha), was applied, seed emergence increases of 50 plants per square meter were obtained in wheat. Nevertheless the effect on yield was not determined (Reis, 1982).

Interference with Water and Mineral Procurement and Absorption (Crown and Root Rot)

The root rotting disease complex is caused by a wide variety of fungi. They show some advance in parasitism over seedling pathogens because they are able to invade well-formed roots through lignified, mature tissue. Most of them are poorly specialized as plant parasites but they are aggressive pathogens. The routes of invasion are usually through wounds created by emergence of secondary roots, abrasion with soil particles, nematode penetration, or insect damage. Roots may be preconditioned to invasion by excess moisture or nutrient conditions such as are found in the brown root rot of wheat (Wiese, 1991).

In any evolutionary scheme, the root pathogens might be classified as the first strongly aggressive plant invaders. They include destructive fungi species, such as *Gaeumannomyces graminis* (Sacc.) von Arx and Olivier var. *tritici* Walker, *Bipolaris sorokiniana* (Sacc. In Sorok.) Shoem. and *Fusarium graminearum* Schwabe. Damage to wheat is related to the extent and time of colonization of roots and stem bases. In the case of take-all, when symptoms occur, grain yields normally are less than half of those of healthy plants. Some of these pathogens, such as *F. graminearum,* that are natural soil inhabitants are difficult to control and are chronically destructive, but the soil invaders such as take-all fungus, which are poorly adapted to survival apart from their hosts because of competition with other microorganisms, are controlled by short-term crop rotation with oats (Reis, 1990).

The root rot parasites are poorly specialized even though some have become adjusted to the rhizosphere environment and thrive best on or in roots. They are general cortical invaders with very little specialization in preference for tissue. The hosts have very poorly developed resistance. Most of the resistance exploited agronomically is in the nature of disease-escaping tactics, when a particular variety has ability to develop new roots to replace those destroyed by the pathogen or has a favorable seasonal development that avoids the infection period. This phenomenon has been reported for take-all (McNew, 1960).

Crop rotation and seed treatment are the most important control measures for the soil invaders, but for soil inhabitants the best control would be the development of soil antibiosis. On a worldwide basis yield reduction caused by take-all is about 30 percent, and in Southern Brazil common root rot caused by *B. sorokiniana* and *F. graminearum* may cause losses up to 20 percent (Diehl, Tinline, and Kockhann, 1983).

Interference with Upward Transport (Wilt)

The vascular or wilt disease complex is caused by xylem parasites that interfere with the upward movement of water and mineral nutrients. Some of the fungus component of this complex live in soil and have habits similar to those of the root rot complex. They invade directly through root hairs or natural wounds, traverse the cortical tissues, and become established in the xylem vessels. Although they are only facultative parasites, they show a much higher order of specialization than the root rot in their preference for a specific tissue and for certain varieties of crops (McNew, 1960).

The masses of mycelium cells present in the tracheal tubes interfere with the movement of water and nutrients. These mechanical impediments do not explain fully all the wilting symptoms, so attention is focused upon the metabolic by-products of the parasite (Wiese, 1991).

The only known vascular disease in wheat is cephalosporium stripe, caused by *Cephalosporium gramineum* Nis. and Ika (syn. *Hymenula cerealis* Ell. and Ev.). The organism is a soilborne fungus and is the only true vascular pathogen of wheat. It colonizes and blocks xylem vessels, restricting the transport of water and nutrients to nodes, leaf veins, and interveinal tissues. This disease has not been reported in South America yet. The incidence of disease in the United States may range up to 100 percent in some fields, and yields may be reduced up to 80 percent (Wiese, 1991).

Interference with the Photosynthetic Process

The pathogens that invade green plant parts include three diverse groups ranging from poorly specialized necrotrophs to highly specialized biotroph parasites. Most of the general leaf-blight organisms have the ability to infect and destroy stems and leaves. Thus they may cause substantial loss in the food factory by reducing the leaf area index. This overlapping of parasitic activity on different organs is not surprising because the tissues involved are morphologically analogous.

The primary mechanism of pathogenesis is destruction of the food-synthesizing potential of the plant. This results from local necrosis of the leaf blade, which may become very extensive in some diseases such as spot blotch, yellow spot, septoria leaf spot, and speckled leaf spot. Extensive rupture of the cuticle by the sporulating fungus may disturb the water economy of the plant with serious consequences such as often occur in wheat fields affected with *Puccinia* spp. (McNew, 1960).

Leaf Blights

The leaf blights are caused by necrotroph parasitic fungi that cause lesions, causing tissue death. In their parasitic action they produce toxins, making them more destructive. Taxonomically they are Imperfecti, and Ascomycetes fungi. The most important leaf-blights in wheat are yellow spot: *Pyrenophora tritici-repentis* (Died.) Drechs., anamorph *Drechslera tritici-repentis* (Died.) Shoem.; septoria leaf and glume blotches: *Leptosphaeria nodorum* Müller, anamorph *Septoria nodorum* (Berk.); *Micosphaerella graminicola* (Fuckel) Schroeter, anamorph *Septoria tritici* Rob. In Desm.; and spot blotch: *Cochliobolus sativus* (Ito and Kurib) Drechs. Ex Dastur, anamorph *Bipolaris sorokiniana* (Sacc. in Sorok) Schoem. *Bipolaris sorokiniana* is also a root rot fungus (Wiese, 1991).

The control of this disease by genetic resistance strategies has been found to be difficult because the pathogens do not show a high degree of specialization. Hence, integrated disease management including crop rotation, seed treatment, and chemical control has been used by growers (Reis et al., 1992).

In the southern zone of South America the damage caused by leaf-blights has been as high as 80 percent (Metha, 1993). The economic threshold recommended to chemically control leaf blights is 5 percent foliar severity (Recomendação, 1996). For this complex of diseases severity (S) may be estimated through incidence (I) by the equation (Reis, Blum, in press):

$$S = 13.4494 + 13.9728\, I \;(R^2 = 0.82)$$

Downy Mildew

Downy mildew is caused by much more specialized parasites than the general leaf blights. Members of the Peronosporaceae are bitrophic parasites that establish a compatible relationship with leaf tissue and maintain it until the fungus begins to sporulate.

They usually have rather specific host requirements and there are sharp differences in the varietal reaction of plants to their invasion. It is possible to breed for disease resistance. The use of haustoria by the pathogen for food searching with a minimum of injury to the host appears for the first time in this group. In general, this group of pathogens has a much more highly developed type of parasitism than any of those in the preceding group, not only host specificity but also a restrained type of feeding that avoids immediate destruction of the invaded tissue (McNew, 1960).

Downy mildews are caused by fungi called "water molds" because they depend on free moisture for at least part of their life cycle. On wheat, however, the sign of downy fungal growth is rarely visible. The causal organism is *Sclerophthora macrospora* (Sacc.) T. S. and N. The downy mildew pathogen may attack barley, rice, corn, oats, sorghum, and about 140 species of perennial and annual grasses. Damage has not been quantified (Wiese, 1991).

Powdery Mildew

Powdery mildew of wheat is caused by the biotrophic parasite (Erysiphaceae) *Erysiphe graminis* Dc. f.sp. *tritici* E. Marschal (syn. *Blumeria graminis*), with well-developed haustoria that permit it to feed on the cell contents without extensive injury to the supporting tissue. The dense growth of mycelium and sporophores on the leaf surface accelerates respiration and undoubtedly decreases photosynthesis by preventing sun rays from reaching the photosynthetizing surface. However, infected leaves persist for weeks and may show only the slightest traces of necrosis. This could be attributable to the fact that relatively little of the parasite's body reaches the interior of the host tissue where its toxic secretions would be added to the cells of the host. Preference for specific hosts is very marked in the powdery mildews. The significant feature of the parasitism of powdery mildew is that the parasite has no saprophytic existence apart from its host and must depend entirely upon specialization for its survival. With this change, there appears to be some evidence of a very compatible establishment in the host even though the host tissue is not stimulated to excessive growth (McNew, 1960).

In southern Brazil, powdery mildew has been reported to cause yield losses up to 39 percent in susceptible cultivars such as Trigo BR 23. An equation to estimate yield damage (Y) as a function of foliar incidence (I) has been established:

$$Y = 2{,}073.12 - 11.39\, I\ (R^2 = 0.72)$$

(Reis, Casa, and Hoffmann, in press). The economic threshold for chemical control of powdery mildew in Brazil is 10 to 15 percent foliar incidence after tillering (Recomendação, 1996).

Rusts

Rusts are caused by parasites (Uredinales) that are very similar to powdery mildews in their parasitic ability. Both may be grouped together as biotrophic leaf parasites except that the rusts are even more specialized. The rusts have achieved remarkable progress in approaching a commensal relationship with the host. A hypersensitive reaction of the host cell that ends in its death is actually an immune reaction in this type of establishment. The more adapted races of rust invade the tissue intercellularly without causing such injury. The haustorium or occasional intracellular mycelium approaches the nucleus or attracts the nucleus in such a way that they often adhere closely to each other. The parasite may be suspected of either finding special nutrients at this center of cell activities or stimulating host metabolism from this location (McNew, 1960).

The susceptible host cell enlarges rapidly and begins to undergo division so that a new type of metabolically active tissue is formed, composed of the fungus and host cells in a very compatible relationship. A change in the rate of respiration and in the respiratory quotient indicates that a new type of metabolism has come into existence. This is different from normal host activity. No one knows exactly which compound is metabolized but there is a definitive change in the citric acid metabolism of the tissue and there is strong indication that succinic acid affects the respiration in uredospores of *P. graminis.* As the rust pustule ruptures the cuticle to expose uredospores or spermogonia, there may be considerable necrosis of the host cells at the base of the lesion in the more resistant varieties and there is an appreciable loss in the water economy of the host (McNew, 1960).

In wheat three important rusts are very well studied: stem rust (*Puccinia graminis* Pers f.sp. *tritici* Eriks and Henn.), leaf rust (*P. recondita* Rob. Ex Desm. f.sp. *tritici*), and strip rust (*P. striiformes* West.). Stem rust is one of the most destructive diseases known.

Barcellos and Ignaczak (1978) determined losses of 50 percent caused by leaf rust in southern Brazil. Burleigh, Roelfs, and Eversmeyer (1972) developed an equation to estimate the damage caused by leaf rust:

$$\text{Damage (\%)} = 5.3788 + 5.5260\, X_2 - 0.3308\, X_5 + 0.5019\, X_7$$

where X_2 is the severity per tiller at stage 10, X_5 is the severity on flag leaf at stage 11.1, and X_7 is the severity on flag leaf at stage 11.2 (Feeks scale modified by Large, 1954).

Wheat rusts have been controlled specially through resistant cultivars, but in those cases where resistance is broken chemical control has been recommended to growers. The recommendation to control leaf rust in southern Brazil is based on a foliar severity of 1 to 5 percent (Recomendação, 1996). Reis (1996) established the correlation to estimate severity (S) from incidence (I) through to the equation:

$$S = -7.948 \cdot \log(1 - I/100)\ (r^2 = 0.62)$$

Thus a foliar severity of 1 to 5 percent corresponds to 35 to 45 percent foliar incidence.

DIVERSION OF FOODSTUFFS TO ABNORMAL USES (BUNT AND SMUTS)

Interference with use of food materials is encountered in many types of parasitism such as those discussed above in the rust fungi. Infected tissues are stimulated to abnormal uses of foodstuffs in synthesizing materials that are not of value to the host, since they do not make any contribution and can be considered secondary to the host's essential functions. There are two groups of parasites in this classification: the smuts and viruses.

Bunts and Smuts

According to Wiese (1991), plants infected with common bunt may be stunted and will be distinguished from the healthy ones by head emergence. Mycelium inhabits the developing ovary and displaces all tissues within the pericarp. Common bunt reduces wheat yield and grain quality. It is caused by *Tilletia tritici* (Bjerk.) Wint. and *T. laevis* Kühn. The state of restrained parasitism may endure for weeks or even months before the parasite becomes aggressively pathogenic. Pathogenesis awaits the proper stage of nutritional development in the host which, in this case, apparently occurs when the leaves begin to supply food materials to the young ovary. The food materials that would have enriched the endosperm and developed an embryo are diverted into nourishing the parasite and overgrown gall tissue.

Loose smut of wheat is easily recognized in the field. The causal fungus, *Ustilago tritici* (Pers.) Rostr., is unique in that it is first incorporated into developing kernels and persists within seed embryos.

This particular disease reduces yield in proportion to its incidence. Yield losses in southern Brazil have been as high as 22 percent in some fields.

The Viral Diseases

The viruses divert essential amino acids and nucleotides into synthesis of virus nucleoprotein. Viral disease damage to plants is not due primarily to depletion of nutrients that have been diverted to synthesis of the virus itself, but to other more indirect effects. These effects are caused by the virus-induced synthesis of new proteins (nucleoproteins), which can interfere with the normal metabolism of the plant. These nucleoproteins can never be digested or converted into host protoplasm again even when the normal host tissue is starved for nitrogen (Agrios, 1988). This loss of organic nitrogen to virus multiplication constitutes a drain on the host's metabolic processes. There is much more to the pathogenic activities of the virus than the drain on nitrogen metabolism by the increased synthesis of nucleoproteins.

Viruses cause a decrease in photosynthesis through a decrease in the amount of chlorophyll in the leaf, chlorophyll efficiency, and in leaf area per plant. Viruses may cause also a decrease in plant growth regulators. Respiration of plants increases immediately after infection with a virus.

The main virus disease in wheat in South America is barley yellow dwarf, probably the most widely distributed and the most destructive virus disease of cereals. Damage to wheat is estimated to reduce yields 5 to 25 percent.

The second virus disease is soilborne wheat mosaic, transmitted by a soil fungus vector, *Polymyxa graminis* (Wiese, 1991). No reports on the losses caused by this virus in South America were found.

SCAB—AN EXCEPTION

The interference of scab in physiological processes does not fit into McNew's classes. Scab is a floral-infecting disease caused by *Gibberella zeae* (Schw.) Petch, anamorph *F. graminearum*. The infection takes place after the deposition of ascospores on the caught anthers. This is the sole infection point of wheat because anthers are rich in chlorine and betaine (Strange, Majer, and Smith, 1974). Green tissues are resistant and the susceptible period of wheat heads extends from the beginning of flowering up to ripening. When anthers are removed infection does not occur. Flowers are destroyed and the pathogen reaches the rachis, blocking the upward nutrition flow. A ten-year period of loss determination showed a grain yield reduction up to 14 percent (Reis, 1988).

Control of scab is very difficult to achieve through breeding, and chemical control is not feasible because the deposition of fungicides on the infection courts is limited.

REFERENCES

Agrios, G.N. (1988). *Plant Pathology.* 3rd ed. San Diego and London: Academic Press.

Barcellos, A.L. and Ignaczak, J.C. (1978). Efeito da ferrugem da folha em diferentes estádios de desenvolvimento do trigo. In *Reunião Anual Conjunta de Pesquisa de Trigo,* 10., Porto Alegre, 1978. Solos e Técnicas Culturais, Economia e Sanidade. Passo Fundo: EMBRAPA/CNPT, pp. 212-219.

Burleigh, J.R., Roelfs, A.P., and Eversmeyer, M.G. (1972). Estimating damage to wheat caused by *Puccinia recondita tritici. Phytopathology.* 62:944-946.

Diehl, J.A., Tinline, R.D., and Kochhann, R.A. (1983). Perdas causadas pela podridão comum de raízes no Rio Grande do Sul, 1979-1981. *Fitopatologia Brasileira.* 8:507-511.

Large, E.C. (1954). Growth stages in cereals: Illustrations of the Feekes scale. *Plant Pathology.* 3:128-129.

McNew, G.L. (1960). The nature, origin, and evolution of parasitism. In Horsfall, J.G. and Dimond, A.E. (eds.). *Plant Pathology, vol. 2.* New York: Academic Press, pp. 2-66.

Metha, Y.R. (1993). *Manejo integrado de enfermedades del trigo.* Londrina, PR, Brazil.

Recomendação da Comissão Sul-Brasileira de Pesquisa de Trigo. (1996). *Reunião da Comissão Sul-Brasileira de Pesquisa de Trigo,* 28, 1996. Passo Fundo, Brazil.

Reis, E.M. (1982). Podridão de sementes. In *Fundação Cargill.* Trigo no Brasil. Campinas. pp. 477-480.

Reis, E.M. (1988). *Doenças do trigo III—Giberella.* 2nd ed. Revista e ampliada.

Reis, E.M. (1990). Control of disease of small grains by rotation and management of crop residues, in Southern Brazil. International Workshop on Conservation Tillage Systems. Passo Fundo: CNPT/EMBRAPA.

Reis, E.M. (1996). Relação entre a severidade e a incidência da ferrugem da folha do trigo, causada por *Puccinia recondita* f.sp. *tritici. Fitopatologia Brasileira.* 21(3):369-372.

Reis, E.M., Blum, M.M.C., Casa, R.T., and Costa, C. (in press). Relação entre a incidência e severidade de manchas foliares, em trigo. *Fitopatologia Brasileira.*

Reis, E.M., Casa, R.T., and Hoffmann, L.L. (in press). Quantificação de perdas causadas por *Erysiphe graminis* f.sp. *tritici. Fitopatologia Brasileira.*

Reis, E.M., Santos, H.P., Lhamby, J.C.B., and Blum, M.M.C. (1992). Effect of soil management and crop rotation on the control of leaf blotches of wheat in

Southern Brazil. In Congresso Interamericano de Siembra Directa, 1, 1992. Villa Giardino. Trabajos presentados. S.I.: Associación Argentina Productores en Siembra Directa/Sociedad de Conservación de Suelos/Clube Amigos da Terra/ Fundação ABC/Associación Uruguaya Pro Siembra Directa, (eds), pp. 217-236.

Strange, R.N., Majer, J.R., and Smith, H. (1974). The isolation and identification of choline and betaine as two major components in anthers and wheat germ that stimulate *Fusarium graminearum* in vitro. *Physiology Plant Pathology.* 4:277-290.

Wiese, M.V. (1991). *Compendium of Wheat Diseases.* 2nd ed. The American Phytopathological Society.

Zadoks, J.C. (1985). On the conceptual basis of crop loss assessment the threshold theory. *Annual Review Phytopathology.* 23:455-473.

Chapter 11

Wheat As a Polyculture Component

Santiago J. Sarandon

INTRODUCTION

One of the main characteristics of modern agricultural systems is their low biodiversity. Of almost 80,000 edible plants that exist, around 200 are commonly used, and only 12 are considered basic food for humanity (FNUAP, 1991). These main crops are cultivated in extensive areas with monocultures of a few successful cultivars, drastically reducing genetic diversity and increasing agroecosystem fragility. Because of concerns about the impact of conventional agriculture, many farmers have begun to adopt alternative practices with the goal of reducing input costs, preserving the resource base, and protecting human health. Sustainable agriculture has emphasized the need for more diversified systems, because they tend to be more stable and resilient, reduce financial risk, and provide a hedge against drought, pest infestation, and other natural factors limiting production (National Research Council, 1989).

One alternative is the use of intercropping. This is a form of multiple cropping where two or more crops are grown together on the same land during all or part of their life cycle. It can include mixtures of different species, or mixtures of cultivars or genotypes of the same species, which can be grown in strip, row, or random mixed intercropping, according to a spatial pattern or arrangement.

In many parts of the world intercropping is a common component of agroecosystem management. This is an old farming practice in many tropical zones, especially in developing countries where 50 to 80 percent of the rainfed crops are planted as intercrops. Almost 98 percent of the cowpeas grown in Africa are intercropped, and 90 percent of the beans in Colombia are grown as polycultures (Arnon, 1972; Gutiérrez, Infante, and

---

Special thanks go to H. Acciaresi, N. Greco, and M. Sisterna for their helpful comments on the manuscript, and to C. Flores for her assistance with the figures.

Pinchinot, 1975, in Vandermeer, 1989). As Francis (1986) pointed out, multiple cropping has its roots in the history of civilization. Though the importance of intercropping as a farming practice has long been recognized, it is only very recently that it has awakened real interest among research workers (Willey, 1981), especially as an appropriate ecological tool for reducing external inputs (Amador and Gliessman, 1990). One reason could be the possibility of increasing productivity of lands and efficient use of resources (Hook and Gascho, 1988).

Some advantages and limitations of multiple cropping systems can be summarized as follows:

Advantages

- Greater biodiversity.
- Greater yield stability over time. Lower economic risk.
- Better performance against climate or weather, pests, diseases, weeds.
- Better nutritional source.
- Better distribution of labor over time.
- Better use of resources (water, light, nutrients).

Limitations

- Requires better understanding of plant interactions.
- Difficult to generalize performance. Site specific.
- Problems (in some cases) with mechanized harvests.
- Requires more intensive labor.
- Problems with commercialization and quality standards.
- More complex evaluation methodology.
- Breeding and methodology for select genotypes with good combining ability more difficult and not fully understood.

Despite the importance of this farming system, intercropping is not seen as a real alternative in temperate zones where small cereal grains are the common crop. Nevertheless, results using intercrops with wheat as a main crop or mixtures of wheat cultivars can be a viable strategy for sustainable productivity. Smithson and Lenné (1997) summarized data from more than 100 studies of mixed crops and the results of comparing the yield of mixtures of varieties with the main yield of their components. They found that the yield of mixtures exceeded that of their components by small but significant amounts on average, being greatest for wheat (5.4 percent), and there were significant positive skews indicating that yield advantage of mixtures tended to be greater than disadvantages. The aim of this chapter is, therefore, to discuss the advantages, limitations, and possibilities of using wheat as a polyculture or intercropping component, especially in sustainable agriculture systems.

HOW TO MEASURE INTERCROPPING PERFORMANCE

In intercropping it is important to evaluate the advantage of the whole system and not only of its components. Several indices have been proposed to measure intercrop performance, but, whatever the method of evaluation, the underlying basis is always a comparison of the performance of the species in intercrop to its performance in monoculture (Vandermeer, 1989). The simplest method is to compare the yield of mixtures with the yield of their components in monoculture weighted according to the composition of the mixture. But this gives little information on the performance of the components, which can be important if its economic value is not the same. Another way of evaluating intercrop performance is using land equivalent ratio (LER) or relative yield total (RYT). LER has been proposed as a measure of an area of land which is required, sowing the components of an intercropping system as a sole crop, to obtain a yield equivalent when grown in intercrop. LER or RYT of a given mixture is the sum of relative yields (RY) of each of its components compared with monoculture (Silvertown, 1982).

LER or RYT are calculated as:

$$RYT = RYij + RYji$$
$$\text{if } RYij = Yij/Yii \text{ and } Ryji = Yji/Yjj$$

where Yij is the relative yield of cultivar (or species) i in mixture with cultivar (or species) j, and Yii is the yield of the cultivar (or species) i when growing in pure stand (alone).

An RYT or LER value greater than 1 indicates an overyielding or a benefit from intercropping. A value equal to 1 indicates no advantage and a value less than 1 indicates that the yield in intercropping is lower than in pure culture.

Experiments to study intercropping or mixture performance utilize several designs. One of the most common is replacement series (de Wit, 1960). This consists of maintaining a fixed density of each component but varying their proportion in the mixture so that the total density is maintained without change. Some theoretical cases of mixture results are summarized in Figure 11.1.

FIGURE 11.1. Some Possible Interactions Between Mixture Components in a Replacement Series Design

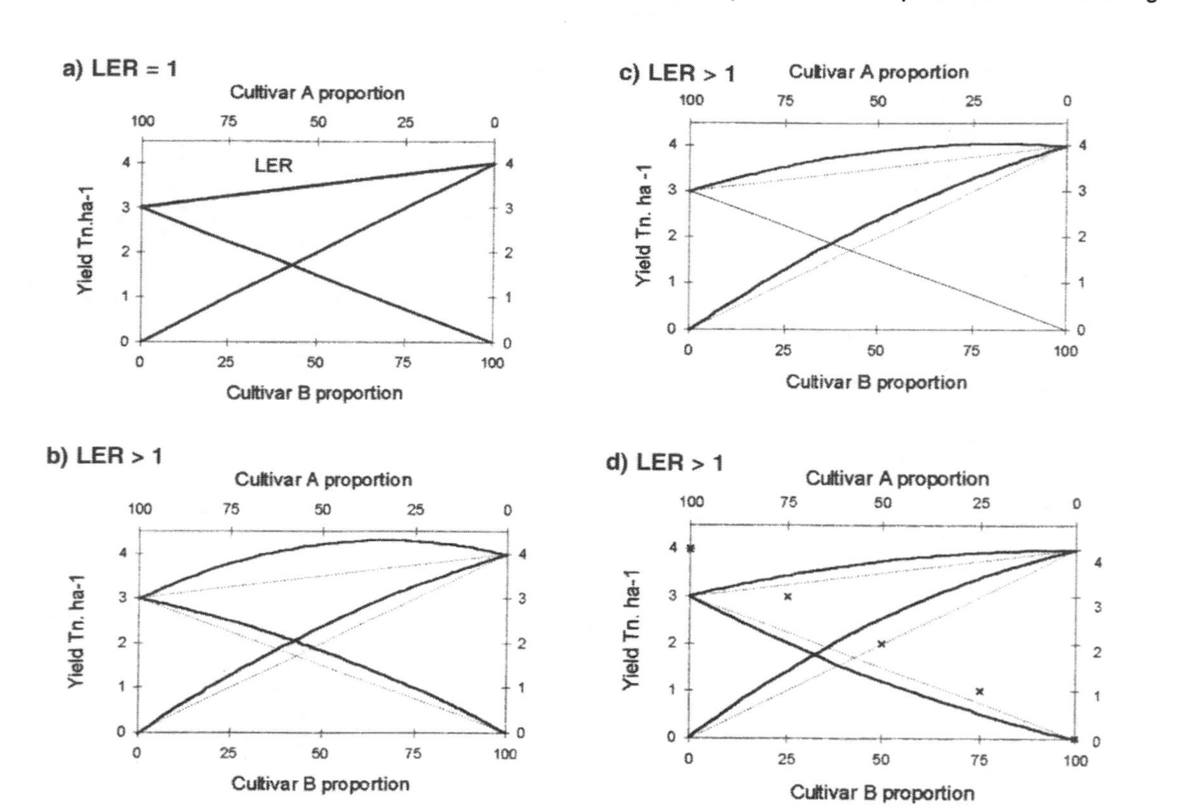

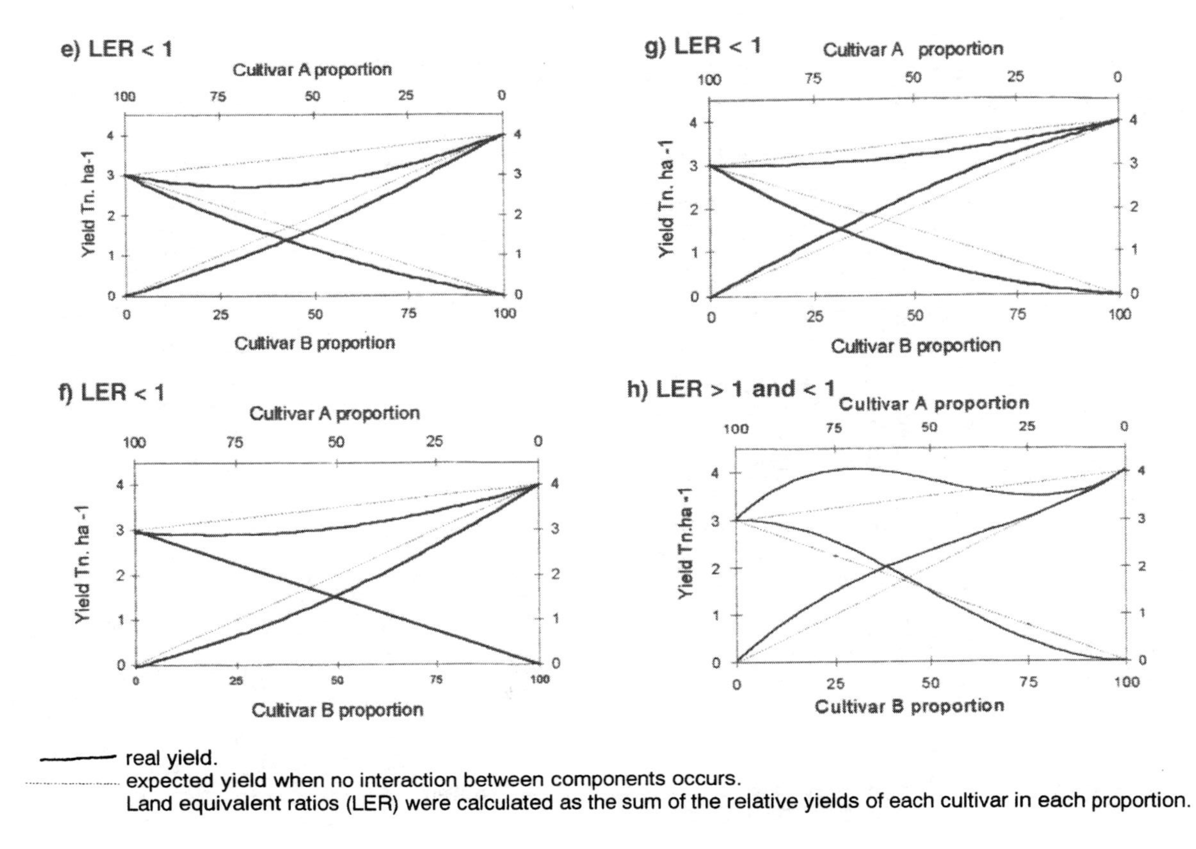

—— real yield.
········ expected yield when no interaction between components occurs.
Land equivalent ratios (LER) were calculated as the sum of the relative yields of each cultivar in each proportion.

When no interaction exists between mixture components, then the relative yield of each one must be proportional to its participation in the mixture. In this case the LER value will be 1 (Figure 11.1a). But when interaction occurs other results can be observed. When LER is higher than 1, there are several possibilities: each component yields more in mixture than in pure culture (Figure 11.1b), one component yields the same and the other more (Figure 11.1c), or one yields less but the other more than proportional (Figure 11.1d). There are similar possibilities when LER is lower than 1 (Figures 11.1e, 11.1f, and 11.1g). Another possibility is that LER value changes according to the proportion between its components. This occurs when one of the components yields more than expected only when it is a minority (25:75 proportion), but yields less than expected when it is a majority (75:25). In this case LER will be higher than 1 in one proportion and lower than 1 in the other (Figure 11.1h). This indicates that mixture advantage can depend in some cases on the proportion of the components. These examples illustrate some possibilities of two components interacting in mixtures and pointed out the difficulties in analyzing experimental results, especially when mixtures have more than two components. Another way to measure intercropping performance takes into account the monetary aspects, the intercropping advantage index proposed by Banik (1996).

EFFECT OF MIXTURES ON PLANT DISEASES

The use of intercropping to improve performance against plant diseases is well documented, and is perhaps the most common research field in cereal intercropping. Moreover, most of the research on the epidemiology of cultivars, mixtures, and multilines has been done with small grains (Browning and Frey, 1969; Mundt and Browning, 1985; Wolfe, 1985). Disease severity reductions up to 97 percent compared with monocrops have been cited in wheat mixtures (Finckh and Mundt, 1992b).

One of the reasons for the interest in the use of cultivar mixtures to control plant disease is that there is a fundamental difference in the ability of mixtures to cope with abiotic and disease stress (Wolfe, 1985). Under an abiotic stress, such as drought or late frost, the mixture cannot directly affect the stress, and only a compensatory mechanism could be expected. In the case of diseases, however, pure stands and mixtures can influence disease progress directly: they control the degree of stress. According to Wolfe's review (1985), mixtures restrain the spread of the pathogen population in three ways:

1. By decreasing the spatial density of susceptible plants.
2. By providing a barrier effect with resistant plants that fill the space between susceptible ones.
3. By inducing resistance by nonpathogenic spores, limiting the productivity of pathogenic ones in the same area.

Diversification as a means of disease control can be achieved in two ways: with multilines or cultivar mixtures. The multiline concept was suggested by Jensen (1952) and consisted of blending a series of pure lines to introduce diversity into a variety. Borlaug (1959) proposed wheat multilines by backcrossing lines that differed in their susceptibility to rusts but maintained uniformity in other agronomic characteristics. On the other hand, cultivar mixtures are composed of different cultivars or genotypes selected for a priori good combining ability, mixed in different proportions. No uniformity is necessary. There are several advantages in using mixtures of cultivars instead of multilines: facility of synthesis (there is no need for breeding mixture components), better ability to use new cultivars as they are released, additional disease control against the target disease through race-nonspecific resistance, reduced selection for super-race (a race virulent to all host components in the mixture), a protection from nontarget diseases and increasing possibilities for positive yield synergism (Finckh and Mundt, 1992b; Akanda and Mundt, 1996).

Five aspects of the performance of cultivar mixtures against disease have been identified:

1. *Size of individual plants and its relationship to lesion size.* Lannou et al. (1994a), in a study including controlled conditions and computer simulation, found that a wheat seedling mixture was less effective against yellow rust (*Puccinia striiformis*), which produced large lesions, than brown rust (*Puccinia recondita* f. sp. *tritici*) over two generations. They suggested that this resulted from the differences between the parasites' lesion size. But in another study Lannou, de Vallavieille-Pope, and Goyeau (1994) found that this was not true in adult plants, where the lesion size could be much more limited in relation to the plant size. Thus, though lesion size has a significant influence on the efficacy of host mixtures for epidemic control, it seems to depend on the relative size of the plant.

2. *Spatial distribution of susceptible host.* In determining the performance of cultivar mixtures, the spatial arrangement of the components has proved to be important. Mundt, Brophy, and Kolar (1996) studied its effects on severity of artificial inoculations of yellow rust (*P. striiformis*) and found that random mixtures of susceptible and resistant wheat cultivars in 1:1 proportion increased grain yield relative to pure stands, but

alternating strips did not. This is related to the effect of the genotype unit area (GUA) on disease control. Genotype unit area is defined as the ground area occupied by an independent, genetically homogeneous unit of host population. Generally, cultivar mixtures with the largest GUAs (such as alternative strips) seem to be less effective in disease control than those with smaller GUAs (random mixtures). Brophy and Mundt (1991) found that disease (*P. striiformis* and *P. recondita*) control in wheat was superior for geometries with small host GUA, so their ranking in order of decreasing performance was random mixture, row, and strip.

Akanda and Mundt (1996) studied the effect of different proportions of wheat cultivars differing in resistance to various races of *P. striiformis* growing in mixtures and suggested that greater disease control could potentially be obtained by increasing the number of cultivars in the mixture to reduce the proportion of plants susceptible to a given race.

3. *Spatial distribution of inoculum in the plot.* The spatial distribution of the inoculum also affects the performance of mixtures. The GUA has no significant effect on the efficacy of mixtures if there is a focal epidemic but, in contrast, increased GUA reduces the efficacy of the mixtures when disease is distributed uniformly. Mundt and Brophy (1988) suggested that the number of genotype units may have a larger influence on mixture efficacy than do the GUAs. Mundt, Brophy, and Kolar (1996) conducted a field experiment with wheat plots of different sizes (0.36, 5.76, and 24.01 m^2) to determine if GUA interacts with number of genotype units in determining the efficacy of host mixtures for controlling wheat yellow rust (*P. striiformis*). They found in only one of the three experiments that in large plots where the number of genotype units was greater the disease severity was reduced compared with small plots. They concluded that there exists little evidence to suggest that genotype unit number is a major influence on the efficacy of host mixtures for disease control. Nevertheless their results suggest that number of genotype units may sometimes interact with GUA to determine mixture efficacy for disease control and that this may be influenced by environmental factors such as wind conditions.

4. *The phenological stage of the plant at which disease is present.* The moment when the infection occurs is very important for a wheat mixture performance. If diseases occur early in the growing season, then compensatory growth via tillering of healthy plants can compensate for the lower tillering of the disease-susceptible component. But sometimes, infections occur too late to allow tillering of healthy plants in response to reduced competition from diseased plants. This suggests that cultivar mixtures could be more important for control of diseases that infect early in the growing season, when the advantage of the mixture could be maximized

compared with pure stands. On the other hand, no greater advantages of mixtures over sole crops can be expected for head or grain diseases.

Finally, cultivar mixtures may perform well against diseases other than those for which they were designed (nontarget diseases). Brophy and Mundt (1991) found that mixtures designed to control stripe rust (*P. striiformis*) also controlled two nontarget diseases: leaf rust (*P. recondita*) and powdery mildew (*Erysiphe graminis* DC) in two seasons.

5. *Relationship between mixture components in absence or presence of the target disease.* Finckh and Mundt (1992a), in an experiment with five wheat varieties differing in their susceptibility to *P. striiformis,* found that only 67 percent of the variation in disease severity could be attributed to the frequency of susceptible genotypes in the mixtures. They suggested that competitive interactions among plant genotypes sometimes appeared to alter susceptibility and obscured the relationship. Then, to understand the effect of disease on interactions among different plant genotypes and the effects of host genetic composition on disease severity, it is necessary to study the population dynamics of host mixtures in the absence and presence of disease (Finckh and Mundt, 1992b) and the competitive abilities of individual cultivars in mixtures (Knott and Mundt, 1990).

In the performance of the mixture one can expect different results based on the presence or absence of the target disease, as two mechanisms occur simultaneously. One is protection against disease and the other is interactions between mixture components. In the presence of the target disease, the major resistance of one component could lead to a better grain yield of the mixture compared to pure stands. But in the absence of the target disease the relation between the components of the mixture could be more competitive and grain yield may be reduced in comparison to pure stands. However, there are several cases where mixtures yielded more than monoculture even in the absence of disease, which has been related to a better use of resources.

EFFECT OF MIXTURES ON RESOURCE USE

One of the advantages attributed to intercropping is better efficiency in resource use (water, light, nutrients) than the monocrops. Generally this has been well documented in mixtures of different species, especially when legumes are one component of the intercrop, but it is also applicable to mixtures of varieties. In wheat, grain yield increase in modern cultivars has been achieved by means of a higher harvest index (HI) with little or no change in biological production (Austin et al., 1980; Perry and D'Antuono, 1989; Slafer and Andrade, 1989). As there exists a maximum HI compatible with normal growth and development of crops (Austin et al., 1980) and it seems that there is not enough genetic variability in the biomass productivity (Austin, Ford, and Morgan, 1989), the genetic improvement of crops by breeding in this direction is reaching its limit (Austin, 1989). This suggests that to obtain high productivity without an increase in HI, new strategies to improve efficiency in resource utilization must be explored.

In a monoculture, especially in self-fertilizing species such as wheat, the plants cannot exploit all the resources available in a determined area, because all the genotypes utilize the same resources at the same time. The spatial and/or temporal availability of light, water, and nutrients may be explored and exploited differently by various genotypes or species. Trenbath (1986) suggested the hypothesis that in many agricultural environments those factors could be more completely absorbed and converted into biomass by a mixed stand than by a pure stand.

Though cultivar mixtures have been considered a potential way to increase crop yield (Harper, 1964), studies on mixtures of genotypes of the same species have focused on the competitive ability of mixture components (Jennings and Aquino, 1968; Jennings and de Jesus, 1968; Mc Kenzie and Grant, 1980), or on the behavior of the mixture against pathogens, but not on the productivity of the mixture itself. Studies in groundnut (*Arachis hypogea*) (Rattunde et al., 1988) and barley (*Hordeum vulgare*) (Stützel and Aufhammer, 1990) have tested mixtures as a means of exploiting morpho-developmental differences among cultivars to increase crop productivity, but they have shown variable results.

Wheat cultivars may differ in morphology, canopy architecture, and physiological attributes as much as different species do. If these differences could result (in some cases) in varying abilities to utilize present resources (de Wit, 1960; Trenbath, 1974), then the total amount of resources used by a mixture of different genotypes may be higher than that of a pure cultivar growing alone. This should lead to a higher biomass production of the mixture, and could eventually result in an increase in grain yield and/or a better grain protein content (Sarandón and Sarandón, 1995).

To take advantage of intercropping systems it is important to understand the ecological mechanisms of yield increase. According to the model proposed by de Wit (1960), and for cultivar mixtures by Stützel and Aufhammer (1990), the biological yield of each component in a mixture is strictly proportional to the share of environmental resources it can acquire. Then the proportional increase in one mixture component will tend to equal the proportional decrease in biomass yield of the other component.

Nevertheless, Smithson and Lenée (1997), analyzing results of several mixture experiments, found higher yields than expected for biomass or grain production in mixtures, indicating that an overcompensatory effect is possible.

Vandermeer (1989) pointed out two principles that can explain the overyielding mechanisms of intercrop as compared with pure stands. These are the *competitive production principle* and the *facilitation principle.* When one species or genotype has an effect on the environment that causes a negative response in the other species that grows with it, even though both can utilize necessary resources more efficiently when grown together than when in pure stands, the competitive production principle applies. There is partial competition between components of the mixture. This means that two or more crops use different resources or they use them in different ways, exploiting different ecological niches (niches are partially overlapped).

The facilitation principle applies when one species modifies the environment in such a way that the second species or genotype benefits. These two principles can act simultaneously and the final performance of the intercrop depends on which one predominates. According to Vandermeer (1989), though both principles act simultaneously, when the relative yield (or partial LER) of one intercrop component yields more than 1, something more than the competitive production principle must be operating. When competition is greater than the benefit of facilitation, then LER will be less than 1 and the performance of intercrop is worse than monoculture. This can be the case for the performance of mixed cultivars against pathogens: in the absence of the target disease the competition between mixture components can be greater than the benefit of disease resistance and the mixture's performance can be worse than pure culture.

The relationship between competition and facilitation can depend also on the density of the components or on agricultural practices such as fertilization and/or irrigation, which suggest the importance of choosing the adequate combination of components for each environmental condition. In an experiment with two wheat varieties grown in mixtures at different proportions, Sarandón and Sarandón (1995) found that the mixture produced more biomass than the best cultivar only when varieties were in one of the proportions (33:67) and not in the others (50:50 or 67:33). This suggests that differences among cultivars may increase the use of resources but, depending on the proportion and on the environmental conditions, the effect of competition may be more important. Thus, to obtain a highly productive mixture, the optimal combination will depend not only on the choice of the components, but also on their relative proportions. This was not taken into account by Trenbath (1974), who reviewed results of biomass productivity only in mixtures with a 1:1 ratio. Neither was considered by by Knott and Mundt (1990), who proposed that the combining ability of cultivars must be tested in mixtures of 1:1 proportion.

In wheat, morphological and/or physiological differences between cultivars suggest that two potential effects in the interaction of plant genotypes within a mixture should be recognized: (1) the competition for common resources, and (2) the increase in potential resource availability (Sarandón and Sarandón, 1995). When the cultivars used in a mixture are agronomically identical it could be expected that they strongly compete for the same resources, and that the mixture will not yield more than the best-yielding cultivar (Simmonds, 1962). This explains why multilines tend not to yield more than sole crops in the absence of the target disease.

Soil resources (nitrogen and water) could be important factors for mixture performance (Trenbath, 1974). Differences in the pattern of root distribution within the soil have been found among wheat genotypes (Lupton et al., 1974), which can be related to differences in HI (Siddique, Bedford, and Tennant, 1990). This suggests that if cultivars differ in their HI a different root growth pattern could be expected between them. Mutual avoidance by adjacent root systems could lead to a late-developing root system occupying deeper soil layers in mixtures than in monocultures (Berendse, 1979). In this case the LER value may be expected to exceed 1, as the resources available for the mixture have increased. Sarandón and Sarandón (1995) observed a better performance of wheat mixture in unfertilized plots, suggesting a better utilization of nitrogen (N) by the mixture when N was the limiting factor. This effect of N availability on mixture performance was also found by Aufhammer et al. (1989), who suggested that under suboptimal conditions for one of the components, ontogenetic differences between components resulted in positive mixing effects on grain yield since the yield proportion of the accompanying partner increased more than proportionally.

Light is another important factor in mixture performance. In the absence of water and nutrient deficit, competition for light can be the principal factor in mixture productivity. Although modeling mixtures of wheat cultivars with contrasting leaf inclinations predicted only a very small advantage in gross photosynthesis (Trenbath, 1972, in Trenbath, 1974), some experiments have been done to exploit morphophysiological differences between wheat genotypes to improve photosynthesis efficiency. Prasad and Reddy (1973) used the height variability in wheat genotypes to create a systematic stand to optimize light utilization. Sharma and Prasad (1978) found that three genotypes of wheat differing in height and grown in different spatial arrangements yielded more in mixtures than in monoculture, which was attributed to better light use.

Prasad and Sharma (1980) conducted field experiments for two seasons in India, comparing the yield potentials of mixed and pure stands of three spring wheat cultivars varying in plant height (60, 80, and 120 cm) when sown at different spatial arrangements. Five canopy stands and two row spacings (15 and 22.5 cm) were evaluated. Canopy structures were denominated as pyramidal or columnar (staircase design). They found that under high rates of N fertilization the columnar canopy stand out-yielded the high-yielding cultivars, suggesting that there was a better N utilization efficiency in columnar canopy stands. They also found in one of the two years that narrower row spacing (15 cm) yielded more than wider spacing (22 cm), showing a significant year × row spacing interaction. Straw yield of all three cultivars was also greater in mixed stands, but it was more evident in tall cultivars in two experiments (Sharma and Prasad, 1978; Prasad and Sharma, 1980), suggesting that in this spatial arrangement tall cultivars had the advantage of height in competition for light.

In spite of the importance of this type of experiment on spatial arrangements or systematic mixed designs to optimize light or nutrient utilization efficiency, these patterns are not appropriate for use by farmers because of difficulties in harvesting due to the different height of mixture components.

Although wheat cultivar mixtures can produce a higher total aerial biomass than a monoculture, this is not necessarily associated with a higher grain yield. Within a mixture, the interaction between cultivars may cause a different dry matter partition than in pure stands. This was confirmed in a field experiment using two contrasting wheat cultivars sown at various proportions (100:0, 67:33, 50:50, 33:67, and 0:100) in mixtures with different N availability (Sarandón and Sarandón, 1995). A decrease in HI of mixtures compared with pure stands was observed, especially in 1:1 proportion in both N conditions, indicating that a higher proportion of resources was allocated to competitive structures (stems and leaves). Because modern high-yielding cultivars have a high HI, and consequently a high grain/straw proportion, a small amount of residues are left in the field after a crop cycle. Due to the importance of soil organic matter in low-input agricultural systems, a greater biomass production of mixtures, while maintaining grain yield and protein content, is another important benefit gained from this alternative agricultural technology (Sarandón and Sarandón, 1995).

Intercropping has also been considered as an alternative practice to control weeds in sustainable agriculture (Liebman and Dyck, 1993). This is based on its capacity to exploit more resources than sole crops. Intercrops or mixtures of varieties may suppress weeds more efficiently because fewer resources will be available for weed growth. Liebman and Dyck (1993) listed a series of experiments in which suppression of weed biomass accumulation by intercrops generally exceeds that of at least one of the component species grown as sole crops. Though these authors suggested that this phenomenon was particularly important in systems in which low-growing weed-suppressive species are sown between rows of main crops, it can extend to cultivar mixtures if they can be more competitive for light, water, and/or nutrients than sole crops. Understanding how intercrops and weeds respond to manipulable environmental and cultural factors should benefit the design of weed-suppressive agronomically productive systems (Liebman and Dyck, 1993).

One concern about mixtures has been their persistence over time. It has become clear that intergenotypic interactions within a mixture cause drastic changes in its composition and distort the yields of its components relative to yields in monoculture in several species studied. This has been demonstrated in barley (Harlan and Martini, 1938), rice (Jennings and de Jesús, 1968), Lolium (Hill and Shimamoto, 1973), and in wheat (Mc Kenzie and Grant, 1980; Shaalan, Heyne, and Lofgren, 1966). It means that mixtures must be reconstituted by farmers every year to maintain the correct proportions of components.

EFFECT OF MIXTURES ON GRAIN QUALITY

Increasing grain yield together with grain protein content is an important goal in wheat production. However, this is not easy due to the inverse relationship between grain yield and grain protein content (Kramer, 1979; Löffler, Rauch, and Busch, 1985). The increase in the dry matter portion of the grain (HI) in modern cultivars has also been associated with a decrease in grain protein content (Austin et al., 1977; Kramer, 1979; Paccaud, Fossati, and Hong,, 1985), since 70 to 90 percent of the N in the grain comes from N previously accumulated in vegetative structures (Austin et al., 1977; Scholz, 1984). As the nitrogen requirements of high-yielding cultivars in low N conditions cannot be supplied by the remobilization of stored N in the vegetative structures (Sarandón and Gianibelli, 1992), grain protein content can be severely affected in low-input environments.

Mixtures of cultivars making a more efficient use of the resources (higher biological yield) can be an efficient way to increase productivity without increasing HI. Then the grain protein content of the mixture could be increased as compared with cultivars grown in pure stands, due to a more favorable source-sink relationship. This was confirmed by Sarandón and Sarandón (1995), who found a higher grain protein content in a mixture composed of two cultivars differing in protein content. This was related to an increase in biomass production and a decrease of HI, which led to the availability of more N for the grain. Then a greater protein content in mixtures could be expected in those cases where a higher biomass production is achieved, maintaining the HI. But this assumption must be confirmed in future experiments.

INTERCROPPING WHEAT WITH OTHER SPECIES

The use of wheat in intercropping with other species, especially legumes, is considered in many countries a way to minimize the risk of crop failure under unfavorable weather conditions and also to stabilize yields and maintain soil health through legume association (Banik, 1996). Reynolds, Sayre, and Vivar (1994) examined the benefits of intercropping wheat with five species of N-fixing legumes at low levels of N input to the soil in the CIMMYT (International Maize and Wheat Improvement Center), Mexico during four years under rainfed conditions. None of the legumes tested reduced yields of the cereal crop in comparison to controls, while the extra aboveground biomass for legumes in some cases was more than double. This led to an intercropped LER as high as 1.54. This performance was attributed to the fact that for wheat growing at suboptimal levels of N fertilization, light is not a limiting factor and can be used by an N-fixing intercrop without detriment to the main crop (wheat).

Some of the benefits of intercropping legumes with cereals have been attributed to the availability of soil N to subsequent wheat crops (Singh, 1983; Danso and Papastylianou, 1992). Brandt, Hons, and Haby (1989) studied the effects of four interseeded subterranean clover (*Trifolium subterraneum* L.) cultivars on soft red winter wheat. They found that interseeding decreased grain yield in the first year, but increased the yield during the second year due to the release of N fixed in the previous year. However, Banik (1996) found that when wheat was grown in mixture with legumes the wheat crop benefited in the first season and yield was increased, which was attributed to the complementary effect of legume association. This is in agreement with the evidence that the intercropping of cereals with N-fixing legumes could permit the lateral movement of fixed N to the cereal (Willey and Reddy, 1981), possibly via mycorrhizal connections as was reported for soybean-maize intercropping (van Kessel, Singleton, and Hoben, 1985). The apparent differences between Banik (1996) and Brandt, Hons, and Haby's (1989) results may be related to the way they measured intercrop performance. There exist two ways of defining wheat performance as a component of intercropping. The first is considering wheat as the main crop. Then the interseeded species must not interfere with wheat production. The objective of this intercropping design is to look for a lateral effect of accompanying species, such as soil cover or protection against erosion. In this way, Brandt, Hons, and Haby (1989) tested the differences between grain yield of wheat (main crop) sown at common plant density as sole crop and yield of wheat when interseeded with legumes.

Another way of evaluating intercrop advantage is looking at the system itself. Though wheat can also be the main crop, the interest is in intercrop performance and not in one of the components. This type of experiment uses LER as one of the main indicators of performance or intercropping advantage index (Banik, 1996), which also takes into account the monetary aspects. Banik used replacement series and focused his experiment on the overall performance of the mixture itself and not only on wheat performance. He found an LER higher than 1, indicating a better resource use than sole crops.

THE EVALUATION OF GENOTYPES FOR INTERCROPPING

From the analysis of mixture performance in the literature it is clear that the success of intercropping or mixture of varieties depends on the availability of appropriate genotypes. One of the main challenges to design mixtures of wheat cultivars that perform well in particular environmental conditions is to choose the appropriate components. But this is not easy because agronomists generally have focused on the performance of the component and not on the system itself.

Knott and Mundt (1990) proposed that the combining ability of cultivars could be calculated based on data derived by mixing cultivars in a 1:1 ratio in all possible two-way combinations. In this way general mixing ability (GMA) and specific mixing ability (SMA) could be detected. In an experiment with mixtures of five wheat cultivars in two-way combinations, and *P. striiformis,* Knott and Mundt (1990) found that the examination of GMA and SMA revealed cultivars and combinations that were statistically better "mixers" than others. In spite of this successful result, this methodology raises the difficulty of testing all possible combinations of genotypes to choose the best. Agronomists must be able to select cultivars (growing in pure stands) that have a potential good combining ability. But genotypes bred for sole crop may not be as good for mixed crop cultivation (Zimmerman, 1996). This agrees with the results of Jennings and Aquino (1968) and Jennings and de Jesus (1968) in rice, which demonstrated that the performance of a component in pure stands does not necessarily indicate a good ability to perform well as a mixture component.

Considerable attention has been given to predicting the performance of genotypes when grown in different environments as sole crops (Willey and Rao, 1981). A sequence of steps to be followed for a successful selection strategy of genotypes for intercropping has been proposed (Wien and Smithson, 1981, modified):

- Definition of intercrop systems for which genotypes are to be selected (climatic, biotic, economic, cultural factors)
- Manipulation of sowing date, spacing, arrangement, and soil nutrient levels
- Screening under defined intercrop conditions of a large number of cultivars growing in pure stands to identify characters of importance in adaptation to intercropping
- Determination of the extent to which the same characteristics are also expressed under monocrop conditions, to allow monocrop selection for intercropping
- Preliminary negative screening for intercrop adaptation using monocrop conditions
- Positive screening of genotypes for adaptation to the selected intercrop system

FUTURE RESEARCH NEEDS

In this chapter the benefits of using wheat as a component of mixtures have been evaluated. Nevertheless there are still some research areas that may need more attention: (1) to identify and screen those characteristics of wheat cultivars under sole crops that indicate a good ability to perform well in mixtures; (2) to search for a plant breeding methodology that allows selection of promising genotypes under conditions where this good combinatory ability can be observed; and (3) to gain a better understanding of the precise environmental conditions in which a mixture can perform better than the pure culture or monoculture.

CONCLUDING REMARKS

In the present review the use of mixtures of cultivars has been discussed as a promising ecological alternative to optimize wheat production, especially in low-input agricultural systems. However, because of the relatively few experiments in this area, the generality of its results should be tested on a large scale, over more environments and with more cultivars. The increase in crop diversity could have multiple positive consequences for agroecosystem functioning, such as a lower ecological fragility or a lower external input dependence, with the same productivity and quality, with similar production and trading cost. Moreover, it could also be a good alternative to maintain in situ diversity.

The existing differences among some wheat genotypes (in morphology, architecture, physiology, etc.) could result in a different exploration and utilization of resources, and to better resistance against diseases. But the performance of a mixture depends also on the proportion of its components, and/or on the environmental conditions (e.g., the presence or absence of the target disease or fertilizer application). A minimum uniformity is needed to sow, harvest, or industrialize wheat and other crops. If the cultivars selected are of similar crop cycle and height, the mixture should not show problems at sowing or harvest. Moreover, there should not be problems in selling mixtures because millers blend them in silos according to their quality.

Finally, an analysis of the prospects and perspectives of cultivar mixtures or the performance of wheat as a mixture component must take into account that for several years agriculture research has focused on selecting genotypes that perform well under monoculture conditions and good environmental conditions. No work has been done on selecting genotypes that perform well in mixtures or in combination with other genotypes of the same cultivars or different species, under low-input conditions. Nevertheless the existence of some successful cases in the literature suggests there is potential for improvement in this research field.

REFERENCES

Akanda, S.I. and Mundt, C.C. 1996. Effects of two-component wheat cultivar mixtures on stripe rust severity. *Phytopathology,* 86(4): 347-353.

Amador, M.F. and Gliessman, S.R. 1990. An ecological approach to reducing external inputs through the use of intercropping. In *Agroecology: Researching the ecological basis for sustainable agriculture.* S.R. Gliessman (Editor), Vol. 78: 146-159, Springer-Verlag, New York.

Arnon, Y. 1972. *Crop production in dry region.* Leonard Hill, London.

Aufhammer, W., Kempf, H., Kübler, E., and Stützel, H. 1989. Effekte der Sorten-(Weizen) und der Arten (Weizen, Roggen) Mischung auf die Ertragsleistung krankheitsfreier Bestände. *Journal of Agronomy and Crop Science,* 163: 319-329.

Austin, R.B. 1989. Genetic variation in photosynthesis (review). *Journal of Agricultural Science, Cambridge,* 112: 287-294.

Austin, R.B., Bingham, J., Blackwell, R.D., Evans, L.T., Ford, M.A., Morgan, C.L., and Taylor, M. 1980. Genetic improvements in winter wheat since 1900 and associated physiological changes. *Journal of Agricultural Science, Cambridge,* 94: 675-689.

Austin, R.B., Ford, M.A., Edrich, J.A., and Blackwell, R.D. 1977. The nitrogen economy of winter wheat. *Journal of Agricultural Science, Cambridge,* 88: 159-167.

Austin, R.B., Ford, M., and Morgan, C.L. 1989. Genetic improvement in the yield of winter wheat: A further evaluation. *Journal of Agricultural Science, Cambridge,* 112: 295-301.

Banik, P. 1996. Evaluation of wheat (*Triticum aestivum*) and legume intercropping under 1:1 and 2:1 row-replacement series system. *Journal of Agronomy and Crop Science,* 176: 289-294.

Berendse, F. 1979. Competition between plant populations with different rooting depths. I. Theoretical Considerations. *Oecologia (Berlin),* 43: 19-26.

Borlaug, N.E. 1959. The use of multilineal or composite varieties to control airborne epidemic diseases of self-pollinated crop plants. *Proceedings First International Wheat Genetics Symposium,* 12-26, Winnipeg, Canada.

Brandt, J.E., Hons, F.H., and Haby, V.A. 1989. Effects of subterranean clover interseeding on grain yield, yield components and nitrogen content of soft red winter wheat. *Journal of Production Agriculture,* 2: 347-351.

Brophy, L.S. and Mundt, C.C. 1991. Influence of spatial patterns on disease dynamics, plant competition and grain yield in genetically diverse wheat populations. *Agriculture Ecosystems and Environments,* 35: 1-12.

Browning, J.A. and Frey, K.J. 1969. Multiline cultivars as a means of disease control. *Annual Review of Phytopathology,* 7: 355-382.

Danso, S.K.A. and Papastylianou, Y. 1992. Evaluation of the nitrogen contribution of legumes to subsequent cereals. *Journal of Agricultural Science, Cambridge,* 199: 13-18.

de Wit, C.T. 1960. On competition. *Verslagen van Landbowkundige Onderzoekinge,* 66: 1-82.

Francis, C.A. 1986. Distribution and importance of multiple cropping. In *Multiple cropping systems,* C.A. Francis (Editor), Macmillan, New York: 1-19.

Finckh, M.R. and Mundt, C.C. 1992a. Plant competition and disease in genetically diverse wheat populations. *Oecologica,* 91: 82-92.

Finckh, M.R. and Mundt, C.C. 1992b. Stripe rust, yield and plant competition in wheat cultivar mixtures. *Phytopathology,* 82: 905-913.

FNUAP. 1991. *La población y el medio ambiente: Los problemas que se avecinan.* Fondo de Población de las Naciones Unidas.

Gutierrez, V., Infante, M., and Pinchinot, A. 1975. *Situación del cultivo de frijol en América Latina.* Centro Internacional de Agricultura Tropical, Cali, Colombia.

Harlan, H.S. and Martini, M.L. 1938. The effect of natural selection in a mixture of barley varieties. *Journal of Agricultural Research*, 57: 189-199.

Harper, J.L. 1964. The nature and consequence of interference amongst plants. In S.J. Geertz (Editor), *Genetics today. Proceedings of the XI International Congress of Genetics* (1964) 2: 465-482. Pergamon Press, New York.

Hill, J. and Shimamoto, Y. 1973. Methods of analyzing competition with special reference to herbage plants. I. Establishment. *Journal of Agricultural Science, Cambridge,* 81: 77-89.

Hook, J.E. and Gascho, G.J. 1988. Multiple cropping for efficient use of water and nitrogen. In W.L. Hargrove (Editor), *Cropping strategies for efficient use of water and nitrogen.* ASA special publication number 51: 7-20. American Society of Agronomy, Madison, WI.

Jennings, P.R. and Aquino, R.C. 1968. Studies on competition in rice. III. The mechanism of competition among phenotypes. *Evolution,* 22: 529-542.

Jennings, P.R. and de Jesus, J. 1968. Studies on competition in rice. I. Competition in mixture of varieties. *Evolution,* 22: 119-124.

Jensen N.F. 1952. Intravarietal diversification in oat breeding. *Agronomy Journal,* 44: 30-34.

Knott, E.A. and Mundt, C.C. 1990. Mixing ability analysis of wheat cultivar mixtures under diseased and nondiseased conditions. *Theoretical and Applied Genetics,* 80: 313-320.

Kramer, T. 1979. Yield-protein relationship in cereal varieties. In J.H.J. Spiertz and T. Kramer (Editors) *Crop physiology and cereal breeding.* Proceedings of a Eucarpia Workshop. Centre for Agricultural Publishing and Documentation, Wageningen, The Netherlands. 1978.

Lannou, C., de Vallavieille-Pope, C., Bias, C., and Goyeau, H. 1994. The efficacy of mixtures of susceptible and resistant host to two wheat rusts of different lesion size: Controlled condition experiments and computerized simulations. *Journal of Phytopathology,* 140: 227-237.

Lannou, C., de Vallavieille-Pope, C., and Goyeau, H. 1994. Host mixture efficacy in disease control: Effects of lesion growth analyzed through computer-simulated epidemics. *Plant Pathology,* 43: 651-662.

Liebman, M. and Dyck, E. 1993. Crop rotation and intercropping strategies for weed management. *Ecological Applications,* 3(1): 92-122.

Löffler, C.M., Rauch, T.L., and Busch, R.H. 1985. Grain yield and plant protein relationships in hard red spring wheat. *Crop Science,* 25(3): 521-525.

Lupton, F.G.H., Oliver, R.H., Ellis, F.B., Barnes, BT., House, K.R., Welbank, P.J., and Taylor P.J. 1974. Root and shoot growth of semi-dwarf and tall winter wheats. *Annals of Applied Biology,* 77: 129-144.

Mc Kenzie, H. and Grant, M.N. 1980. Survival of common spring wheat cultivars grown in mixtures in three environments. *Canadian Journal of Plant Sciences,* 60: 1309-1313.

Mundt, C.C. and Brophy, L.S. 1988. Influence of number of host genotype units on the effectiveness of host mixtures for disease control: A modeling approach. *Phytopathology,* 78: 1087-1094.

Mundt, C.C., Brophy, L.S., and Kolar, S.C. 1996. Effect of genotype unit number and spatial arrangement on severity of yellow rust in wheat cultivar mixtures. *Plant Pathology* 45: 215-222.

Mundt, C.C. and Browning, J.A. 1985. Genetic diversity and cereal rust management. In A.P. Roelfs and W.R. Bushnell (Editors) *The cereal rusts*, Vol. 2. Academic Press, Orlando, FL: 527-560.

National Research Council. 1989. *Alternative Agriculture.* Committee on the role of alternative farming methods in modern production agriculture. National Academy Press, Washington, DC.

Paccaud, F.X., Fossati, A., and Hong Shen Cao. 1985. Breeding for yield and quality in winter wheat. Consequences for nitrogen uptake and partitioning efficiency. *Journal of Plant Breeding,* 94(2): 89-101.

Perry, M.W. and D'Antuono, M.F. 1989. Yield improvement and associated characteristics of some Australian spring wheats introduced between 1860-1982. *Australian Journal of Agricultural Research,* 40: 458-472.

Prasad, R. and Reddy, M.R. 1973. Note on the efficient use of solar energy through a mixed culture of wheat genotypes. *Indian Journal of Agricultural Sciences,* 45:528-529.

Prasad, R. and Sharma, S.N. 1980. Systematic mixed stands of spring wheat cultivars. *Journal of Agricultural Science, Cambridge,* 94: 529-532.

Rattunde, H.F., Ramraj, V.M., Williams, J.H., and Gibbon, R.W. 1988. Cultivar mixtures: A means of exploiting morpho-developmental differences among cultivated groundnuts. *Field Crop Research,* 19: 201-210.

Reynolds, M.P., Sayre, K.D., and Vivar, H.E. 1994. Intercropping wheat and barley with N-fixing legume species: A method for improving ground cover, N-use efficiency and productivity in low input systems. *Journal of Agricultural Science, Cambridge,* 123: 175-183.

Sarandón, S.J. and Gianibelli, M.C. 1992. Effect of foliar sprayings of urea during or after anthesis on dry matter and nitrogen accumulation in the grain of two wheat cultivars of *Triticum aestivum* L. *Fertilizer Research,* 31(1): 79-84.

Sarandón, S.J. and Sarandón, R. 1995. Mixture of cultivars: Plot field trial of an ecological alternative to improve production or quality of wheat (*Triticum aestivum* L). *Journal of Applied Ecology,* 32: 288-294.

Scholz, F. 1984. Some problems and implications in improving cereal grain protein by plant breeding. In K. Muntz and C. Horstmann (Editors), *Genetics and seed proteins.* Proceeding of the 3rd Seed Protein Symposium, Gatersleben, 1983. Die Kulturpflanze vol. 32.

Shaalan, M.I., Heyne, E.G., and Lofgren, J.R. 1966. Mixtures of hard red winter wheat cultivars. *Agronomy Journal,* 58(1): 89-91.

Sharma, S.N. and Prasad, R. 1978. Systematic mixed versus pure stands of wheat genotypes. *Journal of Agricultural Science, Cambridge,* 90: 441-444.

Siddique, K.H.M., Beldford, R.K., and Tennant, D. 1990. Root:shoot ratios of old and modern, tall and semi-dwarf wheats in a Mediterranean environment. *Plant and Soil,* 121: 89-98.

Silvertown, J.M. 1982. *Introduction to plant population ecology.* Longman, London.

Simmonds, N.W. 1962. Variability in crop plants, its use and conservation. *Biological Review,* 37(3): 422-465.

Singh, S.P. 1983. Summer legume intercrop effects on yield and nitrogen economy of wheat in the succeeding season. *Journal of Agricultural Science, Cambridge,* 101: 401-405.

Slafer, G.A. and Andrade, F.H. 1989. Genetic improvement in bread wheat (*Triticum aestivum*) yield in Argentina. *Field Crop Research,* 21(3-4): 289-297.

Smithson, J.B. and Lenné, J.M. 1997. Varietal mixtures: A viable strategy for sustainable productivity in subsistence agriculture. *Annals of Applied Biology,* 128: 127-158.

Stützel, H. and Aufhammer, W. 1990. The physiological causes of mixing effects in cultivar mixtures: A general hypothesis. *Agricultural Systems,* 32: 41-53.

Trenbath, B.R. 1972. The productivity of varietal mixtures of wheat. PhD thesis, University of Adelaide, Australia.

Trenbath, B.R. 1974. Biomass productivity of mixtures. *Advances in Agronomy,* 26: 177-210.

Trenbath, B.R. 1986. Resource use by intercrops. In C. A. Francis (Editor), *Multiple cropping systems,* Macmillan, New York: 57-81.

Vandermeer, J. 1989. *The ecology of intercropping.* Cambridge University Press, New York.

van Kessel, C., Singleton, P.W., and Hoben, H.J. 1985. Enhanced N-transfer from soybean to maize by vesicular arbuscular mycorrizhal (VAM) fungi. *Plant Physiology,* 79: 562-563.

Wien, H.C. and Smithson, J.B. 1981. The evaluation of genotypes for intercropping. In R.H. Willey (Editor), ICRISAT (International Crop Institute for Semiarid Tropics) Proceedings of the International Workshop on Intercropping, Hyderabad, India, 1979. ICRISAT, Patancheru, India: 105-116.

Willey, R.W. 1981. A scientific approach to intercropping research. In R.H. Willey (Editor), ICRISAT (International Crop Institute for Semiarid Tropics) Proceedings of the International Workshop on Intercropping, Hyderabad, India, 1979. ICRISAT, Patancheru, India: 4-14.

Willey, R.W. and Rao, M.R. 1981. Genotypes studies at ICRISAT. In R.H. Willey (Editor), ICRISAT (International Crop Institute for Semiarid Tropics) Proceedings of the International Workshop on Intercropping, Hyderabad, India, 1979. ICRISAT, Patancheru, India: 117-121.

Willey, R.W. and Reddy, M.S. 1981. A field technique for separating above and below ground interactions in intercropping: An experiment with pearl millet/groundnut. *Experimental Agriculture,* 17: 257-264.

Wolfe, M.S. 1985. The current status and prospects of multiline cultivars and variety mixtures for disease resistance. *Annual Review of Phytopathology,* 23: 251-273.

Zimmerman, M.J.O. 1996. Breeding for yield, in mixtures of common beans (*Phaseolus vulgaris* L.) and maize (*Zea mays* L.) *Euphytica,* 92: 129-134.

PART III: WHEAT PRODUCTION SYSTEMS

Chapter 12

Wheat Production in the Great Plains of North America

William R. Raun
Gordon V. Johnson
Robert L. Westerman
Jeffory A. Hattey

INTRODUCTION

The Great Plains encompass portions of 13 states (Iowa, Kansas, Missouri, Nebraska, Oklahoma, Texas, New Mexico, Colorado, Wyoming, North Dakota, South Dakota, Montana, and Minnesota), and three Canadian provinces (Manitoba, Saskatchewan, and Alberta) (http://www.epa.gov/ecoplaces/part1/site9.html). This area is bound by the Rocky Mountains to the west and the Mississippi River valley on the east. The Great Plains area was once the largest grassland in the world and covered over 2.6 million square kilometers.

Importance of Wheat Production, Climate, and Soils

As reported by Peterson (1996), the Great Plains of North America are recognized around the world for their wheat production and fertile Mollisol soils, which dominate landscapes from Canada to Texas. Average annual rainfall in the Great Plains region varies considerably, ranging from 25 cm in the western states (Montana, Wyoming, Colorado, and New Mexico) to as much as 120 cm in the eastern states (Minnesota, Iowa, and Missouri). In general, average annual rainfall is lowest in the western Great Plains states, increases in the central region (North Dakota, South Dakota, Nebraska, Kansas, Oklahoma, and Texas) and is greatest in the eastern states.

Krall and Schuman (1996) reported that soil organic carbon levels have declined on many Great Plains soils since the start of cultivation. This decline has been documented by many soil scientists. Work by Raun et al. (1997) found that when N was applied at rates >90 kg ha^{-1}, surface soil (0-30 cm) organic C was either equal to that of the control (no N applied) or slightly greater following more than 20 years of continuous winter wheat production under conventional tillage. Their work also demonstrated that high rates of applied N had the beneficial effect of increasing soil organic C over the 20-plus-year period evaluated.

Main Characteristics of Wheat Planted in the Great Plains

Five main classes of wheat are grown: hard red winter, hard red spring, soft red winter, white, and durum (Briggle and Curtis, 1987). Most of the hard red winter wheat is grown in the central and southern Great Plains states, while hard red spring is grown in the northern Great Plains. Most of the durum wheat grown in the United States is in North Dakota. In recent years, an increase in soft red winter wheat production has been observed in Illinois and Missouri (Briggle and Curtis, 1987).

DESCRIPTION OF WHEAT PRODUCTION SYSTEMS

Dhuyvetter et al. (1996) reported that the most important dryland cropping system in the Great Plains has been wheat-fallow (WF) rotation (one crop in two years). They also found that between 1991 and 1993, the harvested area of dryland winter wheat in western Kansas, western Nebraska, and eastern Colorado ranged from 2.5 to 2.9 million hectares annually. Most of the 2.8 million hectares of wheat grown in Oklahoma is continuous winter wheat without fallow or rotation. Norwood (1994) noted that the use of grain sorghum in the wheat-sorghum-fallow (WSF) system (two crops in three years) has become increasingly popular in the Great Plains region. Additional cropping systems in this area include sorghum-fallow (SF) and continuous sorghum (SS).

Winter wheat grown solely for grazing cattle has increased in the past two decades. Krenzer, Thompson, and Carver (1992) reported that 35 to 55 percent of the winter wheat is used as a dual-purpose crop in the southern Great Plains, whereby the winter wheat is grazed in the fall and winter followed by grain harvest in late spring. In these winter wheat forage/grain production systems, cattle are generally removed between February and late March, from the southern to northern regions, respectively. Christiansen, Svejcar, and Phillips (1989) reported that straw yields were lowered by grazing; therefore, the harvest index (grain weight/grain + straw weight) was higher for grazed versus nongrazed wheat.

Halvorson (1988) concluded that more intensive cropping systems than the traditional crop-fallow system are needed to make more efficient use of water supplies under dryland conditions in the Great Plains. Later work by Halvorson and Reule (1994) indicated that more intensive dryland cropping systems needed to be adopted in the central Great Plains to increase water use efficiency and better maintain soil quality.

Tillage and Rotation

In order to increase water storage capacity in areas where precipitation was limiting, substantial research has been aimed at evaluating improved management practices. Because of the wind and water erosion associated with soil tillage, alternative sweeps that would leave the land with more cover (versus a plow or disk) even after tillage were developed, thus leading to what is now referred to as stubble mulching (Peterson, 1996). Early work by Staple, Lehane, and Wenhardt (1960) reported that 37 percent of winter precipitation was conserved as stored soil water in untilled wheat stubble, compared with only 9 percent when the soil surface did not have stubble.

Dhuyvetter et al. (1996) noted that the fallow period increases stored moisture and weed control in the central Great Plains. In seven of eight studies, net returns were greater from a more intensive crop rotation than from WF when reduced-tillage (RT) or no-till (NT) were used after wheat harvest but prior to planting in the summer (Dhuyvetter et al., 1996). Cropping systems using more intensive rotations with less tillage had higher production costs than WF, but also had increased net returns and reduced financial risk (Dhuyvetter et al., 1996). Halvorson et al. (1994) reported that if costs or application rates of herbicides could be reduced, NT would be more economically competitive with other tillage systems. Because cost variations between tillage systems are minimal, the added benefits of increased moisture storage and decreased erosion with RT and NT must be considered (Halvorson et al., 1994). Norwood (1994) reported twice as much available water was stored prior to sorghum planting in a wheat-sorghum-fallow production system using no-tillage compared to conventional tillage.

LeMahieu and Brinkman (1990) reported that double-cropping soybean (*Glycine max* L.) after harvesting winter wheat, winter rye (*Secale cereale* L.), or spring barley (*Hordeum vulgare* L.) as forage is feasible in the north central United States. Double cropping soybeans and/or sorghum following wheat has become increasingly popular in the eastern portions of the Great Plains where precipitation is sufficient for an added summer crop. Black and Unger (1987) reported sustained success of dryland agriculture depends upon a unique blend of agricultural sciences, all based on the underlying principle of conserving the soil while using available water efficiently. In this regard, successful implementation of soil conservation practices (fallow, reduced tillage, and rotation) in the Great Plains has had a significant impact on present and long-term productivity in the region.

Planting Dates

Winter wheat in the Great Plains is generally planted from early September to late November. Earlier planting dates have been encouraged when wheat is grown primarily for forage, although early planting dates significantly increase the risk of frost damage in the spring and increased severity of some soilborne diseases. Work conducted in Kansas indicated that the optimum planting date for winter wheat produced for grain ranged from September to early October in the central Great Plains (Witt, 1996). This work also reported that March 1 was the last planting date for winter wheat where heads and grain could be produced, although grain yield levels were significantly reduced when compared to fall planting dates. Winter and Musick (1993) indicated that the planting date of winter wheat grown in the southern High Plains of Texas varied from August to November. They further reported that planting in August compared to October did not increase soil water extraction at anthesis and grain yield was reduced. Dahlke et al. (1993) suggested that for the northern region of the Great Plains, maximum grain yield occurred when planting winter wheat on September 3 using 301 seeds per m^2. Delaying planting to late September required a seeding rate of 449 to 599 seeds per m^2 to maximize yield. Grain yield decreased with a planting date into October, and kernel weight and heads per unit area decreased as well.

Fertilizer Management

Immobile and Mobile Nutrients

For immobile nutrients such as phosphorus, plants can only extract the nutrient from soil close to the root surface. Very little of the nutrient is moved to the root by capillary water movement because soil solution concentrations are small (<0.05 $\mu g\ g^{-1}$ for phosphate compared to as much as 100 $\mu g\ g^{-1}$ for nitrate-N). As a plant grows and roots extend out into the soil, roots come in contact with new soil from which they can extract phosphate. The amount extracted is limited by the concentration at (or very near) the root-soil interface. If the concentration of phosphate available to the plant at the root-soil interface is inadequate, then the plant will be deficient in P throughout its development. Plant growth and crop yield will be limited by the degree to which the immobile nutrient is deficient. Another, perhaps more common way of expressing this limitation is to state that yield will be obtained according to the sufficiency of the nutrient supply. When this is expressed as a percentage of the yield possibility, the term "percent sufficiency" may be applied. Whenever the percent sufficiency is less than 100, plant performance is less than the yield possibility provided by the growing environment. Consequently, it does not matter whether the yield possibility is 2 or 3 Mg ha^{-1}, if the percent sufficiency is 80, then the actual yield obtained (theoretically) will only be 80 percent of the yield possibility.

The soil test for mobile nutrients is an indicator of the total amount available. If this amount is enough to produce 2 Mg ha^{-1}, more N would have to be added to the total pool to produce 3 Mg ha^{-1}. With immobile nutrients such as P and K, an index is developed that is independent of the environment. If the crop year was good, roots would expand into more soil that had the same level of nutrient-supplying capacity. Sufficiency is independent of the environment since increased root growth will expand into areas where diffusion transport and contact exchange uptake is the same (total amount present in the soil is not greatly reduced by more growth).

Yield Goals and Soil Testing

Yield goals apply to all mobile nutrients, while the sufficiency approach is used for immobile nutrients. Yield response to immobile nutrients is not related to the total amount of the available form present in the soil, but instead is a function of the concentration of available form at, or very near, the root surface.

For wheat production in the Great Plains, nitrogen is often the most limiting nutrient. Westerman (1987) indicated that 33 kg N Mg^{-1} (2 lb N/bu) is required for grain yield goals up to 3360 kg ha^{-1} (50 bu/Ac) and an additional 30 kg N Mg^{-1} (60 lb N/ton) of dry matter forage removed by pasturing. The 33 kg is calculated from the protein or N content (on average) of a megagram of wheat, with the added assumption that measured soil nitrate-N and added fertilizer N will be only 70 percent utilized.

Halvorson, Alley, and Murphy (1987) reported that in most states, residual NO_3-N in the surface 60 to 120 cm of soil is considered 100 percent available and if soil testing is available, this amount is subtracted from the N fertilizer recommended for a particular yield goal. Westfall et al. (1996) reported that the quantification of residual soil NO_3-N is the main component of soil testing for determining N needs. Some states have recently considered the addition of NH_4-N to the soil test. As a result the sum of NH_4-N and NO_3-N is subtracted from the N fertilizer requirement determined for a given yield goal. Soil organic matter is also considered in some laboratories whereby the recommendation is altered based on predicted N mineralization and N availability over the growing season (Follett et al., 1991).

Soil test potassium levels in the Great Plains are generally sufficient. Alternatively, phosphorus deficiencies can be commonly found where wheat is grown in this region. Response to applied sulfur has been observed on soils with low soil organic matter and coarse texture, although this nutrient is not considered to be deficient to a large extent. Other work has documented increased wheat grain yields as a result of chloride applications; however, much of the response has been associated with reduced incidence of take-all (*Gaeumannomyces graminis*) (Christensen and Brett, 1985), powdery mildew (*Erysiphe graminis*), and leaf rust (*Puccinia recondita*) (Engel, Eckhoff, and Berg, 1994). Alternatively, Brennan (1993) indicated that wheat grain yield losses from take-all were most severe where plants were N deficient. He further noted that chloride-containing fertilizers are unlikely to control take-all disease.

Halvorson, Alley, and Murphy (1987) reported that water is the factor most often limiting dryland wheat yields in the semiarid regions; therefore, estimating potential grain yield requires consideration of both the amount of soil water at planting and the expected growing season precipitation. They further noted that a number of factors can affect the yield goal, including availability of capital to spend on fertilizer. However, the yield goal should probably be close to the maximum yield obtained in the area. The main reason for fertilizing for maximum yields is that unused nutrients can be available for subsequent crops if not used (Dahnke, Swenson, and Goos, 1983). Maximum yields may require slightly more N to produce a unit of grain and maintain quality due to decreased N-use efficiency with increasing yield level (Halvorson, Alley, and Murphy, 1987).

How the wheat is used will also affect N requirements, and an example of this change is the common practice of grazing wheat in the central Great Plains. Removal of forage can increase N needs and should be considered when formulating N recommendations (Halvorson, Alley, and Murphy, 1987). Oklahoma recommends 30 kg of N for every 1000 kg of forage produced. In the past, 7 kg N/animal/month has been used as a rough estimate for N requirements in a wheat forage production system (Billy Tucker, personal communication, 1997). Halvorson, Alley, and Murphy

(1987) indicated that on an animal gain basis, about 1 kg of N will be required to replace that removed in 3 kg of animal gain.

Nitrogen Source and Timing

The common N source for winter wheat production in the Great Plains is anhydrous ammonia. AA (Anhydrous ammonia) is generally injected prior to planting at a depth of 15 cm with a shank spacing of 75 cm. The majority of farmers who inject AA preplanting, apply all of the seasonal N requirement at this time. However, occasionally weather conditions restrict N applications at planting (extremely wet or dry conditions in the fall that prevent good soil closure behind knife applicators). Because of this, alternative midseason topdress N applications have become popular. Topdress N has generally been applied as urea ammonium nitrate (UAN) in winter months. Recently, Boman et al. (1995a) injected AA into established wheat stands and found no significant grain yield reduction when compared to UAN topdress. Because the cost of AA is half that of UAN, this method/source of midseason N application may become increasingly popular. Other work by Boman et al. (1995b), evaluating time of application, found that N topdressed from December to January resulted in equivalent forage yields when compared to N broadcast and incorporated before planting. Their work further noted that the optimum time of N application for maximum spring forage and wheat grain yields was mid-November and early January, respectively. It should be noted that delayed N applications generally have a greater effect on grain protein than on yield.

Halvorson, Alley, and Murphy (1987) noted that N fertilizers placed with the seed at planting should not exceed 20 kg N ha^{-1}. They further reported that rates of N applied with the seed should be lower on sandy soils when soil moisture conditions at seeding are poor or when soil pH is greater than 7.3. Westfall et al. (1996) indicated that while producers appear to have greater flexibility in choosing fertilizer N placements, N rate is the more critical management decision.

Urea fertilizers should not be applied on the surface of soils without incorporation, especially when soil pH is >7.0 and/or when surface residues are present, due to the increased potential for ammonia volatilization. Ammonia volatilization losses from surface-applied urea without incorporation will also occur when soil surfaces are wet, temperatures are high, and there is little chance for rain soon after application

Environmental Impact of Nitrogen Fertilizers in Winter Wheat

Work by Johnson and Raun (1995) and Raun and Johnson (1995) reported that applying more fertilizer N than that required for maximum wheat grain yield did not immediately pose a risk to groundwater quality. This conclusion was the product of four long-term winter wheat experiments grown under dryland conditions and where average annual rainfall ranged from 765 to 1057 mm. Their data suggested that the soil-plant system was able to buffer against (prevent) soil accumulation of inorganic N. The major buffering mechanisms included documented research where reports of increased plant protein, plant N volatilization, denitrification, and immobilization were found when N rates exceeded that required for maximum yield. Much of the central Great Plains region where wheat is grown is nonirrigated and precipitation generally controls yield level. Because of this, NO_3-N leaching from non-point source fertilizer applications in these rainfed production areas has had little impact on groundwater quality. However, it should be noted that soil testing, especially subsoil analysis for NO_3-N, should be an important input for managing fertilizer N since high subsoil NO_3-N may be an indication of past excess N input and a signal that future inputs of N should be decreased (Johnson and Raun, 1995).

Straw Residue

Westerman (1987) discussed the importance of straw removal of bases on soil pH. Wheat straw contains significant quantities of bases, therefore, when it is removed, the potential exists for increasing soil acidity. This is also true when grazing is employed considering the cation content in the forage. For a grain-only production system, compared to grain + straw or forage + grain, it is apparent why the potential for increased acidity is much lower.

Decomposition of wheat residue results in soil acidification. Carbon dioxide is released from organic residues during decomposition and combines with water to form carbonic acid (H_2CO_3). Carbonic acid dissociates into H^+ and HCO_3^-, resulting in another source of H^+ for increasing soil acidity. Root activity and metabolism may also serve as a source of CO_2 (Westerman, 1987).

Crop Protection Strategies Against Weeds

Lyon, Miller, and Wicks (1996) reported that herbicides have played an important role in dryland agriculture since their introduction in the late 1940s. They have significantly reduced the amount of tillage required for crop production. As a result, reduced-tillage systems have been possible (largely because of the availability of herbicides). Recent trends, however, include declining introduction of new herbicides, potential loss of older herbicides, increased herbicide resistance in weeds, and rising public concern about the effects of pesticides in the environment. All of these may lead to a reduction in herbicide use in the future (Lyon, Miller, and Wicks, 1996). Although atrazine has proven successful for preemergence control of some annual grasses and broadleaf weed species in maize and sorghum production systems, its persistence has been a problem in winter wheat-fallow rotations.

In recent years, cheat (*Bromus secalinus* L.) has become a significant problem in continuous winter wheat production systems. Wheat grain yield losses can exceed 40 percent in fields heavily infested with cheat. A consistent suppression of cheat in wheat has not been found using narrow row spacing (7-14 cm); however, in some years this practice can result in a significant reduction in cheat yield (John B. Solie, 1997, personal communication). The decreased use of rotations has created problems with winter annuals such as cheat in continuous winter wheat production systems. This weed problem signals the need for alternative fallow and/or rotation production practices. Lyon, Miller, and Wicks (1996) noted that the more dissimilar the crops and their management practices are in a rotation, the less opportunity an individual weed species has to become dominant.

Holt and LeBaron (1990) reported that many herbicide-resistant biotypes are resistant to the triazine herbicides (atrazine and simazine). Because of this, some of the newer, environmentally safe herbicides may become ineffective due to weed resistance to herbicides (Lyon, Miller, and Wicks, 1996). This work further noted that to retain herbicides as effective tools in sustainable crop production, strategies need to be developed that include the use of crop rotation, fertilizer placement, tillage, biological control, and precision application technology. Precision application technologies (e.g., spraying only those weeds present using sensor-based systems) have the potential of reducing from $^{1}/_{10}$ to $^{1}/_{1000}$ the total amount applied when using fixed rates.

TECHNOLOGICAL FACTORS THAT HAVE INCREASED GRAIN YIELDS IN THIS CENTURY

Briggle and Curtis (1987) noted that greater use of fertilizer (25 percent increase in global fertilizer N use from 1974 to 1981) coupled with a significant increase in irrigation in some areas has contributed significantly to yield increases in this century. Their work also highlights the widespread adoption of improved high-yielding, semidwarf wheat cultivars, which have greatly increased global genetic potential for higher wheat yields. Breeding efforts for tolerance to acid soils, heat, drought, and certain insects have played an important role in increasing wheat grain yields. However, efforts in these areas cannot replace a sound fertility program, regardless of the wheat production system.

The development of dwarf spring wheats via an intensive shuttle breeding program by Dr. Norman Borlaug stands as one of the most important agricultural achievements of the century. Easterbrook (1997) stated, "Norman Borlaug is responsible for the fact that except in sub-Saharan Africa, food production has expanded faster than the human population, averting the mass starvations that were widely predicted." Pandey et al. (1996) reported that one-fifth of the total U.S. wheat acreage was sown with varieties with CIMMYT (International Maize and Wheat Improvement Center in Mexico, founded by Norman Borlaug and others in cooperation with the Rockefeller Foundation) ancestry by the early 1990s. Dwarf lines from the spring wheat breeding efforts in Mexico were ultimately incorporated into the hard red winter wheat lines in the United States. In the late 1970s, the variety 'Vona' was released by Colorado State University. This was the first hard red winter wheat released in the United States that included the semidwarf gene (from CIMMYT germplasm) and high yield potential (Art Klatt, 1997, personal communication).

TRANSFORMATION OF WHEAT PRODUCTION SYSTEMS IN THE FUTURE

Wheat production in the twenieth century has seen dramatic changes in variety, tillage, rotation, fertilization, and weed-disease-insect management. Because of this, we expect to see more and more specialized diversity for wheat production systems in the future, although adoption of improved practices will continue to be a problem. Dhuyvetter et al. (1996) inferred that even though production benefits of alternative dryland cropping systems have been documented, producers have been slow to adopt these technologies. Because of the relatively low labor and management requirements of a wheat-fallow rotation, producers may be hesitant to change to a more intensive cropping system. Consistent with other authors, they further noted that government program rigidity and producers' attitudes have most likely been major reasons for the slow adoption of alternative crop rotations.

Nitrogen use efficiencies (NUE) in virtually all grain crops seldom exceed 50 percent (Olson and Swallow, 1984; Wuest and Cassman, 1992). This means that 50 percent of the fertilizer nitrogen that is applied is not removed in the grain or straw. The unaccounted N is often assumed to be lost to leaching (Jokela and Randall, 1989), denitrification (Olson et al.,

1979) and as NH_3 volatilized from the leaves of senescing plants (Sharpe et al., 1988). Nitrogen that is applied but not used by the crop represents a sizeable economic loss to the farmer as well as a possible source of contamination to the environment. Current environmental and economic concerns identify the need to combine high yields with increased input use efficiency, especially for nitrogen, which represents, in many cases, the most expensive input used by farmers for their wheat production. In this regard, we feel that production systems and wheat varieties with improved water and nitrogen use efficiency will be developed to assist us in feeding a global population of 10 billion people expected by the year 2050.

REFERENCES

Black, A.L. and P.W. Unger. 1987. Management of the wheat crop, pp. 330-339. In E.G. Heyne (ed.) *Wheat and wheat improvement.* Second edition. Agron. Monogr. 13. ASA, CSSA and SSSA, Madison, WI.

Boman, R.K., R.L. Westerman, W.R. Raun, and M.E. Jojola. 1995a. Spring-applied nitrogen fertilizer influence on winter wheat and residual soil nitrate. *J. Prod. Agric.* 8:584-589.

Boman, R.K., R.L. Westerman, W.R. Raun, and M.E. Jojola. 1995b. Time of nitrogen application: Effects on winter wheat and residual soil nitrate. *Soil Sci. Soc. Am. J.* 59:1364-1369.

Brennan, R.F. 1993. Effect of ammonium chloride, ammonium sulphate and sodium nitrate on take-all and grain yield of wheat grown on soils in south-western Australia. *J. Plant Nutr.* 16:349-358.

Briggle, L.W. and B.C. Curtis. 1987. Wheat worldwide, pp. 1-32. In E.G. Heyne (ed.) *Wheat and wheat improvement.* Second edition. Agron. Monogr. 13. ASA, CSSA and SSSA, Madison, WI.

Christensen, N.W. and M. Brett. 1985. Chloride and liming effects on soil nitrogen form and take-all of wheat. *Agron. J.* 77:157-163.

Christiansen, S., T. Svejcar, and W.A. Phillips. 1989. Spring and fall cattle grazing effects on components and total grain yield of winter wheat. *Agron. J.* 81:145-150.

Dahlke, B.J., E.S. Oplinger, J.M. Gaska, and M.J. Martinka. 1993. Influence of planting date and seeding rate on winter wheat grain yield and yield components. *J. Prod. Agric.* 6:408-414.

Dahnke, W.C., L.J. Swenson, and R.J. Goos. 1983. Choosing a crop yield goal, pp. 69-70. In *1984 Crop production guide.* North Dakota State Univ., Fargo, ND.

Dhuyvetter, K.C., C.R. Thompson, C.A. Norwood, and A.D. Halvorson. 1996. Economics of dryland cropping systems in the great plains: A review. *J. Prod. Agric.* 9:216-222.

Easterbrook, G. 1997. Forgotten benefactor of humanity. *Atlantic Monthly.* Jan.: 75-82.

Engel, R.E., J. Eckhoff, and R.K. Berg. 1994. Grain yield, kernel weight, and disease responses of winter wheat cultivars to chloride fertilization. *Agron. J.* 86:891-896.

Follett, R.H., P.N. Soltanpour, D.G. Westfall, and J.R. Self. 1991. *Soil test recommendations in Colorado.* Colorado Agric. Ext. Serv. XCM-37.

Halvorson, A.D. 1988. Role of cropping systems in environmental quality: Saline seep control, pp. 179-191. In W.L. Hargrove (ed.) *Cropping strategies for efficient use of water and nitrogen.* ASA Spec. Publ. 51. ASA, CSSA, and SSSA, Madison, WI.

Halvorson, A.D., M.M. Alley, and L.S. Murphy. 1987. Nutrient requirements and fertilizer use, pp. 345-383. In E.G. Heyne (ed.) *Wheat and wheat improvement.* Second edition. Agron. Monogr. 13. ASA, CSSA and SSSA, Madison, WI.

Halvorson, A.D., R.L. Anderson, N.E. Toman, and J.R. Welsh. 1994. Economic comparison of three winter wheat-fallow tillage systems. *J. Prod. Agric.* 7:381-385.

Halvorson, A.D. and C.A. Reule. 1994. Nitrogen fertilizer requirements in an annual dryland cropping system. *Agron. J.* 86:315-318.

Holt, J.S. and H.M. LeBaron. 1990. Significance and distribution of herbicide resistance. *Weed Technol.* 4:141-149.

Johnson, G.V. and W.R. Raun. 1995. Nitrate leaching in continuous winter wheat: Use of a soil-plant buffering concept to account for fertilizer nitrogen. *J. Prod. Agric.* 8:486-491.

Jokela, W.E. and G.W. Randall. 1989. Corn yield and residual soil nitrate as affected by time and rate of nitrogen application. *Agron. J.* 81:720-726.

Krall, J.M. and G.E. Schuman. 1996. Integrated dryland crop and livestock production systems on the Great Plains: Extent and outlook. *J. Prod. Agric.* 9:187-191.

Krenzer, E.G. Jr., J.D. Thompson, and B.F. Carver. 1992. Partitioning of genotype × environment interactions of winter wheat forage yield. *Crop Sci.* 32: 1143-1147.

LeMahieu, P.J. and M.A. Brinkman. 1990. Double-cropping soybean after harvesting small grains as forage in the north central USA. *J. Prod. Agric.* 3:385-389.

Lyon, D.J., S.D. Miller, and G.A. Wicks. 1996. The future of herbicides in weed control systems of the Great Plains. *J. Prod. Agric.* 9:209-215.

Norwood, C. 1994. Profile water distribution and grain yield as affected by cropping system and tillage. *Agron. J.* 86:558-563.

Olson, R.V., L.S. Murphy, H.C. Moser, and C.W. Swallow. 1979. Fate of tagged fertilizer nitrogen applied to winter wheat. *Soil Sci. Soc. Am. J.* 43:973-975.

Olson, R.V. and C.W. Swallow. 1984. Fate of labeled nitrogen fertilizer applied to winter wheat for five years. *Soil Sci. Soc. Am. J.* 48:583-586.

Pandey, P.G., J.M. Alston, J.E. Christian, and Shenggen Fan. 1996. *Hidden harvest: U.S. benefits from international research aid.* International Food Policy Research Institute, Washington, DC.

Peterson, G.A. 1996. Cropping systems in the Great Plains. *J. Prod. Agric.* 9:179.

Peterson, G.A., A.J. Schlegel, D.L. Tanaka, and O.R. Jones. 1996. Precipitation use efficiency as affected by cropping and tillage systems. *J. Prod. Agric.* 9:180-186.

Raun, W.R. and G.V. Johnson. 1995. Soil-plant buffering of inorganic nitrogen in continuous winter wheat. *Agron. J.* 87:827-834.

Raun, W.R., G.V. Johnson, S.B. Phillips, and R.L. Westerman. In press. Effect of long-term nitrogen fertilization on soil organic carbon and total nitrogen in continuous wheat. *Soil and Tillage Res.*

Sharpe, R.R., L.A. Harper, J.E. Giddens, and G.W. Langdale. 1988. Nitrogen use efficiency and nitrogen budget for conservation tilled wheat. *Soil Sci. Soc. Am. J.* 52:1394-1398.

Staple, W.J., J.J. Lehane, and A. Wenhardt. 1960. Conservation of soil moisture from fall and winter precipitation. *Can. J. Soil Sci.* 40:80-88.

Westerman, R.L. 1987. Soil reaction-acidity, alkalinity, and salinity, pp. 340-344. In E.G. Heyne (ed.) *Wheat and wheat improvement.* Second edition. Agron. Monogr. 13. ASA, CSSA and SSSA, Madison, WI.

Westfall, D.G., J.L. Havlin, G.W. Hergert, and W.R. Raun. 1996. Nitrogen management in dryland cropping systems. *J. Prod. Agric.* 9:192-199.

Winter, S.R. and J.T. Musick. 1993. Wheat planting date effects on soil water extraction and grain yield. *Agron. J.* 85:912-916.

Witt, M.D. 1996. Delayed planting opportunities with winter wheat in the central great plains. *J. Prod. Agric.* 9:74-78.

Wuest, S.B. and K.G. Cassman. 1992. Fertilizer-nitrogen use efficiency of irrigated wheat: I. Uptake efficiency of preplant versus late-season application. *Agron. J.* 84:682-688.

Chapter 13

Wheat Cropping in Australia

Ralph A. Fischer

INTRODUCTION

Wheat has always been the major grain crop in Australia, currently contributing about 16 million tons or 60 percent of national grain production, followed by barley with 17 percent. Wheat farming began in the late eighteenth century soon after European settlement in order to meet domestic needs. In the latter part of the nineteenth century (after 1860) wheat area expanded rapidly as railways were developed and exporting assumed major importance; today around 75 percent of the crop is exported. However, since wool production was the most important rural enterprise in the early years, the wheat industry developed in conjunction with sheep grazing and nowadays most wheat in Australia is produced in the wheat-sheep zone. The dominance of exporting and the presence of sheep are unique and important features of wheat farming in Australia. This chapter will describe physical aspects of the Australian wheat farming environment, before returning to the evolution of this diverse wheat farming system, its current physical and economic status, and its future challenges.

WHEAT CLIMATE AND SOILS

The wheat climate of Australia is described in detail by Nix (1975). Wheat growing is almost entirely rainfed. Annual average rainfall in the wheat regions ranges from 275 to 700 mm (Figure 13.1). The climate of the western grain region (Western Australia [WA]) is Mediterranean with a strong winter rainfall pattern. Summer rainfall increases moving east to the southern region (South Australia [SA], Victoria [Vic], and southern New South Wales [NSW]), and predominates in the northern region

FIGURE 13.1. Distribution of Wheat Growing in Australia in the Late 1980s and Average Annual Rainfall

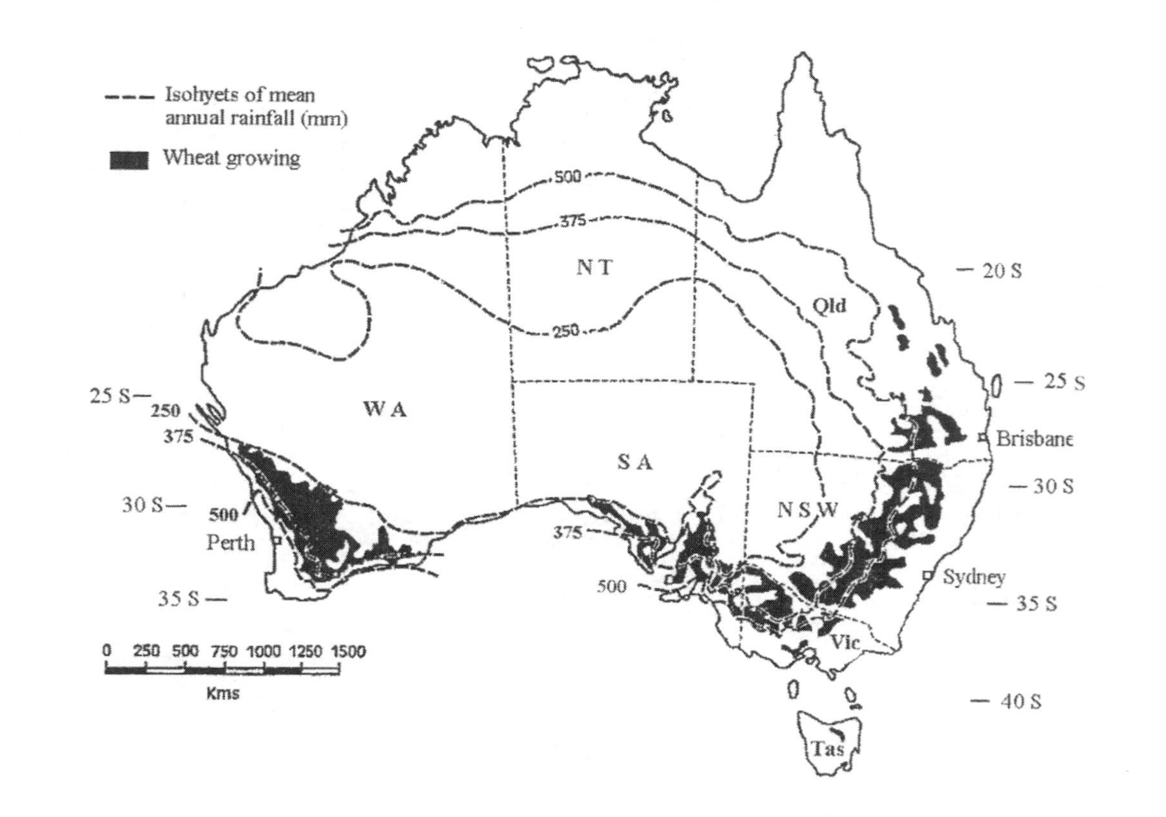

(northern NSW and southeast and central Queensland [Qld]). The wheat-growing season is the winter and spring, generally from May to October when rainfall ranges from 170 to 400 mm. Soil water stored at planting can significantly supplement this rainfall, such that average crop evapotranspiration is around 300 mm. Such storage is especially critical in the north where rain on the crop is least. A feature of NSW and Qld wheat areas is that rainfall anomalies are correlated with various indices of the ENSO (El Nino-Southern Oscillation), determined up to 6 to 12 months beforehand (McBride and Nicholls, 1983; Hammer, Holzworth, and Stone, 1996).

Ranging in latitude from 23° to 38° S (and 42° S in Tasmania where very little wheat is grown), with an altitude usually below 600 m, the wheat areas of Australia have mild winters (mean July temperatures between 7° and 15°C). Thus generally spring-type wheats are planted from May to early July. However, an exception to this is the seeding of facultative (March to April) and true winter (February to March) wheats in the cooler southeastern fringe of the wheat belt for the dual purposes of grazing and grain; there is potential to increase these currently minor plantings as better varieties become available.

The optimum date for flowering for wheat in Australia ranges from late September in the north and in drier areas to late October in the cooler areas of the south and along the wetter margins. Late frost damage, after stem elongation and especially around ear emergence, is a risk particularly in the north, where again anomalies show significant correlations with prior ENSO indices. Flowering must come as soon as frost risk permits to reduce the risk of grain-filling heat or drought damage later in the spring; attention to this dilemma has been a major preoccupation of Australian wheat farmers and researchers alike. Mean grain-filling temperature ranges from 16°C in the southeast to 20°C in the north (Nix, 1975).

Wheat farming occupied the relatively favorable red-brown earths of the south region (the great soil groups of Australia are described in Stace et al., 1968). Red-brown earths have a loamy surface texture overlying moderate to heavy clay subsoils with calcium carbonate at depth. In the drier margins of this region, wheat is now also grown on solonized brown soils or mallee soils, mallee being a unique form of dense *Eucalyptus* woodland. Such soils are characterized by sandy surfaces merging into finer-textured subsoils, with large amounts of calcareous material throughout. Both these soil groups are important also in the west region, where more recently extensive tracts of solodized solonetz and solodic soils have also come under wheat; sandy or loamy surface horizons, often quite deep (>50 cm), sharply contrast with dense clay subsoils having moderate salinity. All the above soil groups were originally occupied by open or dense woodland of *Eucalyptus, Callitris,* and/or *Acacia* species, or in the case of poorer soils, shrublands of the same species. Finally, wheat is grown on cracking clay soils of limited texture contrast with depth, namely gray brown and red clays, and the black earths. These predominate in the last to-be-developed north region, where their ability to store presowing moisture is critical for wheat growing. Heavy gray clays are also found in the south, most notably in the Wimmera district of western Victoria. Originally these soils were usually covered with perennial grassland, but in parts of the north region, open woodland of *Eucalyptus* and *Casuarina* and even dense *Acacia* woodland had to be cleared. With the exception of the last-mentioned clayey soils, the native fertility of the Australian wheat soils was relatively poor, compared for example to the prairie-derived wheat soils of the Northern Hemisphere or Argentina. Soils are low in N, and especially P, and often micronutrients, because of the absence of base-rich parent material and because of a long history of extreme weathering. This is especially so in WA and SA. Another feature of many southern wheat soils is moderate subsoil salinity.

Because the Australian wheat crop is entirely rainfed, and soils generally have poor water-holding capacity, not only is the national average wheat yield low (currently around 1.7 t/ha, see later) but its coefficient of variability (20 percent) is high for countries with a similar area of wheat (Singh and Byerlee, 1990). Even in the past 20 years average yield has fallen below 1.0 t/ha twice due to drought.

HISTORIC CHANGES IN WHEAT YIELD

In 1860 the area of wheat was 0.3 million ha; it reached 2 m ha in 1900 and peaked at 7 m ha in 1930 (Callaghan and Millington, 1956). Wheat area then declined to a low of 4 million ha in the mid-1950s, before rising again to an all-time peak of 13 million ha in 1983; currently the area is steady at around 10 million ha. The falls and rises in wheat growing in the past 60 years are largely a function of changes in wheat prices relative to those of wool.

To look at wheat yield change since 1860, Figure 13.2 takes 10-year means in order to reduce the considerable impact of weather variability. It needs to be noted that in the periods of area expansion this century, drier inland regions with poorer soils tend to be occupied, with consequent downward pressure on expected yields. Nonetheless Donald (1982), who developed Figure 13.2 up until 1970, was able to point to other factors driving three major phases of yield change. The first phase, from 1860 to 1900, was

FIGURE 13.2. Variation in Australian Average Wheat Yield (Ten-Year Mean) from 1860 to 2000

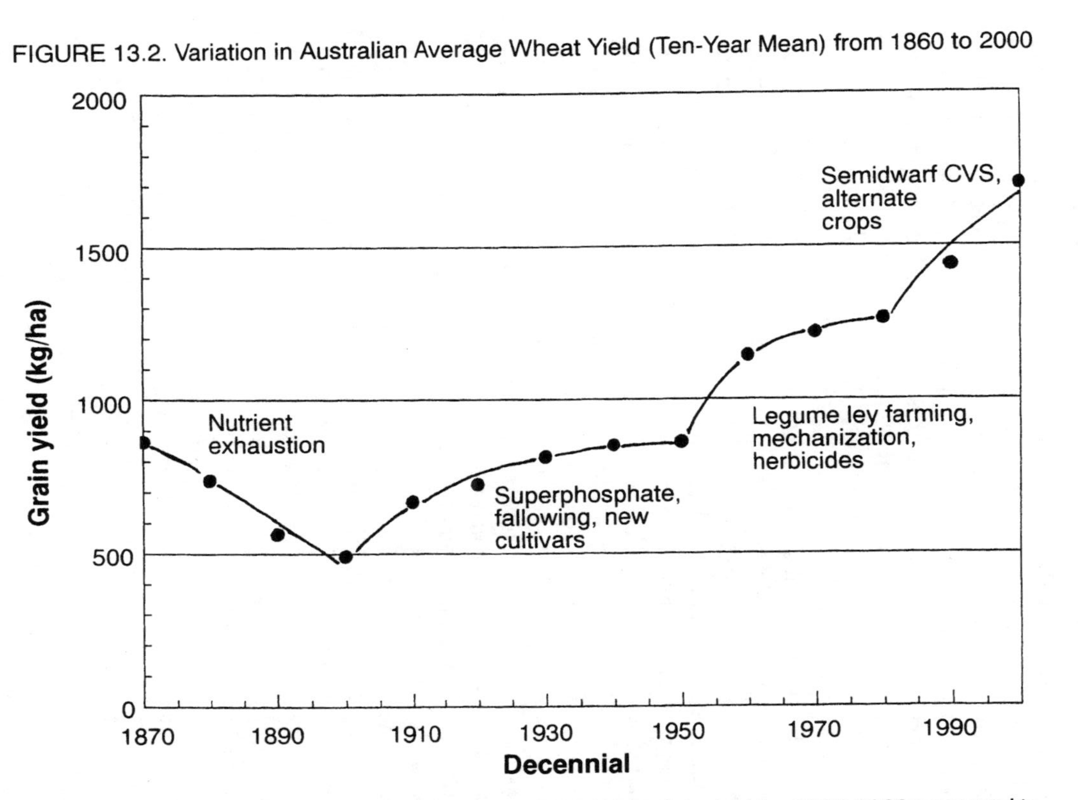

Source: Donald, 1982. Yield plotted at the end of the decennial; yield for 1997-2000 assumed to average 1.71 t/ha.

one of steady yield decline as nutrients were depleted in the initially poor soils and were not replaced by manuring. Following the turn of the century the second phase began as major factors reversed the yield decline. Superphosphate fertilizer was introduced and rapidly adopted, immediately boosting yields by 50 percent or more. Furthermore, bare cultivated fallowing (usually a long fallow of 8 to 10 months or more prior to seeding) was adopted, which in particular enhanced soil mineral nitrogen accumulation through more rapid and prolonged soil organic matter breakdown. The third important factor was the release of the first variety derived from hybridization in 1901, the beginning of a steady stream of improved wheat varieties developed by active breeding programs, which were established in the five wheat-growing states around this time.

By 1930 soil fertility, in particular nitrogen deficiency, was again limiting Australian wheat yields; also, soil erosion encouraged by the bare fallowing was taking its toll, so much so that because of declining wheat area subsequently, yields stagnated. In the early 1950s, however, a third phase commenced with sharp yield rises (Donald, 1982). Up until 1930 Australian wheat farmers were practicing a rotation largely made up of fallow, then wheat, other cereal crops (barley and some oats), and finally volunteer pasture, using the ubiquitous sheep to graze fallow weeds, crop residues, and the pasture. Along with some soil conservation measures, such as contour banks and stubble retention, improved pasture based on the annual legume subterranean clover (*Trifolium subterraneum* L.), and later *Medicago* species, began to be adopted from then on. Sheep numbers rose steadily in the wheat belt over the next 40 years, with the high wool prices around 1950 giving improved pastures and sheep a major boost. By the 1950s farmers were realizing the benefits from the use of these legume leys in terms of improved soil nitrogen status and higher wheat yields: yield increases of 50 percent were not uncommon and soil structure was greatly improved. Yield progress at this time was also increased with the completion of the tractorization of wheat growing, bringing benefits through more timely sowing and harvesting, and the introduction of the first herbicides. Timely sowing is especially important, with yield penalties of 4 to 5 percent per week's delay beyond the optimum date (Kohn and Storrier, 1970; Doyle and Marcellos, 1974).

By 1970 the average wheat farm had almost 2.0 ha of improved pasture and seven sheep for every 1 ha of cereal grown, and the legume ley system of farming was firmly established in southern Australia (Puckridge and French, 1983), with soil N increases ranging from 30 to 150 kg N/ha/year under pasture, depending on legume content, vigor, and rainfall. In the north region such a system was not established, partly because the soils were inherently richer and they had a shorter history of cropping; the need was less urgent and fewer legumes were suited for the summer rainfall area. Cultivated long fallowing remained the commonest practice in 1950 (84 percent of Victoria's wheat grew after such fallow then), but thereafter the practice began to decline in the south as legume leys took care of nitrogen needs and weed control was taken over by herbicides. However, fallowing, short rather than long or for as long as was necessary to build up 100 mm or so of available stored moisture, remained particularly important in the north.

Yield increase over the last two periods of progress identified by Donald (1982), from 1900 to 1970, was around 1.1 percent pa (per annum). Genetic gain from variety improvement over this period was estimated to be 0.41 percent pa for cultivars released between 1886 and 1977 in Victoria (O'Brien, 1982), 0.57 percent pa for cultivars released between 1860 and 1979 in WA (Perry and D'Antuono, 1989), and 0.42 percent pa for cultivars released between 1916 and 1974 in southern NSW (Brennan, 1984). Thus the average for variety improvement is around 0.5 percent pa, fairly typical for world dryland situations, and would therefore have contributed about half of the above yield progress. Breeders over this period produced earlier and shorter cultivars (Loss and Siddique, 1994), with higher harvest indices (even before the Norin 10 semidwarf genes reached Australia). Considerable attention was also given to improved resistance to stem and leaf rust, and to baking quality of grain.

RECENT CHANGES IN WHEAT YIELD

Wheat yield growth slowed again in the 1970s (Figure 13.2), undoubtedly in part because wheat area almost doubled as prices moved to favor wheat; more marginal areas were again farmed. Rotations were shortened as well; by the mid-1980s there was only 1.0 ha of sown pasture and 4.5 sheep per ha of cereal (Tucker and Davenport, 1985). Nevertheless Figure 13.2 suggests that a new phase in Australian wheat yield increase commenced around 1980 and may still be continuing. It was driven first by the introduction of semidwarf varieties based initially on CIMMYT germplasm. The first such varieties were released in 1973, there was 40 percent adoption nationally by 1980, and 85 percent by 1990 (Brennan and Fox, 1995). This was not associated, as it was elsewhere in the world, with a substantial increase in the use of nitrogen fertilizer, which only averaged 5 kg N/ha in 1983-1984 (Tucker and Davenport, 1985): N use on wheat in Australia had always been minimal for economic reasons. Nevertheless the semidwarf varieties performed especially well where pasture leys had boosted soil N, with yield increases of over 25 percent. However, the average yield jump across Australia with the switch to semidwarf wheats was calculated at 8.6 percent by Brennan and Fox (1995), with WA showing a noticeably lower figure than the eastern states and with the recognition that these wheats are more sensitive to poor management such as inadequate weed control (Lemerle et al., 1996) or delayed sowing (Anderson and Smith, 1990).

A second recent positive factor for yield from the mid-1980s onward was the introduction of measures to combat wheat root diseases and nematodes. Grassy ley pastures especially favor root disease: grass control in the pastures (through herbicides) and break crops of noncereals (winter pulses, principally lupins [*Lupinus angustifolius* L.], peas [*Pisum sativum* L.] and chick peas [*Cicer arietinum* L.], and especially oilseed rape or canola [*Brassica napus* L.]) combat root diseases effectively. Wheat yields after canola can be boosted 30 percent or more. The area of these alternate crops in the wheat belt has risen steadily from 0.2 million ha in 1980 to 2.5 million ha in 1996 as improved cultivars and their optimal management elucidated. Progress is expected to continue, for the alternate crop area is still only 20 percent of the cereal area, and nematode problems have not yet been greatly reduced.

CURRENT SITUATION OF WHEAT FARMING

The current situation (1992-1994) of the average wheat farm is summarized in Table 13.1. These statistics are from a sample of around 600 of the 30,000 farmers who produce 98 percent of Australia's wheat (ABARE, 1995, 1996). The surveys also show that 99.5 percent of the cropping area, and presumably of the wheat, is in family farms. Most farms are combined grain and livestock enterprises, the figures indicating the degree of integration. The diversity in cropping is also evident, with a notable increase in noncereal area compared to earlier years. Data was not collected on the actual crop rotation followed. Experience suggests this would now consist largely of one crop per year and would range from two-year rotations of wheat-lupins on deep sandy soils in WA and wheat-*Medicago* pasture in parts of SA, to eight-year rotations with four years legume pasture followed by canola, wheat, grain legume, and barley in the south generally, to fallow-wheat fallow-summer crops in the north, with many variations of these combinations and strong elements of opportunistic cropping throughout. Additional general information is given in Squires and Tow (1991) and detailed physical and biological parameters for a typical northeastern Vic. wheat-sheep farm are given in Loomis and Connor (1992).

TABLE 13.1. Key Mean Statistics for an Australian Farm Producing Wheat; Mean of 1992-1993 and 1993-1994 (ABARE, 1995, 1996)[a]

Description		Average value
Farm area		1482 ha (100%)
Wheat area		255 ha
Barley area		91 ha
Grain legume area		57 ha
Other crop area		65 ha
Total crop area		468 ha (32%)
Improved pasture[b]		500 ha (34%)
Unimproved pasture, woodland, etc.[b]		500 ha (34%)
Sheep equivalents carried		2381 animals
Wheat produced		522 t
Wool produced		9550 kg
Sheep equivalents sold		994 animals
Labor input		109 weeks
Fertilizer input to wheat:	nitrogen (N)[c]	12 kg/ha
	phosphorus (P)[b]	12 kg/ha

[a] Weighted average of wheat and other crops industry (ANZSIC class 0121) and mixed livestock-crops industry (ANZSIC class 0122).

[b] Estimated by author.

[c] Hamblin and Kyneur (1993) data for 1989.

The low labor and fertilizer inputs of the wheat farming enterprise are clear (Table 13.1). Seed costs are also low because most crops are sown with farmer-retained seed and seeding densities tend to be low (e.g., for wheat in the range 25 to 75 kg/ha). Physical data is not available on herbicide use but it has increased markedly and its cost in 1992-1994 approximately equalled that of fuel and of fertilizer, having been less than 20 percent of their cost in the mid-1970s; this cost was about A$30 per ha for each input, or the farm gate value of about 200 kg of wheat. Herbicides are now often used during any fallow phase, as well as pre- and postemergence, and they are used against both grass and broadleaf weeds. By replacing mechanical weed control prior to sowing, achievement of optimal sowing dates has been markedly facilitated. Fungicides or pesticides used in wheat growing in Australia are negligible, and dominated by the almost universal use of seed dressings to control bunt and smut (Brennan and Murray, 1988). Disease control is largely achieved through resistance breeding, and secondly via crop rotation; Brennan and Murray (1988) estimated that yield losses due to disease and pests averaged 240 kg/ha, with *Septoria* sps dominating aboveground, and *Gaumenomyces* and nematodes below ground.

Australia-wide data is not collected on fertilizer use or tillage practices, but the estimate of phosphorus use in Table 13.1 is sound as this has changed little with time. It needs to be noted that P is also applied to the other crops and sometimes to the legume pasture. Thus overall P usage probably continues to exceed P removal in most soils by around 50 to 100 percent. The estimate for average N use on wheat is more uncertain because this figure, having always been <10 kg N/ha before the mid-1980s, appears to have risen significantly in the past 10 years as a result of growing recognition of N deficiency, evidenced for example by declining grain protein contents. One response has been price premiums in the marketplace for wheat with a higher protein content. Another response has been vigorous campaigns for increased N usage, to be guided by target yield, paddock history, soil sampling ahead of planting, and by plant analysis for tillering stage tactical dressings. The exact figure today for N use on wheat in Australia could be as high as 25 kg N/ha.

Anecdotal evidence points to a substantial reduction in the amount of tillage used to grow wheat, the long-cultivated fallow has almost disappeared (although the short one of one to four months is common), disc implements have been replaced with tined ones, tillage is always shallow (<7.5 cm), and conservation tillage practices with residue retention, use of knock-down herbicides for weed control prior to seeding, and/or direct drilling are becoming more widespread. Although burning of the wheat stubble ahead of the next crop is still common where direct drilling is practiced, it is often delayed until just before seeding of the next crop to increase the period of soil protection. The major area of innovation has undoubtedly been in seeding machinery adapted to uncultivated and often trash-covered situations, and much of this appears to be farmer driven.

A recent study by ABARE has looked closely at total factor productivity (TFP) gain in Australia's broadacre farms over the period 1977 to 1993 (Knopke, Strapazzon, and Mullen, 1995). TFP for wheat farms increased at about an impressive 4 percent pa, associated according to the authors with higher-yielding varieties, better crop rotations, greater use of N fertilizers (presumably very late in the period), reduced tillage, and the substitution of materials, especially herbicides, for capital expenditure. New technologies and declining terms of trade for grain farming (−5 percent pa) on the world market, the prices of which have prevailed in Australia since the 1970s, have driven these TFP changes. Another factor is clearly the increase in farm size and in crop area, which was only 1260 and 350 ha, respectively, in 1973-1975 (Bureau of Agricultural Economics, 1979). Such area consolidation has been going on since the land was first opened to wheat farming, when farm size tended to be around 250 ha.

Although Australia may now have on average the largest wheat farms in the world, with low per-hectare costs, the present returns to the family operators' skilled labor and capital are poor relative to that in the nonrural economy. This is because only a modest proportion of the land is cropped annually, and because yields and prices are low, and livestock enterprises depressed. For farms represented in Table 13.1, the average over the two-year period for annual farm cash income (before depreciation and operator's labor deductions) was $A55,000 and the average rate of return on capital 2 percent. These averages conceal huge variation within the sample: it appears that 12 percent of the farms had a negative cash income, and almost 60 percent negative business profit after depreciation and operator labor deductions. This difficult situation is representative of the past decade, with smaller farms generally performing more poorly than larger ones (ABARE, 1995, 1996; Knopke, Strapazzon, and Mullen, 1995). Farms that combine more livestock with cropping have had lower income variability, but less average income at prices prevailing in the 1980s (Stanford, Proctor, and Wright, 1994).

CHALLENGES TO PRODUCTIVITY GAIN

Notwithstanding the current economic difficulties, the picture presented so far suggests one of ongoing generation and adoption of new technologies and steady productivity progress. However, there is much recent evidence to indicate that progress on an industry-wide basis is well below what should be possible, and that some components of the system are particularly weak. One popular analysis of yields that uses a simple correction for the noise created by rainfall variation has been particularly revealing (French and Schultz, 1984). Potential water-limited yield in kg/ha with the current cultivars is considered to be at least 15 to 20 times the crop water supply in mm (soil-stored moisture plus rainfall on the crop, less soil evaporation losses in crop, estimated usually to be 100 mm). Thus a crop evapotranspiration of 300 mm should conservatively yield 3000 kg/ha ($15 \times [300 - 100]$), whereas the current Australia average yield is barely half this figure. French and Schultz (1984) pointed to a host of largely agronomic reasons why this yield gap existed on wheat farms in

SA, and Cornish and Murray (1989) reached similar conclusions in southern NSW, with wheat yields falling below potential especially in the favorable wetter years. The approach has now been used widely as an extension tool, confirming limitations at the paddock level primarily due to inadequate soil N fertility, root diseases, and soil nematodes arising from poor rotations, poor weed control, untimely seeding, and/or foliar diseases such as *Septoria* and *Helminthosporium*.

Hamblin and Kyneur (1993) have made a thorough study of wheat yield trends and changes in practices between 1950 and 1990 in all 210 wheat-growing shires of Australia. They concluded that the relatively low average rate of yield progress in Australia and the decline in grain protein are due especially to declining soil organic matter and nitrogen levels, arising from an excessive cereal cropping intensity (area of cereals exceeding the area of grain legumes plus sown pastures), poor performance of legumes in the pasture phase, and insufficient N fertilizer inputs to compensate for the consequent imbalance between N fixation and N removal from the system. On-farm fertilizer trials tend to confirm the growing wheat responses to N. In some cases research is needed (pasture legumes for the north, disease-resistant pasture legumes for the south), whereas in others the technology exists for good legume pastures, but very difficult market forces have forced the high cereal-cropping intensity (low livestock-to-wheat price ratios), constrained N fertilizer use (Australia has always had N/wheat price ratios above 5 kg/kg), and restricted investment in more sustainable practices. The low rate of yield progress in this study is possibly exaggerated because the Australian wheat area has almost doubled since 1950, largely due to expansion in the 1960s and 1970s into the drier lower-yielding margins. This does not deny the fact that these drier areas have the greatest shortage of satisfactory legumes and the narrowest profit margins, and that the thesis of Hamblin and Kyneur (1993) is likely correct, especially in the latter half of the study period: that soil fertility has declined in many parts of the wheat belt.

Another approach to understanding recent changes is to compare current yield progress to that to be expected from genetic gain. The 25-period from 1961-1970 to 1986-1995 is taken since it encompasses the period of widespread adoption of semidwarf wheats, and wheat area changed little overall (+16 percent); it can be calculated that Australian yield grew at 0.97 percent pa in this period. Earlier-mentioned experiments point to relatively slow breeding progress (around 0.5 percent pa) until 1973 when the first semidwarf releases occurred, representing a jump of 8.6 percent according to Brennan and Fox (1995). Assuming that after the first semidwarf wheats, progress has again dropped to 0.5 percent pa, and assuming no interaction between relative gains and environment, and no bias, over this 25-year period, which encompasses the almost full adoption of semi-dwarf varieties, we can expect a farm-level genetic gain of 0.83 percent pa (= 0.5 percent, + 0.33 percent, which compounded gives 8.6 percent over 25 years). Considering that the 8.6 percent may be an underestimate, it looks like genetic gain on the farm could explain all the yield gain observed (0.97 percent pa). Thus this analysis also suggests that the aforementioned agronomic advances have been either poorly adopted and/or cancelled by soil fertility decline as argued by French and Shultz (1984), Cornish and Murray (1989), and Hamblin and Kyneur (1993).

Although suboptimal management and soil fertility are probably constraining Australian wheat yield in the short term, it should also be mentioned that there are several other negative factors which, although they probably do not constrain yield now, are threatening yield progress and even industry sustainability in the longer term. Actually these problems have all arisen as a result of earlier successful innovations, but took a long time to become evident. First there is dryland salinization, resulting from the very land clearing that gave birth to the wheat industry. When native perennial vegetation was replaced by annual crops and pastures, the seeds of the salinization problem that appeared decades later were unwittingly sown. Annual vegetation such as wheat and clover pasture has lower seasonal water use, thereby increasing accessions to the water table; the water tables are often saline and their gradual approach to the soil surface in lower parts of the landscape brings salt into the root zone. On the same time scale and also associated with water movement below the root zone is the problem of slow soil acidification due largely to nitrate leaching as fertility builds up under leguminous pastures, especially in certain wetter areas (e.g., Helyar et al., 1997). Application of lime, which until now has been minimal in the wheat-sheep zone of Australia, is an immediate but expensive solution to this problem. Finally we have broad-spectrum herbicide resistance in grass weeds, a result of the continual use of certain grass herbicides over a 10- to 20-year period. The salinization problem is a good example of negative off-farm consequences of agricultural practice, because the problems are usually found downslope in nearby farms, public roads, and watercourses. The solution to these three threats will come from yet further research, probably leading to farming system changes such as the inclusion of more perennials (e.g., lucerne or *Medicago sativa* L.) and the use of integrated weed management.

WHEAT RESEARCH AND EXTENSION

The evolution of the Australian wheat farming system over the twentieth century and the increases in its productivity have relied heavily on both local and introduced technologies, in the latter case often modified to suit the local conditions, as in the case of introduced plant germplasm, upon which the system has been especially dependent. Since late in the nineteenth century, Australia has invested substantially in agricultural research: recently research intensity (research cost as a percentage of farm gate gross value of agricultural production) has been as high as 4 percent and is currently around 3.5 percent. This is funded approximately three quarters by the public sector, the federal and state governments, and the remainder by farmers through a product levy, which in the case of wheat is currently about 0.7 percent of farm gate value. The farmer levy and a significant proportion of the public contribution to wheat research is allocated through the Grains Research and Development Corporation, which takes a farming systems and national view on priority setting in which farmers also have a significant input. Table 13.2 summarizes allocation of the approximately A$50 million annual budget of the GRDC among major research and development areas for grains in Australia; this is funding at the research margin but illustrates current priorities. The emphasis in wheat-related activities includes breeding for the distinct wheat qualities required by our different export markets, improvement of alternate crops (barley, triticale, sorghum, pulses, oilseeds) and pasture legumes, application of molecular biology to the improvement of wheat and other crops in the system, soil disease control, integrated weed management, reduced tillage, and indicators of farming sustainability. Other support for these and more traditional research areas, especially breeding, is provided by CSIRO and state departments of agriculture and universities, but increasingly the GRDC is setting the research agenda. Traditional extension activities of the state departments of agriculture, which played a major role in the transfer of technology to wheat farmers in the past, are gradually changing toward innovative models built around farmer groups, concepts of "best practice" management, decision support systems, and strategies that often involve farmer payment for service and private farm management consultants. An example is the Landcare Movement, the formation of whose groups was catalyzed by the dryland salinization problem, with rural communities coming together usually on the basis of stream catchments and with support from the federal government. They are becoming another channel for information flow, not only on salinization but on all aspects of sustainable farming, and an innovative mechanism by which desirable change for the district, as well as the individual farmer, can be motivated.

TABLE 13.2. Allocation of Grains Research and Development Corporation Funds Over the Period 1994-1997

Type of Activity Supported*	Percentage of Budget
Grain quality, marketing, and storage research	16
Breeding and breeding research	45
Disease, pest, and weed research	13
Sustainability research (soil water, nutrients, rotations)	19
Extension and extension research	7

Source: GRDC, 1996.

*Research targets wheat, barley, coarse-grain cereals, pulses, oilseeds, and some pastures, with wheat receiving, for example, 40 percent of the resources spent on breeding.

For a detailed and comprehensive view of the current issues discussed in the later sections of this chapter, the reader is referred to Coombs (1994). Accessible up-to-date statistics on the wheat industry are provided annually by the Australian Bureau of Agricultural and Resource Economics (e.g., ABARE, 1996).

FUTURE PERSPECTIVES

Wheat continues to dominate arable farming in Australia and livestock, principally sheep, continue to be closely integrated with the arable farming enterprise. Exporting of wheat and its exigencies totally dominate the Australian wheat market. These facts are unlikely to change in the future. The real price of wheat in the world market has been declining for over 100 years, driven by and driving the adoption of new technologies and increases in the productivity of the use of land, labor, and capital in wheat production globally, and especially in Australia. This downward trend has recently faltered but most observers anticipate further, albeit slower, price declines (e.g., Rosegrant, Agcaoili-Sombilla, and Perez, 1995). Pressures to remain globally competitive are therefore likely to continue, leading to the need for new ways of increasing productivity. There is little scope to develop new arable lands, but the scale of wheat farming is likely to continue increasing through farm consolidation, with the disappearance of less profitable (usually smaller) farms. Land and capital productivity must also increase and clearly more research will be essential; the GRDC research agenda (Table 13.2) gives an idea of where Australia believes the payoff will come. A new concern, however, is the funding of this research and availability of its products, as concepts of user-pays and privatization become fashionable, and the threat of global patenting of needed technologies looms. Another concern is the increasing operator skill level required for profitable wheat farming.

Finally, a new factor in the future is the growing farmer and community awareness of environmental and sustainability issues. This will add to the research agenda. The Australian wheat farming system may seem to already have valuable elements of sustainability by world standards (reliance on legumes for nitrogen input, diverse rotations, integration of livestock, low levels of fungicide and pesticide use and of tillage) but as we have seen, new threats are emerging, and there will likely be others as yet unanticipated. Besides, the natural resource endowment is inherently poor, and profitability levels are low: there can be no investment in new more sustainable technologies without cash surpluses on the farms. There is a real danger that the open world competition, which Australian wheat farmers have endured for over 20 years and to which most other world wheat farmers are now being exposed, will drive wheat production into unsustainability worldwide, for the market does not reward sustainable production. This is a global problem that will require attention in the future.

REFERENCES

ABARE. (1995). *Farm Surveys Report 1995.* Australian Bureau of Agricultural and Resource Economics. Canberra, Australia.

ABARE. (1996). *Farm Surveys Report 1996.* Australian Bureau of Agricultural and Resource Economics, Canberra, Australia.

Anderson, W.K. and Smith, W.R. (1990). Yield advantage of two semi-dwarf compared to two tall wheats depends on sowing time. *Aust. J. Agric. Res.*, 41: 811-826.

Brennan, J.P. (1984). Measuring the contribution of new varieties to increasing wheat yields. *Rev. Marketing and Ag. Economics,* 52: 175-195.

Brennan, J.P. and Fox, P.N. (1995). *Impact of CIMMYT Wheats in Australia.* Economics Res. Report No. 1/95, NSW Agriculture, Wagga Wagga, Australia.

Brennan, J.P. and Murray, G.M. (1988). Australian wheat diseases: Assessing their economic importance. *Ag. Sci.,* 1(7): 26-35.

Bureau of Agricultural Economics. (1979). *The Australian Wheatgrowing Industry: An Economic Survey; 1973-74 to 1975-76.* Aust. Gov. Publ. Service, Canberra, Australia.

Callaghan, A.R. and Millington, A.J. (1956). *The Wheat Industry in Australia.* Angus and Robertson, Sydney.

Coombs, B. (1994). *Australian Grains: A Complete Reference Book on the Grains Industry.* Morescope Publ., Camberwell, Australia.

Cornish, P.S. and Murray, G.M. (1989). Low rainfall rarely limits wheat yields in southern New South Wales. *Aust J. Exp. Agric.*, 29:77-83.

Donald, C.M. (1982). Innovation in Australian agriculture. In D.B. Williams (Editor), *Agriculture in the Australian Economy,* Second Edition. Sydney University Press, Sydney, pp. 55-82.

Doyle, A.D. and Marcellos, H. (1974). Time of sowing and wheat yield in northern New South Wales. *Aust. J. Exp. Agric. Anim. Husb.*, 14: 93-102.

French, R.J and Schultz, J.E. (1984). Water use efficiency of wheat in a mediterranean-type environment I. The relation between yield, water use and climate. *Aust. J. Agris. Res.*, 35: 743-764.

GRDC. (1996). *Information Paper 1997-98.* Grains Research and Development Corporation. Canberra, Australia.

Hamblin, A. and Kyneur, G. (1993). *Trends in Wheat Yields and Soil Fertility in Australia.* Australian Government Publishing Service, Canberra, Australia.

Hammer, G.L., Holzworth, D.P., and Stone, R. (1996). The value of skill in seasonal climate forecasting to wheat crop management in a region with high climatic variability. *Aust. J. Agric. Res.*, 47: 717-737.

Helyar, K.R., Cullis, B.R., Furniss, K., Kohn, G.D., and Taylor, A.C. (1997). Changes in the acidity and the fertility of a red earth soil under wheat-annual pasture rotations. *Aust. J. Agric. Res.*, 48: 561-586.

Knopke, P., Strapazzon, L., and Mullen, J. (1995). Productivity growth: Total factor productivity on Australian broadacre farms. *Australian Commodities,* 2: 486-497.

Kohn, G.D. and Storrier, R.R. (1970). Time of sowing and wheat production in southern New South Wales. *Aust. J. Agr. Res. Anim. Husb.*, 10: 604-609.

Lemerle, D., Verbeek, B., Cousens, R.D., and Coombes, N.E. (1996). The potential for selecting wheat varieties strongly competitive against weeds. *Weed Res.*, 36: 505-513.

Loomis, R.S. and Connor, D.J. (1992). *Crop Ecology: Productivity and Management in Agricultural Systems.* Cambridge University Press, Cambridge, UK.

Loss, S.P. and Siddique, K.H.M. (1994). Morphological and phenological traits associated with wheat yield increases in mediterranean environments. *Adv. Agron.*, 52: 229-276.

McBride, J.L. and Nicholls, N. (1983). Seasonal relationships between Australian rainfall and the Southern Oscillation Index. *Monthly Weather Rev.*, 111: 1998-2004.

Nix, H.A. (1975). The Australian climate and its effects on grain yield and quality. In A. Lazenby and E. M. Matheson (Editors), *Australian Field Crops Vol I. Wheat and Other Temperate Cereals.* Angus and Robertson, Sydney, pp. 183-226.

O'Brien, L. (1982). Victorian wheat yield trends, 1898-1977. *J. Aust. Inst. Agric. Sci.*, 48: 163-168.

Perry, M.W. and D'Antuono, M.F. (1989). Yield improvement and associated characteristics of some Australian spring wheat cultivars introduced between 1860 and 1982. *Aust. J. Agric. Res.*, 40: 457-472.

Puckridge, D.W. and French, R.J. (1983). The annual legume pasture in cereal-ley farming systems of southern Australia. *Agric. Ecosystems and Env.*, 9: 229-267.

Rosegrant, M.W., Agcaoili-Sombilla, M., and Perez, N.D. (1995). *Global Food Projections to 2020: Implications for Investment.* Food, Agriculture and Environment Discussion Paper 5, International Food Policy Research Institute, Washington, DC.

Singh, A.J. and Byerlee, D. (1990). Relative variability in wheat yields across countries and over time. *J. Agric. Econ.*, 41: 23-32.

Squires, V. and Tow, P. (Editors) (1991). Dryland Farming: A Systems Approach. Sydney University Press, Sydney.

Stace, H.C.T., Hubble, G.D., Brewer, R., Northcote, K.H., Sleeman, J.R., Mulcahy, M.J., and Hallsworth, E.G. (1968). *A Handbook of Australian Soils.* Rellim Tech. Publications, Glenside, South Australia.

Stanford, L., Proctor, W., and Wright, J. (1994). Wheat farming risk. *Farm Surveys Report 1994,* ABARE, Canberra, pp. 72-80.

Tucker, J. and Davenport, S. (1985). Broadacre industries. *Quart. Rev. Rural Economy,* 7: 42-46.

Chapter 14

Wheat Production in Mediterranean Environments

Edmundo H. Acevedo
Paola C. Silva
Hernán R. Silva
Boris R. Solar

THE MEDITERRANEAN CLIMATE

The Mediterranean climate is generally found between latitudes of 30 and 40 degrees on the western coasts of continents. It is characterized by a dry, hot summer alternating with a humid and temperate winter. The dry season may span from one to eight months and it occurs during the summer (the long-day season). It rains in winter with typical mean rainfall values of 500 mm. The vapor pressure deficit (vpd) decreases from mid-autumn to winter, reaching the lowest values during winter. It increases later toward the summer (Strahler, 1986). Globally, the Mediterranean climate is found in five regions (see Figure 14.1): the coasts of the Mediterranean sea, the center and southern coasts of California and northern coast of Mexico, central Chile, the southern tip of Southern Africa, and southwest Australia. Two major variants of the Mediterranean climate can be distinguished according to mean summer temperature: dry, hot summer (mean temperature of the hottest month as high as 22 to 28°C); and marine regions of temperate, humid zones (mean temperature of the hottest month from 14 to 22°C). The hot summer Mediterranean climate is typical of the coasts of the Mediterranean Sea, inland California and southwest Australia. The cool summer type is found in South Africa and the coasts of California and Chile (Rodríguez, 1972).

WHEAT IN MEDITERRANEAN ENVIRONMENTS

The Mediterranean climate presents problems for the adaptation of plants due to its rainless summer. The mean temperature of the coldest

FIGURE 14.1. World Areas with Mediterranean Climate

month goes from 3 to 18°C, never too cold for plant growth (Strahler, 1986). Therefore, winter cereals grow well in these environments. It rains during winter and plants grow vigorously as soon as temperature goes up in spring. The rain is stored in the soil, which supplies the spring and summer water requirements. The soil water is usually not enough to supply the crop water requirements toward maturity and crops suffer postanthesis water stress (Santibáñez, 1994). Seventy to 90 percent of wheat grain yield is produced by postanthesis photosynthesis (Austin et al., 1977), therefore water stress during grain filling reduces carbon assimilation, hampering grain filling and yield (Johnson and Moss, 1976).

Around 10 percent of wheat world production comes from Mediterranean environments. The difficulties for wheat yield improvement in those environments are great (Table 14.1). The year-to-year variation in rain is high (Figure 14.2) and it determines wheat yield, although improved drought-resistant varieties have higher and more stable yield in those environments.

Year-to-year rainfall variability makes it difficult to establish definite agronomic practices in Mediterranean environments. To assure crop success, soil management and fallow practices that store water in the soil profile are required. The seasonal water availability and crop water requirements, as determined by planting date, crop density, and soil fertility, must be balanced (Loomis and Connor, 1992). Much of the wheat agronomy for these environments deals with practices that maximize water use.

TABLE 14.1. Percentage of World Wheat Production in Various Megaenvironments and Estimation of the Difficulty of Improving Yields

Environments	Wheat Surface (%)	Wheat Production (%)	Difficulty*
Irrigated	32	40	+
High rainfall	10	12	++
Mediterranean	10	4	+++
High temperatures	6	6	++++
Acid soils	1.7	1.2	+++
High latitudes	4.9	6.3	++
Low temperatures	12.5	18	++++

Source: Modified from Acevedo and Fereres, 1993.

*Difficulty associated with yield improvement using agronomy or breeding

FIGURE 14.2. Monthly Rainfall Distribution Throughout Four Seasons in Northern Syria

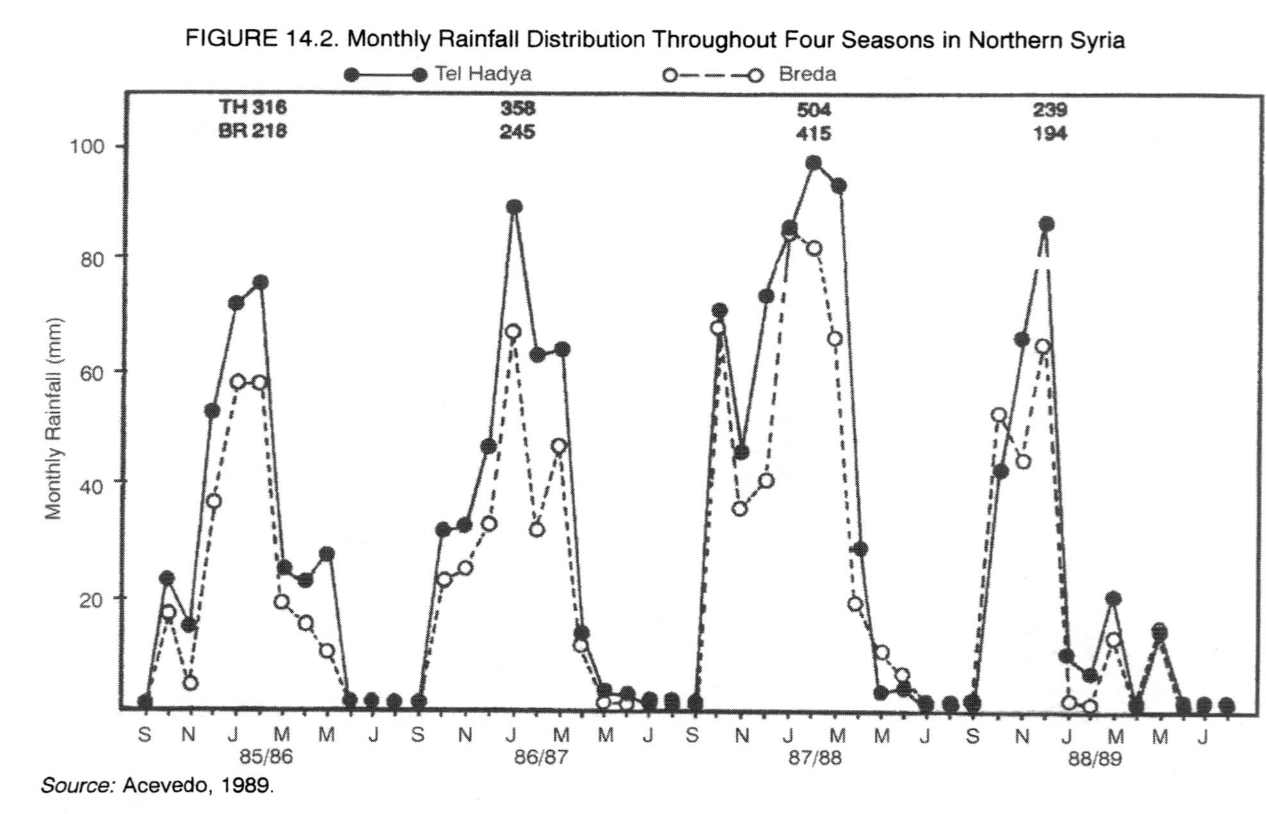

Source: Acevedo, 1989.

Tel Hadya 36°01′N, 36°56′E, 284 m elevation, 328 mm long-term rainfall.
Breda 35°40′N, 37°10′E, 300 m elevation, 281 mm long-term rainfall.

Water Use Efficiency in Mediterranean Environments

Water must be used as efficiently as possible by crops in Mediterranean environments. Crop water use efficiency (WUE) is defined by the ratio of grain yield to water use:

$$\text{WUE} = \frac{\text{Grain Yield (kg ha}^{-1}\text{)}}{\text{Water used (mm)}} \quad (1)$$

Water use is commonly expressed in terms of total water supply (rain plus irrigation or transpiration plus soil evaporation), which allows the evaluation of crop management practices that would increase the use of available water (Bolton, 1991; Cooper et al., 1987).

The yield of a crop grown under dryland conditions can be expressed in terms of its transpiration (T), its transpiration efficiency (TE), and the harvest index (HI) (Passioura, 1977):

$$Y = T \cdot TE \cdot HI \quad (2)$$

If the WUE is measured not only in terms of T but total water use, the water balance equation must be considered:

$$E + T + R + D + I - P = 0 \quad (3)$$

where E is soil water evaporation, T is transpiration, R is runoff, D is drainage water, I is the water intercepted by the canopy, and P is rainfall. It follows that equation 1 can be expressed as:

$$\text{WUE} = \frac{T \cdot TE \cdot HI}{E + T + R + D + HI} \quad (4)$$

$$\text{WUE} = \frac{TE \cdot HI}{1 + \frac{(E + R + D + HI)}{T}} \quad (5)$$

Equation 5 shows the ways to improve WUE:

- Through an increase in genotype *TE*
- By increasing the fraction of ET (evapotranspiration) that is transpired
- WUE is inversely related to the vapor pressure deficit (vpd) experienced by the crop during the transpiration period, such that if vpd decreases, WUE increases (Cooper, 1983)
- By agronomical practices that decrease soil evaporation

The use of some principles set forth by Loomis (1983) may also improve WUE. First, crop evapotranspiration (ET) must be balanced by water supply. Crop growth is closely related to pan evaporation, *Ep*, such that the water use efficiency is inversely proportional to vpd $(e^* - e)$, such that:

$$\mathrm{WUE} = \frac{B}{Ep} = \frac{K}{(e^* - e)} \quad (6)$$

where B is biomass, e^* is the saturation vapor pressure, e is the actual vapor pressure, and K is a crop-specific factor (Tanner and Sinclair, 1983). For a given crop, water is used more efficiently where soil evaporation is minimized and the crop grows when $(e^* - e)$ is small while temperature and radiation are adequate for growth.

In Mediterranean climates the vpd changes markedly through the season, being low during winter and early spring, and increasing rapidly by the end of spring and summer. It changes from 0.3 kPa in winter to 1.2 kPa toward the end of spring and summer, which coincides with the grain-filling period of winter cereals (Gregory, 1991; Cooper et al., 1987; Loomis and Connor, 1992). Higher yields are obtained by stimulating growth during the cold winter months.

Fischer (1981) estimated WUE for wheat grown in south Australia using the following equation:

$$\mathrm{WUE}\ (\mathrm{kg\ Ha^{-1}\ mm^{-1}}) = 10.2 - 1.3\,Ep + 0.053\,Ep^2 \quad (7)$$

when *Ep* is evaporation in mm d^{-1}.

With an *Ep* of 1 mm d^{-1} in winter the WUE would be 9 kg ha^{-1} mm^{-1}. During spring *Ep* goes up to 5 mm d^{-1} and WUE decreases to 5 kg ha^{-1} mm^{-1}. Winter growth is therefore in terms of water use.

During the rainfall period in winter, temperatures and radiation are low while air humidity is high. Much of the rainfall in excess of ET during this period is stored in the soil profile. This stored water is used by the crop during the growing cycle (Duratan et al., 1991).

Soil and crop management alter WUE by affecting soil water infiltration and changing the proportion of water transpired to that lost by direct soil evaporation. The last fraction is an important component of total water lost. Cooper et al. (1987) measured direct soil evaporation as high as 35 to 55 percent for wheat and barley crops, while Hamblin, Tennant, and Perry, (1987) and Gregory (1991) estimated values from 34 to 61 percent for wheat growing in west Australia. French and Schultz (1984) estimated a 33 percent soil water loss via direct soil evaporation in South Australia. The potential to improve WUE is, therefore, high.

Direct soil water evaporation is related to the amount of solar energy reaching the soil surface, therefore it is inversely related to the amount of radiation intercepted by the crop canopy. Any strategy improving the establishment of the crop and the canopy growth and related ground cover will reduce direct soil evaporation and improve WUE.

The total biomass produced (B) is equal to the product of the dry matter produced per intercepted radiation by the canopy (e) times the amount of incident radiation per unit area (S) and times the fraction of intercepted radiation (f) (Monteith, 1973).

Total biomass produced is:

$$B = e * f * S \tag{8}$$

f being the main determinant of crop growth rate, which is turn related to the crop leaf area and leaf arrangement and indirectly to plant density, water, and nutrient availability. The radiation intercepted can be estimated using the following equation:

$$f = 1 - \exp^{-K * \mathrm{LAI}}$$

where K is the extinction coefficient of the crop and LAI is the leaf area index (Cooper, 1983).

Sometimes the question about which is the most limiting factor (water availability or solar radiation) for wheat production in Mediterranean environments is asked. Assuming that solar energy is the only limiting factor to barley yield and that all the incoming radiation is intercepted by the crop, the radiation use efficiency (e) of barley grown in northern Syria was found to fluctuate between 1.3 to 1.6 g MJ^{-1} (grams per megajoule) total radiation. Considering e equal to 1.4 g MJ^{-1} and f (intercepted radiation) equal to 1, the potential dry matter production would be 37.8 t ha^{-1}. On the other hand, assuming water to be the only limiting factor, equation 2 could be used to estimate maximum yield if all the rainfall is transpired by the crop. For the season under study, the rainfall was 284 mm and using a value of K of 2.9 kPa, the maximum dry matter production would be 15.2 t ha^{-1}. Comparing the two values, clearly water is the most limiting factor in these environments (Gregory, 1991).

To improve WUE and yield in these areas, agronomic practices leading to an increase of the dry matter produced during the time of low vpd should be used, that is, growth should be during the cold winter months. Obviously, the total amount of soil water available to the crop should be maximized (Cooper, 1991; Gregory, 1991; Pala, 1991).

The reduction of direct soil water evaporation can be achieved through the following agronomic practices: early planting, modifications of soil plant density and spatial arrangement, selection of varieties with fast early growth, optimum fertilization, and use of straw mulch. To maximize the total amount of available water, other agronomic practices may be used: selection of genotypes having deep roots, correct use of fertilizers, improved soil water infiltration through straw mulch, use of fallow, and reduction of runoff.

Seed Bed Preparation and Planting System

Traditional tillage and no-tillage are the two major ways of planting wheat in Mediterranean environments. The traditional tillage is the most widely used system.

Traditional Tillage

This tillage method is characterized by the use of disc or moldboard plows followed by harrowing. The time of plowing is mainly determined by the date of sowing, which can occur in autumn or spring (Faiguenbaum, 1986). Soil water content dictates when plowing can be done. In clay soils it is common that in autumn and winter the excess water precludes tillage. Disc harrowing is usually the second seedbed preparation activity, which must be done with appropriate soil moisture to avoid the formation of clods (Faiguenbaum, 1986; Aguila, 1987). The labor required to achieve a good tilth varies according to the soil type, the type of machinery, and soil moisture conditions. In Chile farmers usually use one or two plowings and two or more harrowings.

No-Tillage (Direct Drilling)

A new and different approach to growing annual crops has gained researchers' attention and farmers' interest in the past 40 years. The idea is to avoid the tillage associated with traditional seedbed preparation. The seed is drilled directly into the soil with an appropriate planting machine (Crovetto, 1992). The direct drilling technique involves the management of crop residues, which are left on top of the soil as a mulch. Direct drilling has shown advantages over traditional planting methods: it markedly decreases wind and water erosion (Crovetto, 1992), and saves energy, reducing oil needs. In a wheat crop oil savings amount to 60 to 75 percent compared to traditional systems. The no-tillage method of planting gives greater flexibility to enter the crop fields with machinery for sowing, to apply agrochemicals, or for harvesting. It also allows cropping of soils with a higher slope (above 15 percent) and increases water infiltration, decreases direct soil evaporation, and as a result allows for a higher soil water availability.

The soil temperature is lower under a straw mulch. Occasional frosts cause more damage to crops in Mediterranean environments that have been sown directly as compared to conventional tillage. One of the major difficulties for growing crops without tillage is weed control. Adequate herbicides are needed and crop rotations are necessary. The incidence of pests and diseases is usually higher in directly drilled soils with residues left on top.

Sowing Date

In Mediterranean environments, where rainfall occurs mainly in winter, early sowing has a substantial effect on WUE. Early sowing assures that the wheat crop will grow during a period of low vpd and high soil moisture. The higher early autumn temperatures assure a more uniform and better crop emergence, an earlier crop canopy development, and a faster covering of the soil surface, which assures higher WUE and grain yield (Cooper, 1983; Cooper et al, 1987; Jabbour and Naji, 1991; Pala, 1991).

Planting date and genotype should be chosen such that flowering occurs after the risk of spring frosts is over and when the photothermal quotient is at maximum. Planting as early as possible not only decreases the incidence of terminal drought in Mediterranean environments but it also diminishes wheat crop exposure to high temperatures during grain filling. Acevedo (unpublished data) has estimated yield losses of 1 percent of grain yield of dryland wheat per day delay in planting after November 1 in Aleppo (northern Syria).

The optimum wheat planting date in the interior drylands of the regions of Maule and Bio-Bio in central Chile (600-800 mm winter rainfall) is May either for spring or facultative wheat (Ovalle and del Pozo, 1994).

Earlier sowings, in April, are more susceptible to Septoria and plantings done in June experience terminal drought.

The optimum sowing date was found to be during the first 15 days of May in southwest Australia (200-320 mm winter rainfall) (Shackley and Anderson, 1995) for various wheat genotypes. After May 15, all genotypes tested decreased their yield at a mean rate of 20 kg/ha day. French and Shultz (1984) reported a grain yield decline of 200 to 250 kg/ha for each week of delay in planting after the optimal planting date in south Australia. In a different site having 307 mm annual rainfall, delaying sowing by two and a half months decreased biomass production from 12.2 t ha − 1 (optimum sowing date) to 4.6 t ha^{-1}. The WUE decreased from 29.4 to 18.9 kg ha^{-1} mm^{-1}.

Acevedo, Harris, and Cooper (1991) reported grain yield losses of as much as 22 kg/ha per day delay in planting after the end of October (April) for rainfed barley grown in northern Syria.

Early sowing is usually done prior to the first rain. In this case sowing depth has to be increased to avoid a false start of the crop (enough rain to germinate the seeds but not enough to carry the germinated seed to the next rainfall event). The main problem of deep sowing wheat is the decrease in plant emergence and plant vigor (Valverde, 1982). Genotypes having long coleoptiles must be selected for deep planting. Acevedo, Harris, and Cooper (1991) used barley genotypes varying in their coleoptile length in a sowing depth experiment in northern Syria. At the driest location (Breda) they observed that deep plantings (7 to 10 cm) had a better crop establishment in those cases where long-coleoptile genotypes were used.

Population Density and Plant Distribution

Dryland plantings of winter cereals must be uniformly done. This is best achieved with a seeder, or grain yield and WUE are reduced either due to an increase in direct soil evaporation at lower plant densities or by an increase in water stress in areas with a higher population density. Overcast hand plantings have the problem of distribution of the seeds, some of them being exposed to desiccation, pests, and diseases and others planted too deep.

As mentioned above, ground cover has a strong effect on WUE and grain yield in low-rainfall Mediterranean environments. Acevedo, Harris, and Cooper (1991), working with barley, changed row spacing from 10 to 40 cm keeping the same sowing rate (100 kg/ha). The decreased row spacing increased ground cover, markedly improving WUE and grain yield. A shorter distance between rows is beneficial to the crop, increasing early growth and LAI. The increase in early ground cover intercepts more solar radiation, decreasing the soil surface exposed to radiation and therefore decreasing direct soil evaporation and improving WUE.

The use of soil water by winter cereals depends in part on the rate at which the roots explore the available soil volume. A uniform plant distribution with equal space between plants (hexagonal) allows a better exploration of the soil by the roots. On the contrary, sowings with a high plant density in the row (uneven distribution) induces an intense early competition for soil water, hampering early growth (Silim and Saxena, 1991). In turn, the spatial arrangement of the crop affects its competition with weeds. Plants with a more uniform space distribution compete better against weeds than plants with a higher row distance. Densities of 300 plants m^{-2} in autumn plantings of wheat were considered to be adequate in northern Syria and had grain yields 23 percent higher than sowings with 100 plants m^{-2} (Pala, 1991).

Spring sowings of wheat are exposed to decreasing rainfall and increasing vpd. The wheat crop grows essentially on stored soil water. Under these conditions, sowings with low plant densities, even decreasing the transpiration/evapotranspiration ratio, allow better water availability for each plant and better yields due to better soil water content during flowering and grain filling.

Fertilizers

The use of fertilizers induces a better wheat crop growth, higher dry matter accumulation (bigger leaves and higher number of fertile tillers), and a faster canopy expansion, increasing the radiation interception by wheat crops in Mediterranean environments. An increased ground cover due to fertilizer reduces direct soil evaporation, increasing WUE and yield (Gregory, 1991; Kallsen, Sammis, and Gregory, 1984).

Fertilizer use, particularly phosphorus, improves root growth, increasing the chance for the crop to extract water deeper in the soil during spring and summer (Karlsen et al., 1984). Phosphorus hastens wheat development, decreasing the length of the growing cycle of the crop and improving WUE due to lower vpd (Loomis and Connor, 1992; Anderson, 1985).

Phosphorus application to barley grown in areas of very low rainfall and in phosphorus-deficient soils has increased WUE markedly (from 4.7 to 5.4 kg ha^{-1} mm^{-1}). If phosphorus was not used, crop growth was reduced and direct soil evaporation increased to 75 percent of total evapotranspiration (Cooper et al., 1987). Power (1985) observed that nitrogen increased WUE of winter cereals from 6 to 25 kg of grain ha^{-1} mm^{-1} but it also increased water use by 20 mm. Other work also shows higher evapotranspiration and response to fertilizers, but in every case WUE and yield were improved (Mechergui, Gharbi, and Lazaar, 1991).

Loomis and Connor (1992) caution that in areas with low water availability the amount of applied nitrogen has to be decreased, aiming at balancing water use between vegetative and reproductive growth. If excess nitrogen is present in the soil, early growth may be promoted with too much vegetative growth, restricting reproductive growth.

Legumes and fallow increase the amount of available nitrates in the soil. The level of soil nitrates in spring after a crop of grain legumes was 28 percent higher than after a crop of wheat fertilized with N but 43 percent lower than after fallow. The wheat yield of an unfertilized crop following grain legumes was higher than a wheat crop following wheat fertilized with 75 kg N ha^{-1} and similar to a wheat crop following fallow. The nitrogen use efficiency for wheat following legumes was 32 percent higher than for wheat following fallow and 21 percent higher than for continuous wheat.

ROTATIONS

Crop rotations in Mediterranean environments usually include wheat legumes and fallow. The fallow, either short or long, usually precedes the crop and is intended mainly to store water in the soil. The most common legume crops are fava bean in the wettest and lentils in the driest environments. The wheat crop rotation not only increases the amount of available soil N through the legume crop but it also has an effect on weed control, diseases, and soil water content (Fischer, 1981).

Crops of the Rotation

The rotation of cereals with legumes in Mediterranean environments with 300 to 400 mm annual rainfall is an old traditional practice. In areas of North Africa with relatively high rainfall, the main legume in the rotation is fava bean (*Vicia faba*). Lentils (*Lens culinaria*) are usually grown in west Asia (Harris, Cooper, and Pala, 1991).

The Turkish government has promoted the use of legume forages in the rotation such as *Vicia* and *Lathyrus* spp. as an alternative to fallow replacement in the wheat rotations in low rainfall areas (<300 mm), where small ruminants are raised.

In the interior dryland of central Chile the wheat production system usually includes a rotation with clean fallow, wheat, and natural pasture or covered fallow with chickpeas (*Cicer arietinum*) or cowpeas (*Lathyrus* sp.), wheat, lentils, and pastures (del Canto, 1991). In south central Chile, when rainfall is higher the wheat rotation includes rape (*Brassica napus*), oats (*Avena sativa*), and lupin (*Lupinus* sp.). The most common rotation includes pasture, rape or lupin, wheat, and oat. Rape is usually preferred to control grassy weeds of the wheat crop. Lupin is used to increase soil nitrogen and sometimes is even used as a green manure crop. Oats break the cycle of *Gaeumannomyces graminis,* which causes take-all, a very important wheat disease.

The main purpose of crop rotations with wheat is disease control. Sometimes, however, nematodes of the genus *Pratylenchus* infecting roots may reduce wheat yield after chickpea compared to wheat after fallow (Beck, Wery, and Ayady, 1990). The use of a cover crop during winter rains rather than fallow helps to avoid water erosion. The main effect of the legumes in the wheat rotation is to increase the soil nitrogen levels (Reeves, 1991; Crovetto, 1992) and even though they also contribute to weed control WUE is higher after legumes (Reeves, 1991).

Fallow

The cereal-fallow model is an ancient system of cropping. The purpose of the fallow is to increase soil water and nitrogen availability in the soil and to control weeds and soil pathogens (Schultz, 1980). In the past few years fallow has been replaced by grain legumes in areas with rainfall above 400 mm (Durutan et al., 1989).

Fallow, however, continues to be important in areas with limited rainfall where a clean fallow in the crop rotation accumulates annual and sometimes biannual rainfall for crop growth. In areas in which rainfall is concentrated in a short period of the year, the water stored during the fallow period allows the extension of the crop growth period.

Fallow also helps to stabilize yield, as shown in a study where continuous wheat rotation was compared to a fallow-wheat rotation for 30 years. The long-term average wheat yield was 0.7 t ha^{-1} for the continuous wheat system compared to 2.5 t ha^{-1} of the fallow-wheat rotation. In 37 percent of cases of continuous wheat, crop failures occurred (yields below 0.4 t ha^{-1}) while the lowest yield in the fallow-wheat rotation was 1.2 t ha^{-1} (Bolton, 1991).

The cereal-fallow rotation integrated with animals is a common system in west Asia and North Africa (Pala and Mazid, 1989). Weeds and volunteer cereals grow in the fallow land and provide valuable and cheap pasture for sheep. In many areas, however, clean fallow is practiced.

There are two main types of fallow according to the area, local soil, and weather conditions: (1) short-duration fallow, used in areas with rainfall above 300 mm/year (in these areas cereal grain-legume rotations are grown in periods of 9 months); (2) long-duration fallow, used in areas with lower rainfall (less than 300 mm/year). Animals help to control weeds in such areas. The fallow accumulates from 60 to 100 mm of water in the soil.

Long-duration fallow is being replaced by crops due to grain production pressures. Overexploitation of the soil resources, however, may induce soil degradation unless straw is kept on the soil surface and direct seeding replaces tillage as crop intensity is increased (Harris, Cooper, and Pala, 1991).

Water Storage

Fallow may be considered as an agricultural practice used to increase WUE by increasing water availability to the crop. The principles of fallow management derive from the water retention characteristics of the soil and from direct soil water evaporation.

To allow the penetration of rain through the soil surface and hence to become less exposed to direct soil evaporation it is necessary that the rain wets the surface layer of the soil to field capacity, usually from an air-dry state. Each rainfall event requires either 6, 10, or 12 mm of rain for sandy, loamy, and clay soils to wet the surface layer. As a result, the fraction of water lost by evaporation increases with the frequency of rains and is bigger in clay soils with lower infiltration rates. Even though rainfall infiltrates quickly into sandy soils it is also quickly lost due to their low water retention capacity, around 50 mm/m compared to 150 mm/m in loam soil and 240 mm/m for clay (Connor, 1993). Therefore, optimum fallows that maximize rainfall infiltration into the soil and minimize the water lost by evaporation and deep percolation.

Fallow inevitably loses water either by soil evaporation or by drainage below the root zone. Their efficiency, i.e., the fraction of total rainfall contributing to the production of the next crop, is low and variable from year to year. A comparative study of fallow at nine sites in the U.S. Great Plains shows that short fallow (six months) had an efficiency of 19 percent. Fallow efficiency may also be low in areas having summer rainfall. The mean soil storage efficiency for 14 successive winter-summer periods having a mean yearly rain of 340 mm was 11 percent as compared to 71 percent for winter mean rainfalls of 124 mm. In 5 out of the 14 years studied the fallow lost water, the fallow efficiency being −11 percent (Connors, 1993).

Harris, Cooper, and Pala (1991) found that fallow efficiency changed according to the amount of rainfall. Efficiency increased with higher rainfalls. It also varied with soil type, efficiency being higher in deeper, better-structured soils.

Soil Management and Weed Control

Weed control is one of the most important components of crop rotation management due to the direct competition of weeds with the crop. Weed growth reduces water accumulation in systems with or without fallow. Schultze (1988) studied spring-summer fallow in two contrasting soil types (sandy and clay) during three successive seasons in south Australia. The main effect of the clean fallow was to avoid transpiration water loss by weeds from stored soil water.

Grazing is a cheap way to suppress weeds in the fallow period and increase sheep economic return. The fallow efficiency, however, decreases. An initial crop followed by sheep grazing improved wheat yield by 67 percent, compared to a 137 percent increase in yield after clean fallow (Rooney, Sims, and Touhey, 1966).

In the past the only way of maintaining a weed-free fallow was by tillage after each significant rainfall. McDonald and Fischer (1991) pointed out that the high tillage intensity during the fallow period reduced soil organic matter with a loss of soil structure and fertility. The soil infiltration was reduced and the soil erosion risk was increased. All these negative effects could be avoided by keeping crop residues on the surface of the soil. Even in arid and semiarid zones rainfall may exceed the soil infiltration rate of the fallow. Agronomic practices that improve water infiltration, such as maintaining crop residues on the soil surface, become necessary.

Chemical Fallow

Present evidence shows that chemical fallows are more efficient (Cornish and Pratley, 1987). In terms of water conservation, the chemical fallow is at least as good or better than traditional fallow. The plant residues left on the soil surface in the chemical fallow protect the humid surface from direct evaporation. The soil surface remains wet for a long period of time, increasing the probability that the next rainfall event will infiltrate below the surface. This helps to explain why fallow soils with surface residues retain more water than those in which the crop residues are incorporated into the soil or removed. Furthermore, a humid soil surface increases planting opportunities for the farmer and allows crop establishment at the optimal date, which is critical for effective management in environments with limited rainfall. A comparison among fallow techniques showed that fallow with crop residues and mulch (partial incorporation) retained up to 32 percent of rainfall while those maintained with herbicides retained 42 percent (Smika, 1970).

Chemical fallow contributed 38 mm more moisture than a cultivated fallow in a loam soil (Smika, 1970). Similar results have been obtained in clay soils, with a 29 percent higher water retention and 0.7 t ha^{-1} higher yield than the cultivated fallow.

Conservation tillage and no-till practices reduce soil water erosion, maintain soil moisture, and alter the crop microenvironment. The mean quantity of surface residue prior to wheat planting in spring was 1,371 kg ha^{-1} for chemical fallow, 399 kg ha^{-1} for crop residues and 703 kg/ha^{-1} for minimum tillage over four years of study. The higher quantity of residues when chemical fallow was used increased soil moisture significantly to a depth of 1.7 m. Residue retention and no-tillage fallow, however, decreased wheat yields, probably due to decreased soil nitrate concentration and the presence of yellow leaf spot and *Pratilenchus tomei* (Felton, Marcellos, and Martin, 1995).

The retention of residues on top of the soil reduces soil temperature, causing phytotoxicity in some instances. Residues also favor some pathogens such as *Septoria* sp., yellow leaf spot, and *Pyrenophora tritici repenti.* Microclimatic changes may favor root pathogens such as take-all (*G. graminis* var. *tritici*) and eyespot lodging (*Cercosporella herpoptrichioides*). In spite of these drawbacks residue retention is the most promising soil management technique; therefore plant breeders must pay more attention to the incorporation of genetic resistance to pathogens related to this practice (Fischer, 1981).

GERMPLASM

Higher yields of arid and semiarid Mediterranean drylands are the result of good agronomy and better-adapted genotypes. Breeders have been looking for genotypes with higher WUE for years. The genotype × environment interaction, however, markedly decreases breeding progress in low-rainfall Mediterranean environments. Rainfall variability across years and sites is high and crop management is also variable (Austin, 1989).

Bolton (1991) pointed out that the increase in WUE in low-rainfall areas can be achieved mainly through changes in crop and soil management. Increases in WUE due to new genotypes become more important in areas of higher water availability. Early-flowering genotypes or those that have a higher growing rate and faster grain fill at lower temperatures may have a higher WUE (Fischer, 1981). This is due to the limited potential to produce dry matter after anthesis in Mediterranean areas owing to the high vpd and high temperatures in the period (spring-summer). Furthermore, because water stress has a major effect on grain yield from just before flowering to ten days past anthesis, attention should be given to the preanthesis period, aiming at assuring that the crop will complete its growth or reach maturity before water deficits become important (short season, genotypes) and before the vpd increases too much in spring. Water losses through direct soil evaporation may be decreased through a quick ground cover (should become close to 100 percent as soon as possible). These short-cycle genotypes, however, have a lower yield than longer-cycle genotypes, therefore the best choice would depend on weather variability in relation to the climatic characteristics (Loomis and Connor, 1992).

It has been postulated that genotypes having a high root density close to the soil surface may decrease direct soil evaporation (Fischer, 1981). If the root system is also deep, a higher WUE may be achieved (Brown et al., 1987; Cooper, 1983) by selecting genotypes with lower root resistance to water absorption (higher root length density, higher root conductance, and higher root growing rate during crop growth). The genotypes should also optimize radiation interception during early stages to decrease direct soil evaporation and increase transpiration (Austin, 1989).

Yield Potential

Acevedo and Fereres (1993) defined yield potential as the yield of an adapted genotype to a given environment, growing with adequate nutrients and water and where other stresses are effectively controlled. It has been shown that for a certain degree of physical stress varieties with a high yield potential may yield more than stress-resistant genotypes (Romagosa and Fox, 1993). If the stress goes beyond a certain level the opposite may occur and the high-yield-potential genotype may yield less than other genotypes having lower yield potential. This phenomenon is known as "crossover" (Ceccarelli, 1991).

The increases in grain yield potential have been shown to be mainly due to changes in biomass allocation toward the grain (Austin et al., 1980) instead of diverting it to roots, leaves, stems, or reserves. This fact has decreased the ability of high-yield-potential genotypes to compete with weeds, to recover from pests and diseases, and to survive in stressed environments. This may be at least part of the reason for the existence of a crossover under highly stressed conditions where genotypes with a high yield potential have low production (Acevedo and Fereres, 1993).

Future increases in yield potential could be the result of increases in net photosynthesis and biomass. To date these variables have not essentially changed with crop improvement. The main avenue to improve yield potential has been through changes in harvest index. This avenue, however, is coming close to its theoretical limit and lodging is again becoming an important limiting factor to yield potential increases.

Many attempts have been made to select for higher photosynthetic rate, aiming at improving biomass and yield without achieving the expected result (Nelson, 1988). It appears that yield potential increases are related to higher stomata conductance (Rees et al., 1993) but not to increases in photosynthetic capacity. Austin (1990), however, found a higher photosynthetic capacity in *Triticum urartu*, a diploid grass which is donor of the A genome to hexaploid wheat. He failed, however, to incorporate this characteristic into cultivated hexaploid wheat (Austin, personal communication).

Donald (1968) pointed out that the high-yield wheat idiotype was a communal plant that was small, erect, with very few tillers and having small erect leaves, a plant able to survive and produce in a competitive situation surrounded by plants of a similar form. This idiotype should invest minimum resources in plant structure and should, therefore, maximize the harvest index and be very efficient. Evidence supporting the idea that crop improvement is producing higher-yielding genotypes which are more efficient in the use of resources has been provided by Reynolds et al. (1994).

Sedgley (1991) summarized the characteristics of high-yielding idiotypes having high biomass and high harvest index: (1) tolerant to high planting density, (2) efficient in the use of resources, (3) sensitive to cultural practices, (4) having enough florets to assure that the spike has enough capacity to accept assimilates.

Islam and Sedgley (1981) pointed out that the uniculm trait proposed by Donald was favorable in spring wheat of Mediterranean environments. In these environments late tillers do not reach maturity or are starting to fill grain during a period of high water stress. The probable occurrence of early spring frosts, however, makes this character many times undesirable.

Increasing the length of the spike growing period and a decrease in the senescence rate could allow some increase in grain yield potential. The "stay green" trait or slow senescence of the crop canopy during the reproduction phase could be helpful during the grain-filling period (Acevedo and Fereres, 1993).

Yield Under Stress

Drought

Crop water stress may be conveniently defined as the soil water level at which evapotranspiration falls below its maximum value. Water deficit is a common phenomenon in plants and crops. It is accentuated when drought or lack of sufficient water in the rhizosphere occurs. Drought problems are common in Mediterranean rainfed agriculture. This is due to the irregularity of rainfall and to the largely unpredictable nature of weather within most climatic environments.

Crop evapotranspiration and, more precisely, crop transpiration is linearly and positively related to yield in C_3 and C_4 plants. Water stress inevitably decreases yield. This fact of nature has prompted agronomists, breeders, physiologists, and physical scientists to study the nature of drought, the effects of water stress on plant growth, development, and yield, management practices that would alleviate drought, and to search for drought resistant genotypes. The common aim is to minimize the effects of drought on yield in cropping systems, and when conditions are extreme, avoid crop failures.

Three major points need to be emphasized: (1) drought is a complex problem, (2) several disciplines are dealing with it, and (3) the problem should be looked at from a system perspective. Improving the drought resistance of genotypes is one way to alleviate drought effects, i.e., yield losses due to suboptimal water supply.

The Nature of Drought in Rainfed Agriculture

It is important to distinguish the nature of drought in rainfed agriculture because it bears directly on the strategy adopted to cope with it. Two broad situations can be recognized (Richards, 1982): (1) when a crop grows under current rainfall, i.e., the soil profile undergoes recharge and discharge of water during the growing season; and (2) when the crop grows essentially on soil moisture stored prior to sowing. The first case is typical, but not exclusive, of autumn-sown cereals in Mediterranean environments. The second case is common to cereals grown after the major rainfall period has occurred, such as in spring-sown crops in Mediterranean environments, in monsoon areas, or in areas with summer rainfall. In both cases year-to-year variability of rainfall is high and so is the risk of drought. In general the risk increases as seasonal rainfall decreases (Dennett, Keatinge, and Rodgerds, 1984; Virmani, 1982; Watts and El Mourid,

1988). The critical differences between these two extreme cases is that in case 1, rainfall use efficiency has to be maximized at a moment in which rainfall is occurring, while in case 2, a strategy should be adopted that would allow finishing the crop cycle with an already existing and relatively fixed (and known) amount of water in the soil profile.

The stressful environments are often characterized by the occurrence of more than one physical stress at the same time or through the growing cycle. This complicates improvement either by breeding or crop management. In Mediterranean environments, drought periods during winter may be associated with low temperatures (sometimes freezing temperatures) and suboptimal radiation levels, while terminal drought is generally associated with above-optimal temperatures and excess radiation (depending on latitude). Where crops are irrigated, drought and salinity are commonly associated stresses. The soil may impose additional constraints, such as high or low pH, which induces phosphorus and micronutrient deficiencies or toxicity (e.g., manganese and aluminum toxicity at low pH).

Drought Resistance

Plant growth and yield under drought have been the subject of numerous research efforts. Morphological, physiological, and anatomical characteristics have been identified that enable plants to grow and survive in drought-prone environments (Srivastava et al., 1987; Turner and Kramer, 1980; Shultze, 1988; Ludlow and Muchow, 1988; Taylor, Jordan, and Sinclair, 1983). Usually a complex of attributes are present in cereal species that grow and yield under severe drought. Table 14.2, modified from Acevedo and Ceccarelli (1989), illustrates this point for two-row barley landraces that evolved in the Fertile Crescent of southwest Asia and yield better under stress in those Mediterranean environments. It is probable that a combination of the attributes in Table 14.2 confer the better adaptation to stress environments.

In low-rainfall Mediterranean environments, the crops grow essentially under current rainfall. Even with appropriate management the best-yielding entries attain yields under stress that are much lower than in higher-input, less stressed environments. The high-yielding entries usually show high crop ground cover during late autumn and winter, when water availability and crop water use efficiency is at a maximum (Cooper et al., 1987; Acevedo and Ceccarelli, 1989; Acevedo, Pérez-Marco, and van Oosterum, 1990). They also have an early flowering and fast grain-filling period, which increase yield significantly. The attributes (Table 14.2) may be of importance to cope with intermittent drought or to temporarily alleviate its effects. These may also include osmoregulation (Morgan, Hare, and

TABLE 14.2. Attributes of Two Row Pure Lines Isolated from Landraces*

High grain yield under drought
High harvest index
High grain mass
Early heading
Short grain-filling period
High drought-resistance index
Prostrate growth habit in winter
Dark green leaves before stem extension
Light green leaves after stem extension
Short stature under drought
High tillering
High number of fertile spikes
High discrimination for ^{13}C
Low transpiration efficiency
Long emergence to double ridge period
Short ear initiation and ear growth period

Source: Modified from Acevedo and Ceccarelli, 1989.

*Nursery composition: 14 Syrian pure lines isolated from landraces

Fletcher, 1986; Blum, 1988), accumulation of proline and glycine-betaine (Richards, 1983), and others.

Drought resistance is usually quantified in a crop by its yield under drought stress. There are, however, physiological mechanisms and agronomic practices that may result in a higher yield or stability of yield under stress without being necessarily associated with a true resistance to water stress of a genotype.

Yield under stress depends, among other factors, on yield potential and phenology of a genotype. A genotype with a high yield potential will decrease its yield if submitted to water stress, but it may still yield more than a genotype with a lower yield potential (see Figure 14.3). Semidwarf wheat genotypes produce more grain due to their earliness (photoperiod insensitivity and low vernalization requirement) and higher harvest index when compared to tall wheat (Laing and Fischer, 1977). This may not necessarily imply that semidwarfs tolerate intensive water stress better (yields below ca. 1t ha^{-1}) (Richards, 1982). Conversely, a genotype with a relatively long growing cycle may have drought resistance traits but yield very poorly in a stressed, short-season environment.

FIGURE 14.3. Effect of Bread Wheat Yield Potential on Yield Under Stress

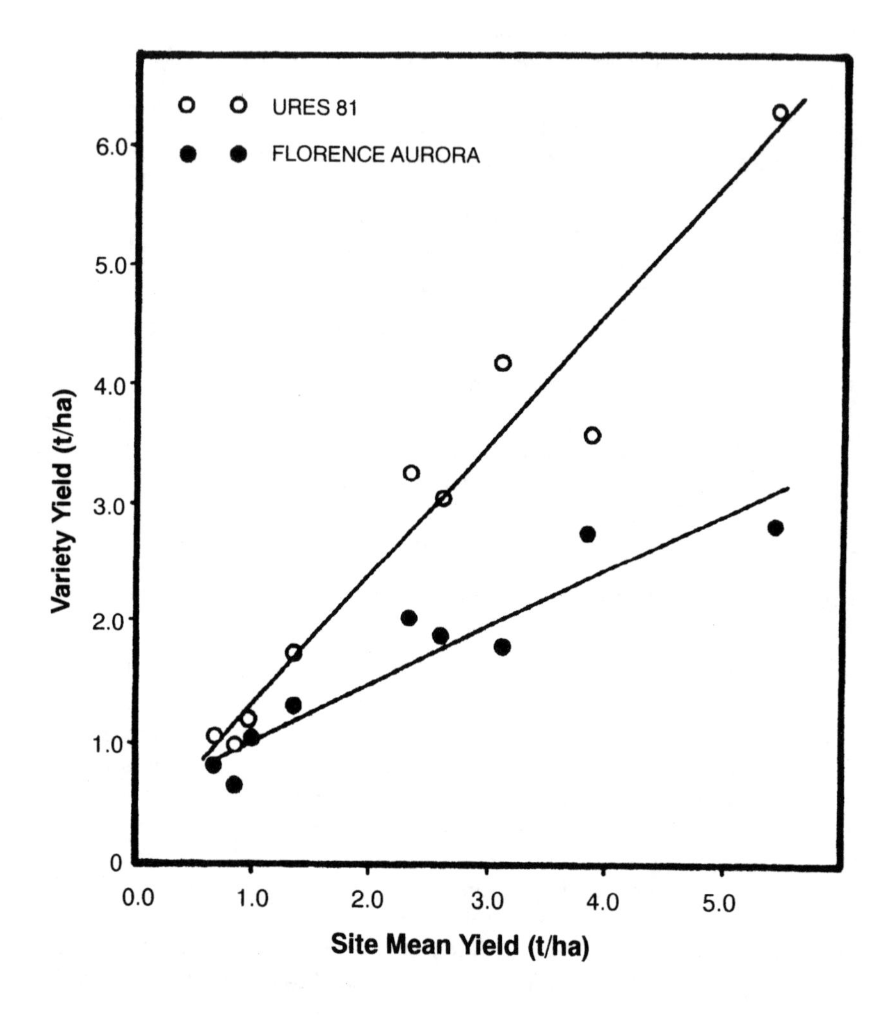

Source: Acevedo, unpublished.

Note: Regression line calculated from data of nine environments. Regression parameters: URES 81 has a regression slope = 1.09 and intercept = 0.23; Florence Aurora has a regression slope = 0.47 and intercept = 0.55.

From the previous paragraphs it may be concluded that yield under drought stress may not be the best definition of drought resistance for the purpose of improving genotypic stress resistance. The contribution made by phenology and yield potential needs to be accounted for (Fischer and Wood, 1979; Bidinger, Mahalakshmi, and Rao, 1987a, 1987b; Acevedo and Ceccarelli, 1989).

The most integrative parameter of plant stresses (they usually occur in combination in a growing cycle) is biomass production. Roots are difficult to measure, therefore assessments of biomass production are generally limited to aboveground biomass, and in the case of wheat, mostly to grain production. It is practical to define stress resistance (avoidance plus tolerance) in terms of grain yield, after accounting for the effects of yield potential and phenology. A convenient indicator of phenology is the number of days from emergence to 50 percent heading (spikes fully emerged from the boot). In a linear regression model, stress resistance would be equivalent to residual yield under stress after accounting for yield potential, phenology, and experimental error (Bidinger, Mahalakshmi, and Rao, 1987a, 1987b). The value may be positive or negative and indicates the relative performance of a genotype with respect to the nursery under evaluation after the yield potential and days-to-heading effects have been removed. This definition of stress resistance would have the advantage of being free of two major contributors to the interaction of genotype and environment on yield, and at the same time, integrates (through the growing cycle) various environmental stresses. If drought is the prevalent stress under consideration, the index may be considered to be a drought resistance index. Morphological and physiological parameters could be assessed by the strength of the correlation with the index. Other yield-derived indices have been proposed and widely used, such as the drought susceptibility index (S) of Fischer and Maurer (1978) and the intercept of the joint regression analysis of genotype yield on site mean yield (e.g., Finlay and Wilkinson, 1963; Blum, 1983, 1988).

In the joint regression analysis part of the genotype × environment accounts for the linear component, but the intercept is highly dependent on yield potential. Furthermore, phenologically unadapted genotypes may present a low regression slope and high intercept without necessarily implying a high level of stress resistance. Similar associations with yield potential have been observed with the drought susceptibility index (S).

Indirect Selection for Yield Under Stress

Because grain yield has low heritability, particularly under stress, indirect selection traits are usually sought by plant breeders. When drought is the major stress under consideration, earliness is an excellent escape mechanism in Mediterranean environments. Remobilization of preanthesis assimilates, rooting depth, and stay green are other usually proposed indirect selection traits (Parlevliet, De Haan, and Schellekens, 1991). The most promising indirect selection traits in wheat for drought tolerance other than growing cycle are, however, osmotic adjustment (Morgan, 1984; Kameli and Lösel, 1995), air-to-canopy temperature difference (Blum, 1988; Rees et al., 1993), and ^{13}C discrimination (Farquhar and Richards, 1984; Austin et al., 1990; Acevedo, 1993). The last two traits are also related to potential wheat yield.

Osmotic Adjustment

Osmotic adjustment consists of the active accumulation of solutes in plant tissues as a response to water shortage. This process lowers the osmotic potential and the total water potential of stems, leaves, and roots (Turner and Jones, 1980; Girma and Krieg, 1992). As a result the plants can absorb water at low soil water potentials and maintain turgor pressure and related physiological activities in plant tissues (Acevedo, 1975; Hsiao et al., 1976; Blum and Sullivan, 1986; Ludlow, Santamaria, and Fukai, 1990). Wheat is a species that has osmotic adjustment and genetic variability for the trait (Morgan, 1984). Wheat yield under water stress is positively correlated with osmotic adjustment. Kameli and Lösel (1995) showed that in wheat the main osmotic factor is glucose, representing 85.5 percent of the total increase in sugars of young plants undergoing moderate water stress. Wheat genotypes with high osmotic adjustment produce higher root biomass, higher root length density, extract more soil water, and have higher transpiration (Morgan, 1984). The higher root growth of genotypes adjusting osmotically is related to turgor maintenance as well as to the amount of carbon fixed which is in turn related to the osmotic adjustment of the apex (Turner, 1986). Osmotic adjustment maintains or even increases the harvest index in wheat (McGowan et al., 1984; Morgan and Condon, 1986). There is little information on the heritability of this trait even though some information indicates that few genes are involved and that the character may be simply inherited (Morgan, 1983; Morgan, Hare, and Fletcher, 1986).

Crop-to-Air Temperature Difference

Wheat genotypes growing under stress differ in their capacity to maintain their water status. This is shown by the canopy temperature in the afternoon hours. Genetic variability has been shown by wheat under stress in this trait (Blum, 1988) and the canopy temperature has become a useful tool to identify drought-tolerant genotypes (Blum, Mayer, and Gozlan, 1982; Blum et al., 1989). Rees et al. (1993) showed that decreased crop temperature is related to potential wheat yield across genotypes (higher flag leaf stomata conductance, lower temperatures, and higher yield).

The association of the crop-to-air temperature difference with leaf conductance indicates that wheat genotypes that maintain cooler leaves under water stress may also have higher photosynthetic rate and higher yield under stress. Crop-to-air temperature differences are therefore related to both wheat yield potential (Rees et al., 1993) and wheat yield under water stress (Blum et al., 1989; Acevedo, unpublished data).

Water Use Efficiency

Wheat genotype WUE can be determined by measuring the discrimination of ^{13}C, which occurs naturally during photosynthesis of C_3 plants. Plants discriminate against ^{13}C of the air. The magnitude of the discrimination (Δ) is negatively and linearly related to transpiration efficiency (Farquhar and Richards, 1984; Hubick and Farquhar, 1989). The value of Δ has been proposed as a selection criterion for genotypes in C_3 species in environments of limited rainfall (Farquhar and Richards, 1984; Austin et al., 1990; Richards, 1991; Acevedo, 1993).

There is genetic variability in wheat Δ (Farquhar and Richards, 1984; Condon, Farquhar, and Richards, 1990; Araus, Reynolds, and Acevedo, 1993) as well as other winter cereals. Heritability estimations of this parameter (broad sense) go from 0.62 to 0.94 (Condon and Richards, 1992; Edhaie and Waines, 1994; Acevedo, 1993). Edhaie and Waines (1994) established dominant and additive effects in Δ expression.

The values of Δ are negatively correlated to WUE. The advantage of the discrimination is that it provides a mean seasonal value weighted by photosynthesis since photosynthetic organs are fixing CO_2 continuously. A selection for low Δ and high yield should provide commercially acceptable genotypes with good WUE (Acevedo et al., 1997). It is worth noting that Δ has a very low interaction with environment (Acevedo, 1993). Al Hakimi et al. (1996) have stressed the possibility of using Δ in crop improvement for WUE and for high yield under drought due to the fact that it can be easily measured, the existence of genetic variability, high heritability, and a good knowledge of associations between Δ and other phenological and morphological characters.

Site for Selection

Selection for stress resistance, using yield or a yield-derived index as a stress indicator, implies an increased vertical stability of yield for a genotype, i.e., a decreased rate at which the genotype decreases yield per unit increase in stress intensity. The variability of stresses across years within location is such that their incidence may be considered random and, therefore, each year for a given site is an environment (Blum, 1988). The assessment of vertical stability comprises a range of environments with a mean of increasing (or decreasing) yield as a result of the major stress(es) under consideration. Test environments within gradients of stresses are required. The gradient should cover the expected variation in stress occurrence and intensity for the target area of the breeding program.

From a physiological standpoint, it is expected that stress resistance traits would have a better expression under high-intensity stresses; therefore, sites of this nature should be included in the testing scheme even at the risk of crop failures. Since yield potential is necessary for quantifying stress resistance, as it plays a role in yield under stress, ideally the nursery under study should be replicated under irrigation at the drier sites. If this is not possible, the yield at the wettest site of the range, located within a narrow range of photo-thermal variation, could be used as the yield potential site.

The most important consideration for choosing the site(s) for selections is their relevance to the environment where the varieties are intended to be grown. Figure 14.4 shows a crossover of wheat genotypes in the relation cultivar yield versus site mean yield; therefore, if the sites for selection fall outside the area, the genotype may be irrelevant for that area.

If a variety yields better than the population mean in a range of testing environments, it is said to have wide adaptation. Yield potential and stress resistance may be combined in such varieties. The question often arises of the range of adaptation of a variety across environments. If the range is extended to cover all possible environments, the existence of a variety with such horizontal stability is doubtful. Many times a trade-off between yield potential and vertical stability has been postulated as shown in Figure 14.3. This has been demonstrated by Ceccarelli and Grando (1989) for barley. They observed a decrease in yield potential across cycles of barley selection at a dry site. This trade-off would have the effect of limiting the range of environments in which the horizontal stability associated with superior yield could be found. Ceccarelli (1989a, 1989b) made a thorough analysis of wide adaptability, showing contrasting behavior of genotypes under stress environments and the need to include stress testing sites in the improvement for stress resistance. Furthermore, the efficiency of selection

FIGURE 14.4. Two Bread Wheat Genotypes Differing in Yield Potential

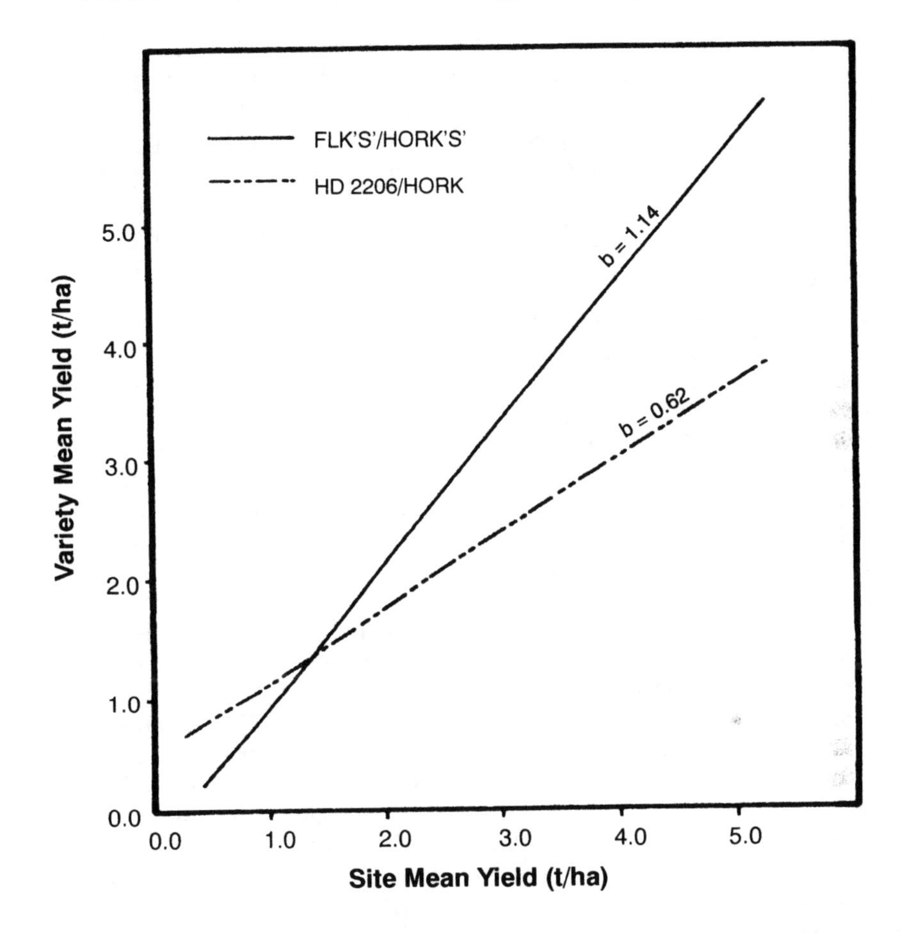

Note: The crossover occurs at about 1.5 (t/ha). Regression lines derived from nine environments in northern Syria ($r^2 = 0.99$ for FLK'S'/HORK'S' and 0.97 for HD 2206/HORK).

for grain yield for stress environment done under stress was 6.9 percent higher for conventional germplasm and 32.4 percent higher with locally adapted germplasm when compared to selections done under favorable environments (Ceccarelli and Grando, 1989).

Ceccarelli et al. (1987) examined breeding strategies for improving cereal yield and stability under drought. The analysis included barley, durum wheat, and bread wheat across environments ranging from 178 to 380 mm of seasonal rainfall for barley and 227 to 600 mm for wheat in northern Syria. A high and negative correlation coefficient was found between the drought susceptibility index and grain yield at the driest sites for the three crops, whereas at the wettest sites, the correlation coefficient was lower and sometimes positive. This was interpreted as an indication that segregating populations and lines with the largest grain yield under favorable conditions are more drought susceptible than those with the largest grain yield under drought. Perhaps this is due to the existence of certain traits that are desirable under drought but undesirable under favorable conditions. What was clear was that selections under nonstress conditions would not provide material adapted to stress environments. It appears that when breeders select based on grain yield for stress environments, they must pay a price in terms of the discriminating power of the environment. Furthermore, selection only in stress environments would result in varieties with narrow adaptation. One possible way to alter this general pattern would be to identify stress resistance traits and conduct selections for yield plus those traits. If the traits are constitutive (always present), selections could be made in environments with high discriminative power for yield. If the traits are largely adaptive (present only under stress conditions), selection under stress would be mandatory. The work by Ceccarelli et al. (1987) and Ceccarelli and Grando (1989) has shown beyond doubt that stress environments are required when breeding for stress. The open question is the extent to which a price in terms of narrow adaptation is being paid. However, Ceccarelli et al. (1987) also presented data indicating that genotypes selected at stressed sites would respond to better environments, particularly in durum wheat. The reverse however was not supported by the data, i.e., it did not appear possible to produce a stress-resistant genotype selecting in high-input environments only.

High Temperature

Temperature increases toward the summer in Mediterranean environments. High temperatures are common along with water deficits in wheat during the grain-filling period. Water stress may worsen the effect of high temperatures; both stresses are closely related through the energy balance of plant organs, transpiration being the most important energy-dissipating mechanism. If the energy dissipation through transpiration is impaired due to water stress, the crop temperature increases.

Heat stress generally affects wheat from anthesis to maturity in Mediterranean environments. Increases in temperature from 15/10°C to 21/16°C (day/night) decreased the grain-filling duration from 60 to 36 days. Increases in temperature from 21/16°C to 30/25°C reduced grain-filling duration even further, from 36 to 22 days (Sofield et al., 1977).

Heat decreases wheat grain weight. Wardlaw, Sofield, and Cartwright (1980) and Wiegand and Cuellar (1981) indicate that in the range of 12 to 26°C of increase in mean temperature during wheat grain filling the grain weight decreases from 4 to 8 percent per degree increase in the mean temperature. Pollen sterility and flower abscission are other problems induced by high temperature.

There is genetic variability for traits related to growth and development of wheat from anthesis to grain filling (Acevedo, Nachit, and Ortíz-Ferrara, 1991) such as glaucousness, grain weight (Wardlaw, Dawson, and Munibi, 1989), photoperiod, and vernalization sensitivity (Midmore, Cartwright, and Fischer, 1982a, 1982b) and grain yield (Acevedo, Nachit, and Ortíz-Ferrara, 1991).

The acceleration of developmental processes by high temperature can be modified by photoperiod and vernalization. Short days prevail at low latitudes, delaying spikelet initiation and prolonging its duration in genotypes sensitive to photoperiod. Genotypes with vernalization requirements also delay flowering at low latitudes. Membrane thermostability is an indirect selection criterion for heat stress resistance (Blum, 1988; Balota et al., 1993). One way to avoid problems due to high temperatures is to improve crop agronomy and water availability such that more energy is dissipated through latent heat flux. All agronomic practices geared to increase the water availability of the crop would reduce the heat stress effects.

CONCLUDING REMARKS

Wheat production in Mediterranean environments is relatively well understood. The major physical constraints to production are terminal drought and terminal heat stress as well as year-to-year weather variability. Drought is probably the major cause of yield loss in these environments. Much work remains to be done to alleviate drought and heat stress, particularly using zero tillage and straw management. This is essential to incorporate sustainability in wheat rotations in these environments. Plant breeding also has a major task in producing varieties that would stabilize yields. Grain yield very much follows the year-to-year variability of rains.

REFERENCES

Acevedo, E. 1975. The growth of maize under field conditions as affected by its water relations. PhD thesis. Department of Water Science and Engineering, University of California, Davis, CA.

Acevedo, E. 1989. Improvement of winter cereal crops in Mediterranean environments. Use of yield, morphological and physiological traits. In Acevedo, E., Conesa, A.P., Monneveaux, P., and Srivastava, J.P. (Eds.), *Physiology—Breeding of Winter Cereals for Stressed Mediterranean Environments* (meeting held at Montpellier, France, July 3-6, 1989).

Acevedo, E., 1993. Potential of carbon isotope discrimination as a selection criterion in barley breeding. In Ehleringer, J., Hall, A., and Farquhar, G. (Eds.), *Stable Isotopes and Plant Carbon-Water Relations.* Academic Press, San Diego, pp. 399-417.

Acevedo, E., Baginsky, C., Solar, B., and Cecarelli, S. 1997. Discriminación isotópica de ^{13}C y su relación con el comportamiento de genotipos silvestres y mejorados de cebada bajo diferentes condiciones hídricas. *Ciencia e Invetigación Agraria* 17: 41-54.

Acevedo, E. and Ceccarelli, S. 1989. Role of physiologist-breeder in a breeding program for drought resistance conditions. In Baker, F.W.G. (Ed.), *Drought Resistance in Cereals.* CAB International, Wallingford, UK, pp. 117-139.

Acevedo, E. and Fereres, E. 1993. Resistance to abiotic stress. In Hayward, M., Bosemark, N., and Romagosa I. (Eds.), *Plant Breeding: Principles and Prospects.* Chapman and Hall, London, pp. 406-421.

Acevedo, E., Harris, H., and Cooper, P. 1991. Crop architecture and water use efficiency in Mediterranean environments. In Harris, H., Cooper, P., and Pala, M. (Eds.), *Soil and Crop Management for Improved Water Use Efficiency in Rainfed Areas.* ICARDA, Syria, pp. 106-118.

Acevedo, E. Nachit, M.M., Ortíz-Ferrara, G. 1991. Selection tools for heat tolerance in wheat: Potential usefulness in breeding. In Saunders, D.A. (Ed.), *Wheat for the Non-Traditional Warm Areas.* CIMMYT, Mexico City.

Acevedo, E., Pérez-Marco, P., and van Oosterom, E. 1990. Physiology of yield of wheat and barley in stressed rainfed Mediterranean environments. In Sinha, S.K., Sane, P.V., Barghava, S.C., and Aggarwal, P.K. (Eds.), *Proceedings of the International Congress of Plant Physiology,* New Delhi, India.

Aguila, H. 1987. *Agricultura general y especial.* Editorial Universitaria, Santiago, Chile.

Al Hakimi, A., Monneveaux, P., and Deleens, E. 1996. Selection response for carbon isotope discrimination in a *Triticum polonicum* × *Triticum durum* cross: Potential interest for improvement of water efficiency in durum wheat. *Euphytica* 36: 248-255.

Anderson, W.K. 1985. Grain yield responses of barley and durum wheat to split nitrogen applications under rainfed conditions in a Mediterranean environment. *Field Crops Research* 12: 191-202.

Austin, R.B. 1989. Maximing crop production in water limited environments. In Baker, F.W. (Ed.), *Drought Resistance in Cereals,* CAB International, Wallingford, UK, pp. 13-25.

Austin, R.B. 1990. Prospects for genetically increasing the photosynthetic capacity of crops. In Zelith, Y. (Ed.), *Perspectives in Biochemical and Genetic Regulation of Photosynthesis,* Alan R. Liss Inc., New York, pp. 395-409.

Austin, R.B., Bingham, J., Blackwell, R.D., Evans, L.T., Ford, M.A., Morgan, C.L., and Taylor, M. 1980. Genetic improvements in winter wheat yields since 1900 and associated physiological changes. *Journal of Agricultural Science, Cambridge* 94: 675-689.

Austin, R., Craufurd, P., Hall, M., Acevedo, E., Da Silveira, B., and Ngugy, E. 1990. Carbon isotope discrimination as a means of evaluating drought resistance in barley, rice and cowpeas. *Bulletin Societé Botanique Francaise* 137, Actualité Botaniques 1: 21-30.

Austin, R., Edrich, J., Ford, M., and Balkwell, R. 1977. The fate of the dry matter carbohydrates and ^{14}C lost from the leaves and stems of wheat during grain filling. *Annals of Botany* 41: 1309-1321.

Balota, M., Amani, Y., Reynolds, M., and Acevedo, E. 1993. Evaluation of membrane thermostability and canopy temperature depression as screening traits for heat tolerance in wheat. Wheat Special Report No.20. CIMMYT, Mexico City.

Beck, D., Wery, J., and Ayady, A. 1990. Dinitrogen fixation and nitrogen balance in cool season food legumes. *Agronomy Journal* 42: 236-242.

Bidinger, F.R., Mahalakshmi, V., and Rao, G.D.P. 1987a. Assessment of drought resistance in Pearl Millet (*Pennisetum americanum* (L.) Leeke). I. Factors affecting yield under stress. *Australian Journal Agricultural Research* 38: 37- 48.

Bidinger, F.R., Mahalakshmi, V., and Rao, G.D.P. 1987b. Assessment of drought resistance in Pearl Millet (*Pennisetum americanum* (L.) Leeke). II. Estimation of genotype response to stress. *Australian Journal Agricultural Research* 38: 49-59.

Blum, A. 1983. Genetic and physiological relationships in plant breeding for drought resistance. *Agricultural Water Management* 7: 195-205.

Blum, A. 1988. *Plant Breeding for Stress Environments.* CRC Press Inc., Boca Raton, FL.

Blum, A., Mayer, J., and Gozlan, G. 1982. Infrared thermal sensing of plant canopies as a screening technique for dehydration avoidance in wheat. *Field Crops Research* 5: 137-146.

Blum, A., Shpiler, L., Golan, G., and Mayer, J. 1989. Yield stability and canopy temperature of wheat genotypes under drought-stress. *Field Crops Research* 22: 289-296.

Blum, A. and Sullivan, C. 1986. The comparative drought resistance of landraces of sorghum and millet from dry and humid regions. *Annals of Botany* 57: 835-846.

Bolton, F.E. 1991. Tillage and stubble management. In Harris, H., Cooper, P., and Pala, M. (Eds.), *Soil and Crop Management for Improved Water Use Efficiency in Rainfed Areas.* ICARDA, Syria, pp. 34-47.

Brown, S.C., Keatinge, J.D., Gregory, P.J., and Cooper, P.J. 1987. Effects of fertilizer, variety and location on barley production under rainfed conditions in northern Syria. 1. Root and shoot growth. *Field Crops Research* 16: 53-66.

Ceccarelli, S. 1989a. Wide adaptation: How wide? *Euphytica* 40: 197-114.

Ceccarelli, S. 1989b. Efficiency of empirical selection under stress conditions in barley. *Journal of Genetics and Breeding* 43: 25-31.

Ceccarelli, S. 1991. Selection for specific environments or wide adaptability. In Acevedo, E., Fereres, E., Gimenez, C., and Srivastava, J. (Eds.), *Improvement and Management of Winter Cereals Under Temperature, Drought and Salinity Stress,* INIA, Madrid, pp. 227-237.

Ceccarelli, S. and Grando, S. 1989. Efficiency of empirical selection under stress conditions in barley. *Journal Genetic and Breeding* 43: 25-31.

Ceccarelli, S., Nachit, M.N., Ferrara, G.O., Mekni, M.S., Tahir, M. van Leur, and Srivastava, J.P. 1987. Breeding strategies for improving cereal yield and stability under drought. In Srivastava, J.P., Porceddu, E., Acevedo, E., and Varma, S. (Eds.), *Drought Tolerance in Winter Cereals.* John Wiley and Sons, Chichester, UK, pp. 101-114.

Condon, A.G., Farquhar, G.D., and Richards, R.A. 1990. Genotypic variation in isotope discrimination and transpiration efficiency in wheat. Leaf gas exchange and whole plant studies. *Australian Journal of Plant Physiology* 17: 9-22.

Condon, A.G. and Richards, R.A. 1992. Broad-sense heritability and genotype x environment interaction for carbon isotope discrimination in field-grown wheat. *Australian Journal of Agricultural Research* 43:921-934.

Connors, D.J. 1993. Managing wheat within the crop livestock system of rainfed agriculture in southeast Australia. In Mickelson, S.H. (Ed.), *International Crop Science Society of America I,* pp. 171-178.

Cooper, P.J. 1983. *Crop Management in Rainfed Agriculture with Special Reference to Water Use Efficiency.* International Potash Institute, Bern, pp. 63-79.

Cooper, P.J. 1991. Fertilizer use, crop growth, water use and WUE in Mediterranean rainfed farming system. In Harris, H., Cooper, P., and Pala, M. (Eds.), *Soil and Crop Management for Improved Water Use Efficiency in Rainfed Areas.* ICARDA, Syria, pp. 135-152.

Cooper, P.J., Gregory, P.J., Tully, D., and Harris, H.C. 1987. Crop water use and water use efficiency in west Asia and North Africa. *Experimental Agriculture* 23: 113-158.

Cornish, P.S. and Pratley, J.E. 1987. *Tillage: New Directions in Australian Agriculture.* Inkata Press, Melbourne, Australia.

Crovetto, L.C. 1992. *Rastrojos sobre el suelo: Una introducción a la cero labranza.* Editorial Universitaria, Santiago, Chile.

Del Canto, P.S. and Mcmahon, M.A. 1991. Fertilizer management on wheat in Chile's Secano interior: An on-farm research case study. In Harris, H., Cooper, P., and Pala, M. (Eds.), *Soil and Crop Management for Improved Water Use Efficiency in Rainfed Areas.* ICARDA, Syria, pp. 186-195.

Dennett, M.D., Keatinge, J.H.D., and Rodgerds, J.A. 1984. A comparison of rainfall regimes at six sites in northern Syria. *Agricultural and Forest Meteorology* 31: 319-328.

Donald, C.M. 1968. The breeding of crop ideotypes. *Euphytyca* 17: 385-403.

Durutan, N.M., Guler, M., Meyveci, K., Karaca, M., Meyveci, A., Avcin, A., and Eyuboglu, H. 1991. The effect of various components of the management package on weed control in dryland agriculture. In Harris, H., Cooper, P., and Pala, M. (Eds.), *Soil and Crop Management for Improved Water Use Efficiency in Rainfed Areas.* ICARDA, Syria, pp. 220-234.

Durutan, N., Pala, M., Karaca, M., and Yesilsoy, M.S. 1989. Soil management, water conservation and crop production in the dryland regions of Turkey. In C.E. Whitman, J.F. Parr, R.I. Papendick, and R.E. Mayer (Eds.), *Soil Water and Crop/Livestock Management System for Rainfed Agriculture in the Near East Region.* Proceedings of Workshop, Amman, Jordan, January, 1986: 60-77.

Edhaie, B. and Waines, J.G. 1994. Genetic analysis of carbon isotope discrimination and agronomic characters in a bread-wheat cross. *Theory Application Genetic* 88: 1023-1028.

Faiguenbaum, H. 1986. *Producción de cultivos en Chile. Cereales-Leguminosas e Industriales.* Publicitaria Yorrelodones LTDA. Santiago, Chile.

Farquhar, G.D. and Richards, R.A. 1984. Isotopic composition of plant carbon correlates with water-use efficiency of wheat genotypes. *Australian Journal of Plant Physiology* 11: 539-552.

Felton, W.L., Marcellos, H., and Martin, R.J. 1995. A comparison of three fallow management strategies for the long-term productivity of wheat in northern New South Wales. *Australian Journal of Experimental Agriculture* 35: 915-921.

Finlay, K.W. and Wilkinson, G.N. 1963. The analysis of adaptation in a breeding program. *Australian Journal of Agricultural Research* 14: 742-754.

Fischer, R.A. 1981. Optimizing the use of water and nitrogen through breeding of crops. *Plant and Soil* 58: 249-278.

Fischer, R.A. and Maurer, R. 1978. Drought resistance in spring wheat cultivars. I. Grain yield responses. *Australian Journal of Agricultural Research* 29: 897-912.

Fischer, R.A. and Wood, J.R. 1979. Drought resistance in spring wheat cultivars. III. Yield associations with morphophysiological traits. *Australian Journal of Agricultural Research* 30: 1001-1020.

French, R.J. and Schultz, J.E. 1984. Water use efficiency of wheat in a Mediterranean type environment. a. Relation between yield water use and climate. *Australian Journal of Agricultural Research* 35: 743-764.

Girma, F. and Krieg, D. 1992. Osmotic adjustment in sorghum. I. Mechanisms of diurnal osmotic potential changes. *Plant Physiology* 99: 577-582.

Gregory, P.J. 1991. Concepts of water use efficiency. In Harris, H., Cooper, P., and Pala, M. (Eds.), *Soil and Crop Management for Improved Water Use Efficiency in Rainfed Areas.* ICARDA, Syria, pp. 9-20.

Hamblin, A., Tennant, D., and Perry, M.W. 1987. Management of soil water for wheat production in western Australia. *Soil Use and Management* 3: 63-69.

Hannah, M.C. and O'Leary, G.J. 1995. Wheat yield response to rainfall in a long-term multi-rotation experiment in the Victorian Wimmera. *Australian Journal of Experimental Agriculture* 35: 951-960.

Harris, H. 1991. Implications of climatic variability. In Harris, H., Cooper, P., and Pala, M. (Eds.), *Soil and Crop Management for Improved Water Use Efficiency in Rainfed Areas.* ICARDA, Syria, pp. 21-34.

Harris, H., Cooper, P., and Pala, M. 1991. *Soil and Crop Management for Improved Water Use Efficiency in Rainfed Areas.* ICARDA, Syria.

Hazel, C.H., Cooper, P.J., and Pala, M. 1991. Soil and Crop Management for Improved Water Use Efficiency in Rainfed Areas. Proceedings of an International Workshop. Ankara, Turkey, May 15-19, 1989.

Hsiao, T., Acevedo, E., Fereres, E., and Henderson, D. 1976. Stress metabolism. *Philosophical Transaction Royal Society of London Series B* 273: 289-296.

Hubick, K. and Farquhar, G. 1989. Carbon isotope discrimination and the ratio of carbon gained to water lost in barley cultivars. *Plant Cell Environment* 13: 795-804.

Islam, T.M. and Sedgley, R.H. 1981. Evidence for a "uniculm effect" in spring wheat (*Triticum aestivum* L.) in a Mediterranean environment. *Euphytica* 30: 277-282.

Jabbour, E. and Naji, M. 1991. Soil and crop management for improved water use efficiency in rainfed areas of Syria. In Harris, H., Cooper, P., and Pala, M. (Eds.), *Soil and Crop Management for Improved Water Use Efficiency in Rainfed Areas.* ICARDA, Syria, pp. 159-167.

Johnson, R. and Moss, D. 1976. Effect of water stress on $^{14}CO_2$ fixation and translocation in wheat during grain filling. *Crop Science* 16: 697-701.

Kallsen, G.E., Sammis, T.W., and Gregory, E.J. 1984. Nitrogen and yield as related to water use of spring barley. *Agronomy Journal* 76: 596.

Kameli, A. and Lösel, D. 1995. Contribution of carbohydrates and other solutes to osmotic adjustment in wheat leaves under water stress. *Journal of Plant Physiology* 145: 363-366.

Laing, D.R. and Fischer, R.A. 1977. Adaptation of semidwarf cultivars to rainfed conditions. *Euphytica* 26: 129-139.

Loomis, R.S. 1983. Crop manipulations for efficient use of water: An overview. In Taylor, H.M., Jordan, W.R., and Sinclair, T.R. (Eds.), *Limitations to Efficient Water Use in Crop Production.* American Society of Agronomy, Madison, WI, pp. 314-334.

Loomis, R.S. and Connor, D.J. 1992. *Crop Ecology: Productivity and Management in Agricultural Systems.* Cambridge University Press, Cambridge, UK.

Ludlow, M. and Muchow, R.C. 1988. Critical evaluation of the possibilities for modifying crops for high production per unit precipitation. In Bidinger, F.R., and Johansen, C. (Eds.), *Drought Research Priorities for the Dryland Tropics.* ICRISAT, Patancheru, India, pp. 179-211.

Ludlow, M., Santamaria, J., and Fukai, S. 1990. Contribution of osmotic adjusment to grain yield in Sorghum bicolor (L.) M. under water-limited conditions. II. Water stress after anthesis. *Australian Journal of Agricultural Research* 41: 67-78.

McDonald, G.K. and Fischer, R.A. 1991. Soil management to maintain crop production in semi-arid environments. In Acevedo, E., Giménez, C., Fereres,

E., and Srivastava, J.P. (Eds.), *Improvement and Management of Winter Cereals Under Temperature, Drought and Salinity Stresses.* Proceedings of the ICARDA-INIA Symposium, Cordova, Spain.

Mcgowan, M., Blanch, P., Gregory, P.J., and Haycock, D. 1984. Water relations of winter wheat. Ch 5. The root system and osmotic adjustment in relation to crop evaporation. *Journal of Agricultural Science* 102: 415-425.

Mechergui, M., Gharbi, A., and Lazaar, S. 1991. The impact of N and P fertilizers impact on root growth, total yield and WUE on rainfed cereals in Tunisia. In Harris, H., Cooper, P., and Pala, M. (Eds.), *Soil and Crop Management for Improved Water Use Efficiency in Rainfed Areas.* ICARDA, Syria, pp. 153-158.

Midmore, D.J., Cartwright, P.M., and Fischer, R.A. 1982a. Wheat in tropical environments: I. Phasic development and spike size. *Field Crops Research* 5: 185-200.

Midmore, D.J., Cartwright, P.M., and Fischer, R.A. 1982b. Wheat in tropical environments: II. Crop growth and grain yield. *Field Crops Research* 8: 202-227.

Monteith, J.L. 1973. *Principles of Environmental Physics. Contemporary Biology.* Edward Arnold, London, UK, p. 241.

Morgan, J. 1983. Osmoregulation as a selection criterion for drought tolerance in wheat. *Australian Journal of Agricultural Research* 34: 607-614.

Morgan, J. 1984. Osmoregulation and water stress in higher plants. *Annual Review of Plant Physiology* 35: 299-319.

Morgan, J. and Condon, A.G. 1986. Water use grain yield and osmoregulation in wheat. *Australian Journal of Plant Physiology* 13: 523-532.

Morgan, J., Hare, R.A., and Fletcher, R.J. 1986. Genetic variation in osmoregulation in bread and durum wheat and its relationship to grain yield in a range of field environment. *Australian Journal of Agricultural Research* 37: 449-457.

Nass, H.G. and Sterling, G.D.E. 1981. Comparisons of test characterizing varieties of barley and wheat for moisture stress resistance. *Canadian Journal of Plant Physiology* 61: 283-289.

Nelson, C.J. 1988. Genetic associations between photosynthetic characteristic and yield: Review of the evidence. *Plant Physiology and Biochemistry* 26: 243-254.

Ovalle, C. and Del Pozo, A. 1994. *La agricultura del secano interior.* Instituto de Investigaciones Agropecuarias (INIA). Ministerio de Agricultura. Cauquenes, Chile.

Pala, M. 1991. The effect of crop management on increased production through improved WUE at sowing. In Harris, H., Cooper, P., and Pala, M. (Eds.), *Soil and Crop Management for Improved Water Use Efficiency in Rainfed Areas.* ICARDA, Syria, pp. 87-105.

Pala, M. and Mazid, A. 1989. *Annual Report for 1988.* Farm Resource Management Program, ICARDA, Aleppo, Syria.

Parlevliet, J., De Haan, A., and Schellekens, J. 1991. *Drought Tolerance Research: Possibilities and Constraints.* Department of Plant Breeding, Agricultural University, Netherlands

Passioura, J.B. 1977. Grain yield, harvest index, and water use of wheat. *Journal of the Australian Institute of Agricultural Science* 43: 117-120.

Power, J.F. 1985. Nitrogen and water use efficiency of several cool season grasses receiving ammonium nitrate for 9 years. *Agronomy Journal* 77: 189-192.

Rees, D., Sayre, K., Acevedo, E., Nava Sanchez, T., Lu, Z., Zeiger, E., and Limon, A. 1993. *Canopy Temperatures of Wheat.* Special Report No. 10. CIMMYT, Mexico, D.F.

Reeves, T.G. 1991. The introduction, development, management and impact of legumes in cereal rotations in southern Australia. In Harris, H., Cooper, P., and Pala, M. (Eds.), *Soil and Crop Management for Improved Water Use Efficiency in Rainfed Areas.* ICARDA, Syria, pp. 274-283.

Reynolds, M.P., Acevedo, E., Sayre, K.D., and Fischer, R.A. 1994. Yield potential in modern wheat varieties: Its association with a less competitive ideotype. *Field Crops Research* 37: 149-160.

Richards, R.A. 1982. Breeding and selecting for drought resistance in wheat. In *Drought Resistance in Crops with Emphasis on Rice.* International Rice Research Institute, Los Banos, Philippines, pp. 303-316.

Richards, R.A. 1983. Glaucousness in wheat, its effects on yield and related characteristics in dryland environments, and its control by minor genes. *Proc. 6th International Wheat Genetics Symposium,* Kyoto, Japan, pp. 447-451.

Richards, R.A. 1991. Crop improvement for temperate Australia: Future opportunities. *Field Crops Research* 41: 141-169.

Rodriguez, E. 1972. *Clasificación climática de Wilhelm Koeppen.* Facultad de Agronomía, Universidad de Chile. Santiago, Chile.

Romagosa, I. and Fox, P. 1993. Genotype X environment interaction and adaptation. In Hayward, M., Bosemark, N., and Romagosa, I. (Eds.), *Plant Breeding: Principles and Prospects.* Chapman and Hall, London, pp. 373-390.

Rooney, D.R., Sims, H.J., and Touhey, C.L. 1966. Cultivation trials in the Wimmera. *Journal of the Department of Agriculture, Victoria* 64: 403-410.

Santibañez, F. 1994. Crop requirements—Temperate crops. In Griffith, J. (Ed.), *Handbook of Agricultural Meteorology,* Oxford University Press, Oxford, UK, pp. 174-188.

Schultze, E.D. 1988. Adaptation mechanisms of noncultivated arid-zone plants: useful lessons for agriculture? In Bidinger, F.R. and Johanson, C. (Eds.), *Drought Research Priorities for the Dryland Tropics.* ICRISAT, Patancheru, India, pp. 159-177.

Schultz, J.E. 1980. Moisture conservation practices on dryland agriculture. In *Proceedings of International Congress on Dryland Farming,* Adelaide, Australia, pp. 440-470.

Sedgley, R.H. 1991. An appraisal of the Donald ideotype after 21 years. *Field Crops Research* 21: 93-112.

Shackley, B.J. and Anderson, W.K. 1995. Response of wheat cultivars to time of sowing in the southern wheatbelt of western Australia. *Australian Journal of Experimental Agriculture* 35: 579-587.

Silim, S. and Saxena, M. 1991. Winter sowing of chickpea: A case study. In Harris, H., Cooper, P., and Pala, M. (Eds.), *Soil and Crop Management for Improved Water Use Efficiency in Rainfed Areas.* ICARDA, Syria, pp. 119-129.

Smika, D.E. 1970. Summer fallow for dryland winter wheat in the semiarid Great Plains. *Agronomy Journal* 62: 15-17.

Sofield, I., Wardlaw, I.F., Evans, L.T., and Lee, S.Y. 1977. Nitrogen, phosphorus, and water contents during grain development and maturation in wheat. *Australian Journal of Plant Physiology* 4: 799-810.

Srisvastava, J.P., Porceddu, E., Acevedo, E., and Varma, S. 1987. *Drought Tolerance in Winter Cereals.* John Wiley and Sons, Chichester, UK.

Strhaler, N.A. 1986. *Geografía Física.* Universidad de Columbia. Ed. Omega, España.

Tanner, C.B. and Sinclair, T.R. 1983. Efficient water use in crop production: Research or research? In Taylor, H.M., Jordan, W.R., and Sinclair, T.R. (Eds.), *Limitations to Efficient Water Use in Crop Production.* ASA, CSSA, SSSA, Madison, WI, pp. 1-27.

Taylor, H.M., Jordan, W.R., and Sinclair, T.R. 1983. Limitations to Efficient Water Use in Crop Production. ASA, CSSA, SSSA, Madison, WI.

Turner, N.C. 1986. Adaptation to water deficit: A changing perspective. *Australian Journal of Plant Physiology* 13: 175-190.

Turner, N. and Jones, M. 1980. Turgor maintenance by osmotic adjustment: A review and evaluation. In Turner, N.C. and Kramer, P.J. (Eds.), *Adaptation of Plants to Water and High Temperature Stress.* Wiley Inter Science, New York, pp. 87-103.

Turner, N. and Kramer, P.J. 1980. *Adaptation of Plants to Water and High Temperature Stress.* Wiley Inter Science, New York.

Valverde, B.J. 1982. Efecto del tipo de suelo y profundidad de siembra sobre cereales de grano pequeño. Tesis Ing. Agrónomo, Facultad de Ciencias Agrarias, Universidad Austral de Chile, Valdivia.

Virmani, S.M. 1982. Rainfall Probability Estimates for Selected Locations of Semi-Arid India. ICRISAT Patancheru, India. Research Bulletin No. 1.

Wardlaw, I.F., Dawson, I.A., and Munibi, P. 1989. The tolerance of wheat to high temperatures during reproductive growth. II. Grain development. *Australian Journal of Agricultural Research* 40: 15-24.

Wardlaw, I.F., Sofield, I., and Cartwright, P. 1980. Factors limiting the rate of dry matter accumulation in the grain of wheat grown at high temperature. *Australian Journal of Plant Physiology* 7: 87-400.

Watts, D.G. and El Mourid, M. 1988. *Rainfall Patterns and Probabilities in the Semi-Arid Cereal Production Region of Morocco.* INRA/MIAC, Morocco.

Wiegand, C.L. and Cuellar, J.A. 1981. Duration of grain filling and kernel weight of wheat as affected by temperature. *Crop Science* 21: 95-101.

Chapter 15

Wheat Production Systems of the Pampas

Emilio H. Satorre
Gustavo A. Slafer

INTRODUCTION

The Pampas is considered one of the most suitable areas of the world for grain crops production and one of the few with great capacity to contribute to future world food supply needs. Wheat is one of the four major grain crops currently produced, together with maize, soybean, and sunflower. Average actual wheat yields are well below estimated potential yields (2.4 and 8.0 t/ha^{-1} respectively; Magrin and Rebella, 1991). This difference suggests that there are opportunities for transforming the present crop production systems into more productive alternatives.

The Pampas extend over nearly 52 million hectares originally covered by grasslands, interrupted only by gallery forests along the main rivers, intrusions of xerophytic forests raised on fossil shell banks in the northeast of Buenos Aires province, and the Tandilia and Ventania hill systems in the south of the same province. (León, Rusch, and Oesterheld, 1984; Hall et al., 1992). The region may be subdivided following the characteristics of the grasslands established in each division. Soriano et al. (1991) recognized five subdivisions named (1) rolling Pampas; (2) inland Pampas (which may be further divided into flat Pampas and western Pampas); (3) flooding Pampas; (4) southern Pampas; and (5) mesopotamic Pampas. Wheat crop production is spread over the whole region (Figure 15.1). However, wheat sown area and production are concentrated in the southern and rolling Pampas; land devoted to wheat in the flat and western Pampas may be 10 percent of arable land, and wheat is only produced in scattered areas in the mesopotamic and flooding Pampas areas.

FIGURE 15.1. (A) Median Rainfall Values for Winter; (B) Distribution of Mollisol Soils (Argiudolls in Black, Hapludolls in Dark Gray, and Haplustolls in Light Gray; and (C) Distribution of Wheat-Producing Areas in the Pampas

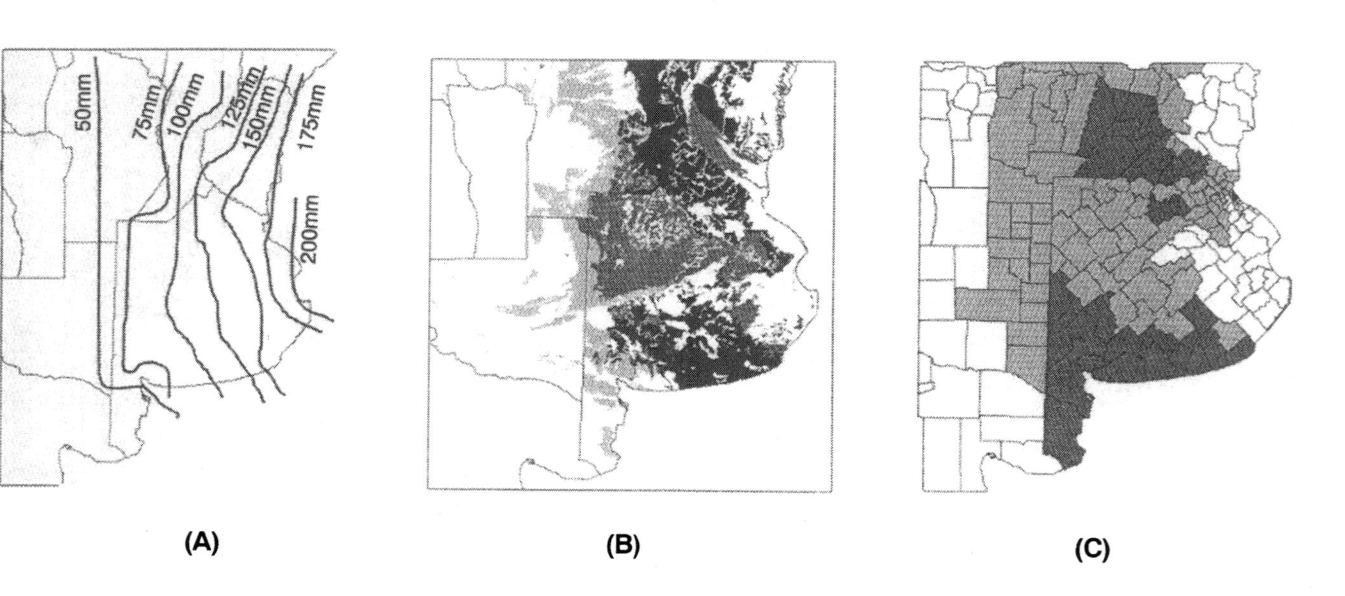

Note: Highly sown and productive areas in black; medium sown and productive areas in gray; and almost unsown areas in white.

A main feature of the region is the mixed (crop-grazing) nature of the farms. The presence of beef cattle raised along with grain crop production is a common denominator in the area, although the proportion of activities may vary among regions. For example, while almost 60 to 80 percent of the land in the rolling Pampas is grain cropped, only 30 to 50 percent is used annually for grain cropping in the western Pampas, the remainder being used mainly for direct grazing of perennial and annual pastures. The area of the Pampas that may be sown with annual grain crops covers approximately 38 million hectares (Mha) but just over 5 Mha are used annually to grow bread wheat (*Triticum aestivum*) and only 40 thousand hectares to grow durum wheat (*Triticum durum*). This chapter will describe climate and soil characteristics and the main attributes of wheat production systems in the most important wheat-producing areas.

CLIMATE AND SOILS: MAJOR CONSTRAINTS FOR WHEAT PRODUCTION

The climate of the Pampas has been defined as "temperate humid without a dry season and with a very hot summer" (Köppen, 1931, cited by Hall et al., 1992). Rainfall and temperature are the main climatic factors for their influence on crop planning and yield.

The general pattern of rainfall distribution indicates that it is minimum in winter and maximum in summer, with decreasing amounts in spring and autumn. Although there is some rainfall during winter in most wheat-producing areas, rainfall distribution is close to monsoonal in the north-west of the Pampas and it tends to an isohigrous pattern southeast of Buenos Aires. In the whole region, the lowest amount of rainfall normally occurs between July and August, when parallel rainfall isohyets decrease from almost 175 mm in the east of the mesopotamic and southern region to 50 mm in the west part of the western and southern region (Figure 15.1). Another important characteristic of rainfall for wheat production is its monthly variability; August and September rainfalls are among the most variable.

The temperature regime for the region shows that June and July are the coldest months and January the hottest. Mean monthly temperatures very rarely fall below 10°C and the period free of frost ranges between 180 and 260 days. However, the interannual variation in the date of last frost can be large, which is of great influence for wheat planning and success. In 50 percent of years late frost may be expected from the end of September in the north to mid-October in the south (Damario and Pascale, 1988) reducing grain set at time of flowering in wheat crops. Temperature indices decrease along a north-south direction, but thermal amplitude also increases from east to west; the frequency and intensity of frosts increase westward. High temperatures are frequent during the end of November and December, which hasten grain filling and reduce the final weight of the growing grains.

The influence of weather on wheat yields has been previously analyzed. A positive relationship between September-October rainfall and wheat yield ($r^2 = 0.42$) has been reported, though excess rain during anthesis can reduce yield in 20 percent of years (Hall et al., 1992). The influence of the El Niño Southern Oscillation (ENSO) on rainfall pattern in the Pampas and wheat yield has been pointed out in recent studies. A strong positive influence of ENSO on wheat crops in the west of the southern Pampas has been demonstrated (Messina, Beltrán, and Ravelo, 1996; Spescha et al., 1997). The combined effect of radiation and temperature patterns on wheat yield potential along the Pampas was described by Magrin et al. (1993), who found a strong association between kernel number per unit area in crops from nursery trials and photothermal quotient calculated for the three weeks prior to crop anthesis, from sites in the northeast to the southeast of the Pampas.

Wheat is cropped in mollisol soils formed over loessic sediments (Fidalgo and Tonni, 1978, cited by Solbrig, 1997) (Figure 15.1); typic argiudolls are the most representative in the rolling and southern Pampas regions (Hall et al., 1992). They have a black A surface horizon of 18 to 35 cm thickness and loam or clay-loam texture followed by a B subsoil layer extended from 35 to 80 cm with high clay content (30 to 57 percent clay, and the predominant mineral is illite), silty-clay to loam-clay in texture (Senigagliesi, Ferrari, and Ostojic, 1996). Topsoil organic matter content may range from 1 to 4 percent with a carbon to nitrogen ratio around 10. In the southern Pampas a petrocalcic subsoil layer may be present in some areas. Typic and entic hapludolls and haplustolls are found in the western Pampas; these soils have a less developed B horizon, with less clay content, and they are deep with predominantly sandy and loamy textures. Pampas soils do not freeze in winter.

The rolling Pampas soils have marked nitrogen and phosphorus deficiencies (Darwich, 1983, 1994); nitrogen deficiency tends to be higher to the west of the region, where organic matter content is low (1.5 to 2 percent). On the other hand potassium levels are high throughout, with exchangeable values well above 10 meq/kg. Some sulphur and magnesium deficiencies have been recently reported (Melgar, 1997).

The Pampas soils have few constraints on wheat cropping. The characteristics of the B clay horizon or the presence of a petrocalcic layer may reduce root penetration. The silty texture of the topsoil makes it prone to crusting and sealing, which reduces water infiltration rate. High-intensity rainfalls increase the risk of soil crusting, particularly when they occur at sowing in crops under conventional tillage systems. The water-holding capacity of Pampas soils varies between 100 to 180 mm of available water in the first meter of the profile. Water held at crop sowing frequently determines wheat success in the rolling and western Pampas, due to the low winter and variable early spring rainfalls.

Overall, weather patterns and soil characteristics described above may greatly influence wheat crop performance. Management practices have evolved in an effort to reduce the negative impact of these constraints, which will be described in the following section.

WHEAT PRODUCTION SYSTEMS

Tillage Systems and Rotations

Wheat in the Pampas is mostly grown within a mixed crop-cattle production system, characterized by a four-year period of pasture and between 6 and 10 years of crops, with only a little area devoted to continuous cropping. During grain production, plant species are rotated following different patterns for each region. In the rolling Pampas, with some of the most productive soils, the predominant rotation includes full-season soybean-wheat double cropped with late-sown soybean-maize, while in the southern and western Pampas the most common rotation tends to be maize-sunflower-wheat. Therefore, soybean and sunflower are the most frequent wheat-preceding crops in the rolling Pampas and in the southern and western Pampas, respectively; although wheat may sometimes follow a maize crop in both regions. Due to its early harvest, which allows soil water replenishment during autumn, sunflower is considered the best crop to precede wheat. It is only in the southwest of the southern Pampas that a rotation having wheat following wheat is usual; it may represent up to 30 percent of the area (Olivero Vila, 1996).

The Pampas region has experienced profound tillage changes in the past ten years, mostly due to the increased interest in maintaining soils covered with plant residues. The chisel plow was introduced in the 1980s and has been quickly replacing the conventional moldboard plow as the deep tillage tool in the western Pampas (Oliverio, 1989). However, the use of a disc harrow after the chisel plow has been usual to improve the sowing with traditional seeders. Following the introduction of double disc seeders, which permitted the proper sowing and establishment of the crop in soils covered with stubble, the number of tillages was effectively reduced. By the early 1990s the performance of no-tilled and direct-drilled crops was evaluated and they started to be used in the region, at present covering nearly 5 percent of the sown area. Tillage sequences for wheat sowing can be summarized as: (1) disc-chisel plow-disc-sowing (conventional with deep [30 cm] soil removal), which is common when maize is the previous crop; (2) disc-disc-sowing (shallow or minimum tillage), which is frequent when either soybean or sunflower are the previous crops. In some cases the disc harrow labor is replaced by the use of a field cultivator.

Usually yields in no-till crops tended to be lower than those under minimum or conventional tillage systems (Ferrari, 1996). The strong influence of fungal diseases in crops sown with no-tillage has been reported; particularly, the incidence and severity of necrotrophic pathogens have increased in wheat crops (Reis and Carmona, 1995; Carmona, 1996). For example, Reis and Carmona (1995) showed a significant increase of yellow spot (*Dreschlera tritici-repentis*) damage on wheat leaves in direct-drilled crops when compared to crops established with some soil removal. Similarly, the incidence of scab (*Fusarium graminearum*) ear disease was greater in no-till crops after maize than in conventional crops (Ivancovich, 1996). In addition, it has been suggested that fertilizer requirements are greater when wheat is sown directly, with no-tillage, than when it is sown after some soil disturbance (Diaz Zorita, 1996).

Sowing Date and Varieties

Wheat crops are sown uninterruptedly from the second half of May to the first half of August. Risk of frost damage at flowering is the main climatic factor determining optimum sowing dates for particular varieties in the various regions.

In the rolling Pampas, long- and intermediate-season cultivars are sown mostly between May 25 and June 20, while short-season cultivars are sown preferably from July 1 to 15. Early planting (early June) of intermediate-long cultivars has proved to be a more productive and stable strategy than the late planting of short cultivars (Savin, Satorre, and Slafer, 1995). Short-season cultivars are frequently sown following double-cropped soybeans due to the short period left from soybean harvest for land preparation. Short-season cultivars are also preferred before double-cropped soybeans mainly because they tend to be harvested early and they produce less stubble, which eases seedbed preparation and sowing. Seeding rates vary from 200 to 250 established plants per m^2 for early to late sowings, respectively.

In the southern and western Pampas, some early-planted wheat crops tend to serve a double purpose for forage and grain production. Optimum planting date for long-season wheat produced for grain ranges from early June to early July, while short-season varieties tend to be sown during July to early August. Seeding rates are slightly lower than in the rolling Pampas for early-sown crops (180 to 220 established plants per m^2). Seeding rates for late-sown varieties vary between 250 and 320 established plants per m^2.

Fertilizer Management

The management of nitrogen and phosphorus is crucial in high-yielding production systems. Although it has been recently reported that some yield response may be obtained from sulfur and potassium application in hapludoll and entic haplustoll soils, the use of this fertilizer source is still very rare (Melgar, 1997). Fertilizers started to be used in crop production very recently; the proportion of land fertilized with either phosphorus or nitrogen in wheat markedly increased from 15 percent in 1987 to almost 60 percent in 1994 (Melgar, 1995).

Fertilizer requirements and decision criteria are based on various approaches. Most of the nitrogen fertilizer is applied before sowing. However, some nitrogen is applied split between sowing and tillering, mostly in the southern and western Pampas. Modified nitrogen balance models are generally used for wheat crops (Remy and Hebert, 1977). In the southern Pampas, González Montaner, Maddonni, and Di Napoli (1991) have proposed a simplified balance model suggesting the application of the difference between the amount of $N\text{-}NO_3$ available in the top 60 cm of the soil before sowing and 125 kg N per hectare. The authors found that this approach tends to optimize the response to fertilizer use in the region. Recently, González Montaner, Maddonni, and Di Napoli (1997) have developed a simple model to predict the yield and yield response to nitrogen fertilizer application for short-season cultivars sown after sunflower. In the western and rolling Pampas much work was done during the 1980s using an approach known as isodose response curve (Barberis et al., 1983). The response to a fixed amount of fertilizer is predicted based upon the availability of $N\text{-}NO_3$ in the top layers of the soil just before sowing. This approach has been slowly replaced by more flexible balance and simulation models. More recently, the crop growth simulation model CERES-Wheat was used to develop nitrogen management criteria for the rolling and western Pampas. In the rolling Pampas that model was used together with field research to predict the performance of commercial wheat crops and develop a decision criteria based on the amount of total N available to the crop at sowing (soil N + fertilizer N) in the top 60 cm (Calderini et al., 1994; Satorre et al., 1997). This decision tool also takes into account the effect of climatic variability on yield responses for various soil and management conditions. This decision criterion was independently validated with data from 140 production fields yielding mean errors lower than 0.1 and 0.4 t/ha on long- and short-season cultivars, respectively. In addition to the approaches described, some research has been carried out on field evaluations of the nitrogen status of plants and on late applications of nitrogen fertilizer to improve protein content and grain quality. Diagnostic methods based on nitrate concentration in the pseudostem tissues have been proposed (Echeverría, 1985; González Montaner, 1987; Viglezzi, Echeverría, and Studdert, 1994). Moreover, research showed that quality of wheat and flour may be improved by postanthesis nitrogen foliar fertilization (Sarandon et al., 1986). However, this technique is only very rarely used in commercial farming.

Phosphorus fertilizer is applied either broadcast before a disc tillage previous to sowing or at sowing with the seeder. It is mostly incorporated in the soil since the efficiency of the fertilizer is markedly reduced when it is left on the soil surface (Loewy and Ron, 1997). Available phosphorus is estimated from soil samples, usually 0 to 20 cm deep, using the Bray-Kurtz No. 1 method. In the case of phosphorus, much effort has been done to estimate the critical thresholds for wheat response under the various soils and regions and to estimate model parameters from yield-dose response curves (Senigagliesi et al., 1983, 1986; Berardo, 1994; Ron and Loewy, 1990; Loewy and Ron, 1997). However, we are still far from reaching a decision tool as that available for nitrogen fertilization.

Crop Protection

Few fungicides and pesticide treatments are used for wheat production in the Pampas. Fungicides are usually applied to the seed to control bunt and smut diseases; however, cases of farmer-retained seed sown with no fungicide treatment are not rare. Scab, take-all and leaf blights and rusts are among the worst fungal diseases frequently observed in the crop. Antonelli (1983) has reported that during 1976 scab damage seriously reduced wheat yield in the southern Pampas, causing 30 percent losses. During 1993 environmental conditions favored scab development in the rolling Pampas, reducing yield and quality. Foliar disease control is intended through selection of resistant cultivars, tillage system, and crop rotation. The importance of and attention paid to foliar diseases has increased in the past ten years. It has been mentioned that foliar diseases may be responsible for up to 10 percent of yield losses every year (Galich et al., 1986). Recently, the use of foliage-applied fungicides has started to increase; when applied during stem elongation, treatments aim to control foliar diseases such as yellow spot, septoria leaf blotch, powdery mildew, and rusts in high-yielding production crops. Some treatments are delayed until ear emergence, intending to protect with a single application the last appearing leaves against foliar diseases and the ear against scab development. Some predictive models that take into account daily temperature, rainfall, and relative humidity are used to determine the occurrence of conditions that favor scab development and the consequent need of fungicide application (Moschini, 1994).

Insecticide use is very limited. If necessary, insecticides are used to control green aphids at crop establishment, but the incidence of this insect has decreased, possibly due to changes in crop and farm management. Russian aphid was recently detected but its influence on yield has not been evaluated yet. Similarly, damage caused by Argentine stem weevil (*Listronotus bonariensis*) on crops from the southern Pampas were recently described (Gallez, Miravalles, and Mockel, 1994). Yield reductions due to its incidence were estimated to vary between 13 and 31 percent in affected crops (Gallez et al., 1994). Insecticides may be also used to prevent damage from *Lepidopterae* species after ear emergence; *Pseudoaletia adultera* and *Faronta albilinea* are the main species affecting the crop during grain filling, but crops are always monitored at this stage and their chemical control is rare.

Weed control during fallow is achieved by means of either mechanical labor or nonselective herbicide application. Weed control strategies during crop growth have been reduced to a single application of a selective postemergence herbicide (or herbicide mix) between the stages when two leaves are unfolded and early stem elongation. Postemergence herbicide use is mostly targeted against broadleaf weeds, since the incidence of grass weeds is very limited. Cruciferae and Poligonaceae species are among the main weed problems in the southern and rolling Pampas. Volunteer sunflower is also considered an important weed in the southern Pampas. Competition studies have pointed out the high competitive ability of species such as *Brassica napus* and *Raphanus sativus* (Guglielmini, García, and Satorre, 1991; Leaden, 1995), while others showed the difficulties of controlling weeds such as *Polygonum convolvulus*, *Polygonum lapathifolium*, and *Polygonum persicaria* (Grosse and Leaden, 1996).

The effect of nitrogen fertilizer application on weeds has been recently demonstrated; the increase of nitrogen availability modifies the competitive balance between Cruciferae and wheat, increasing yield losses, weed biomass, and seed production (Guglielmini et al., 1994, 1995). Some grass weed species such as *Avena fatua* and *Lolium* spp. are important, predominantly in some areas on the southern Pampas. In these cases two weed control strategies are used, one based on the presowing application of trifluralin and the other in the postemergent application of selective graminicides (e.g., diclofop-methyl).

There are no objective and widespread criteria to determine the need of herbicide application. Some weed thresholds have been determined (Satorre and Conterjnic, 1991) but they are rarely used. In all cases decisions are mostly based on subjective aspects of the crop-weed systems, mostly based on the low cost of herbicides and their efficacy and ease of management. Despite the fact that most of the wheat production area is treated with sulfonylurea herbicides every year—80 percent of the area in the rolling Pampas and almost 60 percent in the southern and western Pampas—there have been no reports of the development of herbicide-resistant biotypes of weeds.

Harvest and Double Crops

Wheat crops are harvested by the end of spring, from early to late December, in the rolling and northern part of the western Pampas. In the southern Pampas wheat harvest starts usually by Christmastime, i.e., the end of December, and it extends to January. All wheat in the region is combine harvested. Due to the extent of the crop growth period available before first frost, crops such as soybean, sunflower, or even maize may be sown as a double-crop immediately following wheat harvest in all of the rolling and part of the western Pampas. Although most summer crops may be established after wheat in the northern part of the Pampas, soybean is the most frequently double-cropped species. Double-cropped soybean is only rarely sown in most of the southern Pampas due to the short growth period available for the crop before temperature and radiation diminishes drastically. In this region when a soybean crop is grown in the same season as the previous wheat, it is sometimes intercropped within the wheat at anthesis, in rows left within the fields for that purpose.

In the rolling Pampas soybean crops are sown immediately after wheat harvest. Soybean yield potential is greatly influenced by sowing date; it is reduced even with small delays in planting. Double-cropped soybean is usually sown with no-tillage or stubble treatment. Commonly, weeds are chemically controlled after harvest and the stubble is left standing. In this system, combine harvesters are provided with elements to cut and distribute wheat residues as uniformly as possible and specially adapted direct drillers are used. Some double-cropped soybean is still established after some soil removal, normally using disc harrows and field cultivators. However, due to the chance for timely sowing, the low cost of herbicides, and ease of management, most soybean is directly drilled immediately after wheat harvest. Yield of soybean in this cropping system is quite variable, ranging from 1 to 3 t/ha according to date of sowing, environmental conditions, and crop management. The double-crop system is usually stable and financially convenient, since the wheat crop provides a financial return during December and the soybean crop during June. The possibility of regular double crops is a distinctive characteristic of the northern part of the Pampas that has strongly supported the development of wheat production systems in the region.

FUTURE PERSPECTIVES

Wheat production systems in the Pampas have always been supported by well-established plant breeding programs. Plant breeding has contributed to almost 50 percent of yield increase during this century (Slafer and Andrade, 1991) and breeders are continuously and actively releasing new improved varieties to the market. However, crop production systems were until recently low-input and extensive. At present, agriculture production systems in the Pampas are being intensified and farmers are incorporating technology to help raise grain yields. There will be, therefore, a continuous demand for varieties with characteristics that allow them to respond with high and stable yields to intensified production schemes. At the same time, wheat production in the Pampas is tending to diversify in order to satisfy various market demands. High-quality wheat named "Trigo Plata" was officially promoted through price incentives during 1997. This qualification added to the three grades already existing and the category of forage wheat created two years ago. These policies are expected to promote the development of new crop management skills and plant breeding programs.

New soil management techniques are being adopted, and direct-drill cropped area will increase. This will help to sort out problems such as those derived from wind and water erosion but will build up some new biotic (mainly fungi and weeds) and fertilizer management problems. More complex process-based technologies will need to be incorporated in decision criteria and management techniques. For this reason, a more formal technological and scientific interaction with farmers and their production systems is expected to develop.

Finally, as in other parts of the world, environmental concern and sustainability are topics recently incorporated in the farmer's agenda. Efficient and rational use of agrochemicals has been considered among the technologies required to increase wheat yields while keeping the productive potential and health of the natural resources (Satorre, 1998). Although the new wheat economic and technological scenario is favorable for a continuous grain yield and sown area increase, there is a concern about whether the increase in gross benefits through greater yields is accompanied by a rise in the net benefit received by the farmers. If intensification fails to improve this equation, it may be expected that farmers will tend to slow down the process of technology adoption in the near future, going back to a traditional low-risk, low-input production system. Information and knowledge are crucial elements to continue developing high-yielding and quality wheat production systems in the Pampas, looking for a rational use of energy inputs to allow an increase in farmers' economic returns.

REFERENCES

Antonelli, E. 1983. Principales patógenos que afectan la producción de trigo en Argentina. *Proc. Symp. "Fitomejoramiento y Producción de Cereales,"* Instituto Nacional de Tecnologia Agropecuaria, Marcos Juarez, Argentina, pp. 100-101.

Barberis, L., A. Nervi, H. del Campo, S. Urricarriet, J. Sierra, P. Daniel, M. Vazquez, and D. Zourarakis. 1983. Análisis de la respuesta del trigo a la fertilización nitrogenada en la Pampas ondulada y su predicción. *Ciencia del Suelo* 1(2): 51-64.

Berardo, A. 1994. *Aspectos generales de fertilización y manejo del trigo en el área de influencia de la Estación Experimental INTA Balcarce.* Tech. Bulletin No. 128, Instituto Nacional de Tecnologia Agropecuaria, Balcarce, Argentina, p. 34.

Calderini, D., G. Maddonni, D. Miralles, R. Ruiz, and E. Satorre. 1994. Validación del modelo CERES-Wheat para producciones extensivas de trigo en diferentes situaciones de fertilidad del Norte de la provincia de Buenos Aires. *Proc. III Congreso Nacional de Trigo,* Asociación de Ingenieros Agrónomas del Norte de Buenos Aires, Bahía Blanca, Argentina, pp. 81-82.

Carmona, M.A. 1996. Principales enfermedades del cultivo de trigo. In *Cuaderno de Actualización Técnica No. 56—Trigo* Asociación Argentina de Consorcios Regionales de Experimentación Agrícola (AACREA), Buenos Aires, Argentina, pp. 58-72.

Damario, E.A. and A.J. Pascale. 1988. Características agroclimáticas de la región pampeana. *Revista de la Facultad de Agronomía* (UBA), 9(1-2): 41-69.

Darwich, N. 1983. Niveles de fósforo asimilable en los suelos pampeanos. *IDIA* 409-412: 1-5.

Darwich, N. 1994. Siembra directa y ambiente edáfico. In *Cuaderno de Actualización Técnica No. 54—Siembra Directa* Asociatión Argentina de Consorcios

Regionales de Experimentación Agrícola (AACREA), Buenos Aires, Argentina, pp. 25-28.

Diaz Zorita, M. 1996. Propiedades edáficas y sostenibilidad de los sistemas de producción en la región Noroeste bonaerense. *Proc. Congreso CREA Zona Oeste,* Asociación Argentina de Consorcios Regionales de Experimentación Agrícola (AACREA), Mar del Plata, Argentina, pp. 59-63.

Echeverría, H.E. 1985. Factores que alteran la concentración de nitratos en plantas de trigo. *Ciencia del Suelo* 3(1-2): 115-123.

Ferrari, M. 1996. Los sistemas de labranza en el área de influencia de la EEA INTA Pergamino. Efecto sobre los rendimientos agrícolas y las propiedades de los suelos. *Proc. Congreso CREA Zona Oeste,* Asociación Argentina de Consorcios Regionales de Experimentación Agrícola (AACREA), Mar del Plata, Argentina, pp. 29-33.

Galich, A., M. de Galich, A. Legasa, and G. Musso. 1986. Estimación de pérdidas de rendimiento por enfermedades foliares en cultivares de trigo. *Proc. I Congreso Nacional de Trigo,* Asociación de Ingenieros Agrónomos del Norte de Buenos Aires (AIANBA), Pergamino, Argentina. (4): 41-50.

Gallez, L.M., M.T. Miravalles, M.A. Ecke Follonier, and F.E. Mockel. 1994. Estimación de pérdidas causadas por *Listronotus bonariensis* (Coleoptera: Curculionidae) en trigo. *Proc. III Congreso Nacional de Trigo,* Asociación de Ingenieros Agrónomos del Norte de Buenos Aires (AIANBA), Bahía Blanca, Argentina, pp. 220-221.

Gallez, L.M., M.T. Miravalles, and F.E. Mockel. 1994. Daños causados por *Listronotus bonariensis* (Coleoptera: Curculionidae) en cultivos de trigo. *Proc. III Congreso Nacional de Trigo*, Asociación de Ingenieros Agrónomos del Norte de Buenos Aires (AIANBA), Bahía Blanca, Argentina, pp. 219-220.

González Montaner, J.H. 1987. Deux outils d'analyse de la response du ble a la fertilisation azotee: Les composantes du rendement et la teneur en nitrates des organes vegetaux. Tesis Docteur Ingenieur Sciences Agronomiques, Univ. Paris, Grignon, p. 190.

González Montaner, J.H., G.A. Maddonni, and M.R. Di Napoli. 1997. Modeling grain yield and grain yield response to nitrogen in spring wheat crops in the Argentinean southern Pampas. *Field Crops Research* 51: 241-252.

González Montaner, J.H., G.A. Maddonni, N. Mailland, and M. Posborg. 1991. Optimización de la respuesta a la fertilización nitrogenada en el cultivo de trigo, a partir de un modelo de decisión para la subregión IV (Sudeste de la Provincia de Buenos Aires). *Ciencia del Suelo* 9(1-2): 41-51.

Grosse, R. and M.I. Leaden. 1996. Control de malezas en trigo. In *Cuaderno de Actualización Técnica No. 56—Trigo* Asociación Argentina de Consorcios Regionales de Experimentación Agricola (AACREA), Buenos Aires, Argentina, pp. 73-79.

Guglielmini, A., C. García, and E.H. Satorre. 1991. Efecto de la fertilización con nitrógeno sobre la competencia entre seis cultivares de trigo (*Triticum aestivum*) y *Brassica* sp. *Proc. XII Reunión Argentina sobre la maleza y su control,*

Asociación Argentina para el Control de Malezas, Mar del Plata, Argentina, (2): 79-86.

Guglielmini, A., F. Varela, D. Miguens, and E.H. Satorre. 1994. Competencia entre trigo (*triticum aestivum*) y *Brassica* sp. en ambientes con distinta oferta de nitrógeno. Cambios en la estructura del canopeo. *Proc. III Congreso Nacional de Trigo,* Asociación de Ingenieros Agrónomos del Norte de Buenos Aires (AIANBA), Bahía Blanca, Argentina, pp. 191-192.

Guglielmini, A., F. Varela, D. Miguens, and E.H. Satorre. 1995. Reversión de la habilidad competitiva de una maleza (*Brassica* sp) en cultivos de trigo (*Triticum aestivum*) con y sin fertilización nitrogenada. *Proc. XII Congreso Latinoamericano de Malezas* Asociación Latinoamericana de Malezas (ALAM), Montevideo, Uruguay, p. 59.

Hall, A.J., C.M. Rebella, C.M. Ghersa, and J.P. Culot. 1992. Field-crop systems of the Pampas. In C.J. Pearson (Ed.), *Field crop systems,* pp. 413-450. Ecosystems of the world, Vol. 18, Elsevier, Amsterdam.

Ivancovich, A. 1996. Enfermedades del maíz. In *Cuaderno de Actualización Técnica No. 57—Maíz* Asociación Argentina de Consorcios Regionales de Experimentación Agricola (AACREA), Buenos Aires, Argentina, pp. 62-66.

Leaden, M.I. 1995. Fecha de siembra y balance de competencia entre trigo (*Triticum aestivum*) y nabón (*Raphanus sativus*). Interacciones con la densidad del cultivo y la maleza y la disponibilidad inicial de nitrógeno. Msc Thesis, Univ. Nac. Mar del Plata.

León, R.J.C., G.M. Rusch, and M. Oesterheld. 1984. Pastizales pampeanos—impacto agropecuario. *Phytocoenologia* 12: 201-218.

Loewy, T. and M. Ron. 1997. Fertilización fosfórica del trigo en la región Pampeana. *Revista Fertilizar—Suplemento Trigo,* pp. 10-16.

Magrin, G.O., A.J. Hall, C. Baldy, and M.O. Grondona. 1993. Spatial and interannual variations in the photothermal quotient: Implications for the potential kernel number of wheat crops in Argentina. *Agric. For. Meteorol.* 67: 29-41.

Magrin, G. and C. Rebella. 1991. Proyecto de previsión de cosechas de cereales y oleaginosos. Convenio INTA-JNG; Advance report. Instituto Nacional de Tecnologia Agropecuaria, Buenos Aires, Argentina.

Melgar, R. 1995. Fertilización del trigo. *Revista Fertilizar* (Bol. Div. Tec. No. 107): 1-20.

Melgar, R.J. 1997. Potasio, azufre y otros nutrientes necesarios para considerar en una fertilización. *Revista Fertilizar—Suplemento Trigo,* pp. 17-24.

Messina, C., A. Beltrán, and A. Ravelo. 1996. La variabilidad interanual de los rendimientos de trigo en la región pampeana y su relación con el fenómeno ENSO (El Niño—Southern Oscillation). *Proc. VII Congreso Argentino* and *VII Congreso Latinoamericano e Ibérico de Meteorología,* Centro Argentinode Meteorologos and Federación Latinoamericana e Ibérica de Sociedades de Meteorología, Buenos Aires, Argentina, 55-56.

Moschini, R.C. 1994. Modelos predictivos de la incidencia de fusariosis en trigo basados en variables meteorológicas. *Proc. III Congreso Nacional de Trigo,*

Asociación de Ingenieros Agrónomos del Norte de Buenos Aires (AIANBA), Bahía Blanca, Argentina, pp. 320-326.

Oliverio, G. 1989. Labranza Conservacionista. *Cuaderno de Actualización Técnica No. 37.* Asociación Argentina de Consorcios Regionales de Experimentación Agricola (AACREA), Buenos Aires, Argentina.

Olivero Vila, J.M. 1996. Producción de trigo en la zona semiárida. In *Cuaderno de Actualización Técnica No. 56—Trigo* (AACREA), pp. 115-117.

Reis, E.M. and M.A. Carmona. 1995. Mancha Amarilla de la Hoja de Trigo. Publicación Técnica Bayer, Buenos Aires, Argentina.

Remy, J.C. and J. Hebert. 1977. Le devenir des engrais azotes dans le sol. *Compte Rendu Academie Agricole Francais* 63(11): 700-710.

Ron, M. and T. Loewy. 1990. Fertilización fosfórica del trigo en el SO bonaerense. I. Modelos de respuesta. *Ciencia del Suelo* 8: 187-194.

Sarandon, S.J., M.C. Gianibelli, H.O. Chidichimo, H.O. Arriaga, and C. Favoreti. 1986. Fertilización foliar en trigo (*t. aestivum* L.): Efecto de la dosis y el momento de aplicación sobre el rendimiento y sus componentes, el porcentaje de proteínas y la calidad del grano. *Proc. I Congreso Nacional de Trigo,* Asociación de Ingenieros Agrónomos del Norte de Buenos Aires (AIANBA), Pergamino, Argentina, (2): 242-258.

Satorre, E.H. 1998. Aumentar los rendimientos en forma sustentable en la Pampas argentina. Aspectos generales. In O.T. Solbrig (Ed.), *Hacia una agricultura más productiva y sostenible en la Pampas Argentina.* CPIA Publisher, Buenos Aires.

Satorre, E.H. and S. Conterjnic. 1991. Trigo: Algunas bases para la mas eficiente toma de decisiones en la protección del cultivo. *Revista de los CREA* 148: 49-54.

Satorre, E.H., R.A. Ruiz, D.J. Miralles, G.A. Maddonni, and D.F. Calderini. 1997. Bases de decisión para la fertilización nitrogenada de trigo. Final report Convenio AACREA-Cátedra de Cereales, Univ. Buenos Aires.

Savin, R., E.H. Satorre, A.J. Hall, and G.A. Slafer. 1995. Assessing strategies for wheat cropping in the monsoonal climate of the Pampas using the CERES-Wheat simulation model. *Field Crops Research* 38: 125-133.

Senigagliesi, C.A., M. Ferrari, and J. Ostojic. 1996. La degradación de los suelos en el partido de Pergamino. In J. Morello and O.T. Solbrig (Eds.), ¿Argentina granero del mundo: Hasta cuando? CPIA Publisher, Buenos Aires.

Senigagliesi, C.A., R. García, S. Meira, M. Rivero de Galetto, and M.T. Stornini. 1986. Producción de trigo en el área maicera agrícola-ganadera de influencia de la EEA INTA Pergamino. *Proc. I Congreso Nacional de Trigo,* Asociación de Ingenieros Agrónomos del Norte de Buenos Aires (AIANBA), Pergamino, Argentina, (3): 81-104.

Senigagliesi, C.A., R. García, S. Meira, M.R. Galetto, E. Frutos, and R. Teves. 1983. *La fertilización del cultivo de trigo en el Norte de la Provincia de Buenos Aires y Sur de Santa Fe.* Technical Report No. 191, INTA Pergamino, Pergamino, Argentina.

Slafer, G.A. and F.H. Andrade. 1991. Changes in physiological attributes of the dry matter economy of bread wheat (*Triticum aestivum*) through genetic improvement of grain yield potential at different regions of the world. A review. *Euphytica* 58: 37-49.

Solbrig, O.T. 1997. Towards a sustainable pampa agriculture: Past performance and prospective analysis. A syllabus. Proc. Workshop "Hacia una agricultura productiva y sostenible en la Pampa," Consejo Profesional de Ingeniería Agronómica (CPIA), Buenos Aires, Argentina.

Soriano, A., R.J.C. León, O.E. Sala, R.S. Lavado, V.A. Deregibus, M.A. Cahuepé, O.A. Scaglia, C.A. Velázquez, and J.H. Lemcoff. 1991. Temperate subhumid grasslands of South America. In R.T. Coupland (Ed.), *Natural grasslands.* Ecosystems of the World. Vol. 8, Elsevier, Amsterdam.

Spescha, L., A. Beltrán, C. Messina, and R. Hurtado. 1997. Efectos del ENSO (El Niño—Southern Oscillation) sobre la producción agrícola argentina. Proc. Workshop on "Efectos de El Niño sobre la variabilidad climática, agricultura y recursos hídricos en el sudeste de Sudamérica," Montevideo, Uruguay, pp. 19-22.

Viglezzi, A., H.E. Echeverría, and G.A. Studdert. 1994. Concentración de nitratos en seudotallos de trigo: II Su comportamiento ante prácticas de Manejo. *Proc. III Congreso Nacional de Trigo,* Bahía Blanca, Argentina, pp. 13-14.

PART IV: BREEDING TO FURTHER RAISE WHEAT YIELDS

Chapter 16

Genetic Gains in Wheat Yield and Associated Physiological Changes During the Twentieth Century

Daniel F. Calderini
Matthew P. Reynolds
Gustavo A. Slafer

INTRODUCTION

As already discussed (Chapter 1), noteworthy increases in wheat yields during the present century have been the result of better crop management and the use of cultivars with higher yield potential. To use past experience in future plant breeding efforts we must know more accurately the contribution that breeders made to increase yields; not only the magnitude of the contribution, but also to recognize which attributes of the crop were changed and to what extent the successful criteria used in the past could still be promising in the near future. When physiological traits associated with yield improvements are identified, they can be applied routinely in breeding programs, while the underlying physiological principles can be integrated into our understanding of plant breeding to indicate future selection criteria.

An initial question is how to quantify the contribution of wheat breeding to increases in grain yield. Two main methods have been used extensively. The simplest approach consists of using data from experimental networks extended over long periods of time where the yield of new cultivars is expressed relative to that of known checks. Therefore, cultivars that were never grown together may be compared through their relative yields. The other method is based on experiments in which cultivars released at different times are grown simultaneously in the same environment, and the comparison is made in absolute terms.

Information from experimental networks has the advantage that data are readily available, as these experiments are routinely performed in many experimental stations. But they provide limited information as yield is generally the only variable recorded. This approach does not permit the identification of individual crop attributes responsible for improvements in yield and thus cannot offer alternatives for future breeding. Furthermore, the assessment of genetic gains in yield potential are biased by the sensitivity of the checks to biotic/abiotic constraints that may occur in any particular year, especially diseases. This is especially a problem as checks become older, increasing both the yield penalty produced by them and the likelihood of being affected by new ecotypes of pathogens. In addition, changes in crop management practices over time in these networks may also influence the quantitative assessment of genetic gains.

Most of these biases can be prevented by studies in which the assessment of yield, together with the determination of its major physiological and numerical components, is made by growing cultivars released in different eras simultaneously in replicated experiments. While the latter is accepted as the better of the two approaches (Slafer, Satorre, and Andrade, 1994), its major constraint is the limitations imposed by the growing environment, i.e., crop management and year-to-year variability, since genotype × environment interactions are frequently significant. In addition, environmental factors not only affect the magnitude of expression of genetic gains, but also confound their expression through the complex interaction of genotype with climatic variability (e.g., Perry and D'Antuono, 1989; Sayre, Rajaram, and Fischer, 1997).

In this chapter we attempt to summarize the contributions of wheat breeding for improving yield potential, and the bases on which this increase was possible using a crop physiology approach. To achieve this objective we have mainly reviewed studies reporting experiments with cultivars released at different times, carried out under field conditions in several countries. Initially we will analyze genetic improvement effects on grain yield, and then the associated changes in physiological traits (height, biomass accumulation, and partitioning) and numerical components (grain number, grain weight, and source-sink relationship) will be discussed.

PLANT BREEDING AND GENETIC GAINS IN GRAIN YIELD

Reports showing genetic gains in grain yield potential during this century have been published for different countries such as Argentina (Slafer and Andrade, 1989; Calderini, Dreccer, and Slafer, 1995a), Australia (Perry and D'Antuono, 1989; Siddique, Kirby, and Perry, 1989; Siddique et al., 1989), Canada (Hucl and Baker, 1987), Germany (Feil and Geisler, 1988), India (Sinha et al., 1981; Kulshrestha and Jain, 1982), Italy (Canevara et al., 1994), Mexico (Waddington et al., 1986; Sayre, 1996), New Zealand (McEwan and Cross, 1979), Sweden (Ledent and Stoy, 1988), the United Kingdom (Austin et al., 1980; Austin, Ford, and Morgan, 1989), and the United States (Deckerd, Busch, and Kofoid, 1985; Cox et al., 1988). Most of the studies, evaluating cultivars released at different times under field conditions in which lodging was prevented, revealed that wheat breeding played an important role in the increase of farm yields (Feil, 1992; Loss and Siddique, 1994; Slafer, Satorre, and Andrade, 1994). This genetic contribution to the total gains in wheat yield has been assessed to be ca. 30 percent for Mexico (Bell et al., 1995) and 50 percent worldwide (see Slafer et al., 1994b).

Most results agree that the genetic increase in yield was not linear through the century (Slafer, Satorre, and Andrade, 1994), being generally higher during the second than during the first half of this period. Some authors have even shown that during the 1980s, breeders have been more successful in increasing wheat yield potential than during the previous decade (Austin, Ford, and Morgan, 1989; Calderini, Dreccer, and Slafer, 1995). It could be speculated that during the first part of the century traits other than grain yield per se (e.g., lodging and disease resistance, protein concentration) were prioritized, and that the "critical mass of knowledge" required for supporting a highly efficient breeding program was not reached until the end of the first half of the century. Anyway, it is remarkable that a similar pattern was evident for most plant breeding programs, despite differences in their environments and other economic and political contrasts among countries.

The magnitude of the genetic gains shows differences between countries. Among the extreme values are Mexico, with ca. 71 kg/ha/year and Australia, with ca. 6 kg/ha/year (see Table 16.1). But more useful than these absolute estimates are the relative genetic gains (i.e., grain yield increases relative to the average grain yield of the experiment; Slafer and Andrade, 1991) because the measurement of the impact of genetic improvement on yield (and other characteristics) is affected by the environmental condition of the experiment (see Figure 16.1). Analyzed in relative terms, while the majority of the countries reached values between 0.35 and 0.55 percent/year, very important progress has been reported for Mexico, New Zealand, the latest study of Argentina, and one of the studies in India (ca. 1 percent/year, Table 16.1). In the case of Argentine cultivars, the

TABLE 16.1. Genetic Gain, Average Grain Yield, and Relative Genetic Gain (Genetic Gain to Average Grain Yield Ratio, as Percentage) for Different Countries

Country	Period	Genetic gain (g/m^2/year)	Average yield (g/m^2)	Relative genetic gain (%/year)	Source
Argentina	1912-1980	1.21*	299.83	0.40	Slafer and Andrade, 1989
	1920-1990	5.02***	526.35	0.96	Calderini, Dreccer, and Slafer, 1995
Australia	1860-1982	0.53***	133.63	0.39	Perry and D'Antuono, 1989
	1860-1986	0.61***	169.53	0.36	Siddique, Kirby, and Perry, 1989
Canada	1882-1985	0.50**	285.92	0.50	Hucl and Baker, 1987
India	1901-1980	0.58 (NS)	362.50	0.16	Sinha et al., 1981
	1910-1980	3.82***	361.63	1.05	Kulshrestha and Jain, 1982
Italy	1900-1983	2.94**	417.67	0.50	Canevara et al., 1994
Mexico	1950-1982	5.84**	648.89	0.90	Waddington et al., 1986
	1962-1988	7.09****	803.28	0.88	Sayre, 1996
New Zealand	1935-1973	4.38***	398.18	1.10	McEwan and Cross, 1979
Sweden	1900-1976	1.51**	686.36	0.22	Ledent and Stoy, 1988
UK	1908-1978	2.59**	607.87	0.44	Austin et al., 1980
	1830-1986	2.33**	653.00	0.39	Austin, Ford, and Morgan, 1989
USA	1911-1978	1.04[1]	196.23	0.53	Deckerd, Busch, and Kofoid, 1985
	1874-1987	1.62***	225.00	0.72	Cox et al., 1988

Note: NS, *, **, ***, and **** (no significance, $p \leq 0.10$; $p \leq 0.05$; $p \leq 0.01$ and $p \leq 0.001$, respectively) is the probability at which the regression model fit the data.

[1]Authors provided the genetic gain but not the significance of the relationship.

release of very high-yield potential cultivars (including a commercial hybrid) by the 1990s was suggested as being the cause of the improvement in relative grain yield gain. On the other hand, the introduction of dwarfing genes in India, after a long period in which grain yield was almost unaffected by breeders (Slafer and Andrade, 1991), would be the principal reason for improved performance. The experiments in New Zealand and Mexico did not include cultivars released before 1935 and as the first third

FIGURE 16.1. Relationship Between Genetic Gain in Grain Yield and Average Grain Yield in Each Experiment

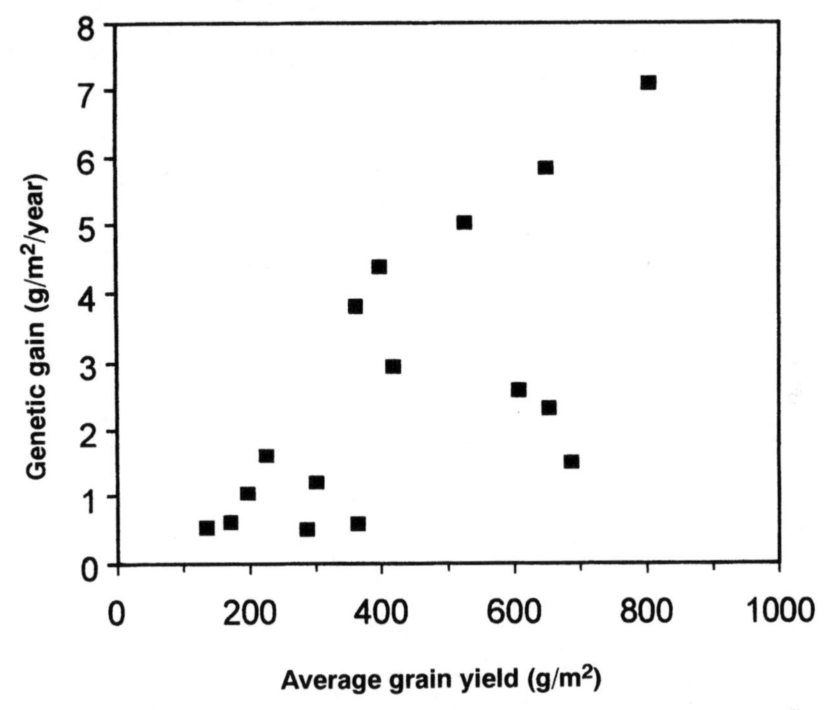

Note: Data from experiments carried out in Argentina, Australia, Canada, India, Italy, Mexico, New Zealand, UK, and USA (see Table 16.1).

of this century represents the less efficient era of plant breeding (see above) these studies are not directly comparable with those using cultivars released very early in this century (or even during the previous century). In the case of Mexico, however, the impact of an international center for crop improvement, namely CIMMYT, most probably contributed significantly to high genetic gains.

CONSEQUENCES OF BREEDING FOR PLANT HEIGHT AND RELATED TRAITS

Results from virtually all countries show that breeding programs produced a clear modification in wheat height. Most modern cultivars are significantly shorter than older cultivars. This has been shown to be the consequence of a continuous reduction in height during the twentieth century. Particular examples of this systematic change can be found for Argentina (Slafer and Andrade, 1989; Calderini, Dreccer, and Slafer, 1995), Australia (Siddique et al., 1989), Italy (Canevara et al., 1994), the United Kingdom (Austin et al., 1980; Austin, Ford, and Morgan, 1989) and the United States (Cox et al., 1988). To illustrate to what extent plant height was modified by breeding, the cases of Argentina (see Figure 16.2a), the United Kingom (Figure 16.2b), and Italy (Figure 16.2c) are shown. These results show a continuous decrease in plant height, which

FIGURE 16.2. Relationship Between Plant Height and Year of Release of Cultivars in Different Countries

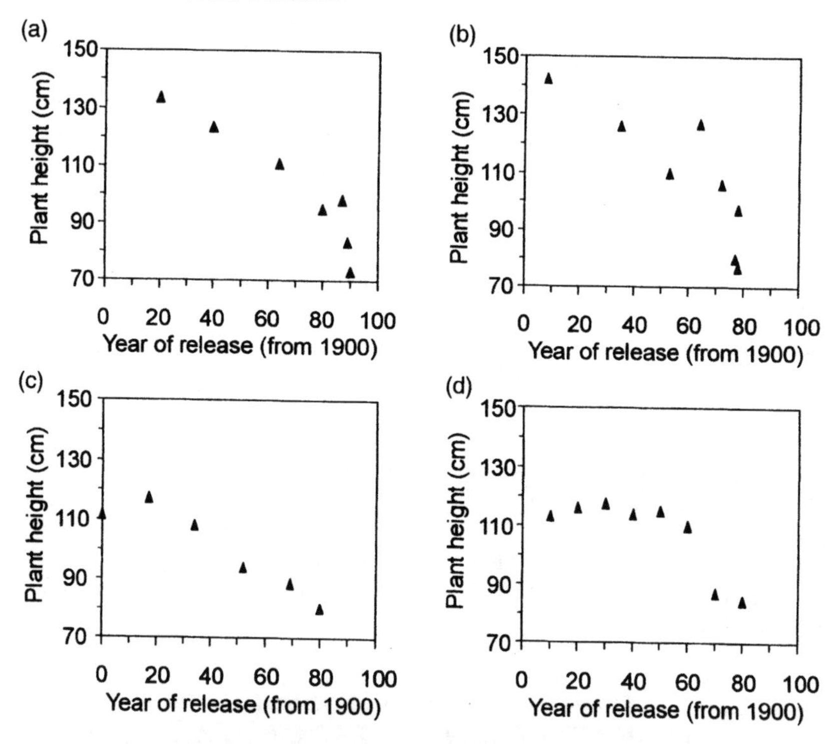

Note: (a) Argentina (Calderini, Dreccer, and Slafer, 1995a); (b) UK (Austin et al., 1980); (c) Italy (Canevara et al., 1994); and (d) India (Kulshrestha and Jain, 1982).

may have resulted from a deliberate selection for lodging resistance, which is known to be positively associated with shorter stems. Stapper and Fischer (1990) estimated that as much as 40 percent of yield can be lost due to lodging during grain filling. In addition, these results confirm a previous suggestion (Calderini, Dreccer, and Slafer, 1995) that for some countries the introgression of major genes for plant height (Rht) represented one more step in the continuos effort for reducing plant height from at least the beginning of the twentieth century.

On the other hand, the study of the effect of breeding on plant height in India showed an exceptional stepwise change, indicating that the reduction in plant height was strongly dependent on the introduction of semidwarfing genes (Figure 16.2d). In addition, it was suggested by Kulshrestha and Jain (1982) that the limited modifications produced in both plant height and grain yield in India until the introduction of cultivars with dwarfing genes could be attributed to the limited genetic variability in wheat available for Indian breeders.

Therefore, grain yield and plant height have been modified by wheat breeders in opposite directions, at least within the range of height shown in Figure 16.2 (70 to 150 cm). The examples analyzed then showed a significant negative relationship between these traits ($r = 0.79$, $p < 0.05$; $r = 0.96$, $p < 0.01$; and $r = 0.81$, $p < 0.05$; for Argentina, Italy and the United Kingdom, respectively). These negative associations agreed with results reported previously for other countries (see Slafer, Satorre, and Andrade, 1994).

PLANT BREEDING AND BIOMASS: EFFECTS ON PARTITIONING AND GROWTH

Trends in Aboveground Biomass and Harvest Index

Analyses of physiological bases for the increase in grain yield showed similarities for most countries in which these studies were conducted. First of all, final biomass was not associated with the year of release of the cultivars (Slafer, Satorre, and Andrade, 1994) and consequently no consistent relationships between grain yield and biomass were found (Figure 16.3). Among all the studies reviewed in this chapter, very few exceptions to this general pattern were found and most of them showed that the increase in biomass was responsible for less than 20 percent of the genetic gain in grain yield, e.g., varieties released by CIMMYT in Mexico (Waddington et al., 1986). The most important increase in biomass was reached

FIGURE 16.3. Relationship Between Grain Yield and Aboveground Biomass in Different Countries

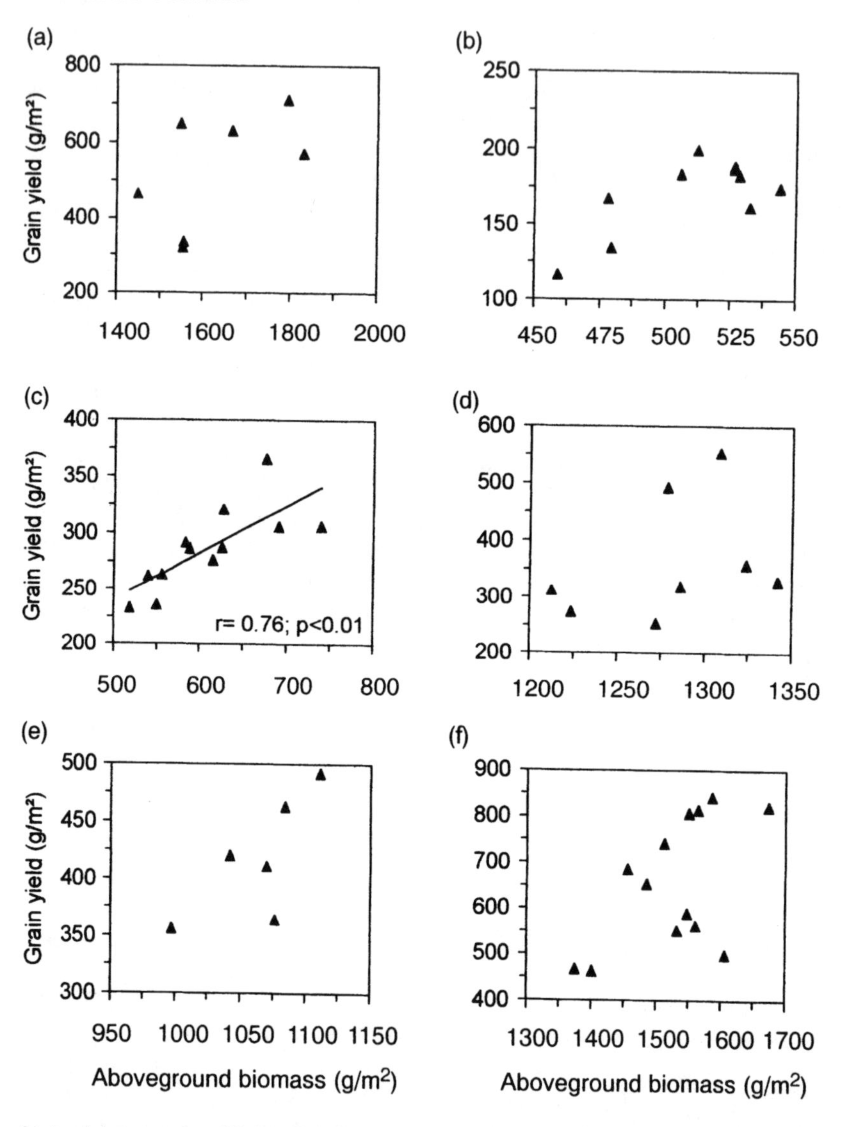

Note: (a) Argentina (Calderini, Dreccer, and Slafer, 1995); (b) Australia (Siddique, Kirby, and Perry, 1989b); (c) Canada (Hucl and Baker, 1987); (d) India (Kulshrestha and Jain, 1982); (e) Italy (Canevara et al., 1994); and (f) UK (Austin, Ford, and Morgan, 1989).

by plant breeding in Canada, where a significant relationship has also been found between grain yield and final aboveground biomass and this genetic gain in biomass has accounted for most of the genetic gain in grain yield (Figure 16.3c). But, independently of particular cases, the increase in grain yield found in the majority of the countries was almost entirely supported by modifications in harvest index (Figure 16.4, and see also reviews by Feil, 1992; Evans, 1993; Loss and Siddique, 1994; Slafer, Satorre, and Andrade, 1994). These modifications in harvest index have probably evolved from reductions in plant height (see above). Most of the research on growth of different organs showed that modern cultivars have higher harvest indexes than their predecessors in accordance with the reduction in culm growth (see Loss and Siddique, 1994; Slafer, Satorre, and Andrade, 1994). Strongly supporting this view, a negative relationship was found between plant height and the biomass allocated to stems in modern cultivars (Austin, Ford, and Morgan, 1989; Slafer and Andrade, 1989; Calderini, Dreccer, and Slafer, 1995). For example, the genetic loss in stem biomass in Argentina was 42 kg/ha/year while the genetic gain in grain yield was 50 kg/ha/year (Calderini, Dreccer, and Slafer, 1995).

The fact that reducing plant height was the main way to increase yields in the past does not imply that we can go further in the same direction. Although this strategy has been extremely successful, it may be expected that further reductions would cause negative consequences in grain yield. For example, an inspection of the relationship between harvest index and the grain yield-to-plant height ratio (Figure 16.5) reveals that further gains at the expense of higher harvest indexes will not be easily obtained by reductions in plant height associated with higher yields. This relationship is curvilinear for most countries, implying that harvest index will hardly be higher than 50 percent, which in fact is quite close to its theoretical upper limit (ca. 60 percent as estimated by Austin et al., 1980). Besides, plant height in modern cultivars ranges between 70 and 100 cm, within which grain yield is optimized (Richards, 1992; Miralles and Slafer, 1995a). The logic behind this optimal height is that further reductions would strongly affect the radiation use efficiency (RUE) of the crop (due to very poor light distribution within the canopy), and taller plants would reduce the partitioning to reproductive sinks.

It is clear that any future quantum leap in grain yield will not come through increases in harvest index simply by reductions in plant height. Therefore, progress would increasingly depend on increasing biomass production while maintaining high values of biomass partitioning (Austin et al., 1980; Kulshrestha and Jain, 1982; Slafer and Andrade, 1991). In this context, the understanding of any changes made by past breeding on

FIGURE 16.4. Relationship Between Grain Yield and Harvest Index in Different Countries

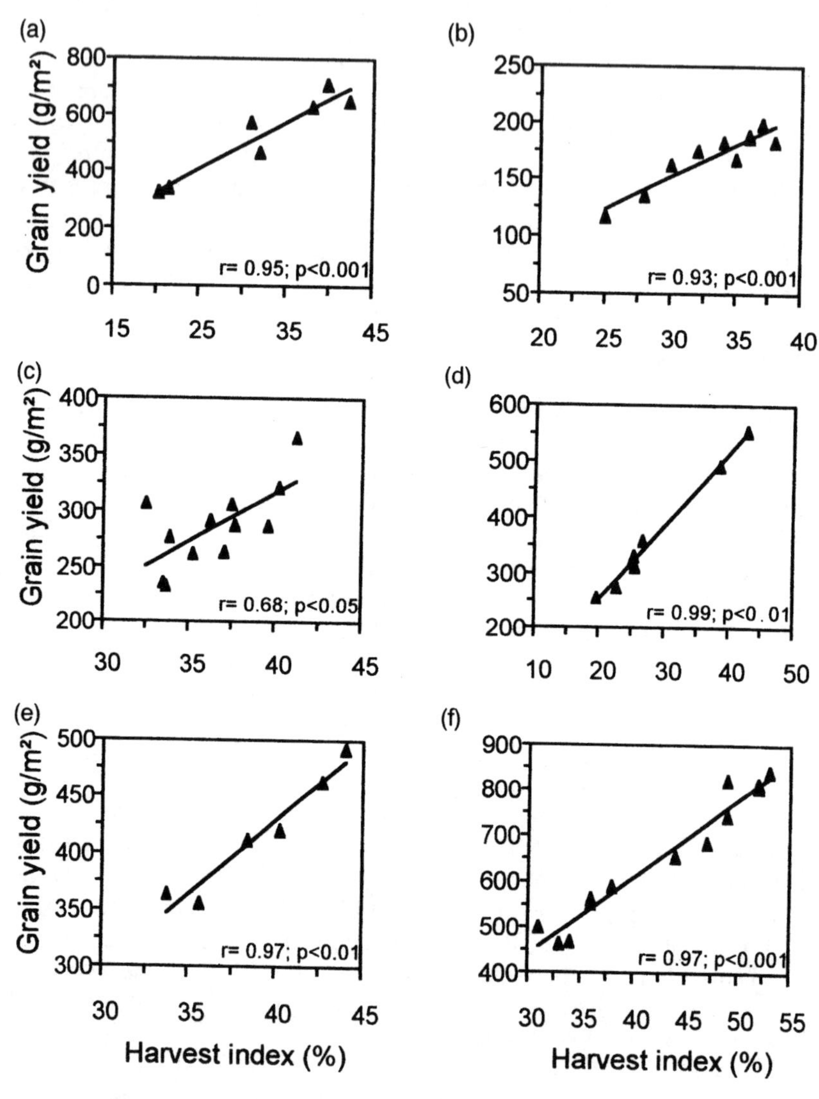

Note: (a) Argentina (Calderini, Dreccer, and Slafer, 1995); (b) Australia (Siddique, Kirby, and Perry, 1989b); (c) Canada (Hucl and Baker, 1987); (d) India (Kulshrestha and Jain, 1982); (e) Italy (Canevara et al., 1994); and (f) UK (Austin, Ford, and Morgan, 1989).

FIGURE 16.5. Relationship Between Harvest Index and Grain Yield-to-Plant Height Ratio in Different Countries

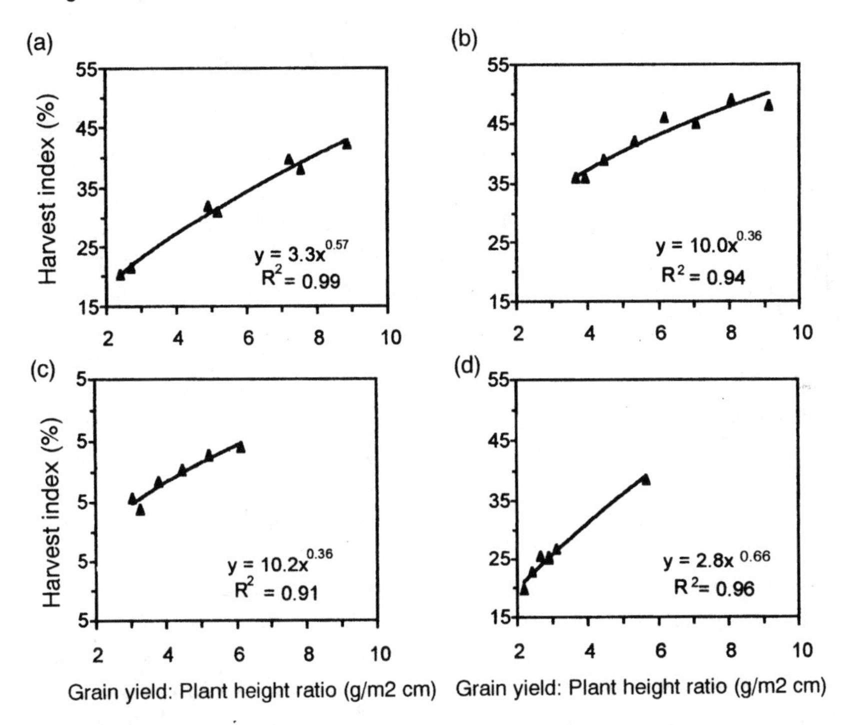

Note: (a) Argentina (Calderini, Dreccer, and Slafer, 1995a); (b) UK (Austin, Ford, and Morgan, 1989); (c) Italy (Canevara et al., 1994); and (d) India (Kulshrestha and Jain, 1982).

physiological components of crop growth is relevant. Although final biomass was negligibly affected by wheat breeders in the past in most cases, the analysis of any effects on its physiological determinants could provide ideas for suggesting alternatives for increasing biomass in future breeding programs.

Physiological Components of Biomass Growth

Results showing that aboveground biomass has not been systematically modified by wheat breeding programs (see Slafer, Satorre, and Andrade, 1994) imply that either the amount of radiation intercepted by the crop and the radiation use efficiency were kept invariable or that these variables were changed in opposite ways. Even if these traits were not systematically changed by breeders, they may become physiological targets for future breeding as genotypic variation in the ability of the canopy to both intercept radiation and convert that energy into new biomass has been documented (e.g., Green, 1989).

As the principal organs for interception and use of radiation are leaf laminae, the leaf area index (LAI) could be an initial trait to be analyzed. Leaf area index of old and modern wheats in the United Kingdom (Austin et al., 1980), the United States (Deckerd, Busch, and Kofoid, 1985), Germany (Feil and Geisler, 1988), and Argentina (Calderini, Dreccer, and Slafer, 1997) were not consistently different. Similarly, Canevara et al. (1994) found only a weak trend toward greater maximum LAI with the year of release of cultivars in Italy. On the other hand, it was reported that in Australia the most recently released cultivars exhibited smaller LAI than the older cultivars during the period between terminal spikelet and booting (Siddique et al., 1989; Yunusa et al., 1993). However, in the Australian studies green area measured included stems, which are more abundant (because of more tillering) and longer in old than in modern cultivars.

When radiation interception is analyzed during the period between seedling emergence and anthesis, two types of results have been found. One of them showed that this trait was not modified by past breeding and that modern and old cultivars are similarly efficient in intercepting the incoming radiation. Examples of this behavior are published for the United States (Deckerd, Busch, and Kofoid, 1985) and Argentina (Figure 16.6a; Slafer, Andrade, and Satorre, 1990; Calderini, Dreccer, and Slafer, 1997). On the other hand, Siddique, Kirby, and Perry (1989) and Yunusa et al. (1993) showed that the most modern Australian cultivars intercepted less radiation through the preanthesis period than old cultivars (Figure 16.6b). An important aspect of the experiments with Australian cultivars was that the cultivars selected to represent particular eras of wheat breeding differed markedly in their rates of development. Therefore, differences in accumulated radiation interception found in Australia were confounded by the length of the sowing-anthesis period. In addition, data of light extinction coefficients (k) calculated for the whole preanthesis period did not show differences among cultivars in Australia (Yunusa et al., 1993) nor in Argentina (Calderini, Dreccer, and Slafer, 1997); however, k values for Australian cultivars were higher than in Argentinian cultivars (ca. 0.7 and 0.4 respectively). The difference was probably attributable to the fact

FIGURE 16.6. Relationship Between Accumulated Radiation Intercepted from Seedling Emergence (SE) to Anthesis and the Year of Release of Cultivars in Argentina (a) and Australia (b), and the Association Between Radiation Use Efficiency and the Year of Release of Cultivars (c)

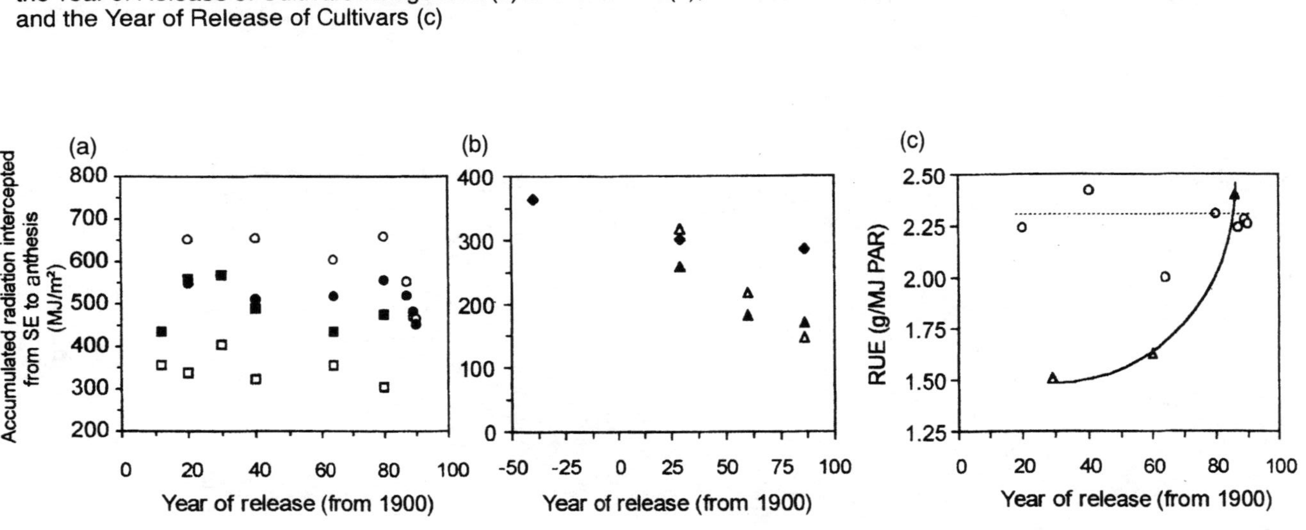

Note: In (a) data from experiments carried out during 1986 (open squares), 1987 (closed squares), 1991 (open circles), and 1992 (closed circles). In (b) data from experiments carried out in 1988 (diamonds), 1990 (open triangles), and 1991 (closed triangles). In (c) data of Argentina are the average of 1991 and 1992 experiments (circles) and data from Australia are the average of 1990 and 1991 experiments (triangles).

that in the Australian study, green area index was used instead of LAI. In addition, it is likely that in a Mediterranean climate, such as that of Western Australia where available water becomes increasingly scarce with time, the better-adapted cultivars were those that cover the soil faster.

As total biomass was generally unchanged, the effect of wheat breeding on RUE during the preanthesis period was positive in Australia (Siddique et al., 1989; Yunusa et al., 1993) and neutral in Argentina (Slafer, Andrade, and Satorre, 1990; Calderini, Dreccer, and Slafer, 1997). One of the possible explanations for the difference found in Australian cultivars has already been suggested by Australian researchers in relation to differences in root biomass, i.e., "efficiency of conversion to total dry matter may show smaller differences between modern and old varieties because of the higher root dry matter of the older varieties" (Siddique, Belford, and Tennant, 1990, p. 96). Another possible explanation for the conflicting results could be the different starting points. It appears that the RUE of old cultivars in Australia was quite low compared to that of old Argentine cultivars, while the most recently released cultivars from both countries possess similar efficiencies (Figure 16.6c). Loomis and Amthor (1996) have recently estimated "a practical potential radiation use efficiency" for wheat of about 3.8 g/MJ PAR (photosynthetically active radiation), which is substantially higher than that found for modern cultivars and suggests that RUE is still a character with potential for genetic improvement.

Bearing in mind the limited information available on radiation interception and RUE during the postanthesis period in cultivars released in different eras (as far as we are aware there is only one experiment), it appears that older cultivars suffered an important decrease in growth rate and RUE after anthesis, while the most recently released cultivars maintain values similar to those of the preanthesis period (Calderini, Dreccer, and Slafer, 1997). The higher RUE during the grain-filling period for more modern cultivars has been suggested previously by Fischer (1984). In addition, Miralles and Slafer (1997), working with Rht isogenic lines of wheat, found higher RUE in a semidwarf than in the tall line and when the number of grains set at anthesis increased, the difference in radiation use efficiency between the pre- and postanthesis periods decreased. This reinforces the idea that at the postanthesis period "crops can respond to an extra demand for photosynthate" (Richards, 1996, p. 142).

Root Biomass

The root system of the crops has been ironically termed "the hidden half" (Waisel and Kafkafi, 1991), emphasizing that it has received far less attention by researchers. This is particularly true for field experiments analyzing the effect of plant breeding. In an initial attempt to evaluate the effect of breeding on root parameters, Siddique, Belford, and Tennant (1990) measured root dry matter of old and modern Australian wheats. Their hypothesis was that harvest index in modern cultivars has been associated with reduced investment in the root system and thus they would possess a lower root-to-shoot ratio during early stages of development. The hypothesis is based on a suggestion by Passioura (1983), who argued that the root system is unnecessarily large and with less investment in the root, more assimilates could be available for shoot and probably for harvested organs. The results obtained by Siddique, Belford, and Tennant (1990) showed that at anthesis the most modern cultivar (Kulin) had lower root biomass than older cultivars (Gamenya and Purple Straw) while their aboveground biomasses were similar. We recently conducted an experiment in Argentina with similar objectives (unpublished data) but did not find differences between old and modern cultivars in root biomass at anthesis.

We clearly need much more information before any conclusion can be reached on the role played by wheat breeders on root growth and particularly before we can speculate about the best way to proceed in the future.

GENETIC IMPROVEMENTS IN NITROGEN AND PHOSPHORUS ECONOMIES

It is widely recognized that major nutrients such as nitrogen and phosphorus directly affect both the rate of crop growth and the quality of grains. However, few studies have attempted to determine the impact of wheat breeding on the economy (mainly absorption and partitioning) of these nutrients.

Several experiments have analyzed the relationships between wheat breeding and nitrogen economy. Although Austin et al. (1980) showed a trend for slightly higher nitrogen uptake in modern cultivars (in only one of the two sites evaluated), most of the studies concluded that wheat breeding did not consistently modify the amount of nitrogen absorbed by the crop (Fischer and Wall, 1976; Paccaud, Fossat, and Cao, 1985; Feil and Geisler, 1988; Slafer, Andrade, and Feingold, 1990; Calderini, Torres León, and Slafer, 1995). Therefore, as modern cultivars outyield their old counterparts, plant breeding increased the nitrogen use efficiency, i.e., the yield produced per unit of nitrogen absorbed.

Despite this lack of genetic improvement effect on nitrogen accumulation, it is important to recognize that genetic variation in total nitrogen uptake has been found in wheat by different authors (Austin et al., 1977;

Cox, Qualset, and Rains, 1985; Heitholt et al., 1990). In addition, studies recently carried out on a historic series of wheat lines released by CIMMYT between 1950 and 1985 (Ortiz Monasterio et al., 1997) show that total nitrogen uptake has increased, suggesting genetic improvement in nitrogen uptake efficiency. Therefore, future breeding programs could exploit this variability to increase the biomass production in new cultivars.

Although nitrogen and phosphorus have substantially different soil dynamics, similar results have been obtained for them; i.e., total nitrogen and phosphorus absorption was similar for all cultivars both at anthesis (Calderini, Torres León, and Slafer, 1995) and at maturity (Figure 16.7a). The lack of trends in aboveground biomass with the year of release agreed with the absence of systematic changes in the ability of new and old cultivars to uptake these macronutrients. However, both nitrogen (see reviews of Feil, 1992 and Slafer, Satorre, and Andrade, 1994) and phosphorus (Calderini, Torres León, and Slafer, 1995) yields (their amounts harvested with the grains) have increased in more modern cultivars, indicating progress in nitrogen and phosphorus harvest indexes (Figure 16.7b). In the future, variability in nitrogen uptake should be exploited as modern cultivars already exhibit high values of nitrogen and phosphorus harvest index. Further increase in partitioning is likely to be difficult since nutrients are used in structural components of vegetative organs that cannot be reallocated to grains during grain filling. Cregan and Van Berkum (1984) indicated that it could be possible to simultaneously increase nitrogen uptake and partition-

FIGURE 16.7. Relationship Between Total Nitrogen (N) Absorbed (Closed Squares) and Total Phosphorus (P) Absorbed (Open Squares) with the Year of Release of Cultivars (a) and Nitrogen Harvest Index (Closed Squares) and Phosphorus Harvest Index (Open Squares) with the Year of Release of Cultivars in Argentina (b)

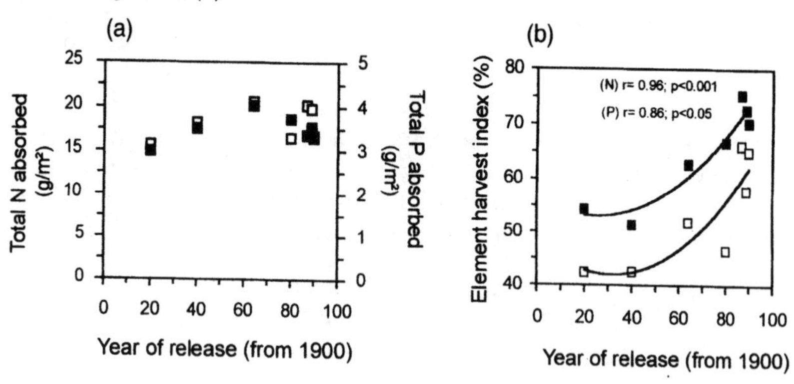

ing due to their physiological independence. The effect of breeding on the partitioning of these nutrients was probably a function of changes in biomass partitioning, and both nitrogen and phosphorus harvest index were significantly associated with harvest index ($r = 0.95$, $p < 0.01$ and $r = 0.93$, $p < 0.01$, respectively; Calderini, Torres León, and Slafer, 1995).

Despite the improvement in nitrogen and phosphorus partitioning, the concentration of both nutrients in the grains tended to decrease with the release of newer cultivars. Grain nitrogen (Austin et al., 1980; Paccaud, Fossati, and Cao, 1985; Feil and Geisler, 1988; Slafer, Andrade, and Feingold, 1990; Canevara et al., 1994; Calderini, Torres León, and Slafer, 1995) and phosphorus (Calderini, Torres León, and Slafer, 1995) concentrations were found to be negatively associated with grain yield. The negative trend for the concentration of these nutrients in the grains, despite the net increase due to increased yields, highlights the dilution effect of increases in grain yield (Figure 16.8). This, in turn, illustrates the principal factor for which breeders have selected, i.e., yield rather than quality. Then the relationship between the ratio of nitrogen or phosphorus harvest index to harvest index (NHI/HI or PHI/HI) and the year of release of the cultivars showed a significant negative slope (Slafer, Satorre, and Andrade, 1994; Calderini, Torres León, and Slafer, 1995). Despite the actual reduction in grain nitro-

FIGURE 16.8. Relationship Between Grain Yield and Grain Nitrogen Yield (a) and Grain Phosphorus Yield (b) in Seven Cultivars Released in Argentina at Different Eras

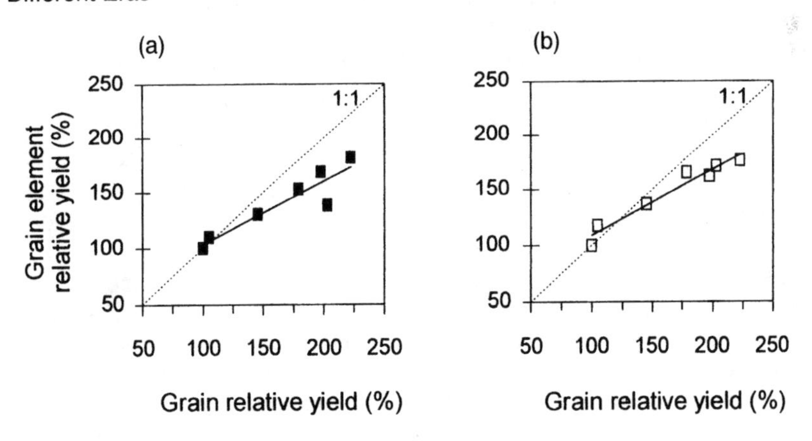

Note: The variables are presented as the ratio between the value observed for each cultivar and that of the oldest cultivar (released in 1920).

gen concentration, it must be highlighted that this is a major, but not the unique, measure of grain quality and that quality of wheat grains may be increased by selecting for genotypes with better protein composition (Canevara et al., 1994).

NUMERICAL YIELD COMPONENTS IN OLD AND MODERN CULTIVARS

Number of Grains

Considering numerical components of grain, the increase in grain number per unit area has been far more important than changes in the other component (Feil, 1992; Loss and Siddique, 1994; Slafer, Satorre, and Andrade, 1994). Therefore, modern cultivars reached higher grain yields mainly because they were able to set a higher number of grains per unit area (Figure 16.9).

In general, but not always, number of grains per m^2 was positively correlated with grain number per spike in studies comparing cultivars released at different eras (e.g., Waddington et al., 1986; Perry and D'Antuono, 1989; Siddique, Kirby, and Perry, 1989; Slafer and Andrade, 1989, 1993). The potential number of grains per spike is, in turn, determined during stem elongation when a relatively small proportion of the floret primordia survives to produce fertile florets at anthesis (Kirby, 1988), most of which set grains thereafter. Only a few studies analyzed the fate of florets in historical series of cultivars. Siddique, Kirby, and Perry (1989) found that modern varieties initiated more floret primordia and possessed more fertile florets per spikelet (only a central spikelet was analyzed), but the higher survival of floret primordia appeared to have played the most important role in differentiating modern and old cultivars (Slafer, Satorre, and Andrade, 1994). Slafer and Andrade (1993) have also observed that the survival of floret primordia (in all the spikelets) immediately before anthesis was a key factor differentiating modern and old Argentine cultivars.

Using an assimilate-based approach to understanding the generation of grain number, such as that proposed by Fischer (1984), it may be expected that the higher number of grains per m^2 of modern wheats was achieved in association with genetic improvement in biomass partitioning to reproductive organs before anthesis. In fact, most modern cultivars showed higher spike dry matter at anthesis than the older cultivars (Siddique, Kirby, and Perry, 1989; Slafer et al., 1990a), although differences in aboveground biomass were not significant. Siddique, Kirby, and Perry (1989) and Slafer

FIGURE 16.9. Relationship Between Grain Yield and Number of Grains per m^2 for Cultivars Released at Different Eras

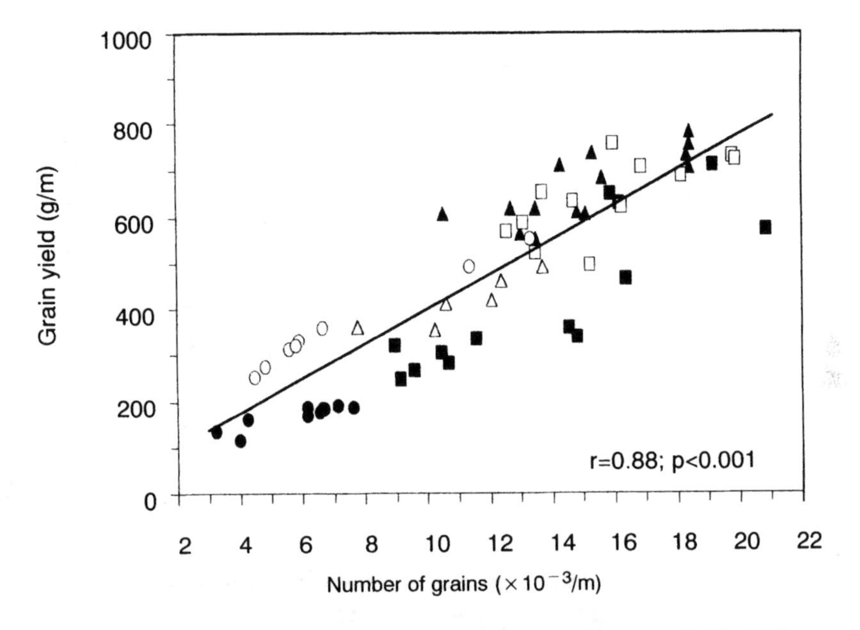

Sources for data from different countries: Argentina (closed squares, Calderini, Dreccer, and Slafer, 1995), Australia (closed circles, Siddique, Kirby, and Perry, 1989), India (open circles, Kulshrestha and Jain, 1982), Italy (open triangles, Canevara et al., 1994), Mexico (closed triangles, Waddington et al., 1986), and UK (open squares, Austin, Ford, and Morgan, 1989).

and Andrade (1993) demonstrated that the heavier spikes of modern cultivars at anthesis, compared with those of old cultivars, was a reflection of their faster rates of dry matter accumulation immediately before anthesis, exclusively due to a more favorable partitioning toward spikes (Figure 16.10).

Individual Grain Weight

Contrasting with what has been described for the number of grains per m^2, and in agreement with studies indicating that individual grain weight is a more conservative attribute than grain number (Fischer, 1985; Savin and Slafer, 1991), it seems that this other major yield component has not

FIGURE 16.10. Relationship Between Spike Dry Weight and Spike-to-Stem Ratio in Cultivars Released at Different Eras

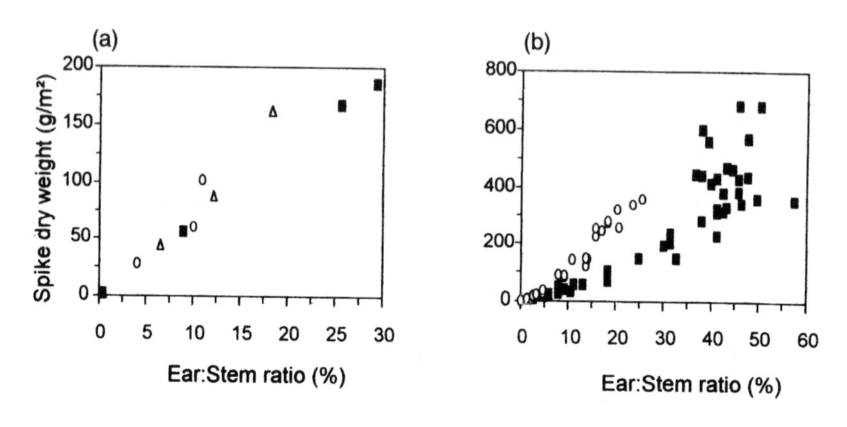

Note: (a) Argentina (1920, circles; 1940, triangles; and 1980, squares) and (b) Australia (1860, circles and 1986, squares).

been markedly modified by wheat breeding (Austin et al., 1980; Waddington et al., 1986) during the twentieth century. Some results even showed that individual grain weight was reduced by genetic improvements (Slafer and Andrade, 1989; Loss et al., 1989). Only few cases have shown some increases in this trait during this century (Cox et al., 1988) or part of it (Calderini, Dreccer, and Slafer, 1995). Therefore, not only was grain weight much less affected than grain number by wheat breeding, but also these small changes were not consistent among studies.

These findings accord closely with evidence from physiological studies of limitation in grain yield potential in wheat. It has been established from these studies in which plants have been stressed (or given more resources) at different stages of crop development that most of the changes in yield were due to changes in the number of grains per m^2 (e.g., Kemp and Whingwiri, 1980; Fischer and Stockman, 1980; Fischer, 1985; Thorne and Wood, 1987; Savin and Slafer, 1991; Magrin et al., 1993). Even when shading, thinning, or defoliations were imposed during grain filling, the impact on grain yield through changes in individual grain weight was small or negligible (e.g., Martinez-Carrasco and Thorne, 1979; Borghi et al., 1986; Mac Maney et al., 1986; Grabau, Van Sanford, and Meng, 1990; Jedel and Hunt, 1990; Savin and Slafer, 1991; Slafer and Miralles, 1992; Nicolas and Turner, 1993; Slafer and Savin, 1994).

The simplest conclusion from these analyses would be that wheat yield is sink limited during grain filling (source would then be in excess for the demands of growing grains), and that genetic improvements in wheat yield have come about largely due to increased sink strength through a higher grain number. There is only a single point that could be interpreted as conflicting with this conclusion: the average individual grain weight is frequently negatively related to the number of grains set per m^2 (Slafer, Calderini, and Miralles, 1996). In fact, this negative relationship could suggest that future increases in number of grains per m^2 might be impeded due to compensating reductions in individual grain weight. The question to answer would then be whether the negative relationship between grain number and grain weight represents an increasing degree of source limitation during grain filling. Only manipulating the source-sink ratio of cultivars released at different eras would provide a proper response, as the negative relationship between major yield components may be noncompetitive in nature (see Miralles and Slafer, 1995b; Slafer, Calderini, and Miralles, 1996).

Source-Sink Relationships

Although there are some exceptions, modern cultivars were frequently reported to possess lower average individual grain weight than their predecessors (Slafer and Andrade, 1989; Siddique, Kirby, and Perry, 1989); negative relationships between these components were commonly reported when cultivars released at different eras are compared (e.g., Waddington et al., 1986; Perry and D'Antuono, 1989; Siddique et al., 1989; Slafer and Andrade, 1989, 1993). However, when the weight of basal grains of the central spikelets was separately evaluated for these cultivars, it was not related to their number of grains per m^2, nor was this individual weight related to their year of release (Loss et al., 1989; Slafer and Miralles, 1993). This suggests that the relatively lower average individual grain weight in modern than in old cultivars was not necessarily a result of stronger competition among the larger number of grains, but that the relative contribution of small apical grains is higher as the number of grains is increased.

Results from studies in which the source-sink ratio had been modified in cultivars released at different eras are scarce, but they constitute the fairest experimental approach to determining the degree of source limitation, if any, modern cultivars may have. To the best of our knowledge, only Koshkin and Tararina (1989) and Kruk, Calderini, and Slafer (1997) analyzed the impact of source-sink manipulations in a set of cultivars released at different eras. Both studies clearly showed that the most modern cultivars were more responsive to source limitation than older cultivars. However, the magnitude of these responses was quite a bit smaller than the manipulation of the source-sink relationship.

The conclusion is that the negative relationship between grain number and grain weight is, in general, mostly due to noncompetitive factors. Thus, it may well still be possible to further increase wheat yields by increasing the capacity of its reproductive sinks; but undoubtedly modern cultivars are exhibiting a growing, though still small, degree of source limitation to completely fill the grains. Therefore, as suggested earlier by Slafer and Savin (1994), modern wheats are likely simultaneously limited by both source and sink strengths.

CONCLUSION

This chapter has addressed the issue of physiological aspects associated with genetic increases in wheat yield over the century. The dominant factor in all environments has been the success of plant breeding for increasing grain yield potential. This achievement was reached, almost exclusively, by changes in biomass partitioning (harvest index). The increase of harvest index was a consequence of the decrease in plant height. Although height reduction was evident throughout the century, the use of major genes such as Rht alleles made it easier to increase the efficiency for simultaneously reducing height and increasing partitioning to reproductive structures.

Higher grain yields and harvest indexes were the consequence of higher grain number per square meter of modern cultivars. This was the grain yield trait that plant breeding consistently modified during the present century while grain weight did not show a clear contribution to grain yield gains worldwide. The increase in biomass partitioning toward spikes during the short period of ca. 20 days immediately previous to anthesis was the physiological change that permitted a higher number of grains per square meter.

Present values of harvest index and plant height are close to their optimum, which suggests that successful traits used in the past will likely not be useful in the future of breeding as we search for higher grain yield potential.

REFERENCES

Austin, R.B., Bingham, J., Blackwell, R.D., Evans, L.T., Ford, M.A., Morgan, C.L., and Taylor, M. (1980). Genetic improvement in winter wheat yield since 1900 and associated physiological changes. *J. Agric. Sci. Camb.*, 94: 675-689.

Austin, R.B., Ford, M.A., Edrich, J.A., and Blackwell, R.D. (1977). The nitrogen economy of winter wheat. *J. Agric. Sci. Camb.,* 88: 159-167.

Austin, R.B., Ford, M.A., and Morgan, C.L. (1989). Genetic improvement in the yield of winter wheat: A further evaluation. *J. Agric. Sci. Camb.,* 112: 295-301.

Bell, M.A., Fischer, R.A., Byerlee, D., and Sayre, K. (1995). Genetic and agronomic contributions to yield gains: A case study for wheat. *Field Crops Res.,* 44: 55-65.

Borghi, B., Corbellini, M., Cattaneo, M., Fornasari, M.A., and Zucchelli, L. (1986). Modification of the sink/source relationship in bread wheat and its influence on grain yield and protein content. *J. Agron. Crop Sci.,* 157: 245-254.

Calderini, D.F., Dreccer, M.F., and Slafer, G.A. (1995). Genetic improvement in wheat yield and associated traits. A re-examination of previous results and the latest trends. *Plant Breed.,* 114: 108-112.

Calderini, D.F., Dreccer, M.F., and Slafer, G.A. (1997). Consequences of plant breeding on biomass growth, radiation interception and radiation use efficiency in wheat. *Field Crops Res.,* 52: 271-281.

Calderini, D.F., Torres León, S., and Slafer, G.A. (1995). Consequences of wheat breeding on nitrogen and phosphorus yield, grain nitrogen and phosphorus concentration and associated traits. *Ann. Bot.,* 76: 315-322.

Canevara, M.G., Romani, M., Corbellini, M., Perenzin, M., and Borghi, B. (1994). Evolutionary trends in morphological, physiological, agronomical and qualitative traits of *Triticum aestivum* L. cultivars bred in Italy since 1900. *Eur. J. Agron.,* 3: 175-185.

Cox, C.M., Qualset, C.O., and Rains, D.W. (1985). Genetic variation for nitrogen assimilation and translocation in wheat. I. Dry matter and nitrogen accumulation. *Crop Sci.,* 25: 430-435.

Cox, T.S., Shroyer, R.J., Ben-Hui, L., Sears, R.G., and Martin, T.J. (1988). Genetic improvement in agronomic traits of hard red winter wheat cultivars from 1919 to 1987. *Crop Sci.,* 28: 756-760.

Cregan, P.B., and Van Berkum, P. (1984). Genetics of nitrogen metabolism and physiological/biochemical selection for increased grain crop productivity. *Theor. Appl. Genet.,* 67: 97-111.

Deckerd, E.L., Busch, R.H., and Kofoid, K.D. (1985). Physiological aspects of spring wheat improvement. In J.E. Harper, L.E. Schrader, and R.W. Howell (Eds.), *Exploitation of Physiological and Genetic Variability to Enhance Crop Productivity.* Am. Soc. Plant Physiol., Rockland, Maryland, pp. 45-54.

Evans, L.T. (1993). *Crop Evolution, Adaptation and Yield.* Cambridge University Press, New York.

Feil, B. (1992). Breeding progress in small grain cereals—A comparison of old and modern cultivars. *Plant Breed.,* 108: 1-11.

Feil, B. and Geisler, G. (1988). Untersuchungen zur Bildung und Verteilung der Biomasse bei alten und neuen deutschen Sommerweizensorten. *J. Agron. Crop Sci.,* 161: 148-156.

Fischer, R.A. (1984). Wheat. In W.H. Smith and S.J. Banta (Eds.), *Potential Productivity of Field Crops Under Different Environments*. IRRI, Los Baños, Philippines, pp. 129-153.

Fischer, R.A. (1985). Number of kernels in wheat crops and the influence of solar radiation and temperature. *J. Agric. Sci.*, 105: 447-461.

Fischer, R.A. and Stockman, Y.M. (1980). Kernel number per spike in wheat (*Triticum aestivum* L.): Responses to preanthesis shading. *Aust. J. Plant Physiol.*, 7: 169-180.

Fischer, R.A. and Wall, P.C. (1976). Wheat breeding in Mexico and yield increases. *J. Aust. Inst. Agric. Sci.*, 42: 139-148.

Grabau, L.J., Van Sanford, D.A., and Meng, Q.W. (1990). Reproductive characteristics of winter wheat cultivars subjected to postanthesis shading. *Crop Sci.*, 30: 771-774.

Green, C.F. (1989). Genotypic differences in the growth of *Triticum aestivum* in relation to absorbed solar radiation. *Field Crops Res.*, 19: 285-295.

Heitholt, J.J., Croy, L.I., Maness, N.O., and Nguyen, H.T. (1990). Nitrogen partitioning in genotypes of winter wheat differing in grain N concentration. *Field Crops Res.*, 23: 133-144.

Hucl, P. and Baker, R.J. (1987). A study of ancestral and modern canadian spring wheats. *Can. J. Plant Sci.*, 67: 87-97.

Jedel, P.E. and Hunt, L.A. (1990). Shading and thinning effects on multi- and standard-floret winter wheat. *Crop Sci.*, 30: 128-133.

Kemp, D.R. and Whingwiri, E.E. (1980). Effect of tiller removal and shading on spikelet development and yield components of the main shoot of wheat and on the sugar concentration on the ear and flag leaf. *Aust. J. Plant Physiol.*, 7: 501-510.

Kirby, E.J.M. (1988). Analysis of leaf, stem and ear growth in wheat from terminal spikelet stage to anthesis. *Field Crops Res.*, 18: 127-140.

Koshkin, E.I. and Tararina, V.V. (1989). Yield and source/sink relations of spring wheat cultivars. *Field Crops Res.*, 22: 297-306.

Kruk, B.C., Calderini D.F., and Slafer, G.A. (1997). Grain weight in wheat cultivars released from 1920 to 1990 as affected by post-anthesis defoliation. *J. Agric. Sci. Camb.*, 128: 273-281.

Kulshrestha, V.P. and Jain, H.K. (1982). Eighty years of wheat breeding in India: past selection pressures and future prospects. *Z. Pflanzenzücht.*, 89: 19-30.

Ledent, J.F. and Stoy, V. (1988). Yield of winter wheat. A comparison of genotypes from 1910 to 1976. *Cereal Res. Comm.*, 16: 151-156.

Loomis, R.S. and Amthor, J.S. (1996). Limits to yield revisited. In M. Reynolds (Ed.), *Increasing Yield Potential in Wheat: Breaking the Barriers*. CIMMYT Int. Symp., CIANO, Cd. Obregon, Mexico. CIMMYT, Mexico, D.F., pp. 76-89.

Loss, S.P., Kirby, E.J.M., Siddique, K.H.M., and Perry, M.W. (1989). Grain growth and development of old and modern Australian wheats. *Field Crops Res.*, 21: 131-146.

Loss, S.P. and Siddique, K.H.M. (1994). Morphological and physiological traits associated with wheat yield increases in Mediterranean environments. *Adv. Agron.*, 52: 229-276.

Mac Maney, M., Díaz, R., Simon, C., Gioia, A., Slafer, G.A., and Andrade, F.H. (1986). Respuesta a la reducción de la capacidad fotosintética durante el llenado de granos en trigo. In *Proceedings I National Congress of Wheat.* AIANBA, Pergamino, Buenos Aires, Argentina, pp. 178-190.

Magrin, G.O., Hall, A.J., Baldy C., and Grondona, M.O. (1993). Spatial and interannual variations in the photothermal quotient: Implications for the potential kernel number of wheat crops in Argentina. *Agric. For. Meteorol.*, 67: 29-41.

Martinez-Carrasco, R. and Thorne, G.N. (1979). Physiological factors limiting grain size in wheat. *J. Exp. Bot.*, 30: 669-679.

McEwan, J.M. and Cross, R.J. (1979). Evolutionary changes in New Zealand wheat cultivars. In S. Ramanujam (Ed.), *Proceedings Fifth International Wheat Genetics Symposium,* Indian Society of Genetics and Plant Breeding, New Delhi, India, pp. 198-203.

Miralles, D.J. and Slafer, G.A. (1995a). Yield, biomass and yield components in dwarf, semi-dwarf and tall isogenic lines of spring wheat under recommended and late sowing dates. *Plant Breed.*, 114: 392-396.

Miralles, D.J. and Slafer, G.A. (1995b). Individual grain weight responses to genetic reduction in culm length in wheat as affected by source-sink manipulations. *Field Crops Res.*, 43: 55-66.

Miralles, D.J. and Slafer, G.A. (1997). Radiation interception and radiation use efficiency of near-isogenic wheat lines with different height. *Euphytica*, 97: 201-208.

Nicolas, M.E. and Turner, N.C. (1993). Use of chemicals dessicants and senescing agents to select wheat lines maintaining stable grain size during post-anthesis drought. *Field Crops Res.*, 31: 155-171.

Ortiz Monasterio, J.I., Sayre, K.D., Rajaram, S., and McMahon, M. (1997). Genetic progress in wheat yield and nitrogen use efficiency under four nitrogen rates. *Crop Sci., 37: 898-904.*

Paccaud, F.X., Fossati, A., and Cao, H.S. (1985). Breeding for yield and quality in winter wheat: Consequences for nitrogen uptake and partitioning efficiency. Z. *Pflanzenzucht.*, 94: 89-100.

Passioura, J.B. (1983). Roots and drought resistance. *Agric. Water Manage.*, 7: 265-280.

Perry, M.W. and D'Antuono, M.F. (1989). Yield improvement and associated characteristics of some Australian spring wheat cultivars introduced between 1860 and 1982. *Aust. J. Agric. Res.*, 40: 457-472.

Richards, R.A. (1992). The effect of dwarfing genes in spring wheat in dry environments. Y. Agronomic characteristics. *Aust. J. Agric. Res.*, 43: 517-523.

Richards, R.A. (1996). Increasing the yield potential in wheat: Manipulating sources and sinks. In M. Reynolds (Ed.), *Increasing Yield Potential in Wheat:*

Breaking the Barriers. CIMMYT Int. Symp., CIANO, Cd. Obregon, Mexico. CIMMYT, Mexico, D.F., pp. 134-149.

Savin, R. and Slafer, G.A. (1991). Shading effects on the yield of an Argentinian wheat cultivar. *J. Agric. Sci.,* 116: 1-7.

Sayre, K.D. (1996). The role of crop management research in CIMMYT in addressing bread wheat yield potential issues. In M. Reynolds (Ed.), *Increasing Yield Potential in Wheat: Breaking the Barriers.* CIMMYT Int. Symp., CIANO, Cd. Obregon, Mexico. CIMMYT, Mexico, D.F., pp. 203-207.

Sayre K.D., Rajaram S., and Fischer R.A. (1997). Yield potential progress in short bread wheats in northwest Mexico. *Crop Sci.,* 37: 36-42.

Siddique, K.H.M., Belford, R.K., Perry, M.W., and Tennant, D. (1989). Growth, development and light interception of old and modern wheat cultivars in a Mediterranean-type environment. *Aust. J. Agric. Res.,* 40: 473-487.

Siddique, K.H.M., Belford, R.K., and Tennant, D. (1990). Root:shoot ratios of old and modern, tall and semi-dwarf wheats in a mediterranean environment. *Plant Soil,* 121: 89-98.

Siddique, K.H.M., Kirby, E.J.M., and Perry, M.W. (1989) Ear to stem ratio in old and modern wheats: Relationship with improvement in number of grains per ear and yield. *Field Crops Res.,* 21: 59-78.

Sinha, S.K., Aggarwal, P.K., Chaturvedi, G.S., Koundal, K.P., and Khanna-Chopra, R. (1981). A comparison of physiological and yield characters in old and new wheat varieties. *J. Agric. Sci.,* 97: 233-236.

Slafer, G.A. and Andrade, F.H. (1989). Genetic improvement in bread wheat (*Triticum aestivum*, L.) yield in Argentina. *Field Crops Res.,* 21: 289-296.

Slafer, G.A. and Andrade, F.H. (1991). Changes in physiological attributes of the dry matter economy of bread wheat (*Triticum aestivum,* L.) through genetic improvement of grain yield potential at different regions of the world. A review. *Euphytica*, 58: 37-49.

Slafer, G.A. and Andrade, F.H. (1993). Physiological attributes related to the generation of grain yield in bread wheat cultivars released at different eras. *Field Crops Res.,* 31: 351-367.

Slafer, G.A., Andrade, F.H., and Feingold, F.E. (1990). Genetic improvement of bread wheat (*Triticum aestivum* L.) in Argentina: Relationships between nitrogen and dry matter. *Euphytica*, 50: 63-71.

Slafer, G.A., Andrade, F.H., and Satorre, E.H. (1990). Genetic improvement effects on pre-anthesis physiological attributes related to wheat grain yield. *Field Crops Res.,* 23: 255-263.

Slafer, G.A., Calderini, D.F., and Miralles, D.J. (1996). Generation of yield components and compensation in wheat: Opportunities for further increasing yield potential. In M. Reynolds (Ed.), *Increasing Yield Potential in Wheat: Breaking the Barriers.* CIMMYT Int. Symp., CIANO, Cd. Obregon, Mexico. CIMMYT, Mexico, D.F., pp. 101-133.

Slafer, G.A. and Miralles, D.J. (1992). Grain area duration during the grain filling period of an Argentine wheat cultivar as influenced by sowing date, temperature and sink strength. *J. Agron. Crop Sci.,* 168: 191-200.

Slafer, G.A. and Miralles, D.J. (1993). Fruiting efficiency in three bread wheat (*Triticum aestivum*) cultivars released at different eras. Number of grains per spike and grain weight. *J. Agron. Crop Sci.,* 170: 251-260.

Slafer, G.A., Satorre, E.H., and Andrade, F.H. (1994). Increases in grain yield in bread wheat from breeding and associated physiological changes. In G.A. Slafer (Ed.), *Genetic Improvement of Field Crops*. Marcel Dekker, New York, pp. 1-68.

Slafer, G.A. and Savin, R. (1994). Grain mass change in a semi-dwarf and a standard-height wheat cultivar under different sink-source relationships. *Field Crops Res.*, 37: 39-49.

Stapper, M. and Fischer, R.A. (1990). Genotype, sowing date and planting spacing influence on high-yielding irrigated wheat in southern New South Wales. Y. Phasic development, canopy growth and spike production. *Aust. J. Agric. Res.,* 41: 997-1019.

Thorne, G.N. and Wood, D.W. (1987). Effects of radiation and temperature on tiller survival, grain number and grain yield in winter wheat. *Ann. Bot.,* 59: 413-426.

Waddington, S.R., Ransom, J.K., Osmanzai, M., and Saunders, D.A. (1986). Improvement in the yield potential of bread wheat adapted to northwest Mexico. *Crop Sci.,* 26: 698-703.

Waisel, Y. and Kafkafi U. (1991). *Plant Roots. The Hidden Half.* Marcel Dekker, New York.

Yunusa, I.A.M., Siddique, K.H.M., Belford, R.K., and Karimi, M.M. (1993). Effect of canopy structure on efficiency of radiation interception and use in spring wheat cultivars during the pre-anthesis period in a Mediterranean-type environment. *Field Crops Res.,* 35: 113-122.

Chapter 17

Physiological Traits That Increase the Yield Potential of Wheat

Gustavo A. Slafer
Jose L. Araus
Richard A. Richards

INTRODUCTION

The Problem

Future genetic gains in grain yield must be maintained at the same pace as before, or even accelerated, to meet the increased demand for food from an increasing population, estimated to be 6 billion people by 2010 (Rasmuson and Zetterström, 1992).

Success in increasing yield potential in the past has mainly been the result of empirical selection for yield per se (Loss and Siddique, 1994). Although this has been successful there is now some concern that both potential and actual yields are leveling off. This is because genetic increases in wheat yields are becoming harder to achieve (see the Preface by T.G. Reeves in Reynolds, Rajaram, and McNab, 1996a; Reynolds et al., 1996; Sayre, 1996). There is now a strong argument that if we are to ensure further genetic gains a physiological and molecular approach to yield improvement may help target the key traits that are currently limiting yield and may therefore complement conventional breeding programs and hasten yield improvement.

Driven by this concern, CIMMYT recently organized a workshop (*Increasing Yield Potential in Wheat: Breaking the Barriers*) in which we were invited to offer our views on physiological approaches to improving yield. In this chapter we have revised, integrated, and summarized these views.

Can Physiologically Assisted Breeding Accelerate Yield Improvement?

There is little doubt that the more precisely we can target key traits, the more rapid genetic progress will be. For example, once the precise disease resistance and grain quality deficiencies were understood, breeders very effectively made progress in overcoming them. The same progress has not been made in understanding the underlying physiological attributes for grain yield and then targeting them in breeding programs. Nevertheless, there are several excellent examples where precise targeting of traits has resulted in substantial yield improvements. For wheat the best known are the targeting of dwarfing genes and genes that control flowering time. The more precisely we can target a key trait underlying yield the easier it is to identify appropriate parents and efficiently select progeny in early generations (Austin, 1993). The requisite though is that we must (1) target those attributes contributing unequivocally to greater yields (Richards, 1996b; Slafer, Calderini, and Miralles, 1996), and (2) develop reliable screening techniques (Araus, 1996; Richards, 1996b). Another avenue involves the use of molecular biology in breeding programs (discussed by Koebner and Snape in Chapter 19), but the power of these techniques for quantitative traits will also strongly depend upon our understanding of yield-determining crop physiological processes.

There are different approaches by which to identify physiological attributes contributing to increased yield potential. Studies involving historical sets of cultivars have been illuminating as have the use of random sets of lines such as those described by Quail, Fischer, and Wood (1989). These studies have identified some critical physiological attributes associated with increases in yield in the past or in random sets of lines and hence identified those attributes where accelerated improvement may result in further yield increases (Shorter, Lawn, and Hammer, 1991). Results from historical sets of wheat varieties in different countries are summarized by Calderini, Reynolds, and Slafer in Chapter 16 (for more details see also reviews by Loss and Siddique, 1994; Slafer, Satorre, and Andrade, 1994).

Another approach used has been ideotype breeding where, on the basis of experimental and/or theoretical evidence, traits have been identified that are expected to have a significant influence on yield.

Objective

The aim of this chapter is to evaluate the likely physiological attributes that may limit the yield potential of wheat. The characteristics that will be discussed have been chosen because of their direct connection with yield from experimental and/or theoretical evidence. Some are speculative whereas for others there is compelling evidence for their importance. We hope that their discussion may lead to a search for important genetic variation for the characteristics and their evaluation and validation in breeding programs. We begin by identifying the traits we consider important through a yield-components approach and end by discussing possibilities for realistically assessing large numbers of lines in a breeding program for some of the characteristics that may be important in the determination of yield.

IDENTIFYING PROMISING TRAITS

Physiological and Numerical Yield Components

A basic requisite for a trait to be considered for selection is that it must be physiologically linked to yield. A way to explore potential traits is to partition yield into simpler components. Grain yield (GY) can be divided into either physiological or numerical components.

A comprehensive description of yield components was provided in Chapters 2 and 3. Briefly, grain yield is considered, from the physiological point of view, to be determined by two components, the amount of biomass at maturity (BY; in general only the aboveground biomass is regarded) and the proportion of it allocated to grains, namely harvest index (HI) (e.g., Gifford et al., 1984). Biomass at maturity is in turn determined by the amount of incident radiation during the growing season (Q), the fraction of this intercepted by the crop canopy (I), and the efficiency of the crop to convert the intercepted radiant energy into dry matter (RUE):

$$GY = BY * HI$$
$$GY = Q * I * RUE * HI$$

From a simpler, numerical analysis, the two major yield components are the number of grains per unit land area (NG) and the averaged individual grain weight (IGWt):

$$GY = NG * IGWt$$

The following subsections will concentrate on these components and some of their physiological determinants to identify whether they can be considered useful for choosing parents and/or selecting progeny.

Can We Expect Further Increases in Harvest Index to Increase Grain Yield?

It has been shown that increases in wheat yield from previous breeding were almost exclusively associated with parallel increases in harvest index (Chapter 16). Since this single trait has been responsible for dramatic increases in grain yield during the whole scientific breeding era, it raises the question of whether we should simply select for an increased harvest index. This may be appropriate where harvest index is still low but this is rare in traditional wheat growing areas without a terminal drought. The fact is that harvest index cannot exceed certain limits. This is because the crop has to maintain a minimum amount of biomass for assimilation and to provide physical support for the reproductive structures. This threshold harvest index has been estimated to be ca. 60 percent (Austin et al., 1980).

In most regions of the world, present values of harvest index are close to or higher than 50 percent and it is likely that further increases in yield through a higher harvest index will be increasingly difficult to achieve. Increases in harvest index through further reductions in height are likely to occur at the expense of total biomass and thus lead to decreases in yield. As previously suggested by many researchers (see Slafer, Satorre, and Andrade, 1994 and references cited therein) future increases in yield potential must largely be achieved by developing cultivars with greater biomass while maintaining a high harvest index.

For regions where harvest index may not yet have reached, say, 50 to 55 percent there is still potential for increase. However, it is important to determine whether harvest index may be a better selection criterion than yield itself (Whan, Rathjen, and Knight, 1981; Whan, Knight, and Rathjen, 1982) or whether it would be easier to target another trait for selection. Two potential candidates would be the spike-to-stem weight ratio at anthesis (Siddique and Whan, 1990; Slafer, Andrade, and Satorre, 1990), a partitioning trait closely associated with harvest index (Siddique, Kirby, and Perry, 1989) and plant height, which is also correlated with harvest index (Miralles and Slafer, 1995a). The former is physiologically sound (see details in Siddique, Kirby, and Perry, 1989) and in practice it could be determined on the main stem while seed from tillers are used for seed continuity.

Plant height is a character that has been consistently reduced during the twentieth century by breeders (see Chapter 16). Unlike harvest index it has a biological limit which is well beyond that of modern cultivars. The relationship between height and yield is parabolic and tall plants have a low yield due to a low harvest index whereas short plants have a low yield due to a smaller final biomass (discussed later). Most modern cultivars fall within an optimum height range of between 70 and 100 cm (Fischer and Quail, 1990; Richards, 1992; Miralles and Slafer, 1995a). It is therefore unlikely that further increases will be achieved by altering plant height.

Biomass Must Be Increased

If the harvest index of modern cultivars is close to their maximum theoretical value (see Austin et al., 1980; and discussion in Slafer and Andrade, 1991) then to increase yield potential we must increase the yield of biomass. An increase in biomass may be achieved by selecting taller plants, which would result in a better light distribution within the canopy. For example, high-biomass barley lines are taller than the modern cultivars to which they were compared (Hanson et al., 1985). However, increasing biomass production in this way could end up with lower rather than higher-yielding cultivars due to two negative associated effects. First, the taller the plant the smaller its harvest index, and second, selecting for increased height is likely to result in more lodging. Greater biomass must therefore be achieved by maintaining the height within the optimum range.

An increase in biomass production may be achieved by increasing the amount of (photosynthetically active) radiation intercepted and/or the efficiency of the photosynthetic tissues to use the intercepted radiation.

Radiation Interception

The amount of radiation intercepted by the crop during the growing season depends on the amount of incident radiation and the ability of the crop to intercept it. Both factors can be genetically manipulated. The total amount of incident radiation to which the crop is exposed may be increased by changing the developmental response of the crop, so that the duration of the growing season may be longer. The proportion of daily incident radiation intercepted by the crop can be increased by altering leaf area index, particularly during the early developmental phases.

Can we actually improve the total amount of incident radiation? Genetic manipulation of plant development to increase the duration of crop growth and in turn the amount of incident radiation during the growing season has been practiced since plants were first selected. This has primarily been to adjust the time of anthesis to occur at the optimum time for a particular location (Flood and Halloran, 1986; Gomez-MacPherson, 1993). As with harvest index, time to anthesis is a trait that has already been optimized for most regions and further change is unlikely (Slafer, Calderini, and Miralles, 1996) unless new or altered cropping opportunities emerge.

Increasing radiation interception. The alternative is to increase the capacity of the crop to intercept radiation. As the main intercepting organs are leaf laminae, the obvious characteristic to increase is leaf area index. However, well-managed crops have maximum leaf area indexes higher than the critical (i.e., the minimum leaf area index required to fully intercept the incoming radiation), and there would be no point in further increasing this value. The opportunity exists only when leaf area index is below the critical level; this mostly occurs during the early growth phases.

Under favorable conditions, wheat yield has a low sensitivity to environmental or management factors during the earliest developmental phases of the crop, but it increases as the crop approaches anthesis (Fischer, 1985; Savin and Slafer, 1991). Genetic manipulation of leaf area during the early developmental phase is therefore unlikely to affect yield when conditions are favorable. However, in less favorable environments where soil water may be limited, there are soil nutritional problems, or there is a very short growing season, then fast early growth of the crop may be critical to guarantee better performance (Richards, 1996a, 1996b). Under these circumstances it may be rewarding to select for higher early vigor.

Exploring the physiological reasons why barley seedlings are far more vigorous than wheat seedlings so as to understand the early vigor of wheat has been very enlightening (López-Castañeda and Richards, 1994; López-Castañeda et al., 1996). Net assimilation rate per unit leaf area of wheat during the early vegetative stages is significantly higher than that of barley. However, because barley has thinner leaves, its leaf area per unit mass of leaves is greater, which more than compensates for the reduced assimilation rate and results in more leaf area and more growth (López-Castañeda, Richards, and Farquhar, 1995). In addition, barley has a larger embryo than wheat. This results in more cells expanding after imbibition and in a larger leaf area, root mass, and total biomass. There is substantial variation in both these characteristics and they have been successfully incorporated into wheat germplasm (Richards, 1996b; Rebetzke, Condon, and Richards, 1996). Several other factors also contribute to variation in the early growth of wheat, which can also be selected to improve vigor. Coleoptile tillers are uncommon in wheat, yet they can also significantly improve early leaf area and biomass (Liang and Richards, 1994). There is also significant variation in emergence time and rate of leaf elongation that, if selected, should also improve early leaf area and biomass growth. It is notable that gibberellic acid-insensitive dwarfing genes, *Rht1* and *Rht2*, which are widespread in wheats globally, confer reduced emergence and leaf area growth. Breeding for other sources of reduced height may be required to overcome this growth penalty but still retain the shorter stature of higher-yielding wheats (Bush and Evans, 1988; Richards, 1992).

Radiation Use Efficiency

For high-yielding environments where yields are less influenced by early growth, the period immediately preceding anthesis has an overwhelming influence on grain yield (Fischer, 1985). Maximizing growth during this period then becomes critical. Radiation interception in well-managed wheat crops is normally higher than 95 percent during this period and it could be rewarding to genetically improve the radiation use efficiency of the crop to obtain greater biomass immediately prior to anthesis. The following discussion concentrates on some physiological attributes that might be considered to increase this efficiency.

Research to understand which physiological traits may improve the ability of the crop to transform intercepted radiation into new dry matter has occurred at two levels. The most studied has been leaf photosynthesis/respiration and the other we shall consider is canopy architecture.

Improving leaf photosynthesis. The possibility of increasing leaf photosynthesis has received considerable attention. The logic has been that, as almost all the dry matter is produced by leaf photosynthesis, it would be expected that increasing this rate would confer greater biomass and yield. In fact, Austin (1992) suggested that leaf photosynthesis must be increased if we are to further raise wheat yields. Photosynthesis can be increased by increasing stomatal conductance, photosynthetic capacity, or both. However, relationships between photosynthesis and yield have generally been found to be insignificant or even negative (e.g., Evans and Dunstone, 1970; Austin et al., 1982; Johnson et al., 1987; Carver, Johnson, and Rayburn, 1989). The negative association is largely due to compensatory increases in leaf size (Planchon and Fesquet, 1982; Austin, 1989; Evans, 1992), and presumably with specific leaf area, as yield increases. This increase in specific leaf area has in turn resulted in a reduction in leaf photosynthesis. The negative correlation between leaf photosynthesis and leaf area has been frequently reported (e.g., Rawson et al., 1983; LeCain, Morgan, and Zerbi, 1989; Morgan, LeCain, and Wells, 1990), although there are exceptions (Araus, Tapia, and Alegre, 1989), and it has consequently been suggested that selection for increased leaf photosynthesis would imply a concomitant decrease in leaf area (Bhagsari and Brown, 1986). Thus, in general, selection for increased maximum leaf photosynthesis has not resulted in an increase in yield (see Austin, 1989 and references cited therein). Among the several possible reasons suggested by Austin (1989) that we emphasize are:

1. lack of variation in maximum leaf photosynthesis in the gene pool exploited in breeding;
2. pleiotropic consequences of high maximum leaf photosynthesis, which adversely affects crop photosynthesis by reducing LAI and leaf area duration (e.g., smaller or fewer leaves);
3. inability of the plant to utilize carbohydrates for growth ("sink limitation"), which would ultimately reduce maximum leaf photosynthesis by feedback mechanisms (Azcon-Bieto, 1983); and
4. genes for high maximum leaf photosynthesis that may be introduced from wild relatives may not be expressed if linkage groups important for high yield in cultivated wheats are broken.

However, these compensations may not always be pleiotropic and may be broken. Carver and Nevo (1990), for example, have reported leaf photosynthetic rates not related to leaf areas in one population of *Triticum dicoccoides* from Israel. Similarly, cultivars released in the Great Plains of the United States varied significantly in both leaf area and leaf photosynthesis, but these attributes were not significantly correlated (Morgan and LeCain, 1991). Both studies identified some genotypes that combine both high photosynthetic rates and large leaves. Such genotypes might prove useful sources of germplasm for increasing photosynthetic rates in commercial wheat breeding programs.

Recent studies provide evidence that increases in photosynthesis are associated with yield increases in wheat breeding programs. These increases have been attributed to greater photosynthetic capacity (Watanabe, Evans, and Chow, 1994; Fischer et al., in press) and stomatal conductance (Rees et al., 1993; Reynolds et al., 1994; Fischer et al., in press). From a study of leaf traits in eight representative dwarf wheats released by CIMMYT over a 26-year period, stomatal conductance was closely associated with yield evaluated over six years (Fischer et al., in press). Conductance measurements were made using both a steady-state porometer and an air-flow porometer. These measurements take less than 30 seconds per leaf and, because of the significant relationship with grain yield, open up the possibility of rapidly screening large populations of plants. This is in line with the positive relationship between crop yield and carbon isotope discrimination (discussed in more detail later) among wheat and other cereals (Condon, Richards, and Farquhar, 1987; Romagosa and Araus, 1991; Hall et al., 1994; Sayre, Acevedo, and Austin, 1995; Richards, 1996a; Araus, 1996) particularly under favorable growing conditions.

Improving respiration efficiency. An alternative approach to increasing net leaf photosynthesis would be to reduce respiration rates (therefore increasing respiration efficiency) and the calorific value of some plant products. Reducing the relative amount of secondary metabolites, which are often rich in nitrogen and have a high caloric value, should result in more sucrose and nitrogen being available for growth and thereby result in more biomass.

Evidence that a substantial proportion of the total carbon assimilated by wheat crops is used in maintenance respiration (Biscoe and Gallagher, 1977) indicates that respiration efficiency may also be improved by decreasing maintenance, in favor of growth respiration. While this has been shown to be a real possibility for forages (Wilson, 1982; Wilson and Jones, 1982), there has been little research on wheat.

Both photorespiration and cyanide-resistant respiration produce no useful chemical energy for crop growth and these have been considered wasteful processes. However, they could confer adaptation to high radiation intensities and suboptimal temperatures, respectively (see Slafer, Satorre, and Andrade, 1994), and a greater understanding of them is required before we can speculate on the advantages of selecting for or against them.

Improving the distribution of radiation within the canopy. At the crop canopy level of organization, radiation use efficiency can be increased by improving the distribution of incident radiation within the canopy, that is, increasing the proportion of the incident radiation used by the lower layers of the canopy by reducing the coefficient of light attenuation, so that each unit of leaf area would intercept less radiation at a higher quantum efficiency. Genetic variability for radiation use efficiency has been reported (Green, 1989; Kiniry et al., 1989) and it is associated with different patterns of radiation distribution within the crop canopy (Aikman, 1989; Green, 1989). In fact, Green (1989) has shown a negative relationship between radiation use efficiency and the coefficient of light attenuation during the preanthesis period.

A single trait strongly associated with the coefficient of light attenuation and in turn with the pattern of radiation distribution within the canopy is the angle of insertion of leaf laminae. Innes and Blackwell (1983) reported that wheat crops with erect upper leaves produced higher yields than those with predominantly horizontal leaves. However, no advantage in having erect leaves at yield levels up to 6.5 t ha^{-1} were reported in studies using spring wheats in Mexico (Araus, Reynolds, and Acevedo, 1993). The range in light attenuation coefficients among wheat cultivars shown by Green (1989), as well as in a study conducted in Australia with cultivars released during different eras (Siddique et al., 1989), indicates the potential to modify canopy architecture and radiation use efficiency in wheat through breeding. Carvalho and Qualset (1978) showed that the angle of insertion of the flag leaf is controlled by genes having major effects. However, the possible existence of an allometric relationship between leaf erectness and smaller leaves, spikes, and stems probably modifies the relative contribution of the different components to final yield and makes progress through selection of a more erect canopy a complex task (Araus, Reynolds, and Acevedo, 1993).

Considerable variation in flag leaf dimensions has been reported in wheat, and there are many wheats with small flag leaves. A genetic reduction in flag leaf area should improve the radiation distribution within the canopy (Richards, 1996b). This would result in wheat having a canopy more like barley and maize, both of which typically have areas of uppermost leaves less than mid-formed leaves. Smaller flag leaves may also be cooler than larger leaves and less likely to senesce when crops are exposed to high temperatures and/or dry conditions. Due to the timing of flag leaf expansion, smaller flag leaves may also improve the survival of floret primordia and result in more fertile florets and grains (more details on these relationships are provided later). Moreover, flag leaves may compensate for a reduced area not only by an improved light penetration into the canopy but also by a higher stomatal conductance. If there is an allometric relationship between flag leaf size and spike size then this may negate any advantage of small leaves unless more fertile tillers are produced or the relation between leaf size and spike size is broken.

Distribution of nitrogen within the canopy. An important factor affecting radiation use efficiency is the concentration of nitrogen in the leaves or other photosynthesizing surfaces (Araus and Tapia, 1987). Bindraban (1996), for example, assumed a linear effect of leaf nitrogen on radiation use efficiency for values of nitrogen concentration between 2 and 3.2 percent (i.e., radiation use efficiency would be 0 g MJ^{-1} for leaf nitrogen $\leq$2 percent and maximum for leaf nitrogen $\geq$3.2 percent). Theoretical studies have suggested that radiation use efficiency would increase if nitrogen is preferentially allocated to the more illuminated leaves (Field, 1983; Hirose and Werger, 1987) in the upper layers of the canopy.

The distribution of nitrogen in a canopy is not uniform and we do not know whether nitrogen distribution can be modified to maximize radiation use efficiency (Dreccer, Slafer, and Rabbinge, 1998). Whether leaf N distribution of modern wheats limits canopy photosynthesis has not been methodically addressed, as far as we are aware, although some attention has been given to it in summer crops (Sinclair and Shiraiwa, 1993; Sadras,

Hall, and Connor, 1993; Giménez, Connor, and Rueda, 1994; Connor, Sadras, and Hall 1995; Hammer and Wright, 1994; Wright and Hammer, 1994). However, manipulation of nitrogen availability by management is likely to override any genetic modification that may be possible.

Further Improving Number of Grains

The previous section dealt with physiological components, and some of their surrogates, that could be considered as potential selection criteria to improve biomass. We must also consider numerical components to devise new criteria to increase wheat yields. Assuming that increases in number of grains per unit land area would not bring about competitive reductions in grain weight (discussion of this assumption is provided later in this chapter), selecting for more grains per unit land area would directly produce higher-yielding cultivars. This has been, in fact, what most breeding programs have indirectly achieved by selecting for yield (see Chapter 16).

The Determination of Number of Grains per Unit Land Area

Selection for grain number is unlikely to be as successful as selecting for grain yield as the measurement of grain number requires several more steps than the measurement of yield. Neither is there evidence of a higher heritability for grain number than for grain yield, which could make selection for grain number worthwhile. Final grain number is achieved through the components spike number, spikelets per spike, and grains per spikelet. Selection for these components to increase grain number has not been successful because each of the components are negatively correlated (Fischer, 1984). The different components contributing to the final number of grains per unit land area are produced throughout the whole period from sowing to a few days after anthesis (see Slafer and Rawson, 1994). The generation of these subcomponents overlaps and this suggests that the negative relationships reflect feedback compensations between them (Slafer, Calderini, and Miralles, 1996). A more fruitful approach to increase grain number may lie in having a better understanding of the processes that regulate its determination in field-grown wheat.

Fischer (1985) identified a critical phase for the definition of final number of grains per unit land area in wheat of ca. 20 to 30 days immediately before anthesis. This has been confirmed in later independent studies (e.g., Thorne and Wood, 1987; Savin and Slafer, 1991). During this phase, some of the tillers and floret primordia die whereas the last stem internodes, leaves, and spikes are actively growing. It is believed that in this phase immediately before anthesis there are inadequate assimilates (source) to fully satisfy the potential growth of all organs (sinks), and hence there is strong competition for available assimilates especially between growing stems and spikes (Brooking and Kirby, 1981; Fischer and Stockman, 1986). Due to this preanthesis source limitation the younger, smaller tillers die and the survival of floret primordia in the surviving tillers depends on the ability of the spike to compete for the available assimilate; there is a high correlation between spike dry weight at anthesis and number of grains per unit land area (Fischer, 1985; Thorne and Wood, 1987; Savin and Slafer, 1991). These findings suggest that to further increase number of grains per unit land area selection should be directed to increase the dry matter allocated to the spikes just before anthesis. Two nonexclusive alternatives are possible: first, to increase the assimilate supply to the spikes by reducing the competitive ability of stems, and second, to reduce the competition by maximizing the availability of assimilates during this critical period.

Increasing Assimilate Supply to the Growing Spikes by Partitioning

An increased assimilate supply to spikes at the expense of stem growth has been the reason for the greater yields achieved by wheat breeders in the past (see Chapter 16) and was a major reason for the "green revolution." It has been largely due to the introgression of major genes for reduced culm stature (*Rht*) into bread wheat (Richards, 1996b). This is because introducing *Rht* genes directly affected the partitioning of dry matter to the spikes by imposing a restriction on stem growth. This in turn results in an increased availability of photoassimilates to the developing spikes (Youssefian, Kirby, and Gale, 1992) and improved floret fertility (Brooking and Kirby, 1981; Fischer and Stockman, 1986; McClung et al., 1986; Slafer and Miralles, 1993; Richards, 1996b), because of less abortion of floret primordia (Youssefian, Kirby, and Gale, 1992). It is likely that any further reduction in height may not result in increased yield as we may have reached an optimum height below which yields may decline (Fischer and Quail, 1990; Richards, 1992; Miralles and Slafer, 1995a). Although shorter stems could easily be selected, extreme dwarfism has been repeatedly found to reduce final biomass (Allan and Pritchett, 1980; Bush and Evans, 1988; Fischer and Stockman, 1986; Fischer and Quail, 1990; Richards, 1992; Miralles and Slafer, 1995a), probably because of the poor radiation distribution within such a compact canopy. Although a further reduction in height is likely to result in an increase in harvest index this is unlikely to be enough to counter the reduction in biomass.

Targeted selection for organ growth during the critical period before anthesis may result in an increased grain number and yield without sacrificing growth. For example, reducing the length of the peduncle (uppermost internode), but not that of other internodes, would maintain a constant leaf canopy height but may significantly reduce the competition for assimilates between the spike and the elongating stem (Richards, 1996b). This is because the peduncle is the most rapidly growing internode just before anthesis, when number of grains per unit land area is mainly determined (Richards, 1996b). Similarly, reducing the size of the flag leaf could also increase grain number as the flag leaf is the last expanding leaf, and its reduced growth could result in more photoassimilates going to the developing ear (Richards, 1996b). This would only be advantageous if the canopy reaches full radiation interception regardless of the size of the flag leaves. It would also improve the distribution of radiation within the canopy and increase radiation use efficiency (see above). Finally, the availability of assimilates for the growing spike might also increase if there is a reduction in tillering (Richards, 1996b). Most tillers die before anthesis. If fewer tillers were produced more assimilates might be available for fertile tiller growth. This is most easily achieved with a single major gene that inhibits tillering in wheat (*tin*; Richards, 1988). It is noteworthy that the spike size of wheats containing the *tin* gene is spectacularly large, indicating how much spike size and grain number can be enhanced by inhibiting the growth of tillers that are likely to be sterile.

Increasing Assimilate Supply to the Growing Spike by Maximizing Growth

In addition to increasing assimilate partitioning to the growing spike during the preanthesis period to increase grain number and yield, there are also opportunities to increase assimilate supply by increasing growth during this period. Factors discussed earlier for improving radiation use efficiency also apply here.

A further opportunity for improvement is to increase the duration of the spike growth period. For most wheat-growing areas the time of anthesis is already close to the optimum and cannot be altered by more than a few days so that any increase must be within the prevailing period to anthesis. This is possible because the duration of the phases to double ridge and to terminal spikelet initiation can be changed without altering anthesis date (Halloran and Pennel, 1982; Slafer and Rawson, 1994). Extending the critical period during which grain number is determined should result in more biomass accumulation by the growing spike, which in turn could increase the final number of grains. The increase in grain number is likely to come about by a reduction in floret abortion or tiller death (Slafer, Calderini, and Miralles, 1996). The consequence of a longer spike growth duration can be assessed from experiments in which the spike growth period has been shortened. For example, increasing air temperature or extending photoperiod both shorten spike growth duration and result in fewer grains set (Fischer, 1985). The ability to increase the early growth of wheat that was discussed earlier could provide the flexibility required to lengthen the duration of the post-initiation phases without sacrificing earlier vegetative growth.

Slafer and Rawson (1996) found that wheat is still sensitive to photoperiod during the stem elongation phase, and that the magnitude of the responses to photoperiod during pre- and postterminal spikelet initiation phases were independent of each other. This indicates that the length of the critical phases, when the number of grains per unit land area is determined, can be altered in response to photoperiod. Previous results (Allison and Daynard, 1976; Rahman and Wilson, 1977; Fischer, 1985; Manupeerapan et al., 1992; Slafer and Rawson, 1995a) support this conclusion and also provide evidence for genetic variation for this response. The length of this phase could also be genetically manipulated through sensitivity to temperature. Evidence for genetic variation has also been reported (Slafer and Rawson, 1995b; Slafer, 1996).

Spike Nitrogen Content

Most of the research on the determination of grain number has centered on biomass growth and the partitioning of it immediately before anthesis. However, recent evidence suggests that nitrogen content of the spike could be a better estimate of the spike's ability to set grains (Abbate, Andrade, and Culot, 1995; van Herwaarden, 1996) and that nitrogen content may determine the degree of floret abortion and consequently the number of florets setting grains (Richards, 1996b). If these findings are confirmed then manipulation of spike nitrogen content and the allocation of nitrogen to spikes at anthesis may be an important determinant of grain number.

Opportunities to Increase Final Grain Weight

Increases in number of grains per unit land area would directly result in greater yield if final weight of the individual grains is not proportionately reduced. The negative relationship between number of grains and their average weight is common in the literature and has also been found when cultivars released in different eras are compared (Waddington et al., 1986; Perry and D'Antuono, 1989; Siddique, Kirby, and Perry, 1989; Slafer and Andrade, 1989), although there are some exceptions to this (Hucl and Baker, 1987; Cox et al., 1988; Calderini, Dreccer, and Slafer, 1995). Understanding whether this compensation is due to competition will determine if future increases in grain number will be increasingly compensated by reductions in individual grain weight.

Is Individual Grain Weight Determined by Resource Availability?

The simplest and most widely accepted hypothesis for the negative relation between grain number and grain weight is that as the number of grains per unit land area increases, the availability of photoassimilates per grain decreases, and that a lower final individual grain weight results (i.e., there is competition between grains for limited resources). However, detailed analyses from source-sink manipulations and analyses of grain weight at particular spike and floret positions do not support this hypothesis (for details see full discussion in Slafer, Calderini, and Miralles, 1996). Thus, despite the negative relationship between the two principal components of yield, there is a substantial body of evidence suggesting that grain growth in wheat is generally sink rather than source limited (e.g., Rawson and Evans, 1971; Evans, 1978; Borghi et al., 1986; Mac Maney et al., 1986; Siddique et al., 1989; Savin and Slafer, 1991; Labraña and Araus, 1991; Slafer and Savin, 1994; Kruk, Calderini, and Slafer, 1997). In other words, the photosynthetic capacity of the crop in the postanthesis period is generally adequate or maybe in excess of that required to completely fill the grains (Savin and Slafer, 1991; Richards, 1996b).

The most likely explanation for the negative relationship between number of grains per unit land area and average individual grain weight is that increases in the former may lead to a greater proportion of grains from floret positions of reduced grain weight potential (e.g., distal positions in central spikelets and/or in apical or basal spikelets and/or in secondary tiller spikes). As their weight potential is smaller, increasing number of grains would necessarily bring about decreases in average individual grain weight even if the availability of assimilates per grain is in excess of the demands (Richards, 1996b; Slafer, Calderini, and Miralles, 1996).

Are Increased Grain Weight and Increased Yield Achievable?

The earlier discussion suggests that there need not be a negative relationship between grain number and grain weight and that increasing either may result in greater yields. However, direct selection for grain weight may not be successful; it may simply result in fewer distal fertile florets. A more targeted approach to increase grain weight potential may be fruitful.

Little is known about the control of grain weight and hence we can only speculate on how to genetically target the factors limiting grain growth. Structural problems could limit grain growth, such as insufficient vascular linkages, particularly to distal florets, so that even though assimilates are available their transport to grains is inadequate (Nátrová and Nátr, 1993). Most evidence, however, suggests that this is unlikely (Evans, 1993). It is known that endosperm cell number is closely related to final grain weight and this may regulate the demand for assimilate (Brocklehurst, 1977; Cochrane and Duffus, 1982). Little is known about genetic variation for this trait in wheat or whether it can be manipulated.

It is possible that *Rht* genes, which have been used globally in wheat breeding, may impose a structural restriction on grain growth because grain coats are smaller at all floret positions in wheats with *Rht* genes. Richards (1996b), for example, suggested that wheats with major dwarfing genes may have smaller caryopses, limiting the potential growth of the grain. This is based on the fact that the cellular dimension of wheats with the *Rht1* and *Rht2* dwarfing genes is smaller, which results in smaller coleoptiles, leaf laminae and sheaths, and internodes (Nilson, Johnson, and Gardner, 1957; Bush and Evans, 1988; Keyes, Paolillo, and Sorrells, 1989; Pinthus et al., 1989; Borrell, Incoll, and Dalling, 1991; McCaig and Morgan, 1993; Calderini, Miralles, and Sadras, 1996). In support of this possibility, Miralles and Slafer (1995b) observed that average grain weight in lines with *Rht1* and *Rht2* alleles is smaller than in tall isolines. This was in part due to a lower weight of the basal grains, which are typically the biggest, and this difference was not removed by doubling the source-sink ratio after anthesis. The effect also appeared to be related to fewer cells in the grain pericarp in a comparison between a dwarf and a tall line (Miralles, Calderini, Pomar, and D'Ambrogio, personal communication, 1997). Wheats with the *Rht1* and *Rht2* alleles are insensitive to gibberellic acid, which is responsible for the reduced cell dimensions of these wheats. If the cellular dimensions of the caryopsis restricts grain weight in wheats with *Rht1* and *Rht2,* then to overcome this would involve the use of other *Rht* genes that are sensitive to GA or the use of minor genes that regulate plant height. This could result in wheats with an optimum plant height but without the penalty of small potential grain sizes.

A physiological, nonstructural trait that might be responsible for variation in grain weight is grain growth in relation to temperature. Genetic variation in duration of grain growth has been reported for wheat (Housley and Ohm, 1992), and variation in response to temperature can be derived from Marcellos and Single (1972) and Hunt, van der Poorten, and Pararajasingham (1991), as shown in a review by Slafer and Rawson (1994). The different sensitivity to temperature may be due to differences in the rate of grain development in relation to temperature (reciprocal of the thermal time) and/or in the base temperature. Then, grain weight may be increased by manipulating either or both of these parameters (the thermal time to grain maturation and the base temperature for the process to occur). For this to be possible we need a better understanding of the genetic basis of these parameters.

SELECTING FOR PROMISING TRAITS

Identifying the most promising traits to increase wheat yield is not easy. Not only must the traits be related to yield, but they must also be highly heritable, have low genotype × environment interactions, and their expression must not be compensated by other related traits that negate their effect on grain yield. Moreover, when these traits are applied in breeding programs simple techniques are needed to screen for them that are quick, reliable, and relatively inexpensive.

Because of the very complex nature of grain yield particular emphasis should be given to single tests that integrate the function of the crop at its highest level of organization (i.e., the canopy) and/or are important for a relatively long period of the crop's life. Some physiological techniques have been modified and, although not as accurate as more complex methods, they may become useful for routine screening. In the next sections we briefly discuss some newer techniques for the evaluation of the integrative traits related to yield (more details are available in Araus, 1996 and references cited therein). These mainly include remote sensing techniques and the isotopic composition of the dry matter.

Remote Sensing Techniques

Assessments based on remote sensing techniques are convenient as they are noninvasive and relatively cheap, while allowing the study of canopy-level physiological phenomena. Remote sensing is particularly useful to assess, for example, leaf area index, absorbed radiation, total canopy chlorophyll, or the potential for CO_2 uptake, xanthophyll pigments, and canopy temperature. These can be determined with spectral reflectance measurements in the visible, near-infrared, and mid-infrared regions (Field, Gamon, and Peñuelas, 1994) and infrared thermometry (Blum, 1988).

The measurement of canopy temperature by remote sensing is the most widely used of these techniques in breeding programs because of the low cost of the infrared thermometer and its easy operation (Blum, Mayer, and Golan, 1982; Blum, 1988). This technique allows to screen for deeper root systems, increased stomatal conductance, and general drought avoidance (Blum, Mayer, and Golan, 1982; Blum, 1988; Blum et al., 1989) as these traits are related to lower canopy temperature or higher canopy temperature depression. For yield potential, canopy temperature depression seems to be promising as well, as recently reported by several researchers from CIMMYT (Reynolds et al., 1994; Sayre, 1996; Fischer et al., in press).

We concentrate here on less popular techniques that might prove useful in breeding programs. Spectral reflectance techniques have so far received little (if any) consideration by breeders, even though they allow the assessment of several traits. Indexes based on the contrast among crop reflectances of different wavelengths can be used not only for monitoring characteristics conferring higher potential growth but also some related to stress resistance. By combining the reading of several discrete wavelengths the simultaneous evaluation of several traits is possible. Then, in a simple and quick screening of progeny plots, relevant information related to potential growth and/or stress resistance might be obtained. In fact, development of a practical portable spectroradiometer (of a size comparable to infrared thermometers) for this practical purpose is advanced (Peñuelas, personal communication, 1997).

Spectral Reflectance Techniques to Estimate Potential Growth

As green tissues are strong absorbers of red radiation and highly reflective of near-infrared radiation, methods based on the red/near-infrared contrast of reflectances can be used to estimate leaf area index and canopy photosynthetic capacity (Araus, 1996). Because the contrast in reflectance between the near-infrared and red is quite small in soils, the contrast increases as the fraction of radiation intercepted by the canopy grows.

A simple index to measure the contrast in reflectance is the normalized difference vegetation index (NDVI), which may be used as an indirect assessment of canopy biomass, leaf area index, light absorption, and potential photosynthetic capacity:

$$\mathrm{NDVI} = (R_{\mathrm{NIR}} - R_{\mathrm{NIR}} + R_{\mathrm{RED}})^{-1}$$

where R_{NIR} and R_{RED} are the reflectance in the near-infrared (e.g., 680 nm) and red (e.g., 900 nm) wavelengths, respectively.

The validity of NDVI as a quantitative indicator of canopy structure or net primary production is largely based on empirical results with horizontally uniform canopies, such as a crop (Kumar and Monteith, 1981; Bartlett, Whiting, and Hartman, 1990), although it is also a good indicator of photosynthetically active radiation absorption, and thus of the potential photosynthetic capacity of the canopy, even in heterogeneous landscapes (Gamon et al., 1995). Indeed NDVI is probably the most commonly used index for analyzing vegetation on a continental and global scale (see references in Gamon et al., 1995). The relationships between NDVI and canopy parameters follow a semilogarithmic pattern, with the canopy parameters being expressed on a (natural) logarithmic scale. The logic of this transformation lies in the negative exponential attenuation of photosynthetically active radiation with canopy depth (Monsi and Saeki, 1953; Sellers, 1987). In addition, where canopy development and photosynthetic activity are in synchrony (for example, during vegetative stages of cereal crops growing under nonlimiting conditions), instantaneous maximum daily photosynthetic rates during vegetative stages of the crop can be evaluated with NDVI (see Gamon et al., 1995).

The NDVI seems to be particularly powerful with leaf area index ≤ 2 (Sellers, 1987), and thus is ideally suited for detecting subtle differences in cover in sparse canopies, such as young crops and grasslands or crops growing under stressed conditions (Tucker et al., 1979; Gallo, Daughtry, and Bauer, 1985; Fernandez et al., 1994; Gamon et al., 1995; Peñuelas et al., 1996). Effects of calibration changes, atmospheric transmission, solar elevation, and canopy architecture still present major challenges (Field, Gamon, and Peñuelas, 1994). Care must be taken when working in very reflective soils, as near-infrared wavelengths in these circumstances may be insensitive to changes in leaf area index (Curran, 1983), and thus traditional vegetation indexes contrasting visible and near-infrared reflectances may be inadequate (Elliott and Regan, 1993). Nevertheless, Fernandez et al. (1994) pointed out that NDVI appears to be a better estimator of leaf area index than spectral indexes incorporating soil reflectance. Moreover, the contribution of genotypic differences in canopy architecture (tiller density and growth habit), plant height (Elliot and Regan, 1993), and phenology to spectral reflectance is poorly documented and further work is required; meanwhile individual calibration for distinct canopy architectures may be required for accurate assessments.

Spectral Reflectance Techniques to Estimate Responses to Stress

Spectral reflectance techniques may also allow the calculation of some indexes of crop responses to abiotic stresses. The photochemical reflectance index (PRI) and the water index (WI):

$$PRI = (R_{REF} - R_{531})(R_{REF} + R_{531})^{-1}$$
$$WI = R_{970} R_{900}^{-1}$$

are two such indexes potentially useful in breeding programs, and are therefore briefly discussed.

The PRI index is used for quantifying the status of the xanthophyll pigments without canopy-scale photosynthetically active radiation manipulations. The basis of this is that changes in the region of spectral reflectance from about 500 to 560 nm wavelength are the result of the interconversion of the xanthophyll pigments. The xanthophyll cycle is involved in processes of excess radiation dissipation (Demmig-Adams and Adams, 1992). Indeed, some processes of excess radiation dissipation have been associated with changes in leaf reflectance near 531 nm compared with those of a reference wavelength (R_{REF}, e.g., 570 nm). Based on the above evidence, PRI could be used to assess photosynthetic use efficiency at leaf or canopy level in a way similar to the widely used fluorescence indicators of the photochemical efficiency of the photosystem II (Gamon, Peñuelas, and Field, 1992; Filella et al., 1996).

The WI has been closely related to changes in relative water content, leaf water potential, stomatal conductance, and canopy temperature when plant water stress was severe (Peñuelas et al., 1993). Water status can then be monitored at the canopy level using the ratio between the reflectance at 970 nm (one of the water absorption bands) and at 900 nm (reference wavelength). For cereals exposed to a soil salinity gradient, WI measured during grain filling has also been seen to be well correlated with carbon isotope discrimination of mature grains and canopy temperature (Peñuelas et al., 1996).

Stable Isotope Composition

Selection based on remote sensing techniques requires each genotype to be grown in plots; this restricts the techniques to advanced generations of a breeding program. Early generation selection is the most desirable as a greater number of plants/lines can often be handled and fewer lines need to be carried through to the next generation. Carbon isotope discrimination is a promising characteristic for early generation selection. In this section we discuss the use of carbon isotope discrimination as well as promising surrogates that might substantially reduce the number of plants and therefore the cost of carbon isotope discrimination.

Carbon and Oxygen Isotopic Composition

The molar abundance of ^{13}C relative to ^{12}C in plant tissues is less (particularly in C_3 plants) than that naturally occurring in atmospheric CO_2, revealing that plants discriminate against ^{13}C in the photosynthetic process (Farquhar, Ehleringer, and Hubick, 1989); a phenomenon termed carbon isotope discrimination (Δ). Although Δ is partly due to kinetic discrimination in diffusion of CO_2 in air and water, its final value is strongly associated with the discrimination of Rubisco against the heavier isotope (Farquhar, Ehleringer, and Hubick, 1989; Hall, 1990) during carboxylation. It is, in turn, directly related to the intracellular concentration of CO_2 (C_i) relative to that in air (C_a) (Farquhar et al., 1982; Farquhar et al., 1989; Hall, 1990). This indicates that the level of ^{13}C discrimination by Rubisco would decrease as leaf internal CO_2 concentration decreases. Thus the value of Δ correlates positively with the ratio C_i/C_a, and negatively with water use efficiency (WUE), measured at either the leaf or canopy level of organization.

Currently, it is widely accepted that for C_3 plants, such as small-grain cereals, Δ provides an integrated measurement of the water use efficiency, or more correctly transpiration efficiency (Farquhar and Richards, 1984; Hubick and Farquhar, 1989), and that Δ could be used as a powerful technique for screening for improved water use efficiency (Farquhar and Richards, 1984; Farquhar et al., 1989; Hall et al., 1994). However, Δ may be a useful criterion to select for potential yield as well. Several studies have reported a positive correlation between grain yield and Δ for bread wheat (e.g., Condon, Richards, and Farquhar, 1987; Araus, Reynolds, and Acevedo, 1993; Sayre, Acevedo, and Austin, 1995), barley (Romagosa and Araus, 1991), and durum wheat (Araus et al., 1998) not only under drought but also without water stress (Araus, 1996; Fischer et al., in press; Sayre, Rajaram, and Fischer, 1997). A positive relationship between Δ and growth has been reported for seedlings grown under adequate water conditions (Febrero et al., 1992; López-Castañeda and Richards, 1994). Indeed, increased early growth and leaf area development may be inherently linked with decreased transpiration efficiency (Turner, 1993) and thus higher Δ. The positive correlation between early vigor and Δ could be due to factors other than thinner leaves alone. Thus, as Δ reflects the water status of the plant, a higher Δ might be an indirect indicator of greater cell elongation due to better water status.

There are other reasons for the positive correlation between Δ and yield in the absence of water stress (see also Richards, 1996a). A higher Δ is related to a higher C_i/C_a ratio due to a larger stomatal conductance (Farquhar and Richards, 1984), which leads to higher photosynthetic rates. This may in turn be due to a greater demand for photosynthate and a greater sink strength (Richards, 1996b) or to a higher intrinsic stomatal conductance. In addition, giving a constant leaf-to-air vapor pressure difference, Δ may provide an integrated indication of transpiration. Therefore, when grown at supraoptimal temperatures Δ may also be related to heat avoidance.

Emphasizing its usefulness for selection, Δ has been shown to be both quite variable and highly heritable (Richards and Condon, 1993), and the variation is independent of phenological differences among genotypes (Sayre, Acevedo, and Austin, 1995; Araus et al., 1998). For example, Condon and Richards (1992) reported evidence for strong genetic control of Δ in wheat with a broad sense heritability ranging from 60 to 98 percent depending on tissue sampled whereas narrow sense heritability in one environment was 82 percent (Richards and Condon, unpublished). These results indicate the feasibility for selecting for a desirable level of Δ in wheat.

The use of Δ as a selection trait might be more efficient if samples of both leaf tissues (from early stages of the crop; Sayre, Acevedo, and Austin, 1995) and mature grains are taken. The former would provide information about basic genetic variation in the population and maybe early vigor, while Δ in grains would probably be more affected by genotype-by-environment interactions and could be useful to mark genotypes less affected by mild stresses that inevitably occur, even under "optimal" agronomic conditions (e.g., midday stomatal closure and transient photoinhibition). Discrimination against ^{13}C is related to both assimilation rate and stomatal conductance. Evidence that oxygen isotope discrimination ($^{18}O/^{16}O$) in leaves is related to conductance is emerging (Farquhar, Barbour, and Henry, 1998). This measurement offers considerable promise as an indicator of transpiration and stomatal conductance. It has advantages similar to carbon isotope discrimination in that it is an integrative measurement and it is relatively nondestructive. In conjunction with $^{13}C/^{12}C$ measurement, $^{18}O/^{16}O$ offers the potential to distinguish between assimilation rate and stomatal conductance as determinants of yield.

Ash Content and NIRS Analysis

The expense involved in carbon isotope analysis (about US $10 per analysis) has meant the emergence of several surrogates for the measurement of Δ. These include the accumulation of minerals such as K or Si, or the total ash content in the vegetative tissues of cereals and forages (Walker and Lance, 1991; Masle, Farquhar, and Wong, 1992; Mayland et al. 1993; Araus et al., 1998). Several aspects still need to be clarified, particularly the mechanisms underlying the genetic association between mineral accumulation and WUE (Walker and Lance, 1991; Masle, Farquhar, and Wong, 1992). However, it seems evident that the amount of minerals accumulated by plants may be a useful indicator of WUE under field conditions (Masle, Farquhar, and Wong, 1992; Mayland et al., 1993). Total ash (or mineral) content seems to be better (negatively) related to WUE than any one mineral. For this reason, as well as its low cost, ash content might become an alternative criterion to Δ, particularly during the early phases of a breeding program when large populations are usually involved. Later selections could be based on the more precise and accurate, yet costly, Δ analysis (Mayland et al., 1993). Indeed, ash content is positively correlated with yield under well-watered field conditions (Mayland et al., 1993; Araus et al., 1998).

A surrogate of both Δ (Clark et al., 1995) and total ash content (Windham, Hill, and Stuedemann, 1991) has been reported for grasses utilizing near-infrared reflectance spectroscopy (NIRS). Indeed, NIRS might be a very useful technique in the routine screening of early generations. In laboratory settings, NIRS is currently the basis for fast, accurate, nondestructive, and highly repeatable assays of many biological materials, including digestibility, nitrogen, energy content, moisture, ash, crude fats, total reducing sugars, alkaloids, and a number of other compounds and classes of compounds in plant matter (Clark, 1989). Preliminary analyses of barley grains (Catala, Blanco and Araus, unpublished results) report reasonably good correlations between Δ, measured by mass spectrometry, and the NIRS assessment.

CONCLUDING REMARKS

Although yield potential has been successfully increased by empirical selection for yield per se, there is a growing argument that to ensure further gains, conventional breeding has to be complemented with a physiological approach to yield improvement. All increases in yield have a physiological basis and the more precisely we can target the traits physiologically linked to yield then the more successful genetic progress will be. The identification of key yield-determining traits has to be associated with the development of reliable techniques to screen for them. In this chapter we have discussed not only physiological attributes that may limit the yield potential of wheat and identified traits that could be used for choosing parents and/or selecting progeny, but also possibilities for some techniques to be used to assess large numbers of lines in a commercial breeding program.

In agreement with others we conclude that for most regions harvest index is close to its theoretical upper threshold and that future increases in yield potential will require the development of cultivars with greater biomass. Nevertheless, if harvest index is low, and there is evidence it can be increased, then it may be more efficient to select for higher spike-to-stem weight ratio at anthesis than for harvest index itself. We note that plant height is negatively related to harvest index, although further selection for reduction in height is discouraged as modern wheats are already within the height range that optimizes yield. It is unlikely that future increases in yield will be achieved by further reducing plant height.

We consider two very promising opportunities to improve wheat yield to increase (1) the capacity of the crop to intercept radiation during the initial growth phases, and (2) the ability of the spikes to capture assimilates during the period immediately preceding anthesis. The former would impact more on yield in stressful environments or where the growing season is short. Selecting for greater leaf area immediately after emergence would produce more vigorous cultivars that intercept more of the incoming radiation during the initial phases. Early leaf area development depends upon the size of the embryo and on specific leaf area. The breadth of the early seedling leaves has been shown to integrate both embryo size and specific leaf area and this can be used to screen thousands of seedlings cheaply and quickly for fast early vigor. Remote sensing techniques also offer promise to screen minicanopies for high early vigor.

For environments where grain number is consistently related to yield, maximizing spike weight per unit land area at anthesis should result in greater yields. The traits we consider most likely to result in heavier spikes at anthesis include: (1) increasing radiation use efficiency and total biomass in the period immediately preceding anthesis, (2) further improving partitioning to the spike without altering the distribution of leaf layers, and (3) lengthening the spike growth period. An increased radiation use efficiency could be achieved by several means. The most promising could be to improve the distribution of incident radiation within the canopy (reducing the coefficient of light attenuation) through a reduction in flag leaf dimensions. The development of a small flag leaf may also release some assimilates to be used by the simultaneously growing spike. An increase in assimilate supply to the growing spike could also be achieved by reducing the length of the peduncle, but not that of the other internodes (so that plant height is reduced without penalties in radiation distribution within the canopy) or by a reduction of tillering, so that more assimilates may be available for fertile tiller growth. Extending the period during which spikes grow and the number of grains are determined could also be altered.

Finally, genetic manipulation of grain weight should be worthwhile. However, we do not yet understand the physiological determinants of individual grain weight potential and until then we can only speculate on some attributes that could be useful. Manipulating grain growth responses to temperature (i.e., the thermal time to grain maturation and the base temperature for the process to occur) are the most obvious traits. It is also possible that the widely used *Rht* genes (*Rht1* and *Rht2*) impose a structural restriction on the growth of the seed coat. This could be overcome by using *Rht* genes sensitive to GA or minor genes that regulate plant height. In both examples, we need more knowledge of the physiological and genetic basis of these parameters before we can more specifically target grain size.

If physiological traits are used as selection criteria in breeding programs, then simple techniques are needed to screen for them that are quick and reliable. Not many traits fulfill these requirements. However, a number of potentially important traits can be easily selected. Some are visual, such as flag leaf dimensions, leaf erectness, and peduncle length. Others require simple instruments, such as vigor (a ruler), ear-to-stem ratio (a balance), or stomatal conductance (an air flow porometer [Rebetzke, Condon, and Richards, 1996] or infrared thermometry), whereas spectroradiometers offer potential to screen for canopy structure and light interception. Yet the measurement of other traits may be more expensive, such as the determination of $^{13}C/^{12}C$ and $^{18}O/^{16}O$, although these have other advantages in that they integrate growth characteristics over time and are often highly repeatable. Nevertheless, even traits considered expensive or difficult to measure often can be simplified, such as the measurement of ash or the use of NIR instead of mass spectrometry to determine $^{13}C/^{12}C$, and the use of an air-flow porometer rather than a steady-state porometer to measure stomatal conductance. Combinations of measurements can also be used. For example, in the CSIRO breeding program for carbon isotope discrimination, the air flow porometer is being used to first cull a large population of lines so that only a subset needs to be analyzed for $^{13}C/^{12}C$ discrimination. Furthermore, for difficult-to-measure traits or traits with low heritability or high genotype $\times$ environmental interactions, then molecular markers offer considerable potential. The final choice of the trait and its measurement will depend on the magnitude of the heritability, the relationship with yield, and the cost and speed of measurement.

The ability to transform wheat opens up opportunities to increase wheat yields using molecular biology. Although there is no compelling evidence that the introduction of a new gene into wheat will increase its yield, we would like to suggest areas of research where molecular biology could greatly increase our understanding of yield physiology and may result in yield increases. A first area is the regulation of sucrose metabolism and the transport of sucrose in leaves to developing apices and in the conversion of sucrose to starch in the grain. Another is the regulation of gibberellic acid biosynthesis, which may regulate germination, leaf and stem growth, and grain growth. The third area is the regulation of cytokinins and genes that control cell division to better understand how meristematic zones and developing grains may be modified.

It is now important to explore the use of these traits in breeding programs, to explore the diversity of genetic variation in them and whether they are amenable to selection, and also the association between the trait and yield, possible pleiotropic effects, and the importance of the genetic background. The technology to rapidly and efficiently develop recombinant inbred lines is now available. This will help to unambiguously validate the importance of many of these traits and to determine any possible complications associated with them.

REFERENCES

Abbate, P.E., Andrade, F.H., and Culot, J.P. (1995). The effects of radiation and nitrogen on number of grains in wheat. *Journal of Agricultural Science* 124, 351-360.

Aikman, D.P. (1989). Potential increase in photosynthetic efficiency from the redistribution of solar radiation in a crop. *Journal of Experimental Botany* 40, 855-864.

Allan, R.E., and Pritchett J.A. (1980). Registration of 16 lines of club wheat germplasm. *Crop Science* 5, 5-8.

Allison, J.C., and Daynard, T.B. (1976). Effect of photoperiod on development and number of spikelets of a temperate and some low-latitude wheats. *Annals of Applied Biology* 83, 93-102.

Araus, J.L. (1996). Integrative physiological criteria associated with yield potential. In *Increasing Yield Potential in Wheat: Breaking the Barriers* (Eds. M.P. Reynolds, S. Rajaram, and A. McNab), pp. 150-166. (CIMMYT: Mexico, DF).

Araus, J.L., Amaro, T., Casadesús, J., Asbati, A., and Nachit, M.M. (1998). Relationships between ash content, carbon isotope discrimination and yield in durum wheat. *Australian Journal of Plant Physiology* 25, 835-842.

Araus, J.L., Reynolds, M.P., and Acevedo, E. (1993). Leaf posture, grain yield, growth, leaf structure and carbon isotope discrimination in wheat. *Crop Science* 33, 1273-1279.

Araus, J.L., and Tapia, L. (1987). Photosynthetic gas exchange characteristics of wheat flag leafblades and sheaths during grain filling. The case of a spring crop grown under Mediterranean climate conditions. *Plant Physiology* 85, 667-673.

Araus, J.L., Tapia, L., and Alegre, L. (1989). The effect of changing sowing date on leaf structure and gas exchange characteristics of wheat flag leaves grown under Mediterranean conditions. *Journal of Experimental Botany* 40, 639-646.

Austin, R.B. (1989). Genetic variation in photosynthesis. *Journal of Agricultural Science* 112, 287-294.

Austin, R.B. (1992). "Can we improve on nature?" *Abstracts of the First International Crop Science Congress,* Ames, Iowa. (Eds. D.R. Buxton, R. Shibles, R.A. Frosberg, B.L. Blad, K.H. Asay, G.M. Paulsen, and R.F. Wilson) (Crop Science Society of America: Madison, WI).

Austin, R.B. (1993). Augmenting yield-based selection. In *Plant Breeding: Principles and Prospects* (Eds. M.D. Hayward, N.O. Bosemark, and I. Romagosa), pp. 391-405. (Chapman and Hall: London).

Austin, R.B., Morgan, C.L., Ford, M.A., and Bhagwat, S.C. (1982). Flag leaf photosynthesis of *Triticum aestivum* and related diploid and tetraploid species. *Annals of Botany* 49, 177-189.

Austin, R.B., Morgan, C.L., Ford, M.A., and Blackwell, R.D. (1980). Contributions to grain yield from preanthesis assimilation in tall and dwarf barley phenotypes in two contrasting seasons. *Annals of Botany* 45, 309-319.

Azcon-Bieto, J. (1983). Inhibition of photosynthesis by carbohydrates in wheat leaves. *Plant Physiology* 73, 681-686.

Bartlett, D.S., Whiting, G.J., and Hartman, J.M. (1990). Use of vegetation indices to estimate intercepted solar radiation and net carbon dioxide exchange of a grass canopy. *Remote Sensing of Environment* 30, 115-128.

Bhagsari, A.S., and Brown, R.H. (1986). Leaf photosynthesis and its correlation with leaf area. *Crop Science* 26, 127-131.

Bindraban, P. (1996). Quantitative understanding of wheat growth and yield for identifying crop characteristics to further increase yield potential. In *Increasing Yield Potential in Wheat: Breaking the Barriers* (Eds. M.P. Reynolds, S. Rajaram, and A. McNab), pp. 230-236. (CIMMYT: Mexico, DF).

Biscoe, P.V., and Gallagher, J.N. (1977). Weather, dry matter production and yield. In *Environmental Effects on Crop Physiology* (Eds. J. Landsberg and C. Cutting), p. 75. (Academic Press: London).

Blum, A. (1988). *Plant Breeding for Stress Environments.* (CRC Press: Boca Raton, FL).

Blum, A., Mayer, J., and Golan, G. (1982). Infrared thermal sensing of plant canopies as a screening technique for dehydration avoidance in wheat. *Field Crops Research* 5, 137-146.

Blum, A., Shpiler, L., Golan, G., and Mayer, J. (1989). Yield stability and canopy temperature of wheat genotypes under drought stress. *Field Crops Research* 22, 289-296.

Borghi, B., Corbellini, M., Cattaneo, M., Fornasari, M.A., and Zucchelli, L. (1986). Modification of the sink/source relationship in bread wheat and its influence on grain yield and protein content. *Journal of Agronomy and Crop Science* 157, 245-254.

Borrell, A.K., Incoll, L.D., and Dalling, M.J. (1991). The influence of the *Rht1* and *Rht2* alleles on the growth of wheat stems and ears. *Annals of Botany* 67, 103-110.

Brocklehurst, P.A. (1977). Factors controlling grain weight in wheat. *Nature* 266, 348-349.

Brooking, I.R., and Kirby, E.J.M. (1981). Interrelationship between stem and ear development in winter wheat: The effects of a Norin 10 dwarfing gene. *Gai/Rht2. Journal of Agricultural Science* 97, 373-381.

Bush, M.G., and Evans, L.T. (1988). Growth and development in tall and dwarf isogenic lines of spring wheat. *Field Crops Research* 18, 243-270.

Calderini D.F., Dreccer, M.F., and Slafer, G.A. (1995). Genetic improvement in wheat yield and associated traits. A re-examination of previous results and latest trends. *Plant Breeding* 14, 108-112.

Calderini, D.F., Miralles, D.J., and Sadras, V.O. (1996). Appearance and growth of individual leaves as affected by semidwarfism in isogenic lines of wheat. *Annals of Botany* 77, 583-589.

Carvalho, F.I.F., and Qualset, C.O. (1978). Genetic variation for canopy architecture and its use in wheat breeding. *Crop Science* 18, 561-567.

Carver, B.F., Johnson, R.C., and Rayburn, A.L. (1989). Genetic analysis of photosynthetic diversity in hexaploid and tetraploid wheat and their interspecific hybrids. *Photosynthesis Research* 20, 105-118.

Carver, B.F., and Nevo, E. (1990). Genetic diversity of photosynthetic characters in native populations of *Triticum dicoccoides. Photosynthesis Research* 25, 119-128.

Clark, D.H. (1989). History of NIRS analysis of agricultural products. In *Near Infrared Reflectance Spectroscopy (NIRS): Analysis of Forage Quality,* (Eds. G.C. Marten, J.S. Shenk, and F.E. Barton II), pp. 7-11. U.S. Department of Agriculture (Agriculture Handbook 643: Washington DC).

Clark, D.H., Johnson, D.A., Kephart, K.D., and Jackson, N.A. (1995). Near infrared reflectance spectroscopy estimation of 13C discrimination in forages. *Journal of Range Management* 48, 132-136.

Cochrane, M.P., and Duffus, C.M. (1982). Endosperm cell number in cultivars of barley differing in grain weight. *Annals of Applied Biology* 102, 177-181.

Condon, A.G., and Richards, R.A. (1992). Broad sense heritability and geotype × environment interaction for carbon isotope discrimination in field-grown wheat. *Australian Journal of Agricultural Research* 43, 921-934.

Condon, A.G., Richards, R.A., and Farquhar, G.D. (1987). Carbon isotope discrimination is positively correlated with grain yield and dry matter production in field-grown wheat. *Crop Science* 27, 996-1001.

Connor, D.J., Sadras, V.O., and Hall, A.J. (1995). Canopy nitrogen distribution and the photosynthetic performance of sunflower crops during grain filling—a quantitative analysis. *Oecologia* 101, 274-281.

Cox, T.S., Shoroyer, J.P., Ben-Hui, L., Sears, R.G., and Martin, T.J. (1988). Genetic improvement in agronomic traits of hard red winter wheat cultivars from 1919 to 1987. *Crop Science* 28, 756-760.

Curran, P.J. (1983). Multispectral remote sensing for the estimation of green leaf area index. *Philosophical Transactions of the Royal Society,* London (Series A) 309, 257-270.

Demmig-Adams, B., and Adams, W.W. III. (1992). Carotenoid composition in sun and shade leaves of plants with different life forms. *Plant, Cell and Environment* 15, 411-419.

Donald, C.M. (1968). The breeding of crop ideotypes. *Euphytica* 17, 385-403.

Dreccer, M.F., Slafer, G.A., and Rabbinge, R. (1998). Optimization of vertical distribution of canopy nitrogen: An alternative trait to increase yield potential in wheat. *Journal of Crop Production* 1, 47-77.

Elliott, G.A., and Regan, K.L. (1993). Use of reflectance measurements to estimate early cereal biomass production on sandplain soils. *Australian Journal of Experimental Agriculture* 33, 179-183.

Evans, L.T. (1978). The influence of irradiance before and after anthesis on grain yield and its components in microcrops of wheat grown in a constant daylength and temperature regime. *Field Crops Research* 1, 5-19.

Evans, L.T. (1992). Processes, genes and yield potential. *Abstracts of the First International Crop Science Congress,* Ames, Iowa. (Eds. D.R. Buxton, R. Shibles, R.A. Frosberg, B.L. Blad, K.H. Asay, G.M. Paulsen, and R.F. Wilson) (Crop Science Society of America: Madison, WI).

Evans, L.T. (1993). *Crop Evolution, Adaptation and Yield.* (Cambridge University Press: New York).

Evans, L.T., and Dunstone, R.L. (1970). Some physiological aspects of evolution in wheat. *Australian Journal of Biological Science* 23, 725-741.

Farquhar, G.D., Ball, M.C., von Caemmerer, S., and Roksandic, Z. (1982). Effects of salinity and humidity on ^{13}C value of halophytes—evidence for diffusional isotope fractionation determined by the ratio of intracellular/atmospheric partial pressure of CO_2 under different environmental conditions. *Oecologia* 52, 121-124.

Farquhar, G.D., Barbour, M.M., and Henry, B.K. (1998). Interpretation of oxygen isotope composition of leaf material. In *Stable Isotopes* (Ed. H. Griffiths), pp. 27-48. (Bios Scientific Publishers: Oxford, UK).

Farquhar, G.D., Ehleringer, J.R., and Hubick, K.T. (1989). Carbon isotope discrimination and photosynthesis. *Annual Review of Plant Physiology and Plant Molecular Biology* 40, 503-537.

Farquhar, G.D., and Richards, R.A. (1984). Isotopic composition of plant carbon correlates with water-use-efficiency of wheat genotypes. *Australian Journal of Plant Physiology* 11, 539-552.

Farquhar, G.D., Wong, S.C., Evans, J.R., and Hubick, K.T. (1989). Photosynthesis and gas exchange. In *Plants Under Stress: Biochemistry, Physiology and Ecology and their Application to Plant Improvement* (Eds. H.G. Jones, T.J. Flowers, and M.B. Jones), p. 47. (Cambridge University Press: Cambridge, UK).

Febrero, A., Blum, A., Romagosa, I., and Araus, J.L. (1992). Relationships between carbon isotope discrimination in field grown barley and some physiological traits of juvenile plants in growth chambers. *Abstracts Supplement of the First International Crop Science Congress*, Ames, IA, (Eds. D.R. Buxton, R. Shibles, R.A. Frosberg, B.L. Blad, K.H. Asay, G.M. Paulsen, and R.F. Wilson), p. 26 (Crop Science Society of America: Madison, WI).

Fernandez, S., Vidal, D., Simon, E., and Sole-Sugranes, L. (1994). Radiometric characteristics of *Triticum aestivum* cv. Astral under water and nitrogen stress. *International Journal of Remote Sensing* 15, 1867-1884.

Field, C. (1983). Allocating leaf nitrogen for the maximisation of carbon gain: Leaf age as a control on the allocation program. *Oecologia* 56, 341-347.

Field, C.B., Gamon, J.A., and Peñuelas, J. (1994). Remote sensing of terrestrial photosynthesis. In *Ecophysiology of Photosynthesis* (Eds. D. Schulze and M.M. Caldwell). Ecological studies, Volume 100. (Springer-Verlag: Berlin).

Filella, I., Amaro, T., Araus, J.L., and Peñuelas, J. (1996). Relationship between photosynthetic radiation-use efficiency of barley canopies and the photochemical reflectance index (PRI). *Physiologia Plantarum* 96, 211-216.

Fischer, R.A. (1984). Wheat. In *Symposium on Potential Productivity of Field Crops Under Different Environments* (Eds. W.H. Smith and S.J. Banta), pp. 129-153. (IRRI: Los Baños, Philippines).

Fischer, R.A. (1985). Number of kernels in wheat crops and the influence of solar radiation and temperature. *Journal of Agricultural Science* 100, 447-461.

Fischer, R.A., and Quail, K.J. (1990). The effect of major dwarfing genes on yield potential in spring wheats. *Euphytica* 46, 51-56.

Fischer, R.A., Rees, D., Sayre, K.D., Lu, Z., Condon, A.G., Larque Saavedra, A., and Zeiger, E. (in press). Wheat yield progress is associated with higher stomal conductance, higher photosynthetic rate and cooler canopies. *Crop Science*.

Fischer, R.A., and Stockman, Y.M. (1986). Increased kernel number in Norin 10-derived dwarf wheat: Evaluation of the cause. *Australian Journal of Plant Physiology* 13, 767-784.

Flood, R.G., and Halloran, G.M. (1986). The influence of genes for vernalization response on development and growth in wheat. *Annals of Botany* 58, 505-508.

Gallo, K.P., Daughtry, C.S.T., and Bauer, M.E. (1985). Spectral estimation of absorbed photosynthetically active radiation in corn canopies. *Remote Sensing of Environment* 17, 221-232.

Gamon, J.A., Field, C.B., Goulden, M.L., Griffin, K.L., Hartley, A.E., Joel G., Peñuelas, J., and Valentini R. (1995). Relationships between NDVI, canopy structure, and photosynthesis in three Californian vegetation types. *Ecological Applications* 5, 28-41.

Gamon, J.A., Peñuelas, J., and Field, C.B. (1992). A narrow-waveband spectral index that tracks diurnal changes in photosynthetic efficiency. *Remote Sensing of Environment* 41, 35-44.

Gifford, R.M., Thorne, J.H., Hitz, W.D., and Giaquinta, R.T. (1984). Crop productivity and photoassimilate partitioning. *Science* 225, 801-808.

Giménez, C., Connor, D.J., and Rueda, F. (1994). Canopy development, photosynthesis and radiation use efficiency in sunflower in response to nitrogen. *Field Crops Research* 38, 15-27.

Gomez-MacPherson, H.A. (1993). Variation in phenology and its influence on growth, development and yield of dryland wheat. Ph.D. Thesis. Australian National University, Canberra, Australia.

Green, C.F. (1989). Genotypic differences in the growth of *Triticum aestivum* in relation to absorbed solar radiation. *Field Crops Research* 19, 285-295.

Hall, A.E. (1990). Physiological ecology of crops in relation to light, water and temperature. *Agroecology* (Eds. C.R. Carroll, J.H. Vandermeer, and P.M. Rosset) p. 191 (McGraw-Hill: New York).

Hall, A.E., Richards, R.A., Condon, A.G., Wright, G.C., and Farquhar, G.D. (1994). Carbon isotope discrimination and plant breeding. *Plant Breeding Reviews* 12, 81-113.

Halloran, G.M., and Pennel, A.L. (1982). Duration and rate of development phases in wheat in two environments. *Annals of Botany* 49, 115-121.

Hammer, G.L., and Wright, G.C. (1994). A theoretical analysis of nitrogen and radiation effects on radiation use efficiency in peanut. *Australian Journal of Agricultural Research* 45, 575-589.

Hanson, P.R., Riggs, T.J., Klose, S.J., and Austin, R.B. (1985). High biomass genotypes in spring barley. *Journal of Agricultural Science* 105, 73-78.

Hirose, T., and Werger, M.J.A. (1987). Maximizing daily canopy photosynthesis with respect to the leaf nitrogen allocation pattern in the canopy. *Oecologia* 72, 520-526.

Housley, T.L., and Ohm, H.W. (1992). Earliness and duration of grain fill in winter wheat. *Canadian Journal of Plant Science* 72, 35-48.

Hubick, K., and Farquhar, G.D. (1989). Carbon isotope discrimination and the ratio of carbon gained to water lost in barley cultivars. *Plant Cell and Environment* 12, 795-804.

Hucl, R., and Baker, R.J. (1987). A study of ancestral and modern Canadian spring wheats. *Canadian Journal of Plant Science* 67, 87-91.

Hunt, L.A., van der Poorten, G., and Pararajasingham, S. (1991). Postanthesis temperature effects on duration and rate of grain filling in some winter and spring wheats. *Canadian Journal of Plant Science* 71, 609-617.

Innes, P., and Blackwell, R.D. (1983). Some effects of leaf posture on the yield and water economy of winter wheat. *Journal of Agricultural Science* 101, 367-376.

Johnson, R.C., Kebede, H., Mornhinweg, D.W., Carver, B.F., Rayburn, A.L., and Nguyen, H.T. (1987). Photosynthetic differences among Triticum accessions at tillering. *Crop Science* 27, 1046.

Keyes, G.H., Paolillo, D.J., and Sorrells, M.E. (1989). The effects of dwarfing genes *Rht1* and *Rht2* on cellular dimensions and rate of leaf elongation in wheat. *Annals of Botany* 64, 683-690.

Kiniry, J.R., Jones, C.A., O'Toole, J.C., Blanchet, R., Cabelguenne, M., and Spanel, D.A. (1989). Radiation-use efficiency in biomass accumulation prior to grain-filling for five grain-crop species. *Field Crops Research* 20, 51-64.

Kruk, B., Calderini, D.F., and Slafer, G.A. (1997). Source-sink ratios in modern and old wheat cultivars. *Journal of Agricultural Science* 128, 273-281.

Kumar, M., and Montheith, J.L. (1981). Remote sensing of crop growth. In *Plants and the Daylight Spectrum* (Ed. H. Smith), pp. 133-144. (Academic Press: London.)

Labraña, X. and Araus, J.L. (1991). Effect of foliar applications of silver nitrate and ear removal on carbon dioxide assimilation in wheat flag leaves during grain filling. *Field Crops Research* 28, 149-162.

LeCain, D.R., Morgan, J.A., and Zerbi, G. (1989). Leaf anatomy and gas exchange in nearly isogenic semidwarf and tall winter wheat. *Crop Science* 29, 1246-1251.

Liang, Y.L., and Richards, R.A. (1994). Coleoptile tiller development is associated with fast early vigour in wheat. *Euphytica* 80, 119-124.

López-Castañeda, C., and Richards, R.A. (1994). Variation in temperate cereals in rainfed environments. I. Grain yield, biomass and agronomic characteristics. *Field Crops Research* 37, 51-62.

López-Castañeda, C., Richards, R.A., and Farquhar, G.D. (1995). Variation in early vigour between barley and wheat. *Crop Science* 35, 472-479.

López-Castañeda, C., Richards, R.A., Farquhar, G.D., and Williamson, R.E. (1996). Seed and seedling characteristics contributing to variation in seedling vigor among temperate cereals. *Crop Science* 36, 1257-1266.

Loss, S.P., and Siddique, K.H.M. (1994). Morphological and physiological traits associated with wheat yield increases in Mediterranean environments. *Advances in Agronomy* 52, 229-276.

Mac Maney, M., Diaz, R., Simon, C., Gioia, A., Slafer, G.A., and Andrade, F.H. (1986). Respuesta a la reduccion de la capacidad fotosintetica durante el llenado de granos en trigo. In *Proceedings I National Congress of Wheat*, pp. 178-190. (Pergamino.) (AIANBA: Buenos Aires, Argentina).

Manupeerapan, T., Davidson, J.L., Pearson, C.J., and Christian, K.R. (1992). Differences in flowering responses of wheat to temperature and photoperiod. *Australian Journal of Agricultural Research* 43, 575-584.

Marcellos, H., and Single, W.V. (1972). The influence of cultivar, temperature, and photoperiod on post-flowering development of wheat. *Australian Journal of Agricultural Research* 23, 533-540.

Masle, J., Farquhar, G.D., and Wong, S.C. (1992). Transpiration ratio and plant mineral content are related among genotypes of a range of species. *Australian Journal of Plant Physiology* 19, 709-721.

Mayland, H.F., Johnson, D.A., Asay, K.H., and Read, J.J. (1993). Ash, carbon isotope discrimination and silicon as estimators of transpiration efficiency in crested wheatgrass. *Australian Journal of Plant Physiology* 20, 361-369.

McCaig, T.N., and Morgan, J.A. (1993). Root and shoot dry matter partitioning in near-isogenic wheat lines differing in height. *Canadian Journal of Plant Science* 73, 679-689.

McClung, A.M., Cantrell, R.G., Quick, J.S., and Gregory, R.S. (1986). Influence of the *Rht* semidwarf gene on yield, yield components and grain protein in durum wheat. *Crop Science* 26, 1095-1098.

Miralles, D.J., and Slafer, G.A. (1995a). Yield, biomass and yield components in dwarf, semidwarf and tall isogenic lines of spring wheat under recommended and late sowing dates. *Plant Breeding* 114, 392-396.

Miralles, D.J., and Slafer, G.A. (1995b). Individual grain weight responses to genetic reduction in culm length in wheat as affected by source-sink manipulation. *Field Crops Research* 43, 55-66.

Monsi, M., and Saeki, T. (1953). Über den Lightfaktor in den Pflanzengesellschaften und seine Bedeutung für die Stoff-produktion. *Japanese Journal of Botany* 14, 22-52.

Morgan, J.A., and LeCain, D.R. (1991). Leaf gas exchange and related leaf traits among 15 winter wheat genotypes. *Crop Science* 31, 443-448.

Morgan, J.A., LeCain, D.R., and Wells, R. (1990). Semidwarfing genes concentrate photosynthetic machinery and affect leaf gas exchange of wheat. *Crop Science* 30, 602-608.

Nátrová, Z., and Nátr, L. (1993). Limitation of kernel yield by the size of conducting tissue in winter wheat varieties. *Field Crops Research* 31, 121-130.

Nilson, E.R., Johnson, V.A., and Gardner, C.O. (1957). Parenchyma and epidermal cell length in relation to plant height and culm internode in winter wheat. *Botanical Gazzette* 119, 38-43.

Peñuelas, J., Filella, I., Biel, C., Serrano, L., and Save, R. (1993). The reflectance at the 950-970 nm region as an indicator of plant water status. *International Journal of Remote Sensing* 14, 1887-1905.

Peñuelas, J., Isla, R., Filella, I., and Araus, J.L. (1996). Visible and near-infrared reflectance assessment of salinity effects on barley. *Crop Science* 37, 198-202.

Perry, M.W., and D'Antuono, M.F. (1989). Yield improvement and associated characteristics of some Australian spring wheat cultivars introduced between 1860 and 1982. *Australian Journal of Agricultural Research* 40, 457-472.

Pinthus, M.J., Gale, M.D., Appleford, N.E.J., and Lenton, J.R. (1989). Effect of temperature on gibberellin (GA) responsiveness and on endogenous GA_1 content of tall and dwarf wheat genotypes. *Plant Physiology* 90, 854-859.

Planchon, C., and Fesquet, J. (1982). Effect of the D genome and of selection on photosynthesis in wheat. *Theoretical and Applied Genetics* 61, 359-365.

Quail, K.J., Fischer, R.A., and Wood, J.T. (1989). Early generation selection in wheat. I. Yield potential. *Australian Journal of Agricultural Research* 40, 1117-1133.

Rahman, M.S., and Wilson, J.H. (1977). Determination of spikelet number in wheat. I. Effects of varying photoperiod on ear development. *Australian Journal of Agricultural Research* 28, 565-574.

Rasmuson, M. and Zetterström, R. (1992). World population, environment and energy demands. *Ambio* 21, 70-74.

Rawson, H.M., and Evans, L.T. (1971). The contribution of stem reserves to grain development in a range of wheat cultivars of different height. *Australian Journal of Agricultural Research* 22, 851-863.

Rawson, H.M., Hindmarsh, J.H., Fischer, R.A., and Stockman, Y.R. (1983). Changes in leaf photosynthesis with plant ontogeny and relationships with yield per ear in wheat cultivars and 120 progeny. *Australian Journal of Plant Physiology* 10, 503-514.

Rebetzke, G.J., Condon, A.G., and Richards, R.A. (1996). Rapid screening of leaf conductance in segregating wheat populations. *Proceedings of the 8th Assembly.* (Eds. R.A. Richards, C.W. Wrigley, H.M. Rawson, G.J. Rebetzke, J.L. Davidson, and R.I.S. Brettell), pp. 130-133. (Wheat Breeding Society of Australia: Canberra).

Rees, D., Sayre, K., Acevedo, E., Nava-Sanchez, T., Lu, Z., Zeiger, E., and Limon, A. (1993). Canopy temperatures of wheat: Relationships with yield and potential as a technique for early generation selection. Wheat Special Rep. No. 10. (CIMMYT: Mexico, DF).

Reynolds, M.P., Balota, M., Delgado, M.I.B., Amani, I., and Fischer, R.A. (1994). Physiological and morphological traits associated with spring wheat yield under hot, irrigated conditions. *Australian Journal of Plant Physiology* 21, 717-730.

Reynolds, M.P., Rajaram, S., and McNab, A. (1996). *Increasing Yield Potential in Wheat: Breaking the Barriers.* (CIMMYT: Mexico, DF).

Reynolds, M.P., van Beem, J., van Ginkel, M., and Hoisington, D. (1996). Breaking the yield barriers in wheat: A brief summary of the outcomes of an international consultation. In *Increasing Yield Potential in Wheat: Breaking the Barriers* (Eds. M.P. Reynolds, S. Rajaram, and A. McNab), pp. 1-10. (CIMMYT: Mexico, DF).

Richards, R.A. (1988). A tiller inhibition gene in wheat and its effect on plant growth. *Australian Journal of Agricultural Research* 39, 749-757.

Richards, R.A. (1992). The effect of dwarfing genes in spring wheat in dry environments. I. Agronomic characteristic. *Australian Journal of Agricultural Research* 43, 517-522.

Richards, R.A. (1996a). Defining selection criteria to improve yield under drought. *Plant Growth Regulation* 20, 157-166.

Richards, R.A. (1996b). Increasing yield potential in wheat—source and sink limitations. In *Increasing Yield Potential in Wheat: Breaking the Barriers*

(Eds. M.P. Reynolds, S. Rajaram, and A. McNab), pp. 134-149. (CIMMYT: Mexico, DF).

Richards, R.A., and Condon, A.G. (1993). Challengers ahead in using carbon isotope discrimination in plant breeding programs. In *Stable Isotopes and Plant Carbon-Water Relations* (Eds. J.R. Ehleringer, A.E. Hall, and G.D. Farquhar), pp. 451-462. (Academic Press: San Diego).

Romagosa, I., and Araus, J.L. (1991). Genotype-environment interaction for grain yield and 13C discrimination in barley. *Barley Genetics* 6, 563-567.

Sadras, V.O., Hall, A.J., and Connor, D.J. (1993). Light-associated nitrogen distribution profile in flowering canopies of sunflower (*Helianthus annuus* L.) altered during grain growth. *Oecologia* 95, 488-494.

Savin, R., and Slafer, G.A. (1991). Shading effects on the yield of an Argentinian wheat cultivar. *Journal of Agricultural Science* 116, 1-7.

Sayre, K.D. (1996). The role of crop management research at CIMMYT in addressing bread wheat yield potential issues. In *Increasing Yield Potential in Wheat: Breaking the Barriers* (Eds. M.P. Reynolds, S. Rajaram, and A. McNab), pp. 203-207. (CIMMYT: Mexico, DF).

Sayre, K.D., Acevedo, E., and Austin, R.B. (1995). Carbon isotope discrimination and grain yield for three bread wheat germplasm groups grown at different levels of water stress. *Field Crops Research* 41, 45-54.

Sayre, K.D., Rajaram, S., and Fischer, R.A. (1997). Yield potential progress in short bread wheats in northwest Mexico. *Crop Science* 37, 36-42.

Sellers, P.J. (1987) Canopy reflectance, photosynthesis, and transpiration. II. The role of biophysics in the linearity of their interdependence. *Remote Sensing of Environment* 21, 43-183.

Shorter, R., Lawn, R.J., and Hammer, G.L. (1991). Improving genotypic adaptation in crops—a role for breeders, physiologists and modellers. *Experimental Agriculture* 27, 155-175.

Siddique, K.H.M., Belford, R.K., Perry, M.W., and Tennant, D. (1989). Growth, development and light interception of old and modern wheat cultivars in a Mediterranean type environment. *Australian Journal of Agricultural Research* 40, 473-487.

Siddique, K.H.M., Kirby, E.J.M., and Perry, M.W. (1989). Ear-to-stem ratio in old and modern wheats; relationship with improvement in number of grains per ear and yield. *Field Crops Research* 21, 59-78.

Siddique, K.H.M., and Whan, B.R. (1990). Ear:stem ratio—a selection criterion for yield improvement in wheat? In *Proceedings of the 6th Assembly* (Eds. L. O'Brien, F.W. Ellison, R.A. Hare, and M.C. MacKay) pp. 67-73. (Wheat Breeding Society of Australia: Tamworth).

Sinclair, T.R., and Shiraiwa, T. (1993). Soybean radiation use efficiency as influenced by non-uniform specific leaf nitrogen distribution and diffuse radiation. *Crop Science* 33, 808-812.

Slafer, G.A. (1996). Differences in phasic development rate amongst wheat cultivars independent of responses to photoperiod and vernalization. A viewpoint of the intrinsic earliness hypothesis. *Journal of Agricultural Science* 126, 403-419.

Slafer, G.A., and Andrade, F.H. (1989). Genetic improvement in bread wheat (*Triticum aestivum*) yield in Argentina. *Field Crops Research* 21, 289-296.

Slafer, G.A., and Andrade, F.H. (1991). Changes in physiological attributes of the dry matter economy of bread wheat (*Triticum aestivum*) through genetic improvement of grain yield potential at different regions of the world. A review. *Euphytica* 58, 37-49.

Slafer, G.A., Andrade, F.H., and Satorre, E.H. (1990). Genetic-improvement effects on pre-anthesis physiological attributes related to wheat grain yield. *Field Crops Research* 23, 255-263.

Slafer, G.A., Calderini, D.F., and Miralles, D.J. (1996). Yield components and compensation in wheat: Opportunities for further increasing yield potential. In *Increasing Yield Potential in Wheat: Breaking the Barriers* (Eds. M.P. Reynolds, S. Rajaram, and A. McNab), pp. 101-133. (CIMMYT: Mexico, DF).

Slafer, G.A., and Miralles, D.J. (1993). Fruiting efficiency in three bread wheat (*Triticum aestivum*) cultivars released at different eras. Number of grains per spike and grain weight. *Journal of Agronomy and Crop Science* 168, 191-200.

Slafer, G.A., and Rawson, H.M. (1994). Sensitivity of wheat phasic development to major environmental factors: A re-examination of some assumptions made by physiologists and modellers. *Australian Journal of Plant Physiology* 21, 393-426.

Slafer, G.A., and Rawson, H.M. (1995a). Development in wheat as affected by timing and length of exposure to long photoperiod. *Journal of Experimental Botany* 46, 1877-1886.

Slafer, G.A., and Rawson, H.M. (1995b). Base and optimum temperatures vary with genotype and stage of development in wheat. *Plant, Cell and Environment* 18, 671-679.

Slafer, G.A., and Rawson, H.M. (1996). Responses to photoperiod change with phenophase and temperature during wheat development. *Field Crops Research* 46, 1-13.

Slafer, G.A., and Savin, R. (1994). Source-sink relationship and grain mass at different positions within the spike in wheat. *Field Crops Research* 37, 39-49.

Slafer, G.A., Satorre, E.H., and Andrade, F.H. (1994). Increases in grain yield in bread wheat from breeding and associated physiological changes. In *Genetic Improvement of Field Crops: Current Status and Development* (Ed. G.A. Slafer), pp. 1-68. (Marcel Dekker, Inc.: New York).

Thorne, G.N., and Wood, D.W. (1987). Effects of radiation and temperature on tiller survival, grain number and grain yield in winter wheat. *Annals of Botany* 59, 413-426.

Tucker, C.J., Elgin, J.H. Jr., McMurtrey J.E. III, and Fan, C.J. (1979). Monitoring corn and soybean crop development with hand-held radiometer spectral data. *Remote Sensing of Environment* 8, 237-248.

Turner, N.C. (1993). Water use efficiency of crop plants: Potential for improvement. In: *International Crop Science* I. (Eds. D.R. Buxton, R. Shibles, R.A.

Frosberg, B.L. Blad, K.H. Asay, G.M. Paulsen, and R.F. Wilson), pp. 75-82. (Crop Science Society of America: Madison, WI).

van Herwaarden, A.F. (1996). Carbon, nitrogen and water dynamics in dryland wheat, with particular reference to haying-off. PhD Thesis. Australian National University, Canberra, Australia.

Waddington, S.R., Ransom, J.K., Osmanzai, M., and Saunders, D.A. (1986). Improvement in the yield potential of bread wheat adapted to northwest Mexico. *Crop Science* 26, 698-703.

Walker, C.D., and Lance, R.C.M. (1991). Silicon accumulation and 13C composition as indices of water-use efficiency in barley cultivars. *Australian Journal of Plant Physiology* 18, 427-434.

Watanabe, N., Evans, J.R., and Chow, W.S. (1994). Changes in the photosynthetic properties of Australian wheat cultivars over the last century. *Australian Journal of Plant Physiology* 21, 169-183.

Whan, B.R., Knight, R., and Rathjen, A.J. (1982). Response to selection for grain yield and harvest index in F_2, F_3, and F_4 derived lines of two wheat crosses. *Euphytica* 31,139-150.

Whan, B.R., Rathjen, A.J., and Knight, R. (1981). The relationship between wheat lines derived from the F_2, F_3, F_4, and F_5 generations for grain yield and harvest index. *Euphytica* 30, 419.

Wilson, D. (1982). Response to selection for dark respiration rate of mature leaves in Lolium perenne L. and its effects on growth of young plants and simulated swards. *Annals of Botany* 49, 303.

Wilson, D., and Jones, J.G. (1982). Effect of selection for dark respiration rate of mature leaves on crop yields of Lolium perenne cv. S_{23}. *Annals of Botany* 49, 313-320.

Windham, W.R., Hill, N.S., and Stuedemann, J.A. (1991). Ash in forage, esophageal, and fecal samples analyzed using near-infrared reflectance spectroscopy. *Crop Science* 31, 1345-1349.

Wright, G.C., and Hammer, G.L. (1994). Distribution of nitrogen and radiation use efficiency in peanut canopies. *Australian Journal of Agricultural Research* 45, 565-574.

Youssefian, S., Kirby, E.J.M., and Gale, M.D. (1992). Pleitropic effects of the GA-insensitive *Rht* dwarfing genes in wheat 2. Effects on leaf, stem, ear and floret growth. *Field Crops Research* 28, 191-210.

Chapter 18

Breeding Hybrid Wheat for Low-Yielding Environments

Jori P. Jordaan

Repeatedly, humans have developed radical new technology for controlling the plants that provided their sustenance; they have modified the genotypes of the plants they cultivated, and at the same time they have modified the way the plants were grown. In modern terminology, plant breeding and plant culture have always been essential agents of change in human society.

Donald N. Duvick, 1996

INTRODUCTION

Wheat breeders around the world have pursued the improvement of wheat and raising the yield level with dedication. Although one of the best adapted crops in the plant kingdom, and is grown on all the continents, there is concern whether wheat production will be able to meet the demands of the ever-growing human population. Wheat scientists are concerned with the genetic gain made in the past decade and are looking at radical new technology, including the exploitation of heterosis, to further improve the crop. Apart from raising the yield potential under favorable growing conditions, an alternative is to address the low-yielding environments that still limit optimal yield expression on a significant percentage of the acreage.

GENOTYPE-STRESS INTERACTION

Selection efficiency is restricted by the extent of the interaction between the environment and the genotype. A large number of factors interact to reduce yield. Some are fixed variables while others are manageable, but of most interest to the breeder are those derived from weather conditions and which cannot be predicted in advance. These include factors such as the amount of precipitation, the distribution thereof, soil moisture, temperature, etc. Attempts by wheat breeders to incorporate genetic stability into these variables were antagonistic to the progress in developing wheat varieties with yield potential. Progress was made developing varieties that perform under inadequate moisture and unfavorable temperatures, even when the plant's response to such conditions was not fully understood or elucidated by physiological and biochemical processes. Conditions of water stress define a stress environment that is usually also characterized by periods of high temperature, resulting in drought.

The Stress Environment

Varying conditions of water stress describe four different kinds of stress environments (Quizenberry, 1981):

1. *The stored moisture environment:* The crop completes its entire life cycle on stored rainfed moisture.
2. *The variable moisture environment:* A rainfed drought stress environment with alternating dry and wet periods.
3. *The reduced irrigation environment:* Suboptimum irrigation causes drought stress.
4. *The optimum moisture environment:* Usually optimum, but occasionally drought stresses can occur during short periods when evaporation greatly exceeds root uptake.

There are areas all over the world that are known to be low yielding. Throughout this chapter, I will refer to a dryland area in Southern Africa where winter wheat is produced, which, with a ten-year average production of 1.18 tons per hectare (South African Wheat Board), can surely be described as one of the lowest yielding environments on which wheat is being grown commercially. This environment can be compared to Quizenberry's environment 1 with elements of environment 2 (Quizenberry, 1981) where wheat is usually grown on only the available moisture in the soil, which was conserved from rains that fell during the preceding summer months. The crop usually grows without any rain up to heading, then receives rain in variable quantities up to maturity. During this period the wheat plant is usually under water stress that might be increased by spells of high temperatures. Factors such as soil depth, soil type, and agronomic practices interact with the availability of water and create an ever-changing environment, which is not predictable in space and time. Specific agronomic systems are prescribed by the water status in the topsoil at planting time. Dry topsoil might be as deep as 15 cm at planting time and requires a specific plant and planter technology. Accordingly, yield conditions will vary from a high potential to a very low potential.

Breeding Strategy

Breeders targeting this area have an integral, although empirical, approach to selecting genotypes that are adapted to multiple site variations. Under these conditions the adaptation of a variety, by definition, is its reaction to soil and climatic conditions, meteorological situations in a specific climatic environment, and specific agronomic systems such as plant nutrition and plant spacing. Breeders differ in their strategy and opinion (Pheiffer, 1988) whether breeding methodology should be variety specific or multipurpose. The strategy of developing varieties that are less responsive to environmental interactions was widely adopted by plant breeders exploring low as well as high-yielding conditions. Van Ginkel (1994) emphasizes CIMMYT's (The International Maize and Wheat Improvement Center) commitment to develop lines that optimize input to output, by alternating environments for early generation selection, which are best selected for under sufficient water, in combination with traits that are best selected for under drought. This very successful approach by CIMMYT has been strategically modified to amalgamate agroecological diversity into megaenvironments (Rajaram, 1994). These environments are transcontinental, defined by similar biotic and abiotic stresses, cropping cultures, and germplasm designated to be exploited in such an environment that accommodate the major stresses but probably not all the secondary stresses to be encountered in that megaenvironment.

Breeders, in general, favor the more optimum environment for development and claim that most progress has resulted from selection that has been done under less environmental stress. Genetic variance, in CIMMYT's International Spring Wheat Yield Nursery, which is replicated in a large number of countries, declines in more stress-associated environments (Pfeiffer, 1988). Experience, however, from breeders targeting specific environments including stress (Hollamby and Bayraktar, 1996) confirm that selection progress depends on a rather more specific strategy and technology.

In Southern Australia they breed for stress tolerance rather than yield potential and argue that since interactions between genotypes, location, and year are not predictable, selection must be carried out across localities and seasons, sampling a production area. These interactions will include response to biotic and abiotic stresses. In such circumstances the priority should be on selection efficiency to ensure that genetic progress can be maintained. If drought stress is sufficiently severe, high yield potential and drought resistance became mutually exclusive (Blum, 1996). Accordingly, knowledge and understanding of the factors involved in an apparent negative association between yield potential and resistance to severe drought stress are important for a more efficient approach to breeding for high yield and yield stability. Consequently, consideration should be given to the possibility of using the competitive plant ideotype, plant phenology, source-sink relationship, and carbon assimilation in the genetic exploitation of growth, yield, and adaptation.

It can be reasoned that most of the yield progress that was achieved in the past resulted from the genetic manipulation of resistance to stress factors with a more simple inheritance such as disease resistance and in particular leaf, stem, and yellow rust resistance. Successful procedures manipulated the genetic complementation between host and pathogen. There are different approaches to resistance breeding, including the utilization of major genes, partial or poligenic resistance, and the role of diversity to buffer small grains against highly epidemic and foliar pathogens. However, whatever the approach, the application thereof has resulted in the continuous improvement of yield potential.

The continuous release of higher-yielding varieties with wider adaptation is proof that there is still enough germplasm available to be exploited. There is, however, concern that the genetic gains realized in recent years and especially in less favorable weather conditions (Bell et al., 1995) are in a declining phase. Kronstad (1996) suggests that this tendency can be reversed and a breakthrough accomplished by making better use of germplasm resources. Continuous progress in Rasmusson's barley breeding program was attributed to regarding germplasm as paramount (Rasmusson, 1996) and that future accomplishments would be achieved by the wise management thereof. The *Triticum* genes accommodate vast genetic resources, which in the past have contributed significantly to breeding for resistance against factors limiting yield potential. In the future breeders will certainly be more dependent on wider genetic resources to further counter detrimental interactions with the phenotype.

Hybrid wheat could in future contribute to raising yield potential, although a large number of breeders were disillusioned with the progress that has been made with the breeding wheat hybrids as well as with the expression of heterosis in wheat. Jordaan (1996) has delivered evidence showing that hybrids perform well in dryland stress environments in South Africa.

Although not much has recently been published on wheat hybrids, recent reports by Bruns and Peterson (1997) and Peterson, Moffatt, and Erickson (1997) gave convincing evidence on the performance of hybrids in the Great Plains of the United States. They ascribed the improved performance of hybrids to inherent yield stability. Hybrids have higher regression slopes and lower deviations from regression than pure lines. They attribute this higher stability of performance to factors such as heterosis for green leaf duration, improved heat tolerance, and disease resistance. The agronomic expression was improved by utilizing computer programs to match exclusive traits of the parents in hybrid combination.

THE HYBRID PROFILE

When corn breeders developed procedures to achieve homozygosity they followed in the footsteps of animal breeders to characterize populations as breeding units and introduced biometrical concepts to develop appropriate breeding plans. Recurrent selection procedures were designed to make use of additive as well as nonadditive gene action (Comstock, Robinson, and Harvey, 1949). Together with reciprocal recurrent procedures (Hallauer and Miranda, 1981) the emphasis was put on selection procedures and theory. These theories formed the basis of a highly successful breeding strategy exploiting hybrid vigor in corn improvement. Wheat breeders are once again focusing on hybrids, even those who claim breeding to be more an art than a science. Hollamby and Bayraktar (1996) suggested that in future wheat breeders should also focus on modeling hybrids with emphasis on procedures that have been successful, especially in the development of hybrid corn. This would require a total change in the attitude of the wheat breeder to a role model that was described by Sprague and Eberhart (1976) as a balanced breeding program:

- *Create* two diverse breeding populations with maximum interpopulation cross performance
- *Devise* an effective selection program to improve populations per se
- *Identify* hybrids with every cycle of selection

GENETIC DIVERSITY IS PARAMOUNT

Thus, learning from the maize model, hybrid wheat breeding should evolve from two different populations (see Figure 18.1). To maximize genetic diversity between the two populations care should be taken to, in the first instance, include in each population germplasm that is distinct from the other. Kinship relationships based on the coefficient of parentage could be used to detect relatedness, but more sophisticated biotechnological methods are available to make a reliable determination. It was found

FIGURE 18.1. Model for Hybrid Wheat Breeding

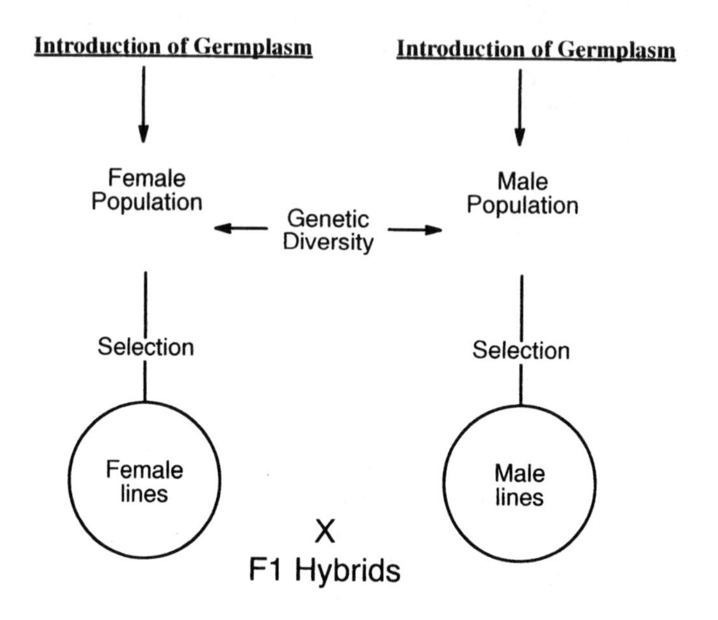

that when parents were assayed for random amplified polymorphic DNA (RAPD), the results indicated that genetic distance could be used to predict hybrid performance (Perenzin et al., 1997). In a more practical, but more time-consuming approach, test crosses could be made to access nicking and placement of germplasm. In pursuing genetic diversity the introgression of alien germplasm could play a significant role in making provision for heterosis based on genomic diversity.

In an unpublished study by Laubscher on wheat in South Africa, 10 lines sampled from the female population were crossed with 10 lines sampled from the male population (Laubscher, 1984). The progeny were tested under a low yield level and at a low seeding rate (5 kg per hectare). The lines were regarded as genetically diverse and originated from the SENSAKO hybrid breeding program. (SENSAKO is a central cooperative involved in hybrid wheat breeding since the middle 1960s, but aggressively so since 1978, and has an established history of pure line releases dating back to 1971.)

For yield the average effect of parents accounted only for 37.2 percent of the hybrid variance while the female × male cross-interaction accounted for 63.8 percent. The results clearly showed that although the genetic variance within the female and male sets was significant, the performance of the hybrids could not be predicted from the average performance of the parents nor from the parent progeny performance. Selection of the 10 best hybrids from this study predicted an advance of 43 percent in yield potential. The assumption was made that the potential of the pure line parents was comparable to the then-conventional commercial varieties. This study presented the basis for the present hybrid development program in South Africa and in retrospect, current hybrid yield potential confirms the prediction that was made then on the cross-performance of the male and female pools (Jordaan et al., 1997).

Most breeders have experienced heterotic F1 combinations in the execution of the conventional crossing programs, but the most convincing evidence of the exploitation of diversity in germplasm utilization in conventional breeding procedures derives from the International Spring × Winter selection program (Kronstad, 1996). Probing the winter and spring gene pools has provided enhanced genetic variability for nearly all the desired agronomic traits. Kronstad suggests that the winter × spring crossing approach may prove interesting for breeders developing hybrid wheat. In limited studies where comparisons were made with winter × spring F1's and those resulting from winter × winter or spring × spring crosses, the former gave a greater expression of hybrid vigor, perhaps reflecting a greater degree of diversity between these two gene pools. In rice, also an autogamous crop, Khush and Peng (1996) reported that the magnitude of heterosis was dependent on the genetic diversity between the two parents. According to Yuan, Virmani, and Mao (1989) there was no gene flow between the indica and japonica germplasm while hybrids between the two groups showed the highest heterosis. The expectation is then that crosses between the recently improved japonicas ideotypes, and the indicas cultivars will show a large expression of heterosis (Khush and Peng, 1996).

Although the introgression of wheat germplasm has proved to be very successful in developing pure self-fertilized lines, it might also be one of the reasons for the ill fate of hybrid wheat. Evidence has also been provided (Pickett, 1993) that hybrids derived from modern, highly bred varieties appear to show less heterosis than hybrids from older, lower-yielding parents.

INTRAPOPULATION IMPROVEMENT

Procedures and technology to develop pure lines are well known and are also applicable in the development of female and male lines to produce hybrids. If the final goal is to produce hybrids that are also adapted to low-yielding environments, then selection in parent development should be ruthlessly done sampling the target environment. To have a meaningful discussion on the selection of parents the structuring or modeling of such a program should be known. Within the female development program (see Figure 18.2) five different levels of selection can be identified.

FIGURE 18.2. A Model of a Female (F) Development Program

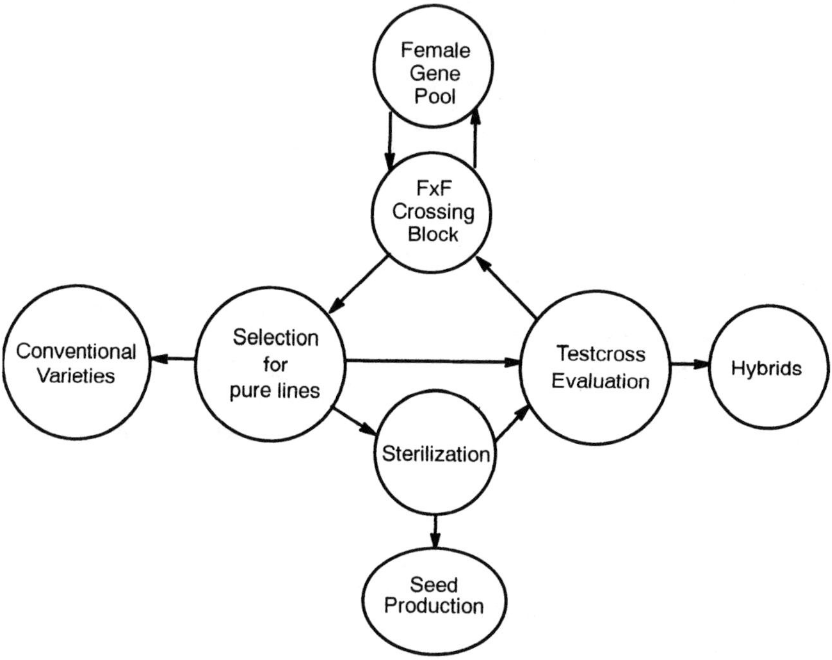

Selection of Germplasm to Be Included in the Female Pool

Selection should be based on diversity from the male pool determined by means of pedigree analyses, marker-assisted selection, and test-cross performance, along with with female characteristics such as duration of receptiveness of the stamen to pollen, quality of the seed, and selection for resistance to biotic and abiotic stresses, which could complement characteristics from the male side.

Selection of Germplasm from the Female Pool to Be Included in the Crossing Block

Emphasis should be placed on recombining new germplasm exhibiting excellence with female lines performing well in hybrid combinations. Focus should be on those crosses that promise to provide the highest frequency of desired progeny.

Developing Pure Female Lines

This should include the normal selection techniques encountered in the development of conventional pure lines. It would, however, be a priority to advance lines through the selection phases as quickly as possible to make an assessment of combining ability with the male lines. Haploid breeding could be a valuable tool to achieve homozygosity as soon as possible before test crossing the lines. If the production area varies in yield potential a shuttle program may be implemented, alternating selection nurseries from high to low yielding potential. In the SENSAKO development program, we aim to raise the level of performance in the female lines to equal the performance of conventional lines at the same level of inbreeding. Pure lines from the female program could also be released as conventional varieties, depending on whether the breeder's release policy allows the commercial release of only hybrid varieties. Female lines from the SENSAKO program are also tested in trials to assess their own performance. A comparison of the consistency of performance (Lin and Bins, 1988) was done on yield data representing low-yielding sites in the target area (less than 2 tons per hectare. Comparing the average of the elite female lines to that of the best-adapted variety under low-yielding conditions, and to that of three new releases of conventional varieties (see Figure 18.3), shows that the female lines on the average have a smaller distance mean square from the maximum response, averaged over all localities, than the conventional varieties, and their performances are comparable to those of the best-performing variety.

Selection for Cross Performance

It is necessary to change the self-fertilizing habit of the wheat plant to produce hybrid seed. This could be done by producing male sterile plants.

For years hand emasculation has been the only method to facilitate crossing. This is a time-consuming activity, limiting the amount of seed that can be produced and also the number of crosses which can be made.

FIGURE 18.3. Comparing the Average Consistency of Performance of Thirteen Elite Females (B) to That of the Newest Released Conventional Varieties (C) to That of the Best Commercial Variety for Low-Yielding Environments (A)

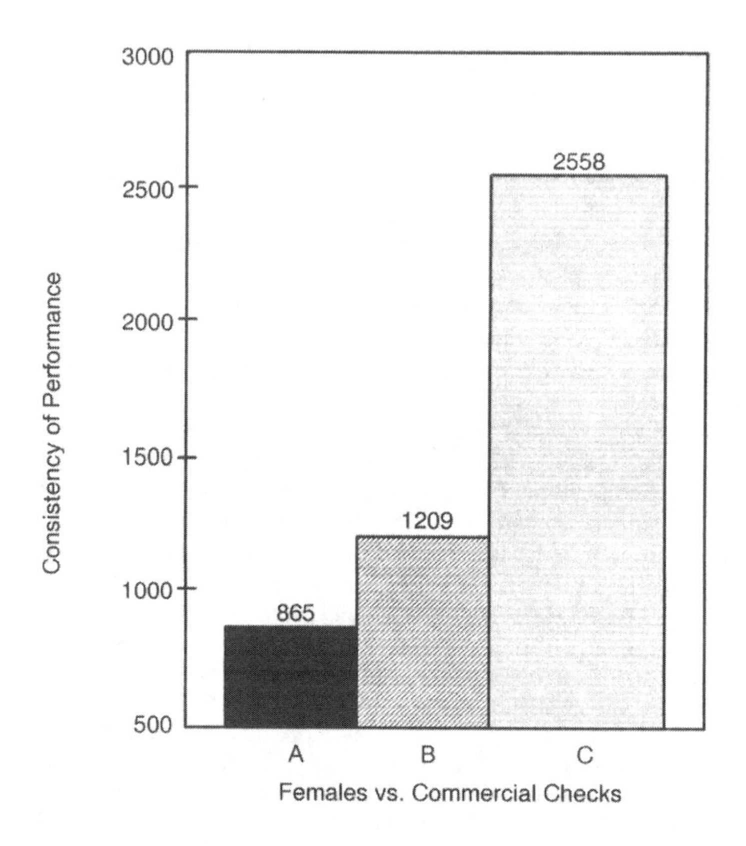

Note: The lowest value is the best performance.

The amount of seed per cross was the major limitation in accessing the value of hybrids in space and time and also disqualified F1 hybrids from commercialization.

Nuclear male sterility was discovered by Wilson and Driscoll in 1983, but did not catch on because of the expensive methods of maintaining seed stocks.

The discovery of a cytoplasmic male sterility system in wheat (Kihara, 1951) made hybridization possible, as in many other crops, and changed the self-fertilized habit of wheat to a breeding system of cross-pollination that could be commercialized. Male sterile hexaploid wheats were produced by substituting the wheat cytoplasm with the *Triticum timpoheevi* cytoplasm, resulting in cytoplasmic induced male sterility (CMS). Finally cross-pollination could be achieved, and male fertile F1 hybrids were produced when a fertility restorer genetic factor was discovered (Schmidt, Johnson, and Maan, 1962).

A significant achievement was the development of chemical hybridizing agents (CHA) to male-sterilize wheat. At first, results were disappointing, but Carver and Nash (1984) reported that the chemicals R.H.007 and WC 84811, which were developed by the companies Rhom and Haas and Shell, respectively, were regarded as successful. Recently, Bruns and Peterson (1997) from the United States, He, Du, and Zuang (1997) from China, and Perenzin et al. (1997) from Italy have reported on the effective use of CHA products to produce hybrids.

CHA products have distinct advantages over CMS mainly because they allow faster development time for hybrids, since female lines need not be sterilized by backcrossing (which must be done by hand) and seed can be multiplied by using the female maintainer line as a pollen parent. It allows the use of technology to facilitate the implementation of recurrent selection procedures, and enough seed for trial purposes could be produced at an earlier generation to do extensive testing throughout the target area. Test cross seed on CMS female lines can be produced from pollen of a male tester in an isolated crossing block. Females having normal cytoplasm can be test crossed in a three-way hybrid by hand crossing the females to the female component (A-line : sterile) of the female tester line and pollinating the resultant sterile F1s with the male tester line (restorer).

Female lines are selected on the basis of the performance of their test crosses. These test crosses are replicated in trials at sites that represent the target area. Selected lines are retained for further selection, backcrossed to CMS or, in the case of CHA, used as parents for producing hybrids, and are finally also recycled in the female pool as parental germplasm.

Selection for Female Characteristics

Female characteristics are those that are of interest in hybrid seed production. Although not important in hybrid performance, there may be critical attributes in the commercialization of hybrids. In the case of the CMS system, the female should also be a good pollinator and shed enough pollen over a long enough period to get a good seed set on the sterile component (A-line). Trapping of anthers is a stress-related character and should be selected against. Furthermore, there are large differences among stamens of female lines in their ability to stay receptive for pollen over several days, especially under conditions of warm dry weather. A higher seed set in the female relates to better seed quality, while a low seed set is usually a distinctive feature of poor seed set or might be a characteristic of the female involved.

SELECTION WITHIN THE MALE POOL

The same rules that apply to the selection of germplasm to be included in the female pool also apply to the male pool. The technology and selection methods that can be used for developing male (M) lines depend on the availability of CHA. When available, recurrent selection producers are possible (Figure 18.4), while for CMS the use of the female tester line is not a practical option since all crosses would have to be made by hand or in isolated cross-pollination blocks for every individual cross. The strategy should rather be to identify the core M lines or germplasm and introduce unique characters by means of backcrossing to a genetic background known for its cross-performance to the F pool (Figure 18.4). The emphasis should be on developing males that restore the CMS in the female. The genetic factor for fertility restoration (R genes) could be included in males by means of backcrossing. However, if the males have the normal cytoplasm it would be necessary to make test crosses on lines having the *Triticum timopheevi* cytoplasm to detect the R genes. The practical option would be to select males segregating for the R genes in the background of the *Triticum timopheevi* (*Tt*) cytoplasm. Hence, every plant is a test cross and allows the strategy for selection of the M germplasm under stress conditions such as insufficient water and high temperatures, which are known to inhibit the expression of fertility restoration.

HETEROSIS AND PLANT POPULATION × ROW WIDTH INTERACTIONS

The use of wider rows relates to the depth of the dry topsoil layer. The deeper the dry topsoil layer, the more dry topsoil has to be pushed away to plant seeds in a wet seedbed, and the wider the rows must be. This technology has already been adopted in practice and is not based on theoretical science. Seeding rate within the rows varies with planting date, maturity of the variety, and soil water status, and very seldom exceeds 30 kg per hectare. Reports on the effect of row width (Paulsen, 1987) and the effects

FIGURE 18.4. Male (M) Development Program Based on CHA System, Normal Cytoplasm (N), Absence of a Restorer Factor (r), and the CMS System Based on *Triticum timopheevi* Cytoplasm (Tt) and the Presence of a Restorer Factor (R)

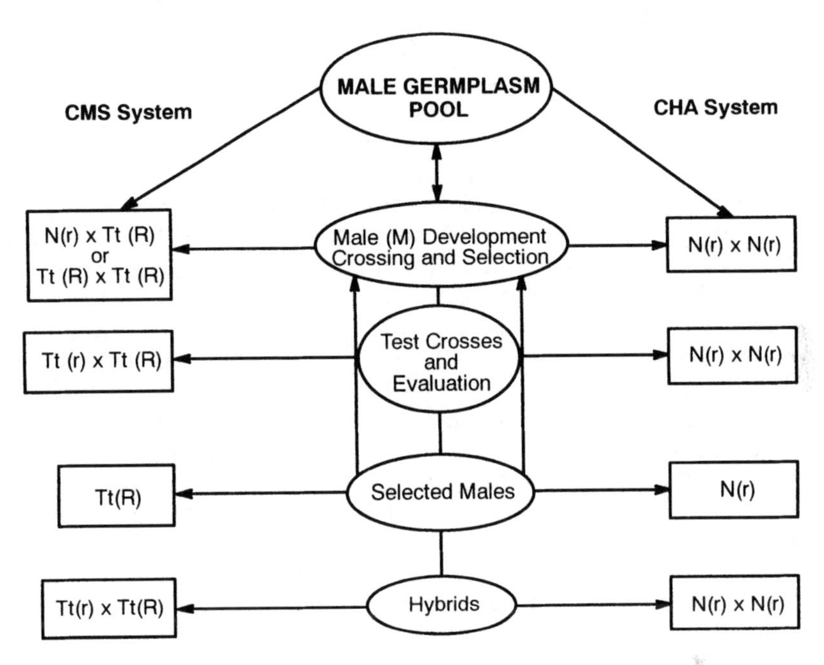

of seeding rate on yield and yield components (Frederick and Marshall, 1985; Bulman and Hunt, 1988; Royo and Ramagosa, 1988) describe interactions and correlated responses in nonstress conditions. Reports on the expression of heterosis for yield and also on the expressions thereof in yield components associated with final yield are inconsistent (Virmani and Edwards, 1983; Pickett, 1993; Borghi and Perenzin, 1994; Martin et al., 1995; Larik, Mahar, and Hafiz, 1995; Uddin et al., 1992; Menon and Sharma, 1994; Winzeler, Schmidt, and Winziler, 1994; Sharma, Smith, and McNew, 1991; Liu and Li, 1994).

The deviation of a hybrid from the mid-parent value (according to the definition of heterosis) was calculated in this specific South African environment, for yield and yield components at different row widths (25, 50, and 75 cm) and three different plant populations (15, 30, and 50 plants per m^2) at different localities and seasons (Engelbrecht, 1991). These row widths and plant densities relate to cultural practices in the specific low-yielding environment. The expression of heterosis, although inferences are limited to one hybrid and its two inbred parents, was significant for all the main variance components and interactions. The biomass production per m^2, number of ears per m^2, and 1,000-kernel weight were highly correlated with yield and significant heterosis was expressed. Interaction between row width and plant population has identified that the hybrid's advantage is specific adaptation to yield potential. In narrow rows the expression of heterosis was higher at the high plant densities and better adapted to high yield potential, while in wider rows the deviation from mid-parent values was the highest at low seeding rates when the mid-parent values were low (low yield potential). Clearly the hybrid's relative performance relates to space arrangements of plants at specific row widths and plant population, supporting the conception that, for the older generation hybrids, the magnitude of heterosis is yield-potential specific and can be optimized by agronomic practices. The hypothesis of wider rows and low seeding rates for low-yielding environments seems to hold in practice.

Under these conditions (wide rows, 75 cm; low plant population, 15 plants m^{-2}) the biggest contribution to hybrid grain yield stems from the number of spikes m^{-2} (rg = 0.57) and the number of spikes per plant (rg = 0.53). In both cases the relationship was not sensitive to narrow rows or higher populations (Engelbrecht, 1991). In the yield-limited environment kernel weight made a significant contribution (rg = 0.43) to higher yields, although it was less in narrower rows and higher populations. The contribution of grain weight, which has showed significant heterosis irrespective of row width and population density, might be an important phenomenon, arguing the importance of heterosis under stress conditions, since the phenological stage when this component is produced (Slafer and Rawson, 1994) coincides with the occurrence of drought (depletion of soil moisture and high temperatures). Thus lower plant density in wider rows together with insufficient soil moisture creates a situation where the environment satisfies the requirements of assimilates for grain growth. The role that the wider rows and low seeding rates play in light inception should be investigated, and also the effect of air flow in cooling the canopy during periods of high temperatures. The underlying processes of physiological traits affecting yield potential have been thoroughly discussed by Slafer, Calderini, and Miralles (1996), Blum (1996), Richards (1996), and Araus (1996).

HYBRID YIELD, STABILITY, AND RESPONSE TO YIELD

Various statistics are available to describe the yield stability of a genotype in environments with varying yield potentials. Trials conducted across localities sampling a major production area are standardized field layouts, and do not make provision for optimizing yield potential at any specific environment by alternating row spacing, seeding rates, or planting dates for individual entries that are included in the trial. When there is a large spectrum of planting dates for winter wheat, as in South Africa, genotypes vary in maturity, resulting from differences in vernalization requirements (true winter types) or daylength sensitivity (intermediate type). Thus, under these conditions varieties exhibiting a vernalization requirement can only be compared to varieties with a photoperiod response at planting dates where the vernalization could be satisfied.

Comparing hybrids to conventional varieties across environmental sites (Jordaan, 1996) was done by comparing the consistency of yield performance (Lin and Bins, 1988). In their comparison yield potential was defined as the distance mean square between the genotype's response and the maximum response within an environment, averaged over all testing sites. Most plant breeders, however, prefer to describe genotype × environment interaction as a linear response to environmental yield potential and the deviations from that response as originally proposed by Eberhart and Russel (1966).

Hybrid and pure line yield stability was compared to their responsiveness to growing conditions in the Great Plains of the United States (Peterson, Moffatt, and Erickson, 1997). Hybrids and pure line yields were regressed on an environmental index based on location mean yield for pure lines. Hybrids showed significantly higher mean yields than pure lines and the yield advantages increased with environmental yield potential. There was no crossover in yield response between hybrids and pure lines at lower yield levels and the deviations from regression were of similar magnitude. Peterson et al. (1997) demonstrated, like Bruns and Peterson (1997), the potential of hybrid wheat for enhanced mean yield and greater yield response to favorable environmental conditions with similar deviations from expected response.

The yield stability of hybrids being developed and grown as winter wheat in South Africa has been defined as a function of the regression slope on an environmental index described by the average of conventional pure lines and cultivars grown in a specific environmental (Jordaan et al., 1997). To eliminate the contribution of aphid (*Duraphuis noxia*) damage to environmental interactions, comparisons were only based on genotypes with adequate genetic resistance to damage caused by the aphid. This limits comparable data to that of the winter season of 1996. Comparisons of the hybrids (9) with the pure lines (16) included in the SENSAKO Performance Trial for winter wheat, exhibiting a vernalization requirement, are summarized at the bottom of Figure 18.5a, while the two newest hybrid releases (SST 966 and SST 936) were compared to the two newest pure line releases (Betta DN and Limpopo) in Figure 18.5b. Genotypes having only a photoperiod response were compared in the SENSAKO Performance Trial for intermediate wheats. The two newest hybrids considered for release are compared to the two newest pure line releases (Gariep and SST 363) in Figure 18.5c, and the provisionally released hybrid (IPT 11) to the average of intermediate pure lines.

The pure line's environmental average was in all cases agreed to be the environmental index. (Thus, the regression slope of 1.00 for pure lines, and an intercept value of 0.00 kg ha^{-1}.) Interpretation of the regression statistics has an astonishing similarity to that of the Great Plains data (Peterson, Moffatt, and Erickson, 1997), showing that hybrids were on the average significantly higher yielding than pure lines and that hybrids showed improved stability to increasing yield potential. Hybrids with an average positive slope deviation from 1.00 do exhibit better adaptation to higher yield potential and do not show crossover at lower yield potential. The hybrids with the best stability proved to be those which are photoperiod sensitive with no vernalization requirement, outyielding the pure lines by a margin of 28 percent when the hybrid with the highest yield potential is considered (see Figure 18.5d). Although not more responsive to lower yield potential, it yielded significantly higher than the pure lines at the lower level of 100 kg ha^{-1} to 1500 kg ha^{-1}). It might be concluded that selection for hybrids in low-yielding, stress environments also results in hybrids exhibiting high yield potential and responsiveness to high-potential environments. Cultivar yield stability, especially at lower yield levels, is an important economic consideration for wheat growers.

THE IDEAL HYBRID

Most plant breeders have a vision of the plant or variety they select for. Visualizing the optimum characteristics of a hybrid wheat phenotype adapted to lower-yielding conditions should not be viewed as unthinkable:

1. An empirical perspective of a phenotype with stability in different environmental conditions targeting the production environment but responsive to higher-yielding conditions.
2. High yield potential might be the ultimate goal, but quality potential should also be regarded as important, and factors that constrain grain marketability eliminated. Hybrids must be bred to satisfy the needs of end consumers or the breeder might find the product exhibiting high yield potential redundant.

3. The physiological routes to greater photosynthesis rates and a faster rate of grain filling are exploited.
4. High levels of resistance to disease and insect damage secure enough healthy green tissue, also, to lower production input costs and raise profitability for the producer.
5. The ideotype plant has strong root development, less foliage in the pretillering stage of phenological development, fewer leaves of a lighter green color and a large number of fertile spikes and spikelets.

FIGURE 18.5a. Winter Hybrids (Solid Lines) versus Pure Lines (Dashed Line)

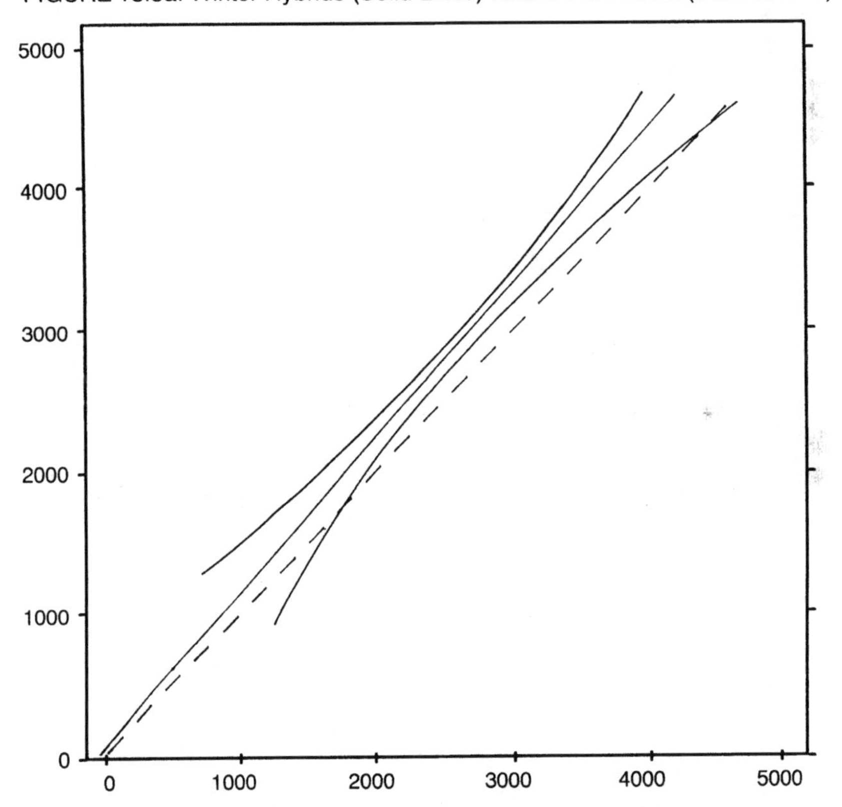

Note: The linear regressions of the mean yields of winter hybrids and pure lines (Figures 18.5a and 18.5b), and intermediate hybrids and pure lines (Figures 18.5c and 18.5d) calculated on an environmental index which was regarded as the mean of the pure lines at each location. The regression coefficient of pure lines was 1.00, intercepting at 0.0 kg ha^{-1}. A 95 percent confidence interval was calculated for the hybrid's regression using a GLM program.

FIGURE 18.5b. SST 966 and SST 936 (Solid Lines) versus Betta DN and Limpopo (Dashed Line)

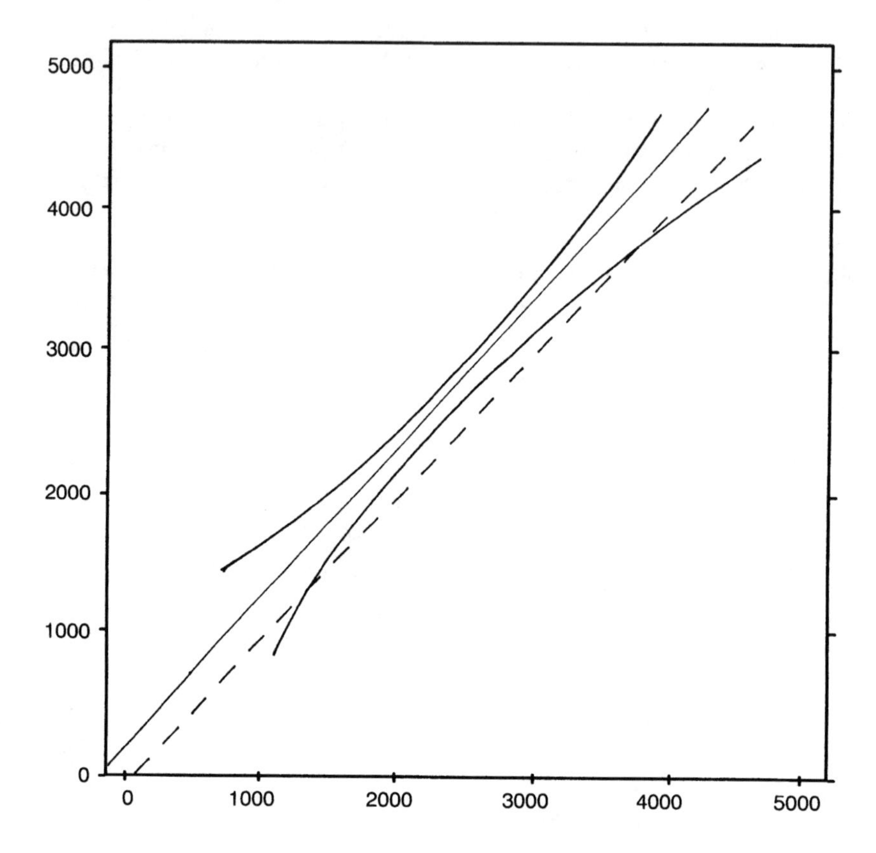

FIGURE 18.5c. Best Hybrids (Solid Lines) versus Best Pure Lines, Intermediate Types (Dashed Line)

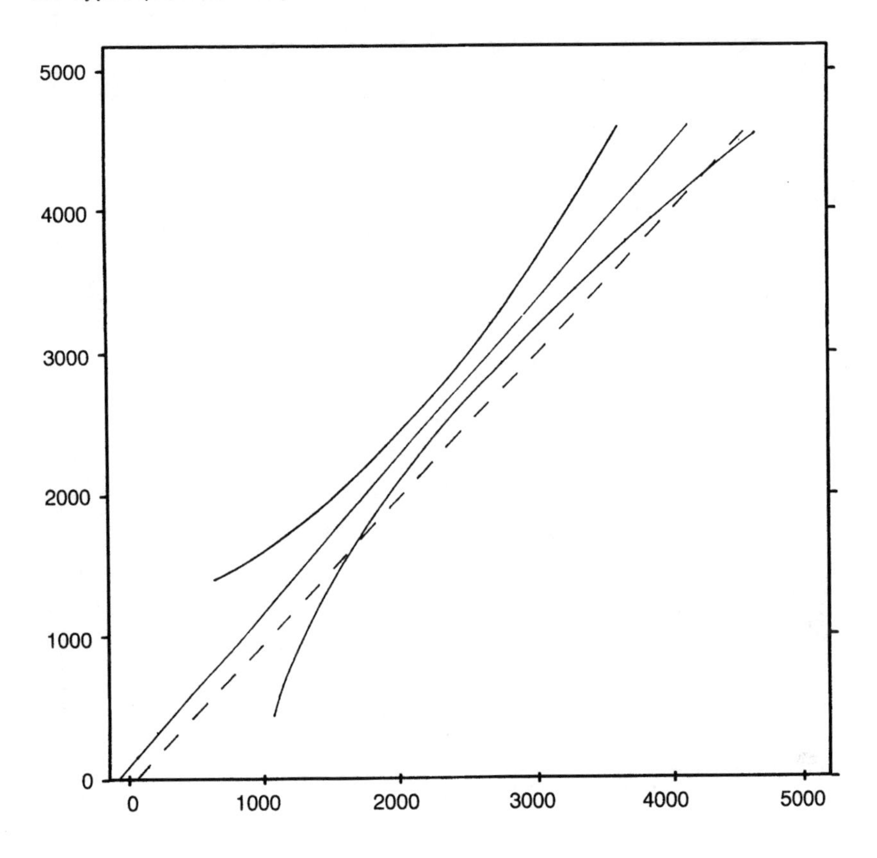

FIGURE 18.5d. IPT 11 (Solid Lines) versus Pure Lines, Intermediate Types (Dashed Line)

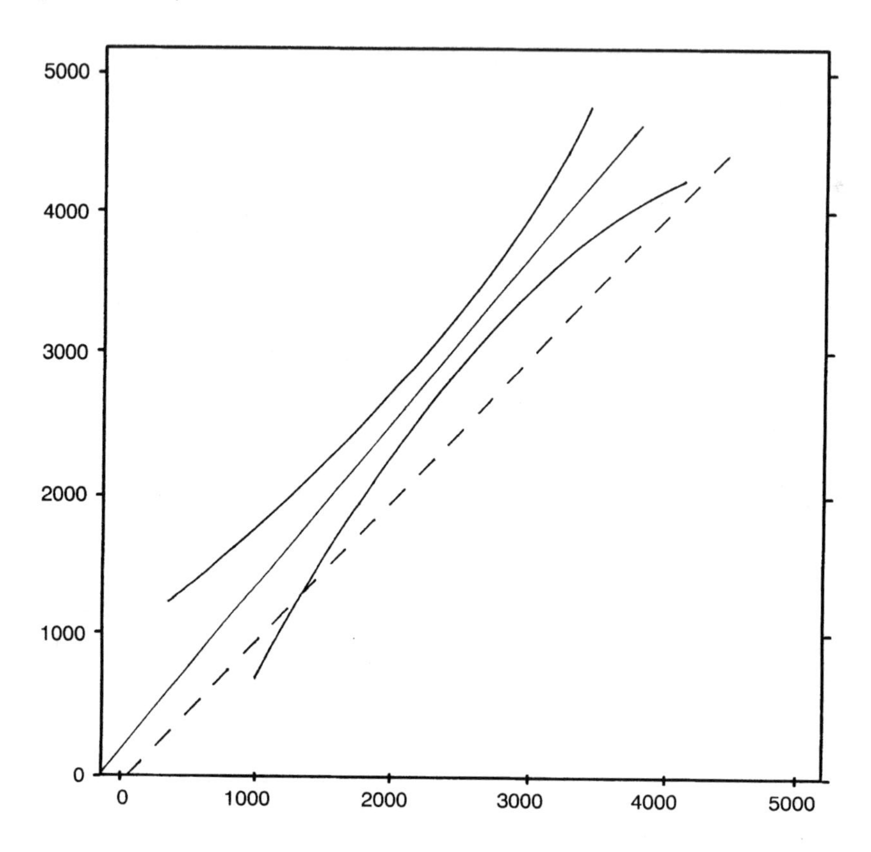

REFERENCES

Araus, J.L. 1996. Integrative physiological criteria associated with yield potential. In: Reynolds, M.P., S. Rajaram, and A. McNab, eds. *Increasing Yield Potential in Wheat: Breaking the Barriers.* Mexico, DF: CIMMYT.

Bell, M.A., Fisher, R.A., Byerlee, D., and Sayre, K. 1995. Genetic and agronomic contributions to yield gains: A case study for wheat. *Field Crops Res.* 44: 55-65.

Blum, A. 1996. Yield potential and drought resistance: Are they mutually exclusive? In: Reynolds, M.P., S. Rajaram, and A. McNab, eds. *Increasing Yield Potential in Wheat: Breaking the Barriers.* Mexico, DF: CIMMYT.

Borghi, B. and Perenzin, M. 1994. Diallel analysis to predict heteroses and combining ability for grain yield, yield components and bread-making quality in bread wheat *(T. Aestivum). Theoretical and Applied Genetics* 89:975-981.

Bruns, R. and Peterson, C.J. 1998. Yield and stability factors associated with hybrid wheat. In: Braun, H.-J., F. Altay, W.E. Kronstad, S.P.S. Beniwal, and A. McNab, eds. *Proceedings of the Fifth International Wheat Conference.* Ankara, Turkey. Dordrecht, Germany: Kluwer Academic Publishers.

Bulman, P. and Hunt, L.A. 1988. Relationship among tillering, spike number and grain yield in winter wheat in Ontario. *Can. J. Plant Sci.* 68:583-596.

Carver, M.F. and Nash, R.J. 1984. The future for hybrid cereals. *SPAN* 27(1): 64-65.

Comstock, R.E., Robinson, H.F., and Harvey, P.H. (1949). A breeding procedure designed to make maximum use of both general and specific combining ability. *Agron. J.* 41:360-367.

Duvick, D.N. 1996. Personal perspective: Plant breeding, an evolutionary concept. *Crop Sci.* 36:539-548.

Eberhart, S.A. and Russell, W.A. 1966. Stability parameters for comparing varieties. *Agron. J.* 6:36-40.

Engelbrecht, S.A. 1991. Stability of heterosis in hybrids for yield and yield components at different plant populations and yield potential. Unpublished PhD thesis in Afrikaans. University of the Free State, South Africa.

Frederick, J.R. and Marshall, H.G. 1985. Grain yield and yield components of soft red winter wheat as effected by management practices. *Agron. J.* 77:495-499.

Hallauer, A.R. and Miranda, J.B. 1981. *Quantitative Genetics in Maize Breeding.* Iowa State Univ. Press, Ames, IA.

He, Z.H., Du, Z.E. and Zuang, Q.S. 1997. Progress of wheat breeding research in China. In: *Proceedings of the Fifth International Wheat Symposium.* Ankara, Turkey, CIMMYT (in press).

Hollamby, G. and Bayrakter, A. 1996. Breeding objectives, philosophies and methods in South Australia. In: Reynolds, M.P., S. Rajaram, and A. McNab, eds. *Increasing Yield Potential in Wheat: Breaking the Barriers.* Mexico, DF: CIMMYT.

Jordaan, J.P. 1996. Hybrid wheat: Advances and challenges. In: Reynolds, M.P., S. Rajaram, and A. McNab, eds. *Increasing Yield Potential in Wheat: Breaking the Barriers.* Mexico, DF: CIMMYT.

Jordaan, J.P., Engelbrecht, S.A., Malan J., and Knobel, H. 1997. Wheat and heteroses. In: Srinivasan, G., ed. *Symposium on the Genetics and Exploitation of Heterosis in Crops.* (In press.) Mexico, DF: CIMMYT.

Khush, G.S. and Peng, S. 1996. Breaking the yield frontier of rice. In: Reynolds, M.P., S. Rajaram, and A. McNab, eds. *Increasing Yield Potential in Wheat: Breaking the Barriers.* Mexico, DF: CIMMYT.

Kihara, H. 1951. Substitution of nucleus and its effects on genome manifestations. *Cytologia* 16:177-193.

Kronstad, W.E. 1996. Genetic diversity and the free exchange of germplasm in breaking yield barriers. In: Reynolds, M.P., S. Rajaram, and A. McNab, eds.

Increasing Yield Potential in Wheat: Breaking the Barriers. Mexico, DF: CIMMYT.

Larik, A.S., Mahar, A.R. and Hafiz, H. 1995. Heterosis and combining ability intermates in diallel crosses of six cultivars of spring wheat. *Wheat Inf. Service* 80:12-19.

Laubscher, M.C. 1984. The evaluation of the cross-performance of probable hybrid breeding parents. In: Afrikaans, unpublished MSc thesis. University of Stellenbosch, South Africa.

Lin, C.S. and Bins, M.R. 1988. A superiority measure of cultivar performance for cultivar x location data. *Can. J. Plant Sci.* 68:193-198.

Liu, Z. and Li, Y. 1994. Heterosis of grain weight in wheat hybrids with *Triticum timopheevi* cytoplasm. *Euphytica* 75:189-193.

Martin, J.M., Talbert, L.E., Lanning, S.P., and Blake, N.K. 1995. Hybrid performance in wheat as related to parental diversity. *Crop Sci.* 35:104-108.

Menon, U. and Sharma, S.N. 1994. Combining ability analysis for yield and its components in bread wheat over environments. *Wheat Inf. Ser.* 79:18-23.

Paulsen, G.M. 1987. Wheat stand establishment. In: E.G. Heyne (ed). *Wheat and Wheat Improvement.* 2nd ed. American Society of Agronomy: Madison, WI.

Perenzin, M., Corbellini, M., Acerbi, M., Vaccino, P. and Borghi, B. 1997. F1 hybrid performance and parental diversity based on molecular marker in bread wheat. In: Braun, H.-J., F. Altay, W.E. Kronstad, S.P.S. Beniwal, and A. McNab, eds. *Proceedings of the Fifth International Wheat Conference.* Ankara, Turkey. Dordrecht, Germany: Kluwer Academic Publishers.

Peterson, C.J., Moffatt, J.M., and Erickson, J.R. 1997. Yield stability of hybrids vs pureline hard winter wheats in regional performance trials. *Crop Sci.* 37:116-120.

Pfeiffer, W.H. 1988. Drought tolerance in bread wheat—analyses of yield improvement over the years in CIMMYT germplasm. In Klatt, A.R. ed. *Wheat Production Constraints in Tropical Environments.* Mexico DF: CIMMYT.

Pickett, A.A. 1993. *Hybrid Wheat—Result and Problems.* Advances in Plant Breeding 15. Berlin: Paul Parey Sc. Publ.

Quizenberry, J.E. 1981. Breeding for draught resistance and plant use efficiency. In: M.N. Christianzen and C.F. Lewis, eds. *Breeding Plants for Less Favourable Environments.* Wiley Interscience, New York, pp. 193-212.

Rajaram, S. 1994. Wheat germplasm improvement: Historical perspectives, philosophy, objectives and missions. In: Rajaram, S. and G.P. Hettel, eds. *Wheat Breeding at CIMMYT: Commemorating 50 Years of Research in Mexico for Global Wheat Improvement.* Wheat Special Report No. 29. Mexico, DF: CIMMYT.

Rasmusson, D.C. 1996. Germplasm is paramount. In: Reynolds, M.P., S. Rajaram, and A. McNab, eds. *Increasing Yield Potential in Wheat: Breaking the Barriers.* Mexico, DF: CIMMYT.

Richards, R.A. 1996. Increasing the yield potential of wheat: manipulating sources and sinks. In: Reynolds, M.P., M.P. Rajaram, and A. McNab, eds.

Increasing Yield Potential in Wheat: Breaking the Barriers. Mexico, DF: CIMMYT.

Royo, C. and Ramagosa, I. 1988. Yield component stability in *Triticum aestivum* L. on *Triticum turgidum* L. var. Durum. *Cer. Res. Com.* 16:77-83.

Schmidt, J.W., Johnson, V.A., and Maan, S.S. 1962. Hybrid wheat. *Nebr. Exp. St. Quart.* 9(3):9.

Sharma, R.C., Smith, E.L., and McNew, R.W. 1991. Combining ability analysis for harvest index in winter wheat. *Euphytica* 55:229-239.

Slafer, G.A., Calderini, D.F., and Miralles, D.J. 1996. Yield components and compensation in wheat: Opportunities for further increasing yield potential. In: Reynolds, M.P., S. Rajaram, and A. McNab, eds. *Increasing Yield Potential in Wheat: Breaking the Barriers.* Mexico, DF: CIMMYT.

Slafer, G.A. and Rawson, H.M. 1994. Sensitivity of wheat phasic development to major environmental factors: A re-examination of some assumptions made by physiologists and modellers. *Australian Journal of Plant Physiology* 21: 393-426.

Sprague, G.F. and Eberhart, S.A. 1976. Corn breeding. In: G.F. Sprague, ed. *Corn and Corn Improvement.* American Soc. of Agron.: Madison, WI.

Uddin, M.N., Ellison, F.W., O'Brien, L. and Latter, B. 1992. The effect of plot type on the estimation of heterosis in bread wheat (*T. Aestivum*). *Aust. J. Agric. Res.* 43:1471-1481.

Van Ginkel, M. 1994. Bread wheat breeding for yield under drought conditions. In: Rajaram, S. and G.P. Hettel, eds. *Wheat Breeding at CIMMYT: Commemorating 50 Years of Research in Mexico for Global Wheat Improvement.* Wheat Special Report No. 29. Mexico, DF: CIMMYT.

Virmani, S.S., and Edwards, I.B. 1983. Current status and future prospects for breeding hybrid rice and wheat. *Adv. Agron.* 36:145-157.

Winzeler, H., Schmidt, J.E., and Winziler, M. 1994. Analysis of the yield potential and yield components of F1 and F2 hybrids of crosses between wheat (*Triticum aestivum* L.) and spelt (*Triticum spelta* L.). *Euphytica* 74:211-218.

Yuan, L.P., Virmani, S., and Mao, C.S. 1989. Hybrid rice: Achievements and future outlook. Pages 219-233. In: *Progress in Irrigated Rice Research.* Int. Rice Res. In.: Los Banos, Philippines.

Chapter 19

Actual and Potential Contributions of Biotechnology to Wheat Breeding

Robert M. D. Koebner
John W. Snape

INTRODUCTION

Advances in wheat breeding, as in other species, depend on three prerequisites. First, new sources of genetic variation to provide desirable alleles for genetic advance; second, technologies to recombine this variation into the generation of new genotypes; and third, technologies for identifying and selecting the phenotypes associated with the new adapted gene complexes created. Several new biotechnologies have been or are in the process of being developed aimed at improving the efficiency or scope of all of these genetic manipulations. The most widely used technologies at present are described in this chapter.

Although the phenotype of the wheat plant is determined by the effects of approximately 30,000 genes, very few (probably only a few hundred) of these have been identified, mapped, and their primary and pleiotropic effects on plant processes elucidated and described. Indeed, in the pedigree selection system normally practiced in wheat, most selection is still carried out only at the phenotypic level without knowing the genotypic constitution of the segregants examined. This restricts options for directed manipulation of the variation by, essentially, selecting desirable gene combinations only by chance. In principle, selection would be much more efficient at the level of genotype, rather than phenotype. However, this requires that the individual genes controlling traits of interest be identified and manipulated singly or in groups. One of the most important contributions of biotechnology to wheat breeding is the development of molecular marker technologies, which are increasingly capable of identification and manipulation of genes controlling both major gene and polygenic traits (so-called quantitative trait loci—QTLs). Once identified, they can be mapped, tagged, and thus followed through breeding programs. There are now several examples where complex traits have been dissected to reveal the numbers and locations of the genes involved. In the future, understanding the physiological bases of wheat adaptation and yield potential will increasingly rely on the identification of their determining major genes and QTLs. The methodologies for molecular marker systems and an insight into their application is described in the next section.

Conventionally, genetic advance in a wheat breeding program has depended almost exclusively on the variation created by intravarietal hybridization. However, recent advances in molecular biology and tissue culture have resulted in the development of new biotechnologies for introducing novel traits by extending the sources of genetic variation available. Thus through interspecific hybridization and chromosome engineering techniques, genes can be introgressed from related species, while by using genetic engineering technologies, genes can be introduced from any biological source, including plants, animals, bacteria, and viruses. The potential of these systems and current uses are described in later sections.

MOLECULAR MARKERS, MAPS, AND THEIR APPLICATION

Molecular Marker Systems in Wheat

The term molecular marker is applied to a variety of techniques that assay variation at the DNA level. Broadly, two classes of these markers can be identified: those based on DNA-DNA hybridization between a defined DNA sequence and the genomic DNA, and those based on the PCR amplification of genomic DNA sequences. In wheat, as in many other major crops, molecular marker systems have revolutionized the process of genetic analysis.

Hybridization-Based Molecular Markers

The dominant technology in this class is restriction fragment length polymorphism (RFLP). The basis of RFLP is the hybridization of a cloned length of DNA ("probe") to one or more restriction fragments, which are generated by the cleavage action of endonucleases on genomic DNA. The fragments can vary in length either due to point mutations at restriction recognition sites, or via insertion/deletion events in the genomic DNA sequences flanking the probe. The restriction fragments are separated one from one another on the basis of their length using gel electrophoresis, and are then immobilized on a nylon membrane and denatured. Probes are obtained from a variety of sources, although the majority come from libraries of genomic plant DNA (gDNA), or of reverse transcriptase products of messenger RNA (cDNA). The probe is labeled, either radioactively or by the attachment of a fluorochrome or an immunological reagent such as biotin, and allowed to bind with the immobilized restriction fragments. Where binding occurs as a result of sequence homology, the detection signal is concentrated in bands, which are visualized by the appropriate detection system. The number of bands generated gives an indication of copy number of the sequence in the genome, thus allowing the classification of RFLP loci as single or multiple copy. Because hexaploid wheat consists of three similar genomes, single-copy sequences in wheat usually generate three bands, a pattern typical of cDNA sequences. Single-copy loci have the advantage that their scoring is unambiguous and that allelism (and therefore codominance) can usually be recognized. In contrast, multicopy loci generate complex banding profiles, which can be useful in the context of DNA fingerprinting, but they are usually avoided in genetic analysis and marker-aided selection applications.

Polymerase Chain Reaction-Based Molecular Markers

PCR (polymerase chain reaction) technology has generated a variety of molecular markers. The common components of the PCR reaction are a thermostable DNA polymerase, the four single nucleotides dATP, dCTP, dGTP, and dTTP, template DNA from the species being analyzed, and one or more (usually two) short synthetic oligonucleotides (primers). The specificity of the reaction to a particular locus or loci depends only on the choice of primers. Amplification is achieved by cycling the reaction temperature a preset number of times, between the melting temperature of DNA (about 90°C) and the annealing temperature of the primer(s) to the template, which depends on both primer length and nucleotide sequence, and varies from about 35°C to 70°C. The reaction products are separated by size on a gel and visualized by staining with either a fluorescent DNA-binding dye or with silver. A critical feature of PCR technology is the low requirement both for sample DNA quantity (ng range, in contrast to the mg range necessary for RFLP assays) and for DNA purity. These features are important in the development of robotic approaches, which will become increasingly necessary for the incorporation of marker technology into plant breeding.

Sequence tagged sites (STS) is a general term applied to any unique genome fragment amplified from known genomic sequences or end-sequenced RFLP probes. Where there is sequence variation at one or both priming sites, polymorphism will be of the on/off type; alternatively, where either part of the amplified sequence is deleted or an extra sequence has been inserted, they will show as length variants. Where no length variants can be detected in a population, internal sequence variation can sometimes be exposed by digesting the PCR product with a range of restriction enzymes before electrophoresis. Where this fails, amplification products from different genotypes are themselves sequenced in a search for internal sequence variation. Any base-pair differences revealed can then be exploited to design primers, which are locus- or allele-specific. A substantial number of barley STS primers have been described by Blake et al. (1996) and a number of these can also be applied in wheat.

An important class of PCR-based markers is represented by microsatellites (also referred to as simple sequence repeats—SSRs). Polymorphism in these markers is based on differences between genotypes in the number of short tandemly repeated sequences (either all completely identical with one another, or including two or more similar, but distinct, sequences) occurring at a particular locus. Specific loci are targeted by designing primers from the sequence of flanking DNA of the satellite locus. Since two genotypes differing in the number of repeat units between the two primer sites will generate a different-sized PCR product, alleles can be detected by gel electrophoresis. SSR markers are generally highly polymorphic, with a probability of discriminating between unrelated genotypes as high as 95 percent in some crop species. Although substantial investment is required to generate SSR markers, their ease of use and information content has ensured their replacement of RFLP-based markers in the human field, where nearly 9,000 SSR loci have been described, of which over 5,000 have been incorporated on the genetic map (Dib et al., 1996). Progress in wheat SSR development is described in Bryan et al. (1997) and Rider et al. (1995).

PCR is possible where no prior knowledge of sequence is necessary. Three such techniques, similar in concept but different in detail, have been described (Welsh and McLelland, 1990; Williams et al., 1990; Caetano-Anolles, 1994). All three techniques usually employ a single primer—its length determines the type of assay. Arbitrarily primed (AP-PCR) utilizes conventional primers of ca. 20 base pairs (bp) and uses annealing temperatures of around 55°C. For random amplified polymorphic DNA (RAPD) the primers are ca. 10 bp and the annealing temperature is ca. 40°C; for DNA amplification fingerprinting (DAF), the primer length is as short as

5 bp, the concentration of primer is high, and the annealing temperature is ca. 35°C. Amplification products are obtained wherever the complementary nucleotide sequence is present in opposite orientation on each DNA strand, and separated by a maximum of 2 kbp. The number of PCR products is small for AP-PCR, is in the range of one to ten in RAPD, and is large for DAF. Of the three techniques, RAPD has proved the most popular compromise between specificity and number of products. RAPD assays are rapid, simple (and therefore potentially can be automated), and a large number of primers are commercially available. As a result, the technique has been applied to a wide range of plant species, with varying success. The major problems experienced in the application of the technique to wheat have been the lack of both reproducibility and polymorphism (Devos and Gale, 1992).

Amplified fragment length polymorphism (AFLP) is the most recently developed and potentially the most powerful of all the PCR-based marker technologies (Vos et al., 1995). The principle of AFLP is to selectively amplify a defined subset of restriction fragments. Following double restriction by a rare-cutting and a common-cutting enzyme, a short oligomer of known sequence is ligated to each cut end of every fragment. This procedure ensures that fragments cut at both ends with the common-cutting enzyme will have the same linker at each end; those with different ends will have a different linker ligated to each end; those with the rare-cutting site at each end are not expected to be present at appreciable frequencies. PCR amplification products are produced using primers complementary to the ligated sequence plus the remainder of the restriction site, with the addition of one to three "selective" bases. These additional bases determine the spectrum of restriction fragments that will be amplified. The profile of amplified products (up to 100 in number) is therefore specific to both the restriction enzyme combination used and the selective bases, and is visualized on a sequencing (denaturing polyacrylamide) gel. The wheat genome is characterized by a high content of repetitive (Flavell, Rimpau, and Smith, 1977), highly methylated (Moore et al., 1993) DNA. To avoid swamping wheat AFLP profiles with this DNA fraction, a methylation-sensitive restriction enzyme is used as the rare cutter. This strategy results in informative and reproducible AFLP profiles (Donini et al., 1997).

Genetic Maps

The major use of molecular marker systems is in the construction of genetic maps. Genetic maps are constructed by analyzing the patterns of cosegregation of molecular and other markers and major genes in defined populations. Increasingly, in wheat as in the human field, these maps will be based on SSRs (e.g., Korzun et al., 1997), but to date the major effort has focused on RFLP-based variation. Wheat maps have been generated from various crosses including an F_2 population between the wide cross, Chinese Spring × Synthetic (Devos and Gale, 1993), recombinant inbred lines such as Opata × Synthetic (Nelson et al., 1995), or doubled haploid populations such as Chinese Spring × Courtôt (Cadalen et al., 1997). The relative frequencies of parental and recombinant types for all pairwise combinations of polymorphic markers allows the detection and calculation of genetic linkages. If the recombination frequency is significantly less than 0.5, linkage can be assumed. Then, from the relative values from two-point and multipoint analyses, loci can be arranged into linkage groups, and then further ordered within these linkage groups. Several software packages are available for carrying out these mapping calculations, the most widely used being MAPMAKER (Lander, 1993) and JoinMap (Stam, 1993). A feature peculiar to wheat is that linkage groups or individual loci can be directly located on individual chromosomes using DNA taken from either the various aneuploid stocks of the reference variety Chinese Spring, or relevant single chromosome intervarietal substitution lines.

A large amount of mapping information on wheat is now available from different laboratories, almost all based on wide crosses, rather than those of agronomic significance. Nevertheless, the information can be applied to map genes of interest in crosses of interest by exploiting the available map and probe information. The development of bioinformatics is making this resource accessible to the wheat genetics community on-line via the Graingenes database, managed by the USDA National Library of Agriculture (http://wheat.pw.usda.gov/graingenes.html). Graingenes contains a considerable and growing amount of information covering populations, probes, maps, genes, and traits emanating from laboratories worldwide.

Dissecting the Genetics of Complex Traits—QTL Analysis

One of the major uses of molecular-marker-based maps is to identify genes of interest, such as those controlling quantitative traits including yield, adaptation, and quality, and to locate these accurately on the genetic map. The principle of marker-mediated location of genes is to derive an association between the segregation of known marker alleles with differences in phenotypic expression of the trait. This involves a four-step procedure. The first step is the establishment of an appropriate recombinant population between the genotypes chosen as differing in phenotype for the character(s) of interest. Second is the characterization of the individual lines in this population for marker loci dispersed throughout the genome and identified as being polymorphic in the cross from examination of the parents. In practice, this will be done by reference to published maps and the availability of particular probes/SSRs. Ideally, the target is to generate markers spaced about 20 to 30 cM apart on each chromosome arm, spanning the centromere. This requires, approximately, at least 6 polymorphic markers per chromosome (given an average genetic distance per chromosome of 150 cM [Devos, personal communication]), i.e., a minimum of 42 and a maximum of 126 loci. The former figure is achievable if single-copy RFLP probes are used which deliver polymorphism at each of the homoeoloci present on the three homoeologues—an unlikely situation! The latter figure where each locus requires its own assay, which will be the norm for SSR analysis, would be unfortunate where RFLP is being employed. In adapted crosses, only about 5 percent of polymorphic RFLP probes give polymorphisms in all three genomes, 15 percent in two genomes, and 80 percent in one genome (Devos, personal communication).

The third step is to evaluate the recombinant lines in appropriate replicated and randomized experiments in appropriate environments. The environments used, whether in a glasshouse, controlled environment, or field, are chosen to maximize the expression of genotypic variation while minimizing random environmental variation. When a sufficient map and good agronomic data are available, the final step is to use statistical procedures to partition the genotypic variation into components attributable to variation at the individual marker loci, either singly or in linked combinations. The simplest approach, and the preliminary to more detailed analysis, is to use analysis of variance (ANOVA) techniques to partition the variation within and between genotypes for each marker in turn. In an F_2 population, for example, the population is characterized at each codominant marker locus into the two homozygous classes and the heterozygote, and an orthogonal ANOVA carried out to separately detect "additive" and "dominance" effects. The difference between the two homozygotes is clearly the most informative, and if significant, a QTL linked to the marker locus is inferred. Further statistical analysis can define the exact location of the QTL on the assumption that only one QTL is in the vicinity of the marker. Initial approaches used individual marker loci in turn as landmarks to estimate the position of (assumed) single QTL (Luo and Kearsey, 1992; Snape et al., 1985). Recently, however, the analytical approaches have become more sophisticated, and numerous approaches are now available to locate genes relative to individual markers or to flanking markers (Lander and Botstein, 1989; Hyne et al., 1994; Haley and Knott, 1992; Zeng, 1993).

Marker-Assisted Selection (MAS)

Having located a gene and particular alleles of interest to a breeding program, the information on map location can be turned into a selection tool—marker-assisted selection (MAS). The principle of MAS is that indirect selection for a trait can be performed by selecting for a marker closely linked to the trait, rather than directly for the trait itself. Single traits can be targeted without knowledge of their map location by the "bulk segregant" method (Michelmore, Paran, and Kesseli, 1991). For this purpose, DNA from individuals with the two alternative phenotypes is bulked and the marker profiles of the two bulks are then compared, in the search for a marker that differentiates one bulk from the other. Since the "background" of the two bulks is identical, these differences should lie at loci close to that determining the target trait. For traits with more complex genetic control, an adequate complete genetic map is the primary prerequisite for successful MAS.

The potential advantages of MAS are considerable. Selection can be made at very early stages in the pedigree, and often at a juvenile plant age. Using PCR, with its requirement for very little DNA, individual seeds can be screened, and only positive selections ever need to be grown. This has large resource implications for breeding programs, as only relatively small populations of segregating material need to be grown to maturity, since undesirable allelic variation at a number of important genes has been greatly reduced or even eliminated by MAS. Furthermore, both the environment and genotype × environment interaction effects, which can be large in conventional breeding programs where trait measurement is the basis of selection, are irrelevant for those traits selected for by MAS. This means, in some cases, that generations can be advanced in off-season conditions or in locations where a particular environmental limitation (disease or abiotic stress) is absent. A further benefit is that individuals carrying a favorable recessive allele can be selected at each backcross generation, without the need for a progeny test, as is required in conventional backcross programs.

Few current examples of the routine use of MAS in wheat are available, since the marker technology is not as yet in general sufficiently breeder-friendly. In many wheat breeding programs, however, two characters are selected for by MAS in early generations. The first involves resistance to the disease eyespot, which is well marked by a variant of the isozyme endopeptidase (Summers et al., 1988). Pathological methods for assessing resistance to the fungus are available, but are time consuming and inexact; replicative testing is necessary and a result is obtained only after ten weeks of plant growth. In contrast, the isozyme test is conducted on individual, nongerminated seeds, and hundreds of individuals can be screened per person per day. The second example involves defined electrophoretic alleles for the glutenin seed storage proteins, which are associated with favorable breadmaking quality (Payne et al., 1983). In this case these proteins themselves are probably responsible for the quality effect, whereas in the eyespot/endopeptidase case, the isozyme is not involved in the plant response to the pathogen.

Passport Data for Parental Selection and Germplasm Conservation

Parental selection is one of the most critical aspects of breeding. In general, the probability of achieving transgressive segregation from a cross (or maximum heterosis in an F_1 hybrid) increases the more genetically diverse the parents are. DNA fingerprinting techniques can provide an objective method of assessing this divergence, since they can be used to sample polymorphism across the entire genome, rather than at just a few loci, as might be done by phenotypic comparisons. Similar considerations relate to germplasm surveys and the management of gene banks. DNA fingerprinting is becoming increasingly useful to formulate rational strategies of ex situ conservation to minimize duplication, to understand population structures, and to provide objective descriptors of individual accessions. Of particular value to these applications are AFLP and SSR analysis as these assays are both highly efficient at uncovering polymorphism and, as PCR-based technologies, have advantages with respect to DNA extraction and processing.

Comparative Genetics

Most gDNA probes and SSR and STS primers do not successfully assay outside the species in which they have been developed. However, cDNA probes, because they represent gene sequences, which tend to be well conserved across species boundaries, do show good levels of portability. Recent research has been exploiting this property to align the genetic maps of distantly related species by simultaneously mapping with common probes. This has begun to show that locus order is conserved over substantial parts of the genomes of species within wide botanical groups, in particular the Gramineae (Moore et al., 1995). A major implication of this discovery is that genetic information can be shared between species that have in the past been considered to be too distantly related for such a purpose. Comparative mapping is of particular importance to the cereals, which dominate as a human food source, and have thus accumulated massive amounts of data related to genetics and physiology. For wheat in particular, the primary relevance of the comparative approach may be in the opportunities provided by the rice genome. For many biotechnological applications, wheat is a problematic target, due to its polyploidy, its many chromosomes, and its large genome size. In contrast, the rice genome, despite possessing a similar gene content and a similar genetic architecture, as demonstrated by comparative mapping, has a genome less than 3 percent as large. Such analysis has shown, for example, that the major genes controlling vernalization response in wheat, the *Vrn* loci, which are located on the long arm of the homoeologous group 5 chromosomes (Galiba et al., 1995) have homoeologues in barley (*Sh2* locus) and rye (*Sp1*). Further, it can be shown that a region on rice chromosome 3 is homoeologous to the *Vrn* regions and this contains a flowering time locus (Sarma et al., 1998). Comparative genetics is expected to become a major tool for locating, cloning, and understanding the biology of important traits in wheat in the future.

INTROGRESSION OF "ALIEN" GENES INTO WHEAT BY CHROMOSOME ENGINEERING

Many of the wild Triticeae species clearly have genes affecting a wide range of physiological traits, including adaptation to different environments, responses to abiotic and biotic stresses, and even photosynthetic performance, which could be useful in wheat improvement. Such genetic variation could be of great value if transferred into wheat. A well-established technology, now augmented by molecular marker and molecular cytogenetical techniques, is in place for the transfer of whole or segments of chromosomes from a large range of other Triticeae species to wheat, taking advantage of the polyploid nature of wheat, both bread (hexaploid, genomes ABD) and durum (tetraploid, genomes AB).

Extending the Wheat Gene Pool by Interspecific and Generic Hybrids

The polyploidy of the wheat genome buffers the loss or gain of individual chromosomes or even whole genomes. This makes it possible to create sexual hybrids by fertilizing wheat with pollen from an extensive range of related grasses. Crossability is itself under genetic control, with a pair of (probably related) genes, *kr1* and *kr2*, determining the extent of crossability of wheat with rye and other related species (Riley and Chapman, 1967). Where fertilization is successful, hybrid endosperm often fails to develop, but often the hybrid embryos can be rescued by culturing on artificial media. Wide hybrids made in this way are universally self-sterile, as all the chromosomes, both wheat and nonwheat, lack homoeologues. However, treatment with the drug colchicine, which doubles chromosome number, in most cases restores at least partial self-fertility.

Many such wide hybrids have been produced, and successfully chromosome-doubled. The phenotype of these plants is generally intermediate between those of the wheat and the nonwheat parents. Particularly when the latter is a diploid species, the phenotype of the hybrid resembles the wheat more than the nonwheat parent. This is understandable in hybrids between hexaploid wheat and a diploid species, since the hybrid genome is then effectively 75 percent wheat. This characteristic has been demonstrated in triticale (wheat × rye), which is seen as an alternative to wheat, adapted to the lower fertility and input conditions that are more suited to rye than to wheat production. Although triticale is recognizably not wheat, its management closely resembles that of wheat, to the extent that wheat farmers can grow triticale without the necessity for any significant changes in husbandry. The breeding of such hybrids has proved to be less straightforward; the enforced combination of two genomes that have not coevolved has required, in triticale at least, a substantial breeding effort to overcome problems of partial sterility and seed shriveling. Similar problems can be anticipated with other wide hybrids. Thus the value of instigating a breeding program to combine, for example, wheat and *Thinopyrum* spp. to generate a crop adapted to salinity-damaged soil has to be balanced against the high cost that such a major breeding program would entail.

Introgression of Exotic Chromosome and Subchromosomal Segments

Introgression of exotic genetic variation has been a long-held goal of wheat cytogeneticists. Ultimately the aim of these manipulations is to introduce a minimum amount of alien genetic material into the wheat genome, so that the target trait can be expressed in wheat without appreciable negative side effects resulting from linkage drag and/or deficiency of necessary wheat genes. Wide hybrids form the starting point for the reduction of the amount of nonwheat genetic information from a whole genome to single chromosomes or chromosome segments. When the hybrid is crossed and backcrossed with wheat, selections can be made that contain a single nonwheat chromosome in a normal wheat background.

The nonwheat chromosome can become relatively stable by generating a disomic addition line, which carries two doses of the chromosome. Disomic addition lines are useful for the assignation of gene effects to individual chromosomes, but their chromosomal instability has ruled them out for use as a crop in their own right.

The analysis of addition line series involving the chromosomes from a number of species related to wheat has led to the recognition that related species genomes are similar in gene content and genetic organization (homoeologous) to those of wheat. In many cases, this allows the successful substitution of a wheat chromosome pair with a pair of nonwheat chromosomes. The resulting lines are generally cytologically stable. However, because homology is frequently incomplete, the nonwheat chromosome does not completely compensate for the loss of a wheat chromosome. Nonetheless, a number of commercial wheat varieties have been released in which rye chromosome 1R has replaced wheat chromosome 1B. A further reduction in the nonwheat content can be achieved by taking advantage of the occasional misdivision of unpaired monosomes to form telocentrics. When two de novo telocentrics form in a single meiocyte, they can fuse together to form Robertsonian translocations; a chromosome of this type, comprising the short arm of rye 1R and the long arm of wheat chromosome 1B, is very widespread in many wheat-breeding programs worldwide, since the translocation appears to have desirable effects on physiological performance in the field. Numerous techniques are available to identify these 1BL.1RS translocation lines (Javornik et al., 1991; Koebner, 1995).

In order to shorten the nonwheat chromosome segment below a half chromosome, two major strategies have been employed: irradiation and pH mediated recombination. Irradiation induces chromosomal breaks, which can be repaired by endogenous mechanisms; these heal the broken end by capping with telomeric sequences to generate a shortened, but still functional chromosome. If two independent breaks are formed, two sorts of novel chromosome will be generated, depending on whether the breaks occur on separate chromosomes, or whether they occur at different points along the same chromosome. In the first case, a translocation between two different chromosomes is generated; in the second case, an intercalary deletion or an inversion on a single chromosome is formed. The random nature of irradiation-induced chromosomal breakage means that the majority of such mutants are of inferior phenotype, as a result of gene deficiency; nevertheless, if the screening procedure is efficient enough, sufficient numbers of mutant phenotypes can be selected, among which desirable types can be found. A good deal of effort was expended in the

1950s and 1960s using irradiation to introgress nonwheat chromosomal segments. Of these, the only one that has found its way into varieties is the *Sr26* segment from *Agropyron elongatum* (Knott, 1961).

Although the individual genomes of wheat are so similar in gene content, the inheritance of genes is exclusively diploid. Meiotically this is borne out by the rarity of chromosome pairing configurations other than bivalents. To maintain the integrity of the constituent genomes, many allopolyploids have evolved a mechanism of pairing inhibition between homoeologous chromosomes. The genetic control of this effect is surprisingly simple in wheat. A single major gene, *Ph1*, residing on chromosome 5B is responsible for the majority of the effect, with a number of genes giving lesser effects distributed on other chromosomes. In the absence of *Ph1*, achieved either through nullisomy for the entire chromosome 5B, or by deletion mutation, homoeologues can recombine at a frequency dependent on the taxonomic distance separating the donor genome from that of wheat. Thus the nonwheat chromosome present in an introgression line can be shortened by recombination. Unlike the radiation-induced translocations, the break points between the wheat and nonwheat chromosome will not be random, as induced recombination relies on recognition and crossing-over mechanisms similar to those that operate in homoeologous recombination. Such recombinant chromosomes have been induced from a wide range of donor species including barley (Islam and Shepherd, 1992), rye (Koebner and Shepherd, 1986), *Aegilops* spp. (Riley, Chapman, and Johnson, 1968; Miller et al., 1987; Ceoloni et al., 1988), and *Agropyron elongatum* (Sears, 1973).

The analysis of alien introgressions has been revolutionized both by the development of molecular markers and by the application of in situ hybridization. While the former allows the definition of the genetic constitution of the recombinant chromosome, the latter allows direct visualization of the physical chromosomal location of specific loci. This is achieved by hybridization of a probe to cytological preparations made, most frequently, from meristematic tissue. Of particular significance to introgression is the genomic in situ hybridization (GISH) technique, which takes advantage of the substantial level of interspecific divergence of repetitive DNA sequences. Since the genome consists overwhelmingly of such repetitive DNA, when nonwheat genomic DNA is used as a probe, it is effectively specific for any nonwheat chromosomal segment. Thus, from mitotic or meiotic chromosome spreads, GISH analysis identifies whole alien chromosomes in addition and substitution lines, and clearly demonstrates any break point between wheat and nonwheat chromatin in introgression lines (Schwarzacher et al., 1992).

GENETIC ENGINEERING OF WHEAT

Transformation Systems

Genetic engineering—the introduction of isolated individual genes from any biological source—is now a reality in wheat following recent developments in tissue culture and transformation technologies (see Jähne, Becker, and Lörz, 1995). Although a number of different techniques have been attempted over the years, including protoplast electroporation and pollen tube microinjection, undoubtedly the most successful approach at present is via biolistic methods of gene delivery into proliferating scutellum tissue of immature embryos (Weeks, Anderson, and Blechl, 1993). This technique is now established in many laboratories, particularly in industry. The system relies on the introduction of a gene of interest either on the same plasmid as a selectable marker and reporter gene, or on a separate plasmid in cotransformation experiments. The most reliable selectable marker is proving to be the *Streptomyces Bar* gene, conferring resistance to the herbicide bialophos, under the control of the maize ubiquitin promoter. The *Gus* gene is still the most widely used reporter of transient and stable transformation in wheat, although the use of the firefly luciferase gene and the jellyfish green fluorescent protein are being tested as nondestructive reporter systems. Several different "guns" are available, including gunpowder-driven devices, but the most widely used is the BioRad PDS1000/Helium gun.

Attention has also now turned to the use of *Agrobacterium* as a method for wheat transformation. Previously, the monocots were thought to be recalcitrant to infection by *Agrobacterium*. However, recent successes with rice (Hiei et al., 1994) and barley (Tingay et al., 1997) have shown that super-virulent strains can be vehicles for gene transfer in monocots, and, undoubtedly, systems competent for wheat will now be developed. These will have the advantage of allowing more control over the integration process, and also should lead to "clean gene" technology, in which transformation is carried out without the necessity for the insertion of undesirable selectable markers such as antibiotic resistance in the vector sequences.

Manipulation of Traits

Two approaches are being taken to the genetic engineering of wheat aimed at the modification of agronomic traits. The first seeks to modify native genes in vitro, and to reinsert these into wheat. This can be done with a view to increase the expression of a native gene, to modify its product, or to switch a gene off using antisense technologies. With respect to altering expression levels, probably the most intensively investigated example at the present time is the modification of wheat storage proteins, particularly the high-molecular-weight glutenins. The Dy10 and Dx5 subunits encoded by chromosome 1D have been isolated from the variety Cheyenne, since these are known to impart good breadmaking quality (Shewry, Halford, and Tatham, 1989). These are being used to study functionality by being reinserted into wheat under the control of native promoters to study their individual effects on breadmaking quality (Lazzeri et al., 1997). Other targets for antisense technology are genes that promote susceptibility to disease. Worland and Law (1991) have shown that reducing the dosage of chromosome 5D increases adult plant resistance of the susceptible variety Hobbit "sib" to yellow rust and to powdery mildew. Thus, if the particular gene involved could be isolated, an antisense construct could be inserted in this or other varieties to increase levels of resistance. As our knowledge of disease resistance mechanisms increases, other targets for modification of gene expression or of gene product are likely to be identified.

The second use of transformation technologies will be to generate novel germplasm by introducing genes from other biological sources into wheat, be it viral, bacterial, plant, or even animal in origin. Primary targets will be the introduction of novel genes to alleviate pest, disease, and stress problems or to create novel products for new end-uses, such as industrial raw materials. The success of these approaches will obviously depend on the availability of cloned genes of interest, and different sources of plant genes are becoming available for both crop and model plant species, particularly the model dicot *Arabidopsis thaliana*. For example, sources of fungal resistance are now emerging from studies of resistance genes such as the *Cf2* and *Cf9* genes of tomato (Jones et al., 1994). The sequences of these genes have been determined and degenerate PCR primers can be designed to extract potential homoeologues from other species, including wheat. This offers the opportunity to take resistance genes from other Triticeae species, particularly wild species, and introduce them directly into wheat, obviating the need for cytogenetical manipulation. Other genes are being introduced from bacteria, for example, for herbicide resistance. In addition, novel sources of genes for modifying end-use come from studies in other crop species. Peas are an example, where several genes have been identified that control starch synthesis and biochemistry. These include the gene controlling *r*, Mendel's "wrinkled" gene, which encodes a starch branching enzyme (Bhattacharyya, Martin, and Smith, 1993).

This and other genes can be used to modify the starch composition of wheat, which could generate products for industrial uses such as biodegradable plastics. Pea lipoxygenases have also been cloned and could be reinserted into wheat to improve dough rheology and ameliorate the need to add soybean flour during the baking process.

Clearly many opportunities for the genetic modification of wheat are opening up, although it should not be forgotten that there are still technical limitations on what can be achieved at present. For example, the mechanisms of transgene insertion and gene expression are not understood and gene expression can be highly variable between different transgenics developed using the same construct under the same conditions. Also, the phenomenon of transgene silencing can occur for reasons still not understood, as can unstable inheritance patterns. Thus challenges still remain in terms of understanding gene expression, stability, and durability. Plus, farmer and consumer acceptance will need to be taken into account, and these techniques are unlikely to be a universal panacea. At present, they should be regarded as complementary to conventional breeding technologies, and used in an integrated approach to crop improvement.

CONCLUSION

The new biotechnologies now enable a much greater understanding of the biology of the wheat plant, and consequently allow the plant breeder much more scope than has ever been possible before for genotypic, and hence phenotypic, modification. This also opens up the possibility of understanding the physiological responses of the wheat plant to its environment at the genetic level—in other words, the opportunity of translating "traits to genes." Thus, the physiological and biochemical processes underlying yield potential in a particular environment can now be identified in terms of individual genes. These can then be manipulated by marker-assisted selection in conventional breeding programs, or cloned, modified, and reintroduced into wheat with changed expression levels or specificities. Also, the wide adaptability of the many related wild species undoubtedly means that these harbor novel alleles for modifying physiological traits in wheat. The technology is developing so that the potential of this gene pool can be "molecularly farmed" for genetic variation for adaptability and yield potential in the cultivated cereals. Clearly, a greater understanding of the physiological basis of yield potential will lead to novel opportunities for genetic modification to meet the challenges of growing wheats with much greater yield potential and yield stability than was previously possible.

REFERENCES

Bhattacharyya M., Martin C., and Smith A. (1993). The importance of starch biosynthesis in the wrinkled seed shape character of peas studied by Mendel. *Plant Molecular Biology* 22: 525-531.

Blake T.K., Kadyrzhanova D., Shepherd K.W., Islam A.K.M.R., Langridge P.L., McDonald C.L., Erpelding J., Larson S., Blake N.K., and Talbert L. (1996). STS-PCR markers appropriate for wheat-barley introgression. *Theoretical and Applied Genetics* 93: 826-832.

Bryan G.J., Collins A.J., Stephenson P., Orry A., Smith J.B., and Gale M.D. (1997). Isolation and characterisation of microsatellites from hexaploid bread wheat. *Theoretical and Applied Genetics* 94: 557-563.

Cadalen T.C., Boeuf S., Bernard S., and Bernard M. (1997). Intervarietal molecular map in *Triticum aestivum* and comparison with a map originated from a wide cross. *Theoretical and Applied Genetics* 94: 367-377.

Caetano-Anolles G. (1994). MAAP: A versatile and universal tool for genome analysis. *Plant Molecular Biology* 25: 1011-1026.

Ceoloni C., Del Signore G., Pasquini M., and Testa A. (1988). Transfer of mildew resistance from *Triticum longissimum* into wheat by *ph1* induced homoeologous recombination. In T.E. Miller and R.M.D. Koebner (eds.) *Proceedings of the 7th International Wheat Genetics Symposium,* Institute of Plant Science Research, Cambridge Laboratory, Cambridge, UK, pp. 221-226.

Devos K.M. and Gale M.D. (1992). The use of random amplified polymorphic DNA in wheat. *Theoretical and Applied Genetics* 84: 567-572.

Devos K.M. and Gale M.D. (1993). The genetic maps of wheat and their potential in plant breeding. *Outlook in Agriculture* 22: 93-99.

Dib C., Fauré S., Fizames C., Samson D., Drouot N., Vignal A., Millasseau P., Marc S., Hazan J., Seboun E. (1996). A comprehensive genetic map of the human genome based on 5,264 microsatellites. *Nature* 380: 152-154.

Donini P., Elias L., Bougourd S., and Koebner R.M.D. (1997). AFLP fingerprinting reveals pattern differences between template DNA extracted from different plant organs. *Genome* 40: 521-526.

Flavell R.B., Rimpau J., and Smith D.B. (1977). Repeated sequence DNA relationships in four cereal genomes. *Chromosoma* 63: 205-222.

Galiba G., Quarrie S.A., Sutka J., Morgounov A., and Snape J.W. (1995). RFLP mapping of the vernalization (*Vrn1*) and frost resistance (*Fr1*) genes on chromosome 5A of wheat. *Theoretical and Applied Genetics* 90: 1174-1179.

Haley C.S. and Knott S.A. (1992). A simple regression method for mapping quantitative trait loci in line crosses using flanking markers. *Heredity* 69: 315-324.

Hiei Y., Ohta S., Komari T., and Kumashiro T. (1994). Efficient transformation of rice (*Oryza sativa* L.) mediated by *Agrobacterium* and sequence-analysis of the boundaries of the T-DNA. *Plant Journal* 6: 271-282

Hyne V., Kearsey M.J., Martìnez O., Wang G., and Snape J.W. (1994). A partial genome assay for quantitative trait loci in wheat (*Triticum aestivum*) using different analytical techniques. *Theoretical and Applied Genetics* 89: 735-741.

Islam A.K.M.R. and Shepherd K.W. (1992). Production of wheat-barley recombinant chromosomes through induced chromosome pairing. I. Isolation of recombinants involving barley arms 3HL and 6HL. *Theoretical and Applied Genetics* 83: 489-494.

Jähne A., Becker D., and Lörz H. (1995). Genetic engineering of cereal crop plants: A review. *Euphytica* 85: 35-44.

Javornik J., Sinkovic T., Vapa L., Koebner R.M.D., and Rogers W.J. (1991). A comparison of the methods for identifying and surveying the presence of 1BL.1RS translocations in bread wheat. *Euphytica* 54: 45-53.

Jones D.A., Thomas K.E., Hammond-Kossack K., Balint-Kurti P.J., and Jones J.D.G. (1994). Isolation of the tomato *Cf-9* gene for resistance to *Cladosporium fulvum* by transposon tagging. *Science* 266: 789-793.

Knott D.R. (1961). The inheritance of rust resistance. VI. The transfer of stem rust resistance from *Agropyron elongatum* to common wheat. *Canadian Journal of Plant Science* 41: 109-123.

Koebner R.M.D. (1995). Generation of PCR-based markers for the detection of rye chromatin in a wheat background. *Theoretical and Applied Genetics* 90: 740-745.

Koebner R.M.D. and Shepherd K.W. (1986). Controlled introgression to wheat of genes from rye chromsome 1RS by induction of allosyndesis. 1. Isolation of recombinants. *Theoretical and Applied Genetics* 73: 197-208.

Korzun V., Röder M., Worland A.J., and Börner A. (1997). Intrachromosomal mapping of genes for dwarfing (*Rht12*) and vernalization response (*Vrn1*) in wheat by using RFLP and microsatellite markers. *Plant Breeding* 116: 227-232.

Lander E.S. (1993). MAPMAKER/Exp 3.0 and MAPMAKER/QTL 1.1. Whitehead Institute, Cambridge, MA.

Lander E.S. and Botstein D. (1989). Mapping Mendelian factors underlying quantitative variation using RFLP linkage maps. *Genetics* 121: 185-199.

Lazzeri P.A., Barcelo P., Barro F., Rooke L., Cannell M.E., Rasco-Gaunt S., Tatham A., Fido R., and Shewry P.R. (1997). Biotechnology of cereals: Genetic manipulation techniques and their use for the improvement of quality, resistance and input use efficiency traits. *Aspects of Applied Biology* 50: 1-8.

Luo Z.W. and Kearsey M.J. (1992). Interval mapping of quantitative trait loci in an F2 population. *Heredity* 69: 236-242.

Michelmore R., Paran I., and Kesseli R. (1991). Identification of markers linked to disease resistance genes by bulked segregant analysis—a rapid method to detect markers in specific genomic regions by using segregating populations. *Proceedings of the National Academy of Sciences of the USA* 88: 9828-9832.

Miller T.E., Reader S.M., Ainsworth C.C., and Summers R.W. (1987). The introduction of a major gene for resistance to powdery mildew of wheat, *Erysiphe graminis* f.sp. *tritici*, from *Aegilops speltoides* into wheat, *Triticum aestivum*. In: M.L. Jerna and L.A.J. Slootmaker (Eds.), *Cereal Breeding Related to Integrated Cereal Production*—EUCARPIA, Pudoc, Wageningen, the Netherlands, pp. 179-183.

Moore G., Abbo S., Cheung W., Foote T., Gale M., Koebner R., Leitch A., Leitch I., Money T., Stancombe P., Yano M., and Flavell R. (1993). Key features of cereal genome organization as revealed by the use of cytosine methylation-sensitive restriction endonucleases. *Genomics* 15: 472-482.

Moore G., Devos K.M., Wang Z., and Gale M.D. (1995). Grasses, line up and form a circle. *Current Biology* 5: 737-739.

Nelson J.C., Vandeynze A.E., Autrique E., Sorrells M.E., Lu Y.H., Negre S., Bernard M., and Leroy P. (1995). Molecular mapping of wheat—homoeologous group 3. *Genome* 38: 525-533.

Payne P., Holt L., Thompson R., Bartels R., Harberd N., Harris P., and Law C. (1983). The high-molecular-weight subunits of glutenin: Classical genetics, molecular genetics and the relationship with breadmaking quality. In: S. Sakamoto (Ed.), *Proceedings of the 6th International Wheat Genetics Symposium,* Plant Germ-Plasm Institute, University of Kyoto, Kyoto, Japan, pp. 827-834.

Rider M., Plaschke J., Koenig S.U., Börner A., Sorrells M.E., Tanksley S.D., and Ganal M.W. (1995). Abundance, variability and chromosomal location of microsatellites in wheat. *Molecular and General Genetics* 246: 327-333.

Riley R. and Chapman V. (1967). The inheritance in wheat of crossability with rye. *Genetical Research, Cambridge* 9: 259-267

Riley R., Chapman V., and Johnson R. (1968). Introduction of yellow rust resistance of *Aegilops comosa* into wheat by genetically induced homoeologous recombination. *Nature* 217: 383-384.

Sarma R.N., Gill B.S., Sasaki T., Galiba G., Sutka J., Laurie D.A., and Snape J.W. (1998). Comparative mapping of the wheat chromosome 5A *Vrn-A1* region with rice and its relationship to QTL for flowering time. *Theoretical and Applied Genetics* 97: 103-109.

Schwarzacher T., Anamthawat-Jónsson K., Harrison G.E., Islam A.K.M.R., Jia J.Z., King I.P., Leitch A.R., Miller T.E., Reader S.M., and Rogers W.J. (1992). Genomic *in situ* hybridization to identify alien chromosomes and chromosome segments in wheat. *Theoretical and Applied Genetics* 84: 778-786.

Sears E.R. (1973). Agropyron-wheat transfers induced by homoeologous pairing. In: E.R. Sears and L.M.S. Sears (Eds.), *Proceedings of the 4th International Wheat Genetics Symposium,* Agricultural Experiment Station, University of Missouri, Columbia, MO, pp. 191-199.

Shewry P.R., Halford N.G., and Tatham A.S. (1989). The high-molecular-weight subunits of wheat, barley and rye: Genetics, molecular biology, chemistry and role in wheat gluten structure and functionality. In: B.J. Miflin (Ed.), *Oxford Surveys of Plant Molecular and Cell Biology,* Oxford University Press, Oxford, pp. 163-219.

Snape J.W., Law C.N., Parker B.B., and Worland A.J. (1985). Genetical analysis of chromosome 5A of wheat and its influence on important agronomic characters. *Theoretical and Applied Genetics* 71: 518-526.

Stam P. (1993). Construction of integrated genetic linkage maps by means of a new computer package: JoinMap. *Plant Journal* 5: 739-744.

Summers R.W., Koebner R.M.D., Hollins T.W., Förster J., and Macartney D.P. (1988). The use of an isozyme marker in breeding wheat (*Triticum aestivum*) resistant to the eyespot pathogen (*Pseudocercosporella herpotrichoides*). In: T.E. Miller and R.M.D. Koebner (Eds.), *Proceedings of the 7th International Wheat Genetics Symposium,* Institute of Plant Science Research, Cambridge Laboratory, Cambridge, UK, pp. 1195-1197.

Tingay S., McElroy D., Kalla R., Fieg S., Wang M.B., Thornto S., and Brettell R. (1997). *Agrobacterium tumefaciens*-mediated barley transformation. *Plant Journal* 11: 1369-1376.

Vos P., Hogers R., Bleeker M., Reijans M., van de Lee T., Hornes M., Freijters A., Pot J., Peleman J., Kuiper M., and Zabeau M. (1995). AFLP: A new technique for DNA fingerprinting. *Nucleic Acids Research* 23: 4407-4414.

Weeks J.T., Anderson O.D., and Blechl A.E. (1993). Rapid production of multiple independent lines of fertile transgenic wheat (*Triticum aestivum*). *Plant Physiology* 102: 1077-1084.

Welsh J. and McLelland M. (1990). Fingerprinting genomes using PCR with arbitrary primers. *Nucleic Acids Research* 18: 7213-7218.

Williams J.G.K., Kubelik A.R., Livak K.J., Rafalski J.A., and Tingey S.V. (1990). DNA polymorphisms amplified by arbitrary primers are useful as genetic markers. *Nucleic Acids Research* 18: 6531-6535.

Worland A.J. and Law C.N. (1991). Improving disease resistance in wheat by inactivating genes promoting disease susceptibility. *Mutation Breeding Newsletter* 38: 2-5.

Zeng Z.B. (1993). Theoretical basis of separation of multiple effects on mapping quantitative trait loci. *Proceedings of the National Academy of Sciences of the USA* 90: 10972-10976.

Author Index

Subject Index